OFDM for Optical Communications

OFDM for Optical Communications

William Shieh
Department of Electrical and Electronic Engineering
The University of Melbourne

Ivan Djordjevic
Department of Electrical and Computer Engineering
The University of Arizona

AMSTERDAM • BOSTON • HEIDELBERG • LONDON
NEW YORK • OXFORD • PARIS • SAN DIEGO
SAN FRANCISCO • SINGAPORE • SYDNEY • TOKYO
Academic Press is an imprint of Elsevier

Academic Press is an imprint of Elsevier
30 Corporate Drive, Suite 400, Burlington, MA 01803, USA
525 B Street, Suite 1900, San Diego, California 92101-4495, USA
84 Theobald's Road, London WC1X 8RR, UK

Library of Congress Cataloging-in-Publication Data
Application submitted.

British Library Cataloguing-in-Publication Data
A catalogue record for this book is available from the British Library.

ISBN: 978-0-12-374879-9

For information on all Academic Press publications
visit our Web site at www.elsevierdirect.com

Printed in the United States of America.
09 10 9 8 7 6 5 4 3 2 1

Working together to grow
libraries in developing countries

www.elsevier.com | www.bookaid.org | www.sabre.org

ELSEVIER   BOOK AID
             International   Sabre Foundation

*To my wife, Jennifer, my lovely kids Heather and Edmund,
and my parents Zhengfa and Chaorong.*
—William Shieh

*To my parents Blagoje and Verica, brother Slavisa,
and to Milena.*
—Ivan Djordjevic

Preface

We have recently witnessed a dramatic surge of interest in orthogonal frequency-division multiplexing (OFDM) from the optical communications community. This is evidenced by an increase in the number of publications, and many eye-catching "hero" experimental demonstrations of OFDM technology in a wide spectrum of applications. One of the main drivers behind the emergence of OFDM technology in optical communications is the rapid advance of silicon signal processing capability underpinned by Moore's Law. The introduction of sophisticated electronic digital signal processing (DSP) and coding could fundamentally alter optical networks as we see them today. On the other hand, although OFDM has emerged as the leading physical interface in wireless communications over the past two decades, it is still regarded as alien by many optical engineers. The most frequently asked question is whether OFDM can provide a solution to the major problems of today's "fragile" and "rigid" optical networks.

This book intends to give a coherent and comprehensive introduction to the fundamentals of OFDM signal processing and coding, with a distinctive focus on its broad range of applications. It is designed for two diverse groups of researchers: (1) optical communication engineers who are proficient in optical science and interested in applying OFDM technology, but who are intimidated by sophisticated OFDM terminologies, and (2) wireless communication engineers who are content with their DSP skill set, but are disoriented by a perceived huge gap between the optical and radio-frequency (RF) communications worlds. We have attempted to make the individual chapters self-contained while maintaining the flow and connectivity between them.

The book is organized into twelve chapters, and describes various topics related to optical OFDM, starting from basic mathematical formulation through OFDM signal processing and coding for OFDM, to various applications, such as single-mode fiber transmission, multimode fiber transmission, free space-optical systems, and optical access networks. The book also provides fundamental concepts of optical communication, basic channel impairments and noise sources, and optical system engineering process. The book presents interesting research problems in the emerging field of optical OFDM and touches the intriguing issue of standardization of optical OFDM technology.

The authors would like to thank their colleagues, in particular, R. S. Tucker, G. Pendock, R. Evans, B. Krongold, H. Bao, W. Chen, X. Yi, Y. Tang, Y. Ma, Q. Yang, S. Chen, B. Vasic, L. Xu, and T. Wang, whose collaboration on the subject of optical OFDM has contributed directly and indirectly to the completion of the book. We also express our sincere gratitude to C. Xie at Bell Labs, X. Yi at University of California, Davis, K. Hinton, R. S. Tucker, and Y. Tang at the University of Melbourne, and Hussam G. Batshon at the University of Arizona for their careful proofreading of the book chapters. The authors would like to acknowledge the ARC Special Research Centre for Ultra-Broadband Information Networks (CUBIN), National ICT Australia (NICTA), and the National Science Foundation (NSF) for their support of OFDM related research activities.

Finally, special thanks are extended to Tim Pitts, Melanie Benson, and Sarah Binns of Elsevier for their tremendous effort in organizing the logistics of the book, including the editing and promotion that allowed this book to happen.

Author Biography

William Shieh is an associate professor and reader in the Electrical and Electronic Engineering Department at the University of Melbourne, Melbourne, Australia. He received his M.S. in electrical engineering and Ph.D. in physics from the University of Southern California, Los Angeles, in 1994 and 1996, respectively. From 1996 to 1998 he worked as a member of Technical Staff in the Jet Propulsion Laboratory, Pasadena, CA. From 1998 to 2000 he worked as a member of Technical Staff in Bell Labs, Lucent Technologies, Holmdel, NJ. From 2000 to 2003 he worked as a technical manager in Dorsal Networks, Columbia, MD. Since 2004 he has been with the Electrical and Electronic Engineering Department at the University of Melbourne, Melbourne, Australia.

Ivan Djordjevic is an assistant professor of Electrical and Computer Engineering (ECE) at the University of Arizona, Tucson. Prior to this appointment in August 2006 he was with University of Arizona, Tucson, as a research assistant professor; University of the West of England, Bristol, UK; University of Bristol, Bristol, UK; Tyco Telecommunications, Eatontown, USA; and National Technical University of Athens, Athens, Greece. His current research interests include optical networks, error control coding, constrained coding, coded modulation, turbo equalization, OFDM applications, and quantum error correction. He currently directs the Optical Communications Systems Laboratory (OCSL) within the Electrical and Computer Engineering Department at the University of Arizona. Dr. Djordjevic serves as an associate editor for *Research Letters in Optics* and as an associate editor for *International Journal of Optics*. Dr. Djordjevic is an author of about 100 journal publications and over 80 conference papers.

Contents

CHAPTER 1

Introduction

In the virtually infinite broad electromagnetic spectrum, there are only two windows that have been largely used for modern-day broadband communications. The first window spans from the long-wave radio to millimeter wave, or from 100 kHz to 300 GHz in frequency, whereas the second window lies in the infrared lightwave region, or from 30 THz to 300 THz in frequency. The first window provides the applications that we use in our daily lives, including broadcast radio and TV, wireless local area networks (LANs), and mobile phones. These applications offer the first meter or first mile access of the information networks to the end user with broadband connectivity or the mobility in the case of the wireless systems. Nevertheless, most of the data rates are capped below gigabit per second (Gb/s) primarily due to the lack of the available spectrum in the RF microwave range. In contrast, due to the enormous bandwidth over several terahertz (THz) in the second window, the lightwave systems can provide a staggering capacity of 100 Tb/s and beyond. In fact, the optical communication systems, or fiber-optic systems in particular, have become indispensable as the backbone of the modern-day information infrastructure. There has been a worldwide campaign in the past decade to push the fiber ever closer to the home. Despite the fact that the Internet "bubble" fizzled out in the early 2000s, Internet traffic has been increasing at an astounding rate of 75% per year.[1,2] The new emerging video-centric applications such as IPTV will continue to put pressure on the underlying information infrastructure.

Digital modulation techniques can be generally classified into two categories. The first is single-carrier modulation, in which the data are carried on a single main carrier. This is the "conventional" modulation format that has been the workhorse in optical communications for more than three decades. Single-carrier modulation has in fact experienced rapid advancement in recent years, and many variants to the conventional non-return-to-zero (NRZ) format have been actively explored, including return-to-zero (RZ),[3,4] duobinary,[5,6] differential phase-shift keying (DPSK),[7,8,9] and coherent quaternary phase-shift keying (QPSK).[10–12] The second category of modulation technique is multicarrier transmission, in which the data are carried through many closely spaced subcarriers. Orthogonal frequency-division multiplexing (OFDM) is a special class of MCM system that has only recently gained attention in the optical communication community, especially after being proposed as the attractive long-haul transmission format in coherent detection[13] and direct detection.[14,15]

Experiments on coherent optical OFDM (CO-OFDM) transmission at 100 Gb/s by various groups[16–18] have put the optical OFDM in the race for the next generation of 100 Gb/s Ethernet transport.

OFDM has emerged as the leading modulation technique in the RF domain, and it has evolved into a fast-progressing and vibrant field. It has been triumphant in almost every major communication standard, including wireless LAN (IEEE 802.11 a/g, also known as Wi-Fi), digital video and audio standards (DAV/DAB), and digital subscriber loop (DSL). It is not surprising that the two competing fourth-generation (4G) mobile network standards, Worldwide Interoperability for Microwave Access (WiMAX, or IEEE 802.16) from the computing community and Long-Term Evolution (LTE) from the telecommunication community, both have adopted OFDM as the core of their physical interface. Although the arrival of optical OFDM has been quite recent, it does inherit the major controversy that has lingered more than a decade in the wireless community—the debate about the supremacy of single-carrier or multicarrier transmission.[19,20] It has been claimed that OFDM is advantageous with regard to computation efficiency due to the use of fast Fourier transform (FFT), but the single carrier that incorporates cyclic prefix based on blocked transmission can achieve the same purpose.[19,20] Perhaps the advantage of the OFDM has to do with the two unique features that are intrinsic to multicarrier modulation. The first is scalable spectrum partitioning from individual subcarriers to a sub-band and the entire OFDM spectrum, which provides tremendous flexibility in either device-, or subsystem-, or system-level design compared to single-carrier transmission. The second is the adaptation of pilot subcarriers simultaneously with the data carriers enabling rapid and convenient ways for channel and phase estimation. In this book, we do not intend to resolve the debate on the superiority between single-carrier and multicarrier transmission. Instead, we focus on multicarrier modulation related to its principle, design, transmission, and application. Readers who are interested in advanced modulation formats for single-carrier transmission are referred to other excellent reading material that summarizes progress in single-carrier transmission.[21,22]

Optical OFDM bears both similarities to and differences from the RF counterpart. On the one hand, optical OFDM suffers from two well-known problems, namely high peak-to-average power ratio (PAPR) and sensitivity to phase/frequency noise. On the other hand, the optical channel has its own unique set of problems. One of the prominent differences is the existence of fiber channel nonlinearity and its intricate interaction with fiber dispersion, which is nonexistent in the RF systems. Furthermore, in the RF systems, the main nonlinearity occurs in the RF power amplifier, where a bandpass filter cannot be used to cut off the out-of-band leakage due to unacceptable filter loss. However, in optical OFDM systems, the erbium-doped fiber amplifier (EDFA; by far the most prevalent optical amplifier) is perfectly linear regardless of the level of saturation, and it is usually accompanied by a wavelength multiplexor that can remove the out-of-band spectral leakage.

In summary, after reading this book, we expect that readers—whether from an RF or an optical background—will grasp the unique promises and challenges of the optical OFDM systems.

1.1 Historical Perspective of Optical Communications

The use of light as a means of communication is natural and can be traced back to early ages of many civilizations. For instance, along the Great Wall of China is a relatively sophisticated ancient communication system composed of countless beacon towers that in many ways resembles modern-day optical communication systems. Using the color of smoke or the number of lanterns to inform the size of an invading enemy is a crude method of "multilevel" signaling. Analogous to today's repeated communication systems, the beacon towers are positioned at regular intervals along the Great Wall, and guards in each tower, upon seeing a signal from the previous one, would send the same pattern of signal to the next tower. A message could be relayed from one end of the Great Wall to the other, more than 7300 km, in slightly more than 1 hour.

Optical communication systems took a back seat for quite awhile after the advent of telegraphy, telephone, and radio networks in the first half of the 20th century. However, in the late 20th century, such electrical-based systems had reached a point of saturation in terms of capacity and reach. A typical coaxial transport system operated at a rate of 200 Mb/s needs to regenerate every 1 km, which is costly to operate. The natural trend was to study the lightwave communication systems, in which the data rate can be increased dramatically. This was boosted after the invention and the realization of a laser that gives a coherent source for the transmitter.[23] The remaining obstacle is to find an appropriate lightwave transmission medium. In 1966, Kao and Hockman proposed the idea of using the optical fiber as the lightwave transmission medium despite the fact that optical fiber at the time suffered unacceptable loss.[24] They argued that the attenuation in fibers available at the time was caused by impurities, which could be removed, rather than by any fundamental physical effects such as Rayleigh scattering. Their prophetic prediction of 20 dB/km for telecom-grade optical fiber was realized 5 years later by researchers from Corning, and currently a loss of 0.2 dB/km is the routine specification for single-mode fiber.

Despite their extremely low loss compared to that of the RF counterpart, optical systems still need regeneration for spans commonly less than 100 km. In the late 1980s and early 1990s, coherent detection communication systems were introduced to enhance the transmission distance.[25–27] However, this effort faded after the invention of the optical amplifier in the 1990s. The advent of the optical amplifier heralded a new era of optical communications in which a massive number of wavelength-division multiplexing (WDM) signals can be conveyed over thousands of kilometers.[28]

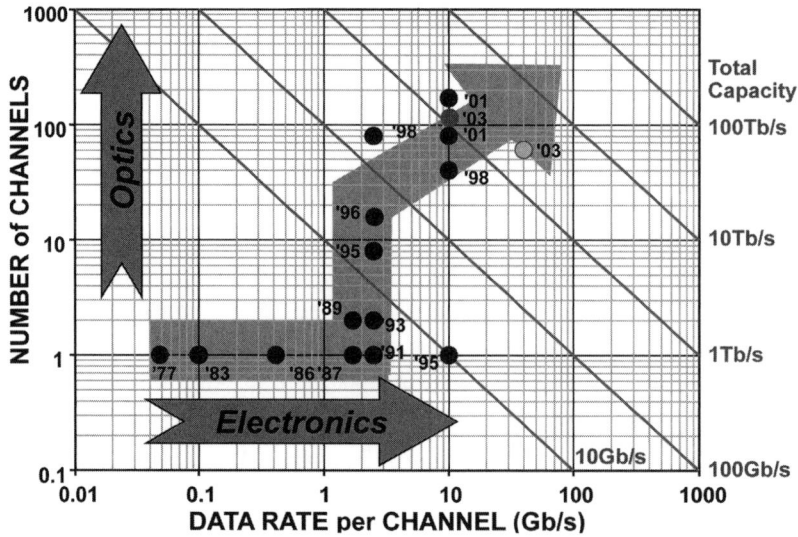

Figure 1.1: Advancement of optical fiber transmission systems in terms of the number of WDM channels and data rate per channel. Each circle represents the event of the commercial availability of the transmission system with a given performance. The year of the event is depicted adjacent to the corresponding event. Courtesy of H. Kogelnik.

Figure 1.1 summarizes the progress in optical communication that has been made during the past three decades. System capacity growth has occurred due to advancement in the line rate (x-axis) and the number of wavelength channels that can be packed into one fiber (y-axis). By the early 1990s, the single-channel system capacity had reached 10 Gb/s. Amplified systems came into use in the early 1990s and instantly boosted the capacity via supporting WDM systems. By 2003, the 40 Gb/s per channel WDM system was commercially deployed. Recent research demonstrates transmission of 21.7 Tb/s, thereby crossing the 10 Tb/s per fiber barrier.[29,30] The fiber capacity of 100 Tb/s poses new challenges to the optical communication community, requiring the innovation of novel schemes of optical amplification, modulation format, or fiber design.

1.2 Trends in Optical Communications

The advent of the Internet fundamentally changed the underlying information communication infrastructure. The phenomenal bandwidth growth led to worldwide telecom build out in the late 1990s and early 2000s, coinciding with the extremely rapid technical advancement in optical communications highlighted by the massive WDM optical amplified systems. Internet traffic has rapidly increased despite the subsequent burst of the so-called "Internet bubble" in the equity market. Many new applications have emerged, such as YouTube and IPTV, which have again continued to drive the bandwidth demand. It does not appear

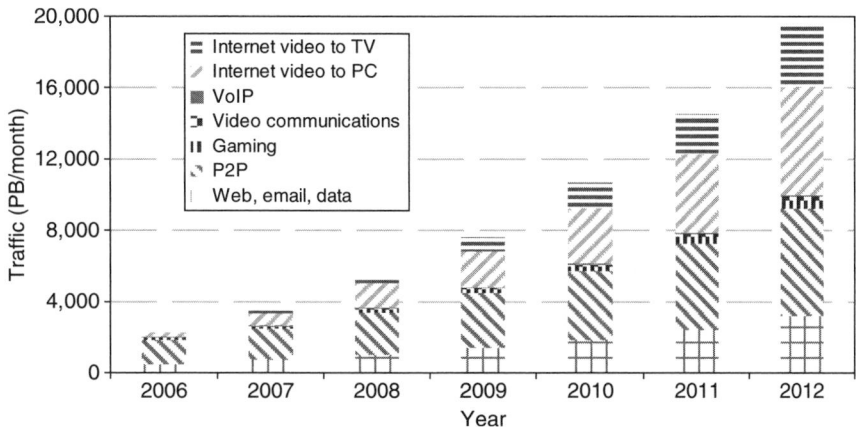

Figure 1.2: Internet traffic growth projected to 2011. Modified from Cisco, Inc.[31]

that the growth of Internet traffic will slow in the foreseeable future.[31] Figure 1.2 shows Cisco's projection of Internet traffic to 2011,[31] which shows bandwidth growth of a factor of 2 every 2 years. This phenomenal growth places tremendous pressure on the underlying information infrastructure at every level, from core to metro and access networks. In the following sections, we identify several trends in optical communication networks arising from the rapid increase in IP traffic and merging applications.

1.2.1 Evolution toward 100 Gb/s Ethernet

During the past 30 years of evolution and development, Ethernet (IEEE 802.3) has expanded from original share-medium LAN technology to a reliable standard across all levels of the network from campus LANs to metro and wide area networks (MANs/WANs). Because of the availability of a wide range of products and applications that conform to the IEEE 802.3 standard, Ethernet has become the undisputed communication technology of choice with regard to cost and reliability. There are two drivers to move the transmission rate beyond the current 10 Gb/s Ethernet (10 GbE). First, the traffic in IP backbones has grown so quickly that in 2007, some large Internet service providers had already reported router-to-router trunk connectivity over 100 Gb/s.[2] New emerging applications, such as Internet video to TV and video communications, will further triple the bandwidth demand by 2011, requiring IP link capacity scalable to 1 Tb/s.[31] Second, migration of the line rate from 10 GbE to 100 GbE is expected to achieve a reduction in capital and operational cost. It is expected that 100 GbE will operate at four or five times 10 GbE, achieving cost reduction per Gb/s. Migration to 100 GbE also leads to fewer larger bandwidth pipes between the IP routers, and it is expected to reduce the traffic planning/engineering and operating cost.

1.2.2 Emergence of Dynamically Reconfiguration Networks

The explosive growth of bandwidth-rich Internet video applications will place tremendous strain on traditional communication networks. Although the link capacity can be enhanced through migration of the transmission speed to 40 Gb or 100 Gb or by filling up more WDM channels, such simplistic augmentation of the optical transport capacity works only when transmitting information between a simple two-node point-to-point network. However, to accommodate the ever-changing pattern of bandwidth demand, optical networks are required to dynamically add, drop, and route the wavelength channels at many individual nodes. This is preferably performed in the optical domain to save unnecessary transponder costs associated with the optical-to-electrical and electrical-to-optical conversion. This form of optical domain bandwidth management can be achieved by the so-called reconfigurable optical add–drop multiplexer (ROADM).[32–34] Figure 1.3 shows a typical configuration of the ROADM-based metro core network that interconnects many distribution networks. The optical reach and the number of ROADM nodes that can be traversed transparently can be more than 1000 km and 20 pass-through nodes, respectively.[32] Since their introduction in 2003, ROADMs have become a mainstream necessity in the core network and are an essential feature in metro DWDM deployments.

The trend toward reconfigurable networks with the transport speed beyond 100 Gb/s poses two major challenges to the network design. First, the signal has become extremely sensitive to the chromatic dispersion, polarization mode dispersion (PMD), ROADM filtering effects, and the imperfection of the optoelectronic components. For instance, the tolerance to the residual chromatic dispersion for the conventional NRZ modulation format needs to be reduced 100-fold when the line rate increases from 10 to 100 Gb/s. As such, it is mandatory

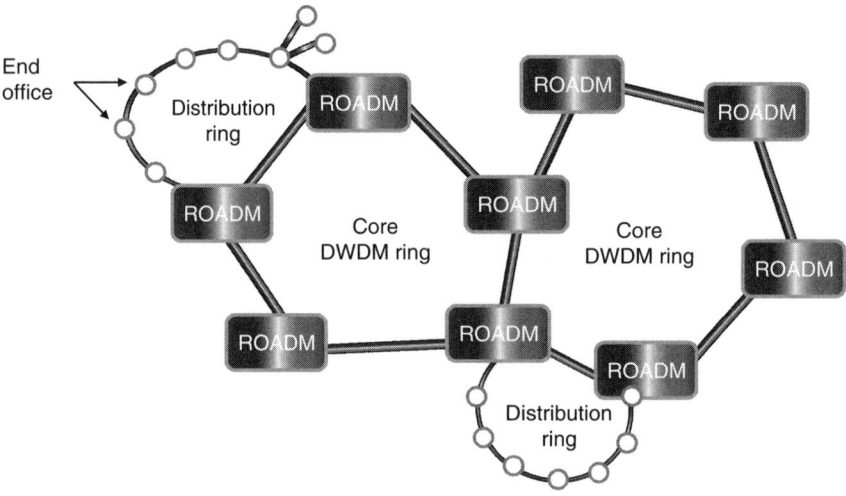

Figure 1.3: Optical networks interconnected with ROADMs.

that the per channel optical dispersion compensation is used. Furthermore, to support long-haul transmission in the range of 1000 km, even the best fiber link requires PMD dispersion compensation for 100 Gb/s and beyond. Optical PMD compensators are lossy, bulky, and expensive.[35,36] It is doubtful that such a brute-force migration to the 100 GbE can achieve the original goal of cost-savings. Second, the current optical network is "inflexible" compared to the RF counterpart—that is, it is more difficult to install, maintain, and administrate—and therefore it is quite costly to operate. As such, an adaptive optical transmission system for an agile and reconfigurable optical network is essential to support high capacity and ever evolving user demand.

1.2.3 Software-Defined Optical Transmission

In response to the emergence of a plethora of analog and digital standards in the 1980s, the concept of software-defined radio (SDR) has been proposed as a practical solution,[37,38] namely a software implementation of the radio transceiver that can be dynamically adapted to the user environment instead of relying on dedicated hardware. SDR does not imply a specific modulation technique; instead, it promotes the mitigation trend from analog to digital in wireless communications. Not surprisingly, a similar challenge arises for modern optical communications, in which multiple advanced modulation formats[10,39,40] have been proposed for the next-generation 100 Gb/s Ethernet transport. This may signal the trend toward software-defined optical transmission (SDOT), in which the transponder can be adapted or reconfigured to multiple standards or multiple modulation formats.[41] In particular, we envisage an intelligent SDOT, a similar concept to SDR, that can

1. dynamically set up the physical link without human intervention—for instance, measuring link loss and dispersion—and set up the dispersion compensation module;

2. assign an optimal line rate for the link with a sufficient margin;

3. perhaps operate at multimode—that is, elect to run in either multicarrier mode or single-carrier mode; and

4. accurately report the channel conditioning parameters, including OSNR, chromatic dispersion, PMD, and electrical SNR, that may identify the fault or predict the alarm before it occurs.

We anticipate that electronic digital signal processing (DSP)-enabled SDOT will lead to a fundamental paradigm shift from the inflexible optical networks of today to robust, "plug-and-play" optical networks in the future. The introduction of SDOT places focus on automation and reconfigurability and will inevitably lower the maintenance and operational costs, all of which are critical to ensure the sustainability of the information infrastructure that can scale up cost-effectively with the explosive bandwidth demand.

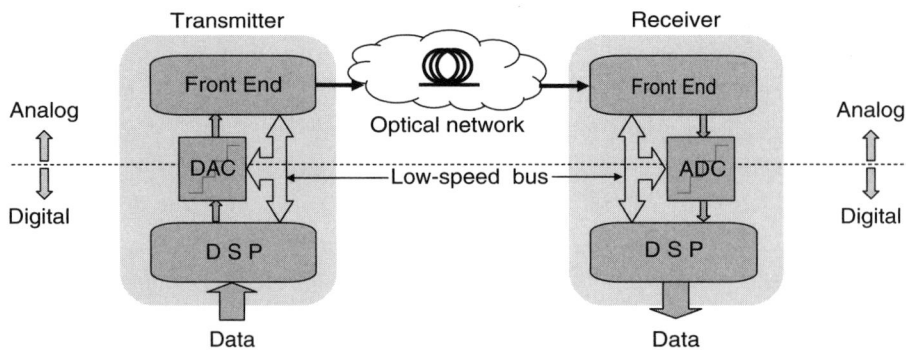

Figure 1.4: Conceptual diagram of software-defined optical transmission (SDOT).

A conceptual diagram of SDOT is shown in Figure 1.4. A salient difference from conventional optical systems is the presence of digital-to-analog/analog-to-digital converter (DAC/ADC) and DSP in the architecture of the SDOT. The entire communication system is partitioned into analog and digital domains by DAC/ADC. For optimization and application purposes, there are low-speed interactions among DSP, DAC/ADC, and the front end. Again, SDOT promotes the migration from analog to digital to enhance the optical transmissions via dynamic adaptation to the transmission channel and reconfiguration to an appropriate modulation format.

Analog-to-digital migration can be better appreciated by a discussion of the development of electronic dispersion compensation (EDC). The very early approaches of EDC were hardware based, including the feed-forward equalizer (FFE) and decision feedback equalizer (DFE), with limited performance improvement.[42] However, the ensuing EDC via DSP has shown much more significant performance improvement.[43] SDOT in essence also provides a generic architecture for various EDC via DSP. For instance, for conventional optical front ends of intensity modulation/direct detection (IM/DD) systems, maximum likelihood sequence estimation (MLSE) can be used[44,45]; for an optical in-phase/quadrature (IQ) modulator and direction detection, precompensation can be used[46,47]; for a coherent detection front end, digital phase estimation can be used to replace conventional optical phase-locked loops[48]; and for an optical IQ modulator and coherent detection front end, CO-OFDM can be realized.[13,49,50] The front ends in these examples are quite distinct, but they all take advantage of DSP to achieve significant performance enhancement of chromatic dispersion tolerance, and they can be described via the generic architecture of the SDOT shown in Figure 1.4.

1.3 Moore's Law and Its Effect on Digital Signal Processing

DSP-enabled SDOT for future optical networks is an enticing prospect. The key question is how current technologies are matched to the concept; for example, will we be able to perform the SDOT functionalities at a data rate of 100 Gb/s? To answer this question, we first need to review the current status of the underlying semiconductor technology.

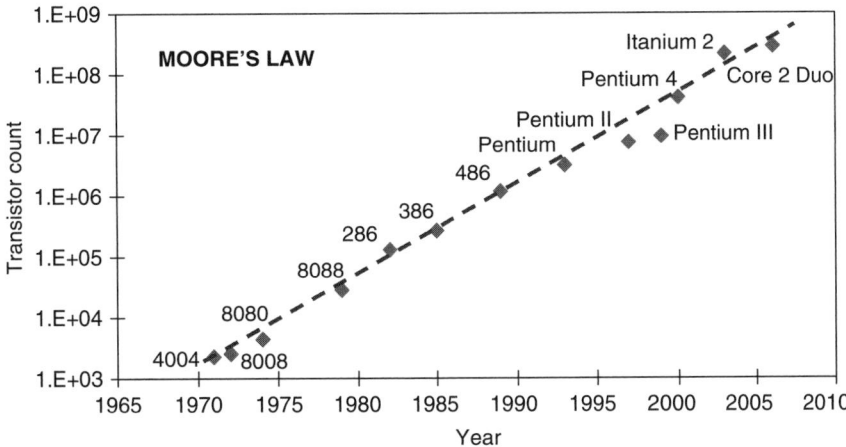

Figure 1.5: Evolution of the transistor counts for Intel's CPU chips.

In 1965, Intel co-founder Gordon Moore published a simple yet profound observation[51] that the number of transistors on a chip will double approximately every 2 years, which later was coined "Moore's law." Moore's law has become prophetic and has been the underlying driving force behind the phenomenal growth in the performance of the semiconductor device and systems. It has fundamentally impacted almost every area of our lives, from mobile phones to video game players, personal computing, and supercomputers for advanced science research. For the past four decades, through sheer creativity of researchers throughout the world as well as heavy capital investment from the industry, semiconductor technology has been kept in line with Moore's law. Figure 1.5 shows the transistor count for the popular CPU chips during the past four decades. The transistor count has increased from less than 10,000 for the 8008 to more than 1 billion for the Titanium CPU—a five order of magnitude improvement. Despite perpetual doubt among many that Moore's law will hit a brick wall in the near future, the popular belief is that this exponential growth rate will remain valid for the next decade if not longer. In fact, the International Technology Roadmap for Semiconductors predicts (as of 2008) that Moore's law will continue for several chip generations through 2029,[52] which implies a potential four or five orders of magnitude performance improvement for the CMOS technology.

1.3.1 Moore's Law Scaling

The driving force for the exponential growth rate of Moore's law is the manifestation of the continuous improvement of two basic CMOS processings[53]:

1. The gate feature—for example, the gate length and the wire pitch in the integrated circuits are shrinking at a rate of 13% per year. Also, the gate length has decreased from 50 µm in the 1960s to 90 nm in 2008.[53]

2. The size of the IC chip, or chip edge, is increasing at a rate of 6% per year,[53] or doubling approximately every decade.

Almost every other measure of the IC is a derivative to these two fundamental parameters, such as IC processing speed, power consumption, and clocking speed. Next, we discuss these important parameters of scaling based on some sensible engineering assumptions. In doing the exercise, we will search for the annual rate of the change of each parameter, or scaling of the parameter, whereas the absolute value of the corresponding parameter is omitted.

1. The number of devices that can be placed on an IC, or Moore's law: Let's denote the gate feature as x equal to 0.87 and the chip edge as y equal to 1.06, which is the annual change rate for these two parameters, as discussed previously. Then the gate counts on the chip will be y^2/x^2, equal to 1.49; that is, the gate counts are increasing by 49% per year. This is an approximate reinterpretation of Moore's law, which states that the number of devices that can be placed on an IC is doubled every 18–24 months.

2. The clock speed of an IC: We make an assumption that the carrier transport velocity in the device is a constant; therefore, the clock speed, a reflection of the switching time of the device, is inversely proportional to the gate feature x. Subsequently, we conclude that the clock speed is increasing exponentially at the rate of 15% per year.

3. Digital signal processing capacity of an IC: The improvement of processing capacity is benefited from the combination of two effects. First, more devices can be placed on the chip at the rate of y^2/x^2. Second, the clock speed is constantly accelerating at the rate of $1/x$. Therefore, the processing capacity of an IC chip equals y^2/x^3, increasing by a whopping 71% per year. This is the fundamental driving force behind ever increasing use of silicon electronics for a broad range of applications, from Apple iPhone and Nintendo Wii to personal computing and supercomputing. The application of DSP, or SDOT, in optical communications is merely a manifestation of this exponential growth trend of DSP capability. Most important, after the first DSP-based transceiver has made inroads in certain applications, it is expected that the DSP-based transceiver can potentially be improved at a stunning rate of 71% per year.

4. Power consumption: Although the leakage current can potentially be a fundamental problem for silicon IC, we assume that the power consumption is a result of moving the charge across a junction, equal to the $C_{gate}V^2$, where C_{gate} and V are the capacitance and voltage across the junction, respectively. The capacity C_{gate} can be approximated as a plate capacitor, and it can be easily shown scaled as the gate length of x. Assuming the field strength is approximately constant, the voltage V is also approximately proportional to the gate length x. Subsequently, the power consumption per gate is equal to x^3, or decreasing approximately 34% per year. This power

scaling is profound, signifying that one of the best options to obtain energy efficient devices is to use a newer generation of silicon platform that has a smaller feature, x.

1.3.2 Progress in Electronic Digital Signal Processing for Optical Communication

Conventional optical systems employ a dispersion management scheme that places a dispersion compensation module (DCF) at the amplifier site to negate the dispersion of the transmission link.[54,55] The DCF could be placed at the optical amplifier site within a double-stage amplifier or be distributed in a dispersion mapped span as shown in Figure 1.6. Such a dispersion management scheme works fine for transmission systems at 10 Gb/s and lower, and it is difficult to perfect at 40 Gb/s and higher. Both the nominal dispersion and the dispersion slope of the DCF need to be matched precisely. Any residual mismatched dispersion needs to be compensated on a per-channel basis with a fixed or tunable optical dispersion module.[56,57] Since the early 1990s, there has been great interest in using electronic equalizers as replacements for optical dispersion compensating modules at the receiver.[58–60] Compared with the optical counterpart, the electron equalizer has the advantages of lower cost, small footprint, and ease of adaption. Electronic equalizers that have adopted the classical equalization approaches include FFE, DFE, and Viterbi equalizer. These early stage equalizers mostly utilized SiGe or InP/AlGaAs technology with the channel length limited to 2 or 3 bits. Nevertheless, they were successfully used in commercial deployment.[61] The major breakthrough in electronic signal processing occurred in 2004 when researchers from Nortel published their predistortion equalizer showing 5000 km transmission over standard single-mode fiber (SSMF) without an optical dispersion compensator.[46] The predistortion equalizer is based on a powerful silicon chip encompassing a digital finite impulse filter with a large number of taps, high-speed DAC, and an optical IQ modulator. This work raised an interesting and fundamental question for the optical community regarding the necessity of the

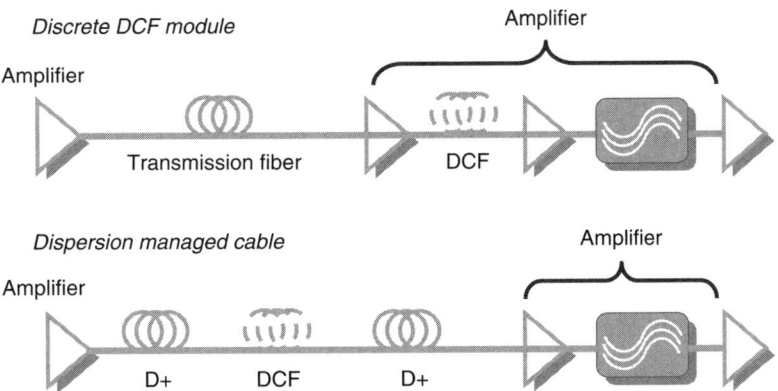

Figure 1.6: Application of the discrete DCF module and dispersion managed span.

dispersion compensated map as shown in Figure 1.6, which has been so deeply ingrained in the optical communications field for decades. The work ushered in an era of electronic DSP enabled optical transmission followed by the subsequent dramatic revival of the coherent optical communications.[11,12,62–66]

Although rapid advances were made in single-carrier transmission systems based on electronic digital signal processing, multicarrier modulation started to emerge as the competitive alternative. In 2006, three groups independently proposed two types of optical OFDM for long-haul application that were also aimed at eliminating the need for dispersion management in optical transmission systems: direct-detection optical OFDM (DDO-OFDM)[14,15] and CO-OFDM.[13] CO-OFDM holds the promise of delivering superior performance in spectral efficiency, receiver sensitivity, and polarization-dispersion resilience,[67,68] but implementation is much more complex than for DDO-OFDM.[14,15] The transmission experiments of CO-OFDM produced in research laboratories have achieved 100 Gb/s transmission over 1000 km SSMF.[16–18] Because both the single-carrier coherent system and CO-OFDM are attractive candidates for 100 GbE transport, the intriguing question naturally arises regarding which one is superior. This is addressed in the next section.

1.4 Single-Carrier or Multicarrier Transmission: An Optical Debate

Although OFDM has gained popularity in the past decade and has widely been implemented in numerous communication standards, there has been ongoing debate regarding whether OFDM or single-carrier frequency domain equalizer (SCFDE) is superior.[19,20] OFDM has two fundamental problems: (1) large PAPR and (2) sensitivity to frequency and phase noise. The debate has not produced a clear-cut answer but, rather, resulted in a split decision even in some standards; for instance, the United States has chosen single-carrier 8-level vestigial sideband modulation (8-VSB) as the digital TV standard, whereas Europe, Japan, and most other countries have elected OFDM. It may be premature to conclude that the debate of single-carrier and multicarrier transmission in the optical domain will emerge in the same manner as in the RF domain. Given the fact that the communication channel, devices, and systems are quite distinct between these two domains, it is imperative that we thoroughly understand the problems at hand and clarify the context in which the debate is being conducted.

The debate regarding single-carrier versus multicarrier transmission may have different meaning in the optical domain. Figure 1.7 shows the transmitter architectures for single-carrier systems and CO-OFDM systems. There are two conspicuous differences:

1. Single-carrier systems employ a relatively "conventional" and simpler architecture, in which discrete digital-level modulation is fed into the two arms of the QPSK modulator. With regard to generating the I and Q component, the QPSK modulation resembles that

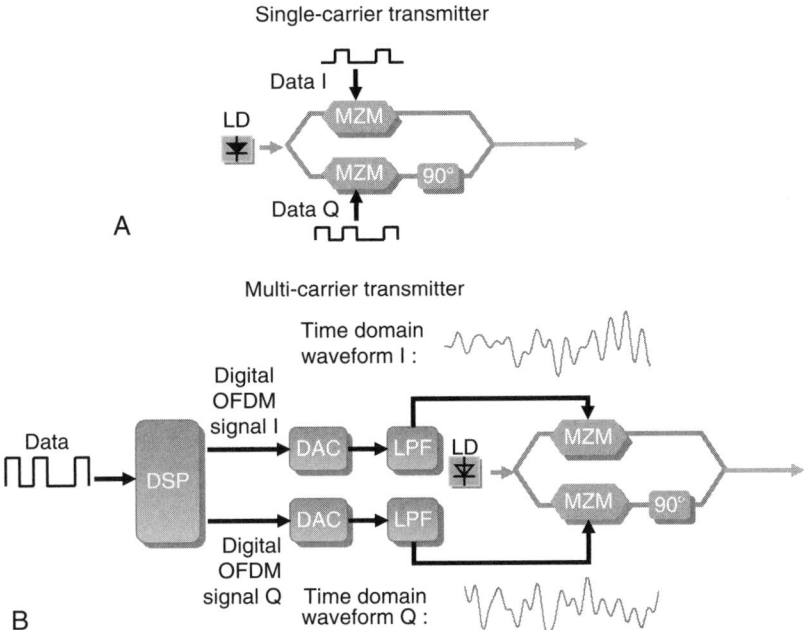

Figure 1.7: Transmitter architecture for (a) single-carrier systems and (b) multicarrier systems.

of conventional BPSK or DPSK modulation. In contrast, the CO-OFDM architecture includes drastic modification from the conventional single-carrier system, in which the electronic DSP module and DAC are required for complex OFDM signal generation at the transmit end. The OFDM transmitter strictly enforces linearity in each component associated with the CO-OFDM transmitter.

2. In the single-carrier systems, the information is coded in the time domain, whereas in CO-OFDM, the information is encoded in the frequency domain, more precisely onto each individual subcarrier.

Based on these two differences, we now make some detailed comparisons of some key properties:

Ease of signal processing: CO-OFDM places the signal processing capability in the transmitter and enables the aforementioned SDOT that brings all the benefit of transmitter adaptability. In particular, we first discuss the two important signal processing procedures for coherent communications: channel estimation and phase estimation. In CO-OFDM-based systems, by using pilot symbols or pilot subcarriers, the channel estimation and phase estimation are made relatively straightforward. In the single-carrier coherent systems as shown in Figure 1.7, the channel estimation has to rely on blind equalization—for instance, using the CMA algorithm—or decision feedback, both of which are prone to error propagation. The phase estimation usually adopts the Viterbi algorithm, which is most effective for the pure phase modulation and less effective for other constellation modulation. Furthermore, differential-phase coding needs to be employed to resolve the intrinsic phase ambiguity for the mth-power law algorithm, resulting in approximately a factor of 2 BER increase.[69]

Higher order modulation: For the commonly used QPSK modulation, the transmitter complexity of CO-OFDM is higher, as shown in Figure 1.7, but once the modulation goes beyond 2 bits per symbol, such as 8-PSK or 8-QAM, the CO-OFDM has lower complexity than the single-carrier system, which subsequently reduces the system cost. The reason is that CO-OFDM can be gracefully scalable to the higher order modulation without optical hardware alternation. The only change from 4-QPSK to 8-QAM modulation is enabled via the software to reconfigure the DSP and DAC. In contrast, the higher order single-carrier optical system requires more complicated optical modulator configuration either in a serial or in a parallel manner,[30] which inevitably increases the system complexity and cost. The drive toward more complex constellation for high-spectral efficiency transmission is certainly turning the tide in favor of CO-OFDM.

Tight bounding of spectral components: Because the OFDM spectral shape is tightly bounded, it is more tolerant to the filter narrowing effect. As long as the filter is wider than the OFDM rectangular-like spectrum, the OFDM signal practically suffers no penalty. Even if the edge subcarriers are attenuated by the narrowing filtering, some form of bit and power loading scheme can be employed to mitigate the effect. In contrast, for the single-carrier system, because of the difficulty in reducing the timing jitter at the high clock rate, it is necessary to maintain some additional excess bandwidth for the pulse shaping such that a sufficient margin is allocated for timing accuracy. The filtering narrowing effect not only causes the pulse distortion but also makes single-carrier signal susceptible to the timing jitter. The resilience to the filter narrowing effect makes CO-OFDM particularly fit for the systems employing long cascades of ROADMs.

Bandwidth scalability: Because CO-OFDM spectrum is inherently tighter than the single-carrier one and the CO-OFDM signal is generated in the frequency domain, it is relatively simple to partition the entire OFDM spectrum into multiple bands and process each band separately.[16,18] In particular, if the orthogonality is maintained between adjacent bands, there is no need for the frequency guard band; that is, there is no sacrifice in spectral efficiency for the sub-banding of OFDM spectrum.[16] In doing so, the OFDM transceiver is not limited to the bandwidth constraint of the DAC/ADC. In contrast, the single-carrier encodes the information across the entire spectrum, making it impossible to scale down the bandwidth. It is foreseeable that single-carrier coherent systems solely relying on the timing domain information encoding will hit the brick wall of the electronic DSP speed much sooner than the CO-OFDM-based systems.

Sub-wavelength bandwidth access for performance monitoring and multiaccess/multicast networks: It is a great advantage to place the DSP in the transmitter for CO-OFDM systems. The ability and flexibility to allocate a certain number of subcarriers for channel estimation and performance monitoring will prove to be an attractive feature for CO-OFDM. For instance, this leaves the option of grouping a band of subcarriers for monitoring, which can be easily detected without processing the entire spectrum. Similarly, sub-banding of OFDM allows for the dynamic bandwidth allocation for multiaccess networks using the orthogonal frequency domain multiple-access scheme.[70] All these are difficult to achieve with the single-carrier system.

Computation complexity: The computation complexity is an important factor that affects the chip design complexity and power consumption of the DSP chip. The single-carrier system using IFFT/FFT has the computation complexity scales as the channel length N_x:

$$C_{bit} \propto \log_2(N_{sc}), \; N_{sc} = \alpha \cdot D \cdot B \quad (1.1)$$

where C_{bit} is the computation complexity defined as the number of multiplications required per bit, and N_{sc} is the number of subcarriers in CO-OFDM or number of points used in FFT/IFFT. The computation complexity of the single-carrier system involving DFT and IDFT is the same as that for CO-OFDM, but for the single-carrier time domain equation systems based on FIR equalization, the computation complexity scales as follows[43,71]:

$$C_{bit} \propto D \cdot B^2 \quad (1.2)$$

It can be seen that CO-OFDM outperforms the FIR equalized single-carrier systems,[43,71] and the advantages are dependent on the detailed design.[71] On the other hand, the SCFDE uses block signal processing, and FFT/IFFT can have the computation complexity on par with CO-OFDM.[19,20,71] However, the block-based signal processing is more conveniently performed if the DSP is available in the transmitter, and if so, the advantage of transmitter simplicity for single-carrier systems as shown in Figure 1.7a disappears. More important, once the DSP and DAC are available at the transmitter and the signal processing is performed on a block basis, the distinction between the multicarrier and single-carrier systems is purely pedagogical.

Sampling rate: For single-carrier systems, it is best for the sampling rate to be twice the signal band rate because it is sensitive to sampling phase inaccuracy if lower.[10–12] Although resampling can reduce the oversample factor somewhat, the computation is intensive and thus impractical. For the CO-OFDM system, oversampling is done simply by not filling the edge subcarriers in order to tightly bound the signal spectrum[16,18]; therefore, approximately 10–20% oversampling is sufficient.[64,72] Reduction of the sampling rate of CO-OFDM from that of the single-carrier system will become more attractive when the high-speed ADC/DACs become more difficult to design.

Tolerance to the component imperfection: It is anticipated that transmitter components, including the RF amplifier, ADC/DAC, and optical IQ modulator, will deviate from their perfect form when operating at high speed for the bit rate of 100 Gb/s and beyond. CO-OFDM enforces linearity throughout every stage of the transmitter design; thus, the imperfections, when linear in nature, can be largely estimated and compensated through the transmitter and receiver signal processing. In contrast, the single-carrier system relies on the drive voltage operating at the saturation, making the component imperfection difficult to estimate and mitigate.

Bit and power loading: One of the major advantages of CO-OFDM is the ability to manipulate the frequency domain at the transmitter, which involves bit and power loading along the line of the "water-filling" algorithm.[73] This is a commonly emphasized advantage in the RF communications in which the channel can be in deep fading, or some part of the spectrum may be completely notched out due to the severe multipath interference. How this bit/power loading capability is to be exploited in the optical domain is of great research interest. Furthermore, the channel rate of a conventional optical transmission system is set at the required level throughout its lifetime, whereas the CO-OFDM offers the new functionality of the adaptive data rate according to the channel condition through bit or power loading. The benefits of the adaptive data rate are reduced transponder inventory because one transponder can be used for multiple data rates and the increased channel usage by delivering more data rate when the margin is plentiful.

Based on the previous comparisons, we conclude that CO-OFDM is advantageous, especially in areas that are key to future transmission systems, including scalability to the ever increasing data rate and transponder adaptability. Nevertheless, the CO-OFDM unavoidably inherits the two main problems intrinsic to OFDM: (1) high PAPR that suggests the CO-OFDM is more susceptible to nonlinearity, and (2) sensitivity to the frequency and phase noise. A proper understanding and treatment of these two important issues is critical to the implementation of CO-OFDM and is without doubt an area of rich research potential. It is also extremely meaningful to briefly discuss different circumstances in which the two problems are being investigated for RF and optical OFDM systems, which are addressed in the next section.

1.5 The Difference between RF OFDM and Optical OFDM Systems

It is a fallacy that because RF OFDM has been extensively studied during approximately the past 20 years, optical OFDM will be an effortless one-to-one translation from the wireless domain to the optical domain. As we shall see in what follows, a clear understanding of the uniqueness of the optical channel and the optical systems makes possible the most efficient

Table 1.1: Comparison between Wireless and Optical Channels

	Mathematical Model	Nonlinearity	Speed
Wireless OFDM	Time domain multiple discrete Rayleigh fading	None	Can be fast for mobile environment
Optical OFDM	Continuous frequency domain dispersion	Significant	Medium

design of the CO-OFDM systems. Using the SMF optical communication systems and the wireless systems as examples, we lay out the following differences that have significant ramifications for the OFDM design:

Channel model: Table 1.1 summarizes the differences between wireless and optical communication channels. A typical wireless channel can be modeled as summarization of the multiple paths that each undergoes a Rayleigh process, given by[74,75]

$$h(t,\tau) = \sum_{l=1}^{L} a_l \cdot g_l(t) \cdot \delta(\tau - \tau_l) \tag{1.3}$$

where a_l is a complex constant, $g_l(t)$ is the Rayleigh fading process, and τ_l is the delay for the lth path. A optical SMF can be more conveniently modeled in the frequency domain for the two polarization components in the fiber expressed as[67]

$$H(t,f) = e^{j\Phi_D(f)} \cdot T_k \tag{1.4}$$

$$\Phi_D(f) = \pi \cdot c \cdot D_t \cdot f^2 / f_{LD1}^2 \tag{1.5}$$

$$T_k = \prod_{l=1}^{N} \exp\left(\left(-\tfrac{1}{2}j \cdot \vec{\beta}_l(t) \cdot f - \tfrac{1}{2}\vec{\alpha}_l(t)\right) \cdot \vec{\sigma}\right) \tag{1.6}$$

where $\Phi_D(f)$ is the phase dispersion due to the fiber chromatic dispersion effect, T_k is the Jones matrix for the fiber link representing the polarization-dependent effect including PMD and polarization dependent loss (PDL), N is the number of PMD/PDL cascading elements represented by their birefringence vector $\vec{\beta}_l$ and PDL vector $\vec{\alpha}_l$,[76] and $\vec{\sigma}$ is the Pauli matrix vector.[76] The significance of the channel model is that the majority of the channel dispersion $e^{j\Phi_D(f)}$ can be first estimated and factored out for the channel estimation. Dynamic dispersion comes from the Jones matrix T_k, but it can be reduced effectively to a summation of only a few taps of the FIR model if its mean PMD value is known. Therefore, dependent on the PMD value and the data rate, the channel estimation can be greatly simplified for the optical OFDM systems.[77]

Channel nonlinearity: Wireless channel is in free space and therefore does not possess any nonlinearity. On the other hand, optical fiber is fairly nonlinear. Coupled with the fiber dispersion, PMD, and PDL effects, the optical channel is arguably more complicated than a wireless channel. Most often, there is no closed-form analytical solution for nonlinear transmission in the optical fiber, and subsequently the numerical solutions to the nonlinear Schrödinger equation that describes the nonlinear wave propagation in the fiber are required to analyze the performance.[78] At first glance, OFDM plagued with high PAPR would not fit for the optical fiber with high nonlinearity. Fortunately, the fiber chromatic dispersion serves as the saving grace that tends to mitigate against the nonlinearity,[79] and experiments have shown successful transmission of 100 Gb/s CO-OFDM over 1000 km SSMF fiber.[16–18]

Time variation of the channel characteristics: As important as the frequency dispersion (or frequency selectivity) of the channel, the time selectivity or dispersion is another determining factor.[74,75] Time dispersion is defined as the rate at which the channel characteristics are changing. In wireless systems, time dispersion is characterized as the Doppler frequency from the fast-moving mobile users, whereas in the fiber-optic systems it is characterized as the polarization rotation from the mechanical disturbance of the fiber-optic link. The extent of the time selectivity is characterized by the product of the Doppler frequency in the wireless systems (or polarization rotation rate in fiber-optic systems) and OFDM symbol length, which is approximately 0.04 for Universal Mobile Telecommunications System or wireless LAN environment[75] or 5×10^{-5} for fiber-optic systems (using 50 ns symbol length and 1 kHz polarization rotation rate). Subsequently, the optical channel can be considered quasistatic. The efficient channel estimation can be adopted by taking advantage of this important fact.

Amplifier nonlinearity: This is an important factor that may not have been commonly recognized. In wireless systems, the major nonlinearity takes place in the power amplifier. It is critical to either have a high saturation power RF amplifier or operate at the sufficient back-off. However, in fiber-optic systems, the predominant amplifier is EDFA, which is perfectly linear. This is because the response time of the EDFA is milliseconds, and therefore any nonlinearity faster than milliseconds would practically vanish. This is significant in the sense that in designing CO-OFDM systems, when confronting the trade-off between the optical loss and RF loss, we would choose the former because it is more linear. For instance, in the CO-OFDM transmitter design, we would choose to minimize the RF drive voltage to the optical IQ modulator and optically amplify the signal to compensate for the excess loss of the optical IQ modulator.

Tolerance to the out-of-band emission: In the wireless systems, because of the scarcity of the spectrum, the RF channel is packed as tightly as possible. Therefore, there is a stringent out-of-band emission requirement that is enforced upon the OFDM transmitter. Figure 1.8 shows the transmission mask for a Wi-Fi signal that details the maximum relative intensity to which the transmission emission should be restricted.[80] For instance, for the Wi-Fi signal with channel spacing of 20 MHz, the maximum out-of-band

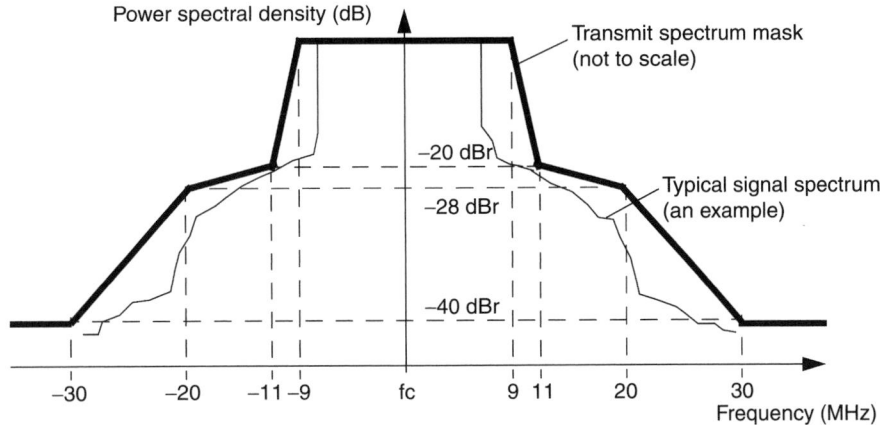

Figure 1.8: Transmission spectrum mask for Wi-Fi signal. After reference 80.

emission intensity at 11, 20, and 30 MHz is, respectively, -20, -28, and -30 dB relative to the in-band density. It would have been an easy task if an RF filter could be used to remove any out-of-band emission at the transmitter in a wireless system. The problem is that the power amplifier is one of the major contributors to the overall chip power consumption, and the introduction of an RF filter following the power amplifier would induce significant loss and decrease the chip power efficiency; therefore, it is commonly avoided in wireless systems. This places a stringent requirement to the OFDM transmitter in terms of the nonlinearity control. In contrast, in the optical systems, wavelength division multiplexing devices are commonly used to combine multiple wavelengths, and any out-of-band emission from the CO-OFDM transmitter is effectively eliminated. As such, the CO-OFDM transmitter is more tolerant to the out-of-band emission. This fact should be taken advantage of when dealing with the PAPR reduction in CO-OFDM systems.

1.6 What Does OFDM Bring to the "Game"?

In Section 1.2, we presented an overview of the trends in optical communications and the evolution toward highly reconfigurable networks at the channel speed of 100 Gb/s and beyond. We also reviewed technological advances in silicon technology indicating that the time has come for SDOT. To capture the gist of the past few sections, here we show that optical OFDM which combines the merits of the coherent detection and OFDM technology has arrived in time for the next-generation optical networks.

1.6.1 Scalability to the High-Speed Transmission

To cope with upgrades to the next transmission speed in optical systems, the WDM equipment needs to be significantly redesigned. Sometimes perhaps a new type of fiber is required to be redeployed. Specifically, the optical transmission systems have advanced from direct modulation in multimode fiber to modulation in single-mode fiber, such as externally modulated NRZ modulation, RZ modulation, DPSK modulation, and single-carrier coherent systems, or perhaps optical OFDM systems. A significant capital investment is required for each technology advance involving high-speed circuit design and testing and also system-level testing and evaluation. There is significant risk regarding whether the investment can be recouped. As such, it is critical to identify a technology that will have a maximum life span and that can be adapted to the next generation of the product with minimum modification in hardware and software. We aim to show that optical OFDM is the ideal future-proof technology that can be gracefully scaled to ever increasing transmission speed.

As we alluded to in Section 1.4, because the information in optical OFDM is managed in the frequency domain, it is relatively simple to partition the entire broad OFDM spectrum into many sub-bands whenever the bandwidth bottleneck occurs for the DAC/ADC signal processing elements. This ensures a smooth migration path of CO-OFDM from 40 to 100 Gb/s, or even 1 Tb/s, by augmenting the spectrum through OFDM band multiplexing.[16,7] In doing so, the hardware and software design developed for 40 Gb/s will very likely be reused at 100 Gb/s, or even Tb/s, without a major design overhaul. Although OFDM modulation does not appear to be as familiar as single-carrier modulation in terms of its transmitter and receiver design as well as the associated signal processing, significant benefit will be gained by developing such a technology that will be reused in the next-generation products for the foreseeable future. This is a much better scenario than adopting a modulation format at 40 Gb/s that may not be applicable to 100 Gb/s or future Tb/s systems.

1.6.2 Compatibility to the Future Reconfigurable Optical Networks

The optical network today is predominately based on point-to-point connections, involving laborious human intervention in installation, maintenance, and bandwidth provisioning. As discussed in Section 1.2, modern optical networks are evolving toward those that are highly adaptive and reconfigurable, similar to the wireless LAN. This concept of plug-and-play optical networks minimizes the cost associated with expensive human intervention and will be critical to the future optical networks that can economically meet the capacity demand from bandwidth-rich Internet video traffic. We identify the following attributes of optical communication systems that are essential to the future optical networks and show how optical OFDM can gracefully meet each challenge that may not be sufficiently addressed with the conventional modulation formats:

Transmission agnostic to the underlying physical link: The optical fiber links deployed throughout the past three decades comprise fibers with differing chromatic dispersion and polarization mode dispersion, as well as varying span distance from 40 to 120 km and varying mechanical stability, dependent on whether the fibers are buried in the ground, hung from poles, or located along disturbance-prone rail tracks. The large divergence in optical dispersion and mechanical stability makes it difficult for conventional modulation formats, such as direct-detection NRZ/RZ or DPSK, to meet these challenges for high-speed transmission at 100 Gb/s and beyond. CO-OFDM, with its intrinsic resilience to any optical dispersion and a simple and convenient method of adaptive equalization based on the pilot subcarriers/symbols, is an ideal candidate for meeting these challenges.

Adaptive data rate provisioning: Because of the vast diversity of physical link, the supported link data rate may not be predetermined or may be too costly to be predetermined. Because CO-OFDM supports digital signal processing at the transmitter and receiver, the optimum data rate can be obtained through the bit and power loading or subcarrier-level manipulation. At the beginning of the life, the carried data rate can be higher than the nominal rate when the margin is available. On the other hand, when the margin is not sufficient, the data rate can be simply reduced through loading less subcarriers or adding more forward error correction (FEC) redundancy. The concept of the link operation at the downscaled rate is beneficial compared to the alternative scenario in which the link is rendered useless if a full rate is enforced. The software provisioning of multiple data rate obviously reduces the inventory counts and its associated cost.

Sub-wavelength granularity bandwidth access: The optical transmission channel data rate is envisioned to advance to 400 Gb/s or even 1 Tb/s within the next decade. CO-OFDM provides a seamless pathway to 1 Tb/s Ethernet from the current generation of 100 Gb/s Ethernet.[81] This enormous data rate of 1 Tb/s may make sense among the major nodes, but it would be beneficial to provide a pathway for a smaller intermediate node to economically access the 1 Tb/s traffic at a lower rate because some of the intermediate nodes may not need 1 Tb/s bandwidth but still need to access the multicast traffic at 1 Tb/s. Band-multiplexed OFDM is ideal for such a purpose. Figure 1.9 shows the 1 Tb/s link between two major nodes, nodes A and B, where the CO-OFDM signal is composed of 12 of 100 Gb/s sub-bands. In the intermediate node C, only the third band is accessed. This is done by simply tuning the receive laser to the center of the third band and passing the down-converted RF signal to a low-pass RF filter. Because the entire receiver signal processing is performed at 100 Gb/s, the cost for the intermediate load can be cheaper than that of the major node at 1 Tb/s. Of course, future bandwidth expansion of the intermediate node can be achieved by adding more circuit cards at 100 Gb/s.

Self-performance monitoring: CO-OFDM is heavily reliant on the channel estimation for equalization, and subsequently various important system parameters can be acquired without resorting to separate monitoring devices.[82,83] Parameters that can be extracted from CO-OFDM receivers include chromatic dispersion, PMD, laser phase noise, OSNR,

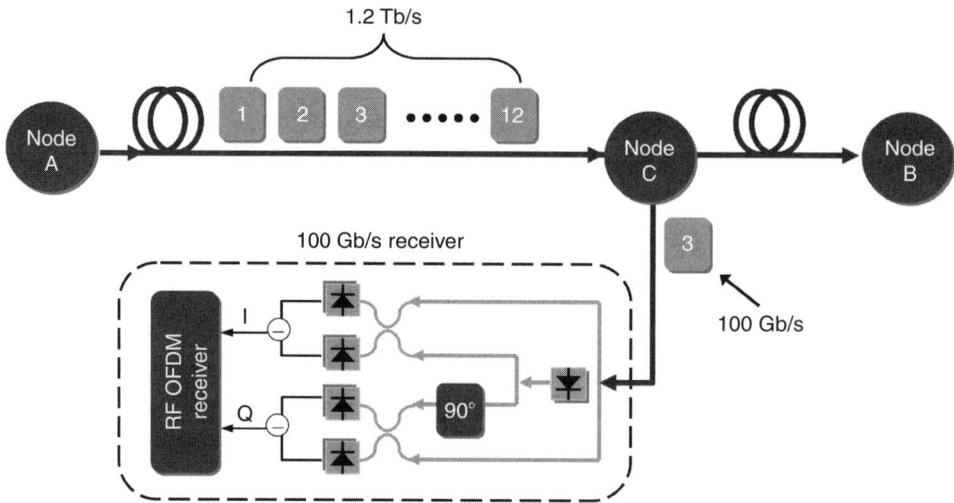

Figure 1.9: Illustration of sub-wavelength bandwidth access of 1 Tb/s CO-OFDM signal for an intermediate node C with a 100 Gb/s receiver. The selectivity is done by tuning the local laser to the center of the desired OFDM band (e.g., the third band in the illustration).

nonlinearity, and Q-factor. These are important parameters for network monitoring and maintenance. For instance, Q-factor statistics can be gathered to give the operator a clear view of the margin for the link, advanced warning for the replacement, or to indicate optical power adjustment before the system collapses into outage.

Energy efficiency awareness: As the channel rate and the overall capacity increase to meet the exponential growth of IP traffic, there is a growing concern about the energy consumption of the telecommunication equipment in the context of its environmental and social impact. It has been recognized that the electrical energy consumption is a significant contributor to greenhouse gasses.[84–86] Conventional transmission systems are designed to operate properly during the worst-case scenario, or all the transponders may be overprescribed. In optical OFDM-based systems, the dynamic adaptation of the bit resolution, RF amplifier power, FEC gain, as well the complexity of the channel estimation will inevitably reduce the unnecessary signal processing and subsequently improve energy efficiency. For instance, most of the time the optical link is quite stable; the rate of the PMD variation is slow (<10 Hz). Therefore, channel estimation can be done less frequently; it only needs to be updated at the faster rate when the fiber link is occasionally disturbed due to human actions at the central office. On average, the computational complexity and its associated power consumption due to the channel estimation are dominated by the slow dynamics of the fiber optics, and therefore energy efficiency is enhanced over that of the design that assumes the fast-varying channel all the time.

1.7 Channel Coding and OFDM

As discussed previously, future Internet traffic growth will require deployment of optical transmission systems with data rates at 100 Gb/s and higher. However, at these data rates, the signal quality is significantly degraded mainly due to the impact of PMD and intrachannel nonlinear effects. By using coded OFDM, we aim to achieve transmission beyond 100 Gb/s while employing commercially available components operating at much lower speeds. Using this approach, modulation, coding, and multiplexing are performed in a unified manner so that, effectively, the transmission, signal processing, detection, and decoding are done at much lower symbol rates. At these lower rates, dealing with the nonlinear effects and PMD is more manageable, while the aggregate data rate is maintained above 100 Gb/s. In this book, we describe different FEC schemes suitable for use in coded OFDM, including different classes of block codes, convolutional codes, and concatenated codes. To deal with bursts of errors introduced by intrachannel nonlinearities and PMD, we describe the use of interleavers and product codes.

Codes on graphs, such as turbo codes and low-density parity-check (LDPC) codes, have revolutionized communications and are becoming standard in many applications. The inherent low complexity of LDPC decoders opens up avenues for their use in different high-speed applications, including optical communications. LDPC codes, invented by Gallager[87] in the 1960s, are linear block codes for which the parity-check matrix has low density of ones. LDPC codes have generated great interest in the coding community, resulting in a better understanding of different aspects of LDPC codes and their decoding process. Moreover, it has been shown that an iterative LDPC decoder based on the sum-product algorithm is able to achieve a performance as close as 0.0045 dB to the Shannon limit.[88] In this book, we describe the large-girth binary LDPC code design and the min-sum-with-correction-term decoding algorithm and its FPGA implementation, and we discuss a class of nonbinary LDPC codes suitable for use in optical communications. We further explain how to combine multilevel modulation and channel coding by using coded OFDM.

1.8 Overview of the Book

This book consists of 12 chapters covering a broad range of topics related to optical OFDM.

In Chapter 2, we give a historical perspective of OFDM and a brief discussion of its emergence in optical communications. We then provide an introduction to OFDM fundamentals, including its basic mathematical formulation, discrete Fourier transform implementation, cyclic prefix, spectral efficiency, and PAPR property. In addition, we provide a detailed analysis of the SNR impairment from the frequency offset and phase noise for OFDM systems.

In Chapter 3, we describe the key optical components used in contemporary optical communication systems: basic signal and noise parameters; major channel impairments, including chromatic dispersion, PMD, and fiber nonlinearities; and the system engineering process.

In Chapter 4, we lay out various aspects of OFDM signal processing, including the three levels of synchronization comprising window synchronization, frequency synchronization and channel estimation, ADC/DAC impact on system performance, PAPR reduction algorithms, and multiple-input multiple-output (MIMO)-OFDM.

Understanding polarization effects is crucial for the study of optical OFDM. In Chapter 5, we describe the basic mathematical formulation of the electrical field in Jones vector representation and its nonlinear propagation through fiber. Then, we discuss the important concepts of PMD and PDL. Both linear and nonlinear polarization effects are analyzed. Finally, the relationship of the polarization effects with MIMO-OFDM is elucidated.

In Chapter 6, we describe coded OFDM as an efficient approach to overcome major channel impairments, including chromatic dispersion and PMD. We describe different channel codes suitable for use in coded OFDM. For access networks and the transmission over multimode fibers, we describe RS codes and concatenated codes as possible candidates. For long-haul transmission and 100 Gb/s Ethernet, we describe iteratively decodable codes, such as turbo-product and LDPC codes, as attractive options for use in coded OFDM. In future optical networks, different destination nodes may require different levels of protection. For those applications, we describe the rate-adaptive channel codes suitable for optical OFDM.

In Chapter 7, we elaborate on the richness of the OFDM variations. We explain the architecture, design, and performance of incoherent and coherent optical OFDM. We also show how these two detection schemes can be employed in various applications. This chapter serves as the transition to subsequent chapters (Chapters 8–11).

In Chapter 8, we show why OFDM is a promising technique for the forthcoming 100 Gb/s Ethernet transport. We elucidate the multiband OFDM as a bandwidth-scalable and spectral efficient modulation and multiplexing scheme, aiming to alleviate the electronic bandwidth constraint from DAC/ADC. We also turn our attention to the higher order modulation transmission, which we consider as one of the critical advantages for optical OFDM.

In Chapter 9, we describe the OFDM as an efficient way to deal with multimode dispersion in MMF links and to overcome bandwidth limitation in plastic optical fiber (POF) links. The coding techniques suitable for use in coded OFDM for MMF and POF links are described as well. The MIMO-OFDM is discussed as a potential candidate to mitigate multimode dispersion in MMF links.

In Chapter 10, we describe the OFDM as an efficient modulation and multiplexing technique that can be used (1) to enable transmission over the strong atmospheric turbulence channel and (2) as an interface between wireless and optical channels. Because the free-space optical (FSO) channel is envisioned as the solution to the connectivity bottleneck problem, and as a supplement to wireless links, the complexity of the transmitter and receiver must be low; therefore, the IM/DD is considered a reasonable choice for FSO links. The optical OFDM with direct detection seems to be a viable option for FSO links. To improve the power efficiency of direct detection OFDM, several power efficient schemes are described, including the clipped OFDM.

In Chapter 11, we explain the basic principles of passive optical network, radio-over-fiber communications, and WiMAX and UWB communications. We also describe how WiMAX and UWB can be integrated with optical OFDM. In the last section of this chapter, we describe the potential use of OFDM in optical LANs.

Finally, in Chapter 12, we illustrate and emphasize why optical OFDM is a vibrant and fast-progressing field that provides great potential for both practical development and research endeavor in the field of optical communications. We present our views on some interesting research problems in the emerging field of optical OFDM as well as touch upon the intriguing issue related to standardization of optical OFDM technology.

References

1. Coffman KG, Odlyzko AM. Growth of the Internet. In: Kaminow IP, Li T, editors. *Optical Fiber Telecommunication IV-B: Systems and Impairments*. San Diego: Academic Press; 2002. p. 17–56.
2. Melle S, Jaeger J, Perkins D, Vusirikala V. Market drivers and implementation options for 100-gbe transport over the WAN. *IEEE Applications Practice* 2007;**45**:18–24.
3. Atia WA, Bondurant RS. Demonstration of return-to-zero signaling in both OOK and DPSK formats to improve receiver sensitivity in an optically preamplified receiver. In: *IEEE Lasers and Electro-Optics Society Annual Meeting*, vol. 1. San Francisco. p. 226–7.
4. Miyamoto Y, Hirano A, Yonenaga K, Sano A, Toba H, Murata K, et al. 320 Gbit/s (8 × 40 Gbit/s) WDM transmission over 367 km with 120 km repeater spacing using carrier-suppressed return-to-zero format. *IET Elect Lett* 1999;**35**:2041–2.
5. Yonenaga K, Kuwano S, Norimatsu S, Shibata N. Optical duobinary transmission system with no receiver sensitivity degradation. *IET Elect Lett* 1995;**31**:302–4.
6. Kaiser W, Wuth T, Wichers M, Rosenkranz W. Reduced complexity optical duobinary 10-Gb/s transmitter setup resulting in an increased transmission distance. *IEEE Photon Technol Lett* 2001;**13**:884–6.
7. Gnauck AH, Winzer PJ. Optical phase-shift-keyed transmission. *J Lightwave Technol* 2005;**23**:115–30.
8. Xu C, Liu X, Wei X. Differential phase-shift keying for high spectral efficiency optical transmissions. *IEEE J Selected Topics Quantum Elect* 2004;**10**:281–93.
9. Ho K-P. *Phase-Modulated Optical Communication Systems*. New York: Springer; 2005.
10. Noé R. Phase noise tolerant synchronous QPSK/BPSK baseband type intradyne receiver concept with feedforward carrier recovery. *J Lightwave Technol* 2005;**23**:802–8.

11. Sun H, Wu K, Roberts K. Real-time measurements of a 40 Gb/s coherent system. *Opt Express* 2008;**16**:873–9.

12. Savory SJ, Gavioli G, Killey RI, Bayvel P. Electronic compensation of chromatic dispersion using a digital coherent receiver. *Opt Express* 2007;**15**:2120–6.

13. Shieh W, Athaudage C. Coherent optical orthogonal frequency division multiplexing. *Electron Lett* 2006;**42**:587–9.

14. Lowery AJ, Du L, Armstrong J. Orthogonal frequency division multiplexing for adaptive dispersion compensation in long haul WDM systems. In: *Opt. Fiber Commun. Conf.*, paper no. PDP 39. Anaheim, CA; 2006.

15. Djordjevic IB, Vasic B. Orthogonal frequency division multiplexing for high-speed optical transmission. *Opt Express* 2006;**14**:3767–75.

16. Shieh W, Yang Q, Ma Y. 107 Gb/s coherent optical OFDM transmission over 1000-km SSMF fiber using orthogonal band multiplexing. *Opt Express* 2008;**16**:6378–86.

17. Kobayash T, Sano A, Yamada E. Electro-optically subcarrier multiplexed 110 Gb/s OFDM signal transmission over 80 km SMF without dispersion compensation. *Elect Lett* 2008;**44**:225–6.

18. Jansen SL, Morita I, Tanaka H. 10 × 121.9-Gb/s PDM-OFDM transmission with 2-b/s/Hz spectral efficiency over 1000 km of SSMF. In: *Opt. Fiber Commun. Conf.*, paper no. PDP2. San Diego; 2008.

19. Falconer D, Ariyavisitakul SL, Benyamin-Seeyar A, Eidson B. Frequency domain equalization for single-carrier broadband wireless systems. *IEEE Commun Magazine* 2002;**40**:58–66.

20. Wang Z, Ma X, Giannakis GB. OFDM or single-carrier block transmissions? *IEEE Trans Commun* 2004;**52**:380–94.

21. Winzer PJ, Essiambre RJ. Advanced optical modulation formats. In: Kaminow IP, Li T, Willner AE, editors. *Optical Fiber Telecommunications V: B: Systems and Networks*. New York: Academic Press; 2008.

22. Liu X, Gill DM, Chandrasekhar S. Optical technologies and techniques for high bit rate fiber transmission. *Bell Labs Technical J* 2006;**11**:83–104.

23. Maiman TH. Stimulated optical radiation in ruby. *Nature* 1960;**187**(4736):493–4.

24. Kao KC, Hockman GA. Dielectric-fiber surface waveguides for optical frequencies. *Proc IEEE* 1966;**113**:1151–8.

25. Kazovsky L. Multichannel coherent optical communications systems. *J Lightwave Technol* 1987;**5**:1095–102.

26. Okoshi T. Heterodyne and coherent optical fiber communications: Recent progress. *IEEE Trans Microwave Techniques* 1982;**82**:1138–49.

27. Kahn JM, Habbab IMI, Giles CR. 1Gbit/s zero-IF DPSK coherent optical system using a single photodetector. *IET Elect Lett* 1988;**24**:1455–7.

28. Gnauck AH, Tkach RW, Chraplyvy AR, Li T. High-capacity optical transmission systems. *J Lightwave Technol* 2008;**26**:1032–45.

29. Sano A, Yamada E, Masuda H, et al. 13.4-Tb/s (134 × 111-Gb/s/ch) no-guard-interval coherent OFDM transmission over 3600 km of SMF with 19-ps average PMD. In: *Eur. Conf. Opt. Commun.*, paper no. Th.3.E.1. Brussels, Belgium; 2008.

30. Yu J, Zhou X, Huang M, et al. 21.7 Tb/s (161 × 114 Gb/s) polmux-RZ-8PSK transmission over 662 km of ultra-low loss fiber using C-band EDFA amplification and digital coherent detection. In: *Eur. Conf. Opt. Commun.*, paper no. Th.3.E.2. Brussels, Belgium, 2008.

31. *Cisco Inc. Approaching the zettabyte era.* Information available at http://www.cisco.com/en/US/solutions/collateral/ns341/ns525/ns537/ns705/ns827/white_paper_c11-481374_ns827_Networking_Solutions_White_Paper.html

32. Basch EB, Egorov R, Gringeri S, Elby S. Architectural tradeoffs for reconfigurable dense wavelength-division multiplexing systems. *IEEE J Selected Topics Quantum Elect* 2006;**12**:615–26.

33. Strasser TA, Taylor J. ROADMS unlock the edge of the network. *IEEE Commun. Magazine* 2008;**46**:146–69.

34. Homa J, Bala K. ROADM architectures and their enabling WSS technology. *IEEE Commun Magazine* 2008;**46**:150–3.

35. Takahashi T, Imai T, Aiki M. Automatic compensation technique for timewise fluctuating polarisation mode dispersion in in-line amplifier systems. *IET Elect Lett* 1994;**30**:348–9.

36. Sunnerud H, Xie C, Karlsson M, Samuelsson R, Andrekson PA. A comparison between different PMD compensation techniques. *J Lightwave Technol* 2002;**20**:368–78.

37. Mitola J. The software radio architecture. *IEEE Commun Magazine* 1995;**33**(5):26–38.

38. Abidi A. The path to the software-defined radio receiver. *IEEE J Solid-State Circuits* 2007;**42**:954–66.

39. Fludger CRS, Duthel T, Van Den Borne D, et al. Coherent equalization and POLMUX-RZ-DQPSK for robust 100-GE transmission. *J Lightwave Technol* 2008;**26**:64–72.

40. Winzer PJ, Raybon G, Duelk M. 107-Gb/s optical ETDM transmitter for 100 G Ethernet transport. In: *Eur. Conf. Opt. Commun.*, paper no. Th4.1.1. Glasgow, Scotland; 2005.

41. Yi X, Shieh W, Ma Y. Phase noise effects on high spectral efficiency coherent optical OFDM transmission. *J Lightwave Technol* 2008;**26**:1309–16.

42. Buchali F, Bulow H. Adaptive PMD compensation by electrical and optical techniques. *J Lightwave Technol* 2004;**22**:1116–26.

43. Bulow H. Electronic dispersion compensation. In: *Opt. Fiber Commun. Conf.*, paper no. OMG5. Anaheim, CA; 2007.

44. Färbert A. Application of digital equalization in optical transmission systems. In: *Opt. Fiber Commun. Conf.*, paper no. OTuE5. Anaheim, CA; 2006.

45. Haunstein HF, Schorr T, Zottmann A, Sauer-Greff W, Urbansky R. Performance comparison of MLSE and iterative equalization in FEC systems for PMD channels with respect to implementation complexity. *J Lightwave Technol* 2006;**24**(11):4047–54.

46. Killey RI, Watts PM, Mikhailov V, Glick M, Bayvel P. Electronic dispersion compensation by signal predistortion using digital processing and a dual-drive Mach–Zehnder modulator. *IEEE Photon Technol Lett* 2005;**17**:714–6.

47. McGhan D, Laperle C, Savchenko A, Li C, O'Sullivan. 5120-km RZ-DPSK transmission over G.652 fiber at 10 Gb/s without optical dispersion compensation. *IEEE Photon Technol Lett* 2006;**18**:400–2.

48. Ly-Gagnon DS, Tsukamoto S, Katoh K, Kikuchi K. Coherent detection of optical quadrature phase-shift keying signals with carrier phase estimation. *J Lightwave Technol* 2006;**24**:12–21.

49. Shieh W, Yi X, Tang Y. Transmission experiment of multi-gigabit coherent optical OFDM systems over 1000 km SSMF fiber. *Elect Lett* 2007;**43**:183–4.

50. Jansen SL, Morita I, Takeda N, Tanaka H. 20-Gb/s OFDM transmission over 4160-km SSMF enabled by RF-Pilot tone phase noise compensation. In: *Opt. Fiber Comm. Conf.*, paper no. PDP15. Anaheim, CA; 2007.

51. Moore E. Cramming more components onto integrated circuits. *Electronics* 1965;**38**:114–7.

52. Intel Corporation. *International Technology Roadmap for Semiconductors*. Available at http://www.intel.com/technology/silicon/itroadmap.htm?iid=tech_mooreslaw+rhc_roadmap.

53. Dally W, Poulton J. *Digital Systems Engineering*. New York: Cambridge University Press; 1998.

54. Li MJ. Recent progress in fiber dispersion compensators. In: *Eur. Conf. on Opt. Commun.*, paper no. Th.M.1.1. Amsterdam, The Netherlands; 2001.

55. Grüner-Nielsen L, Wandel M, Kristensen P, et al. Dispersion-compensating fibers. *J Lightwave Technol* 2005;**23**:3566–79.

56. Willner AE, Feng K-M, Cai J-X, et al. Tunable compensation of channel degrading effects using nonlinearly chirped passive fiber Bragg gratings. *IEEE J Select Topics Quantum Electron* 1999;**5**:1298–311.

57. Eggleton BJ, Ahuja A, Westbrook PS, et al. Integrated tunable fiber gratings for dispersion management in high-bit rate systems. *J Lightwave Technol* 2000;**18**:1418–32.
58. Bulow H, Buchali F, Klekamp A. Electronic dispersion compensation. *J Lightwave Technol* 2007;**26**:158–67.
59. Möller L, Thiede A, Chandrasekhar S, et al. ISI mitigation using decision feedback loop demonstrated with PMD distorted 10 Gbit/s signals. *IET Elect Lett* 1999;**35**:2092–3.
60. Bülow H, Buchali F, Baumert W, Ballentin R, Wehren T. PMD mitigation at 10 Gbit/s using linear and nonlinear integrated electronic equalizer circuits. *IET Elect Lett* 2000;**36**:163–4.
61. Elbers JP, Wernz H, Griesser H, et al. Measurement of the dispersion tolerance of optical duobinary with an MLSE-receiver at 10.7 Gb/s. In: *Opt. Fiber Commun. Conf.*, paper no. OThJ4. Los Angeles; 2005.
62. Ly-Gagnon DS, Tsukarnoto S, Katoh K, Kikuchi K. Coherent detection of optical quadrature phase-shift keying signals with carrier phase estimation. *J Lightwave Technol* 2006;**24**:12–21.
63. Taylor MG. Coherent detection method using DSP for demodulation of signal and subsequent equalization of propagation impairments. *IEEE Photon Technol Lett* 2004;**16**:674–6.
64. Ip E, Lau APT, Barros DJF, Kahn JM. Coherent detection in optical fiber systems. *Opt Express* 2008;**16**:753–91.
65. Noé R. Phase noise tolerant synchronous QPSK/BPSK baseband type intradyne receiver concept with feedforward carrier recovery. *J Lightwave Technol* 2005;**23**:802–8.
66. Charlet G, Renaudier J, Salsi M, et al. Efficient mitigation of fiber impairments in an ultra-long haul transmission of 40Gbit/s polarization-multiplexed data, by digital processing in a coherent receiver. In: *Opt. Fiber Commun. Conf.*, paper no. PDP17. Anaheim, CA; 2007.
67. Shieh W, Yi X, Ma Y, Tang Y. Theoretical and experimental study on PMD-supported transmission using polarization diversity in coherent optical OFDM systems. *Opt Express* 2007;**15**:9936–47.
68. Jansen SL, Morita I, Tanaka H. 16 × 52.5-Gb/s, 50-ghz spaced, POLMUX-CO-OFDM transmission over 4160 km of SSMF enabled by MIMO processing KDDI R&D Laboratories. In: *Eur. Conf. Opt. Commun.*, paper no. PD1.3. Berlin, Germany; 2007.
69. Ly-Gagnon D, Tsukarnoto S, Katoh K, Kikuchi K. Coherent detection of optical quadrature phase-shift keying signals with carrier phase estimation. *J Lightwave Technol* 2006;**24**:12–21.
70. Qian D, Hu J, Yu J, et al. Experimental demonstration of a novel OFDM-A based 10Gb/s PON architecture. In: *Eur. Conf. Opt. Commun.* paper no. 54.1. Berlin, Germany; 2007.
71. Spinnler B, Hauske FN, Kuschnerov M. Adaptive equalizer complexity in coherent optical receivers. In: *Eur. Conf. Opt. Commun.*, paper no. We.2.E.4. Brussels, Belgium; 2008.
72. Shieh W, Yi X, Ma Y, Yang Q. Coherent optical OFDM: Has its time come? [Invited]. *J Opt Networking* 2008;**7**:234–55.
73. Gitlin R, Hayes JF, Weinstein SB. *Data Communications Principles*. New York: Plenum; 1992.
74. Hara S, Prasad R. *Multicarrier Techniques for 4G Mobile Communications*. Boston: Artech House; 2003.
75. Hanzo L, Munster M, Choi BJ, Keller T. *OFDM and MC-CDMA for Broadband Multi-User Communications, WLANs and Broadcasting*. New York: Wiley; 2003.
76. Gisin N, Huttner B. Combined effects of polarization mode dispersion and polarization dependent losses in optical fibers. *Opt Commun* 1997;**142**:119–25.
77. Liu X, Buchali F. Intra-symbol frequency-domain averaging based channel estimation for coherent optical OFDM. *Opt Express* 2008;**16**:21944–57.
78. Agrawal G. *Nonlinear Fiber Optics*. 3rd ed. San Diego: Academic Press; 2001.
79. Nazarathy M, Khurgin J, Weidenfeld R, et al. Phased-array cancellation of nonlinear FWM in coherent OFDM dispersive multi-span links. *Opt Express* 2008;**16**:15777–810.
80. Supplement to IEEE standard for information technology telecommunications and information exchange between systems—local and metropolitan area networks—specific requirements. Part 11: Wireless LAN

medium access control (MAC) and physical layer (PHY) specifications: High-speed physical layer in the 5 GHz band. In: IEEE Std 802.11a-1999. 1999.

81. Shieh W. High spectral efficiency coherent optical OFDM for 1 Tb/s Ethernet transport. In: *Opt. Fiber Commun. Conf.*, paper no. OWW1. San Diego; 2009.

82. Shieh W, Tucker RS, Chen W, Yi X, Pendock G. Optical performance monitoring in coherent optical OFDM systems. *Opt Express* 2007;**15**:350–6.

83. Mayrock M, Haunstein H. Performance monitoring in optical OFDM systems. In: *Opt. Fiber Commun. Conf.*, paper no. OWM3. San Diego; 2008.

84. Gupta M, Singh S. *"Greening of the Internet."* *ACM SIGCOMM*. Karlsruhe, Germany: 2003.

85. Vukovic A. Data centers: Network power density challenges. *ASHREA J* 2005;**47**:55.

86. Baliga J, Hinton K, Tucker RS. Energy consumption of the Internet. In: *32nd Australian Conf. Opt. Fibre Technol.* COIN-ACOFT; 2007.

87. Gallager RG. *Low Density Parity Check Codes*. Cambridge, MA: MIT Press; 1963.

88. Chung SY, Forney GD, Richardson TJ, Urbank R. On the design of low-density parity-check codes within 0.0045 dB of the Shannon Limit. *IEEE Commun Lett* 2001;**5**:58–60.

CHAPTER 2

OFDM Principles

2.1 Introduction

Orthogonal frequency-division multiplexing (OFDM) belongs to a broader class of multicarrier modulation (MCM) in which the data information is carried over many lower rate subcarriers. Two of the fundamental advantages of OFDM are its robustness against channel dispersion and its ease of phase and channel estimation in a time-varying environment. With the advancement of powerful silicon DSP technology, OFDM has triumphed in a broad range of applications in the RF domain from digital audio/video broadcasting (DAB/DVB) to wireless local area networks (LANs). However, OFDM also has intrinsic disadvantages, such as high peak-to-average power ratio (PAPR) and sensitivity to frequency and phase noise. Therefore, a proper understanding of OFDM basics is essential for the study of its applications in the emerging field of optical OFDM.

In this chapter, we present a historical perspective of OFDM and a brief discussion on its emergence in optical communications. We then provide an introduction to the fundamentals of OFDM, including its basic mathematical formulation, discrete Fourier transform implementation, cyclic prefix, spectral efficiency, and PAPR characteristics. Furthermore, we provide the detailed analysis of the signal-to-noise ratio (SNR) impairment from frequency offset and phase noise for OFDM systems.

2.2 Historical Perspective of OFDM

The concept of OFDM was first introduced by Chang in a seminal paper in 1966.[1] The term "OFDM" in fact first appeared in a separate patent of his in 1970.[2] The field of OFDM had long been developed as a peripheral interest in military applications because there was a lack of broadband applications for OFDM and powerful integrated electronic circuits to support the complex computation required by OFDM. However, the arrival of broadband digital applications and maturing of very large-scale integrated (VLSI) CMOS chips in the 1990s brought OFDM into the spotlight. In 1995, OFDM was adopted as the European DAB standard, ensuring its significance as an important modulation technology and heralding a new era of OFDM success in a broad range of applications. Among the important standards that incorporate OFDM modulation technology are the European DVB, wireless local area networks (Wi-Fi; IEEE 802.11a/g), wireless metropolitan area networks (WiMAX; 802.16e), asymmetric digital subscriber line (ADSL; ITU G.992.1), and long-term evolution (LTE)—the fourth-generation mobile communications technology.

The application of OFDM to optical communications occurred surprisingly late and relatively scantly compared with the RF counterpart, although the same acronym of OFDM has long been used to stand for "optical frequency division multiplexing" in the optical communications community.[3,4] The first paper on optical OFDM in the open literature was reported by Pan and Green in 1996,[5] and there was also some intermittent research on optical OFDM over the ensuing years.[6–8] However, the fundamental advantage of OFDM, namely, its robustness against optical channel dispersion was not recognized in optical communications until 2001, when Dixon et al.[9] proposed the use of OFDM to combat modal dispersion in multimode fiber (MMF). Given the fact that the MMF fiber channel resembles that of the wireless channel in terms of multipath fading, it is not surprising that the early body of work on optical OFDM concentrated on MMF fiber application.[9–12] The increased interest in optical OFDM is in large part attributed to independent proposals of optical OFDM for long-haul applications from three groups, including direct-detection optical OFDM (DDO-OFDM)[13,14] and coherent optical OFDM (CO-OFDM).[15] To date, 100 Gb/s CO-OFDM transmission over 1000 km standard single-mode fiber (SSMF) with high spectral efficiency of 2 bit/s/Hz has been demonstrated by various groups.[16–18] One of the major strengths of optical OFDM is that it can be adapted for many different applications, which are discussed in detail in Chapter 7.

2.3 OFDM Basics

2.3.1 Mathematical Formulation of an OFDM Signal

OFDM is a special class of MCM, a generic implementation of which is depicted in Figure 2.1. The structure of a complex multiplier (IQ modulator/demodulator), which is commonly used in MCM systems, is also shown in the figure. The MCM transmitted signal $s(t)$ is represented as

$$s(t) = \sum_{i=-\infty}^{+\infty} \sum_{k=1}^{N_{sc}} c_{ki} s_k (t - iT_s) \tag{2.1}$$

$$s_k(t) = \Pi(t) e^{j2\pi f_k t} \tag{2.2}$$

$$\Pi(t) = \begin{cases} 1, & (0 < t \le T_s) \\ 0, & (t \le 0, t > T_s) \end{cases} \tag{2.3}$$

where c_{ki} is the ith information symbol at the kth subcarrier, s_k is the waveform for the kth subcarrier, N_{sc} is the number of subcarriers, f_k is the frequency of the subcarrier, T_s is the symbol period, and $\Pi(t)$ is the pulse shaping function. The optimum detector for each

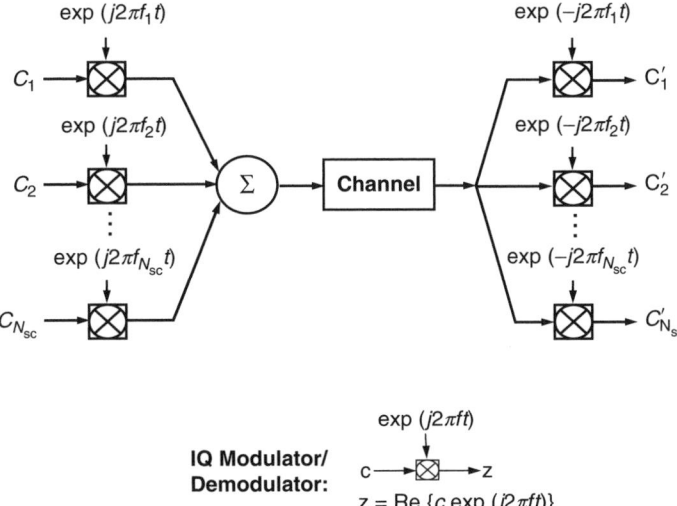

Figure 2.1: Conceptual diagram for a generic multicarrier modulation system.

subcarrier could use a filter that matches the subcarrier waveform or a correlator matched to the subcarrier as shown in Figure 2.1. Therefore, the detected information symbol c'_{ik} at the output of the correlator is given by

$$c'_{ki} = \frac{1}{T_s} \int_0^{T_s} r(t - iT_s) s^*_k dt = \frac{1}{T_s} \int_0^{T_s} r(t - iT_s) e^{-j2\pi f_k t} dt \qquad (2.4)$$

where $r(t)$ is the received time domain signal. The classical MCM uses nonoverlapped band-limited signals and can be implemented with a bank of large numbers of oscillators and filters at both transmit and receive ends.[19,20] The major disadvantage of MCM is that it requires excessive bandwidth. This is because to design the filters and oscillators cost-effectively, the channel spacing has to be a multiple of the symbol rate, greatly reducing the spectral efficiency. A novel approach, OFDM, was investigated by employing overlapped yet orthogonal signal sets.[11] This orthogonality originates from a straightforward correlation between any two subcarriers, given by

$$\delta_{kl} = \frac{1}{T_s} \int_0^{T_s} s_k s^*_l dt = \frac{1}{T_s} \int_0^{T_s} \exp\left(j2\pi (f_k - f_l)t\right) dt = \exp\left(j\pi (f_k - f_l)T_s\right) \frac{\sin\left(\pi (f_k - f_l)T_s\right)}{\pi (f_k - f_l)T_s} \qquad (2.5)$$

It can be seen that if the condition

$$f_k - f_l = m\frac{1}{T_s} \qquad (2.6)$$

is satisfied, then the two subcarriers are orthogonal to each other. This signifies that these orthogonal subcarrier sets, with their frequencies spaced at multiples of inverse of the symbol periods, can be recovered with the matched filters in Eq. (2.4) without intercarrier interference (ICI), despite strong signal spectral overlapping.

2.3.2 Discrete Fourier Transform Implementation of OFDM

A fundamental challenge with OFDM is that a large number of subcarriers are needed so that the transmission channel affects each subcarrier as a flat channel. This leads to an extremely complex architecture involving many oscillators and filters at both transmit and receive ends. Weinsten and Ebert first revealed that OFDM modulation/demodulation can be implemented by using inverse discrete Fourier transform (IDFT)/discrete Fourier transform (DFT).[21] This is evident by studying OFDM modulation (Eq. 2.1) and OFDM demodulation (Eq. 2.4). Let's temporarily omit the index i, re-denote N_{sc} as N in Eq. (2.1) to focus our attention on one OFDM symbol, and assume that we sample $s(t)$ at every interval of T_s/N. The mth sample of $s(t)$ from Eq. (2.1) becomes

$$s_m = \sum_{k=1}^{N} c_k \cdot e^{j2\pi f_k \cdot \frac{(m-1)T_s}{N}} \tag{2.7}$$

Using the orthogonality condition of Eq. (2.6) and the convention that

$$f_k = \frac{k-1}{T_s} \tag{2.8}$$

and substituting Eq. (2.8) into Eq. (2.7), we have

$$s_m = \sum_{k=1}^{N} c_k \cdot e^{j2\pi f_k \cdot \frac{(m-1)T_s}{N}} = \sum_{k=1}^{N} c_k \cdot e^{j2\pi \frac{(k-1)(m-1)}{N}} = \mathfrak{F}^{-1}\{c_k\} \tag{2.9}$$

where $\mathfrak{F}$ is the Fourier transform, and $m \in [1, N]$. In a similar manner, at the receive end, we arrive at

$$c'_k = \mathfrak{F}\{r_m\} \tag{2.10}$$

where r_m is the received signal sampled at every interval of T_s/N. From Eqs. (2.9) and (2.10), it follows that the discrete value of the transmitted OFDM signal $s(t)$ is merely a simple N-point IDFT of the information symbol c_k, and the received information symbol c'_k is a simple N-point DFT of the receive sampled signal. It is worth noting that there are two critical devices we have assumed for the DFT/IDFT implementation: (1) digital-to-analog converter (DAC), needed to convert the discrete value of s_m to the continuous analog value of $s(t)$, and (2) an analog-to-digital converter (ADC), needed to convert the continuous received signal $r(t)$ to discrete sample r_m. There are two fundamental advantages of DFT/IDFT implementation of OFDM. First, because of the existence of an efficient IFFT/FFT algorithm, the number of complex multiplications for IFFT in Eq. (2.9) and FFT in Eq. (2.10) is reduced from N^2 to

$$\frac{N}{2} \log_2(N)$$

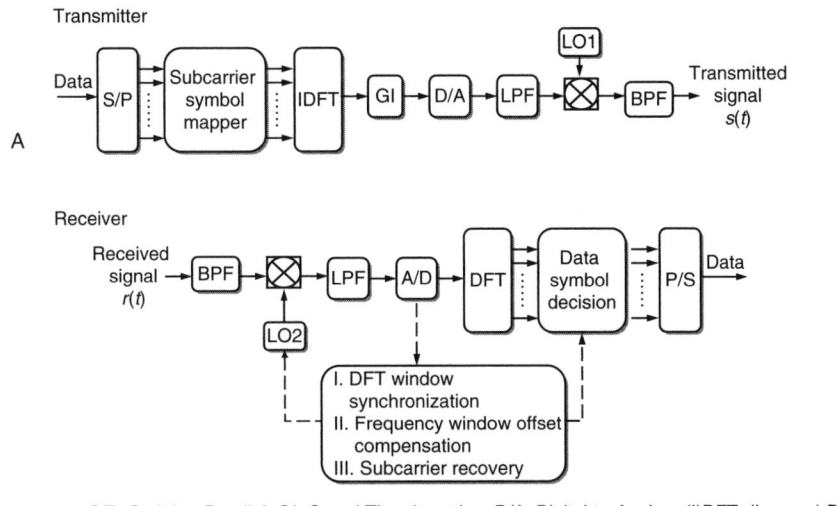

Figure 2.2: Conceptual diagram for (a) OFDM transmitter and (b) OFDM receiver.

almost linearly with the number of subcarriers, N.[22] Second, a large number of orthogonal subcarriers can be generated and demodulated without resorting to much more complex RF oscillators and filters. This leads to a relatively simple architecture for OFDM implementation when large numbers of subcarriers are required. The corresponding architecture using DFT/IDFT and DAC/ADC is shown in Figure 2.2. At the transmit end, the input serial data bits are first converted into many parallel data pipes, each mapped onto corresponding information symbols for the subcarriers within one OFDM symbol, and the digital time domain signal is obtained by using IDFT, which is subsequently inserted with a guard interval and converted into real-time waveform through DAC. The guard interval is inserted to prevent intersymbol-interference (ISI) due to channel dispersion. The baseband signal can be upconverted to an appropriate RF passband with an IQ mixer/modulator. At the receive end, the OFDM signal is downconverted to baseband with an IQ demodulator, sampled with an ADC, and then demodulated by performing DFT and baseband signal processing to recover the data.

Note that from Eq. (2.7), the OFDM signal s_m is a periodical function of f_k with a period of N/T_s. Therefore, any discrete subcarrier set with its frequency components spanning one period of N/T_s is equivalent. Namely, in Eqs. (2.7) and (2.8), the subcarrier frequency f_k and its index k can be generalized as

$$f_k = \frac{k-1}{T_s}, \quad k \in [k_{\min} + 1, k_{\min} + N] \tag{2.11}$$

where k_{min} is an arbitrary integer. However, only two subcarrier index conventions are widely used: $k \in [1, N]$ and $k \in [-N/2 + 1, N/2]$. These two conventions are mathematically equivalent, and both are used in the book.

2.3.3 Cyclic Prefix for OFDM

One of the enabling techniques for OFDM is the insertion of cyclic prefix.[23,24] Let us first consider two consecutive OFDM symbols that undergo a dispersive channel with a delay spread of t_d. For simplicity, each OFDM symbol includes only two subcarriers with the fast delay and slow delay spread at t_d, represented by "fast subcarrier" and "slow subcarrier," respectively. Figure 2.3a shows that inside each OFDM symbol, the two subcarriers—fast subcarrier and slow subcarrier—are aligned upon the transmission. Figure 2.3b shows the same OFDM signals upon the reception, where the slow subcarrier is delayed by t_d against the fast subcarrier. We select a DFT window containing a complete OFDM symbol for the fast subcarrier. It is apparent that due to the channel dispersion, the slow subcarrier has crossed the symbol boundary leading to interference between neighboring OFDM symbols, the so-called ISI. Furthermore, because the OFDM waveform in the DFT window for the slow subcarrier is incomplete, the critical orthogonality condition for the subcarriers (Eq. 2.5) is lost, resulting in an inter-carrier interference (ICI) penalty.

Cyclic prefix was proposed to resolve the channel dispersion-induced ISI and ICI.[23] Figure 2.3c shows insertion of a cyclic prefix by cyclic extension of the OFDM waveform into the guard interval, Δ_G. As shown in Figure 2.3c, the waveform in the guard interval is essentially an identical copy of that in the DFT window, with time shifted forward by t_s. Figure 2.3d shows the OFDM signal with the guard interval upon reception. Let us assume that the signal has traversed the same dispersive channel, and the same DFT window is selected containing a complete OFDM symbol for the fast subcarrier waveform. It can be seen from Figure 2.3d that a complete OFDM symbol for the slow subcarrier is also maintained in the DFT window because a proportion of the cyclic prefix has moved into the DFT window to replace the identical part that has shifted out. As such, the OFDM symbol for the slow subcarrier is an almost identical copy of the transmitted waveform with an additional phase shift. This phase shift is dealt with through channel estimation and will be subsequently removed for symbol decision. Now we arrive at the important condition for ISI-free OFDM transmission, given by

$$t_d < \Delta_G \tag{2.12}$$

It can be seen that to recover the OFDM information symbol properly, there are two critical procedures that need to be carried out: (1) selection of an appropriate DFT window, called DFT window synchronization, and (2) estimation of the phase shift for each subcarrier, called channel estimation or subcarrier recovery. Both signal processing procedures are actively pursued research topics, and they are discussed in both books and journals.[23,24]

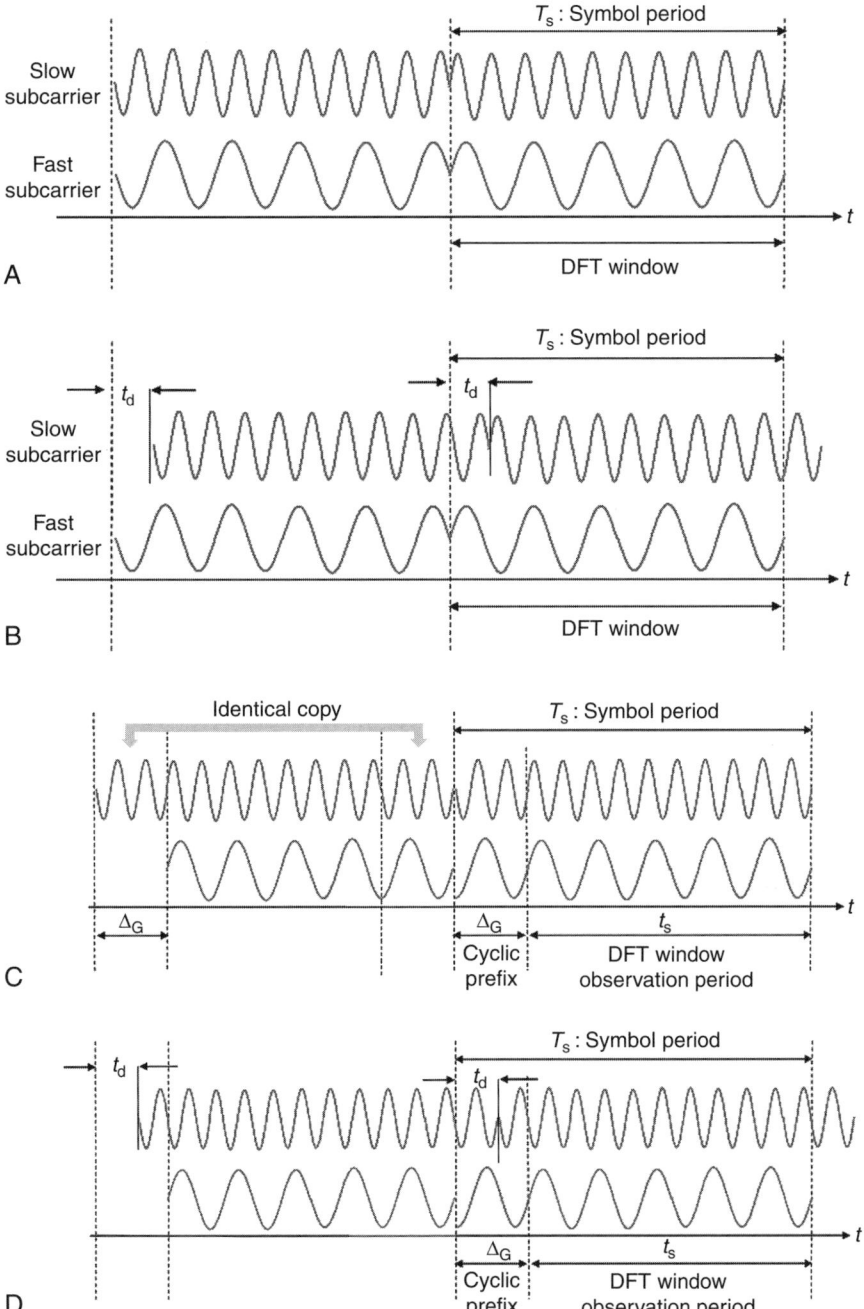

Figure 2.3: OFDM signals (a) without cyclic prefix at the transmitter, (b) without cyclic prefix at the receiver, (c) with cyclic prefix at the transmitter, and (d) with cyclic prefix at the receiver.

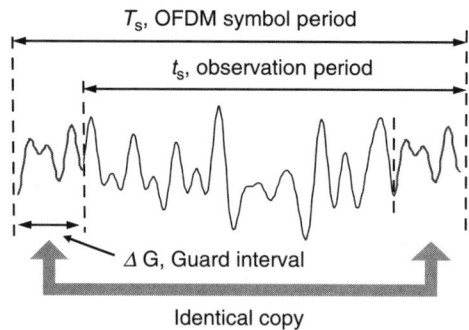

Figure 2.4: The time domain OFDM signal for one complete OFDM symbol.

An elegant way to describe the cyclic prefix is to maintain the same expression as Eq. (2.1) for the transmitted signal $s(t)$ but extend the pulse shape function (Eq. 2.3) to the guard interval, given by

$$\Pi(t) = \begin{cases} 1, (-\Delta_G < t \leq t_s) \\ 0, (t \leq -\Delta_G, t > t_s) \end{cases} \qquad (2.13)$$

The corresponding time domain OFDM symbol is illustrated in Figure 2.4, which shows one complete OFDM symbol composed of an observation period and cyclic prefix. The waveform within the observation period will be used to recover the frequency domain information symbols.

2.3.4 Spectral Efficiency for Optical OFDM

In DDO-OFDM systems, the optical spectrum is usually not a linear replica of the RF spectrum; therefore, the optical spectral efficiency is dependent on the detailed implementation. We turn our attention to the optical spectral efficiency for CO-OFDM systems. In CO-OFDM systems, N_{sc} subcarriers are transmitted in every OFDM symbol period of T_s. Thus, the total symbol rate R for CO-OFDM systems is given by

$$R = N_{sc}/T_s \qquad (2.14)$$

Figure 2.5a shows the spectrum of wavelength-division multiplexed (WDM) channels, each with CO-OFDM modulation, and Figure 2.5b shows the zoomed-in optical spectrum for each wavelength channel. We use the bandwidth of the first null to denote the boundary of each wavelength channel. The OFDM bandwidth, B_{OFDM}, is thus given by

$$B_{OFDM} = \frac{2}{T_s} + \frac{N_{sc}-1}{t_s} \qquad (2.15)$$

where t_s is the observation period (see Figure 2.4). Assuming that a large number of subcarriers are used, the bandwidth efficiency of OFDM η is found to be

$$\eta = 2\frac{R}{B_{OFDM}} = 2\alpha, \quad \alpha = \frac{t_s}{T_s} \qquad (2.16)$$

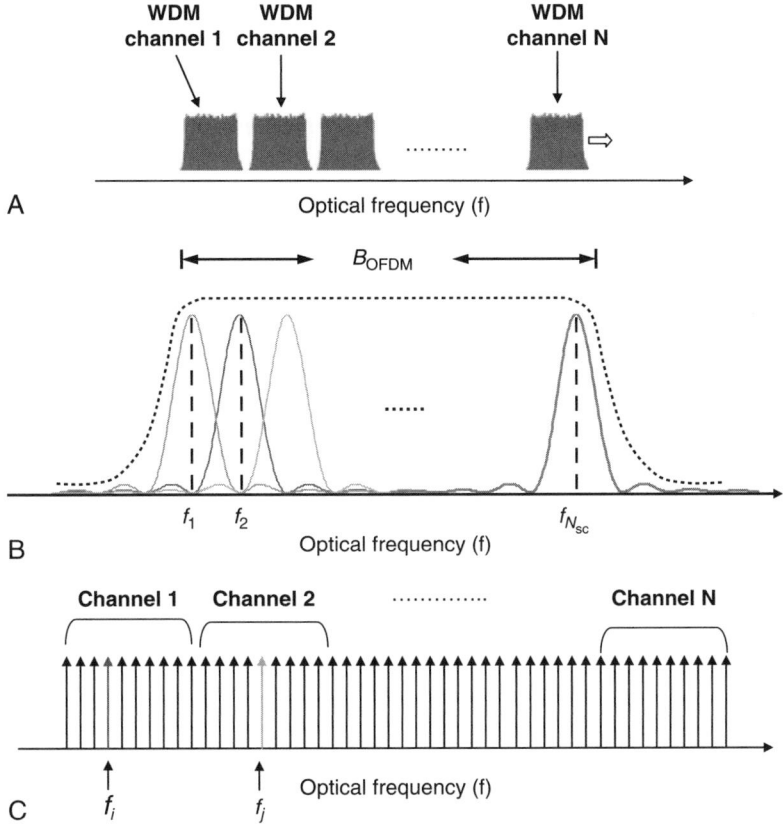

Figure 2.5: Optical spectra for (a) *N* wavelength-division multiplexed (WDM) CO-OFDM channels, (b) zoomed-in OFDM signal for one wavelength, and (c) cross-channel OFDM (XC-OFDM) without guard band.

The factor of 2 accounts for two polarizations in the fiber. Using a typical value of 8/9, we obtain the optical spectral efficiency factor η of 1.8 Bd/Hz. The optical spectral efficiency gives 3.6 bit/s/Hz if quaternary phase-shift keying (QPSK) modulation is used for each subcarrier. The spectral efficiency can be further improved by using higher order QAM modulation.[25,26] To practically implement CO-OFDM systems, the optical spectral efficiency will be reduced due to the need for a sufficient guard band between WDM channels, taking account of laser frequency drift of approximately 2 GHz. This guard band can be avoided by using orthogonality across the WDM channels, which is discussed in the next section.

2.3.5 Cross-Channel OFDM: Multiplexing without Guard Band

The laser frequency drift of WDM channels can be resolved by locking all the lasers to the common optical standard such as an optical comb and directly using the frequency tones from an optical comb.[27] In so doing, all the subcarriers that cross the WDM channels can be orthogonal; that is, the orthogonality condition of Eq. (2.5) is satisfied for any two subcarriers, even from different WDM channels. As shown in Figure 2.5c, the subcarrier f_i in channel 1 is orthogonal to another subcarrier f_j in a different channel (e.g., channel 2). We call this OFDM form XC-OFDM, in which orthogonality applies to the subcarriers from different channels. The term *channel* as discussed here in a broader sense could be either an "RF" one or an "optical" one; that is, cross-channel (XC)-OFDM can be realized through RF subcarrier multiplexing[17] or wavelength multiplexing[18] shown in Figure 2.5c. An optical or electrical filter with bandwidth slightly larger than the channel bandwidth can be used to select the desired channel. The interchannel interference is avoided through the orthogonality of XC-OFDM subcarriers. Consequently, no frequency guard band is necessary in such a scheme. However, this requires DFT window synchronization among the WDM channels, as discussed in Section 2.3.3. To comply with optical add/drop, a small number of such XC-OFDM channels can be grouped/banded together to allow for guard band between individual groups/bands, in a similar manner to all-optical OFDM.[28] Note that to alleviate the bandwidth bottleneck from electronic ADC/DACs, the same concept has been adopted to sub-band the ultrabroad bandwidth for the high-speed optical OFDM signal, and it is called orthogonal-band multiplexed OFDM (OBM-OFDM) to emphasize the bandwidth reduction due to sub-banding.[18]

Another important development is the proposal of no-guard-interval (NGI) CO-OFDM.[16] In such a scheme, each wavelength is modulated with a single-carrier signal with proper pulse shaping, such as the nearly rectangular shape described in Eq. (2.3). The orthogonality condition of Eq. (2.6) is satisfied for the frequency of each subcarrier. The coherent optical detection ensures that the orthogonality between the optical subcarriers is maintained throughout the receive signal processing. This form of optical OFDM has better or similar spectral efficiency as OBM-OFDM, but the multiplexing and demultiplexing of NGI CO-OFDM do not take advantage of the flexibility of cyclic extension in the timing synchronization and the efficient FFT/IFFT algorithm.

2.3.6 Complex and Real Representations of an OFDM Signal

At the very beginning and end of digital signal processing, the baseband OFDM signal is represented in a form of complex value, but during transmission the OFDM signal becomes a real-valued signal. More precisely, frequency upconversion and frequency downconversion are required for this complex-to-real-value conversion, or baseband to passband conversion. Mathematically, such transformation involves a complex multiplier (mixer) or IQ modulator/demodulator, which at the upconversion can be expressed as

$$S_{RF}(t) = \text{Re}\{S(t)e^{j2\pi f_{RF}t}\} = \text{Re}\{S(t)\} \cdot \cos(2\pi f_{RF}t) - \text{Im}\{S(t)\} \cdot \sin(2\pi f_{RF}t) \qquad (2.17)$$

where the passband signal $S_{RF}(t)$ is a real-value signal at the center frequency of f_{RF}, $S(t)$ is the baseband complex-valued signal, and Re and Im are real and imaginary parts of a

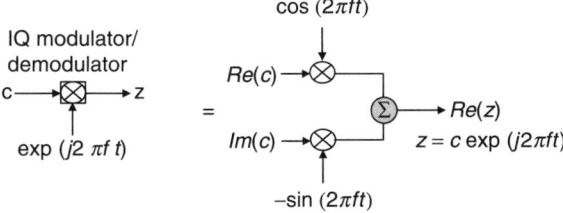

Figure 2.6: IQ modulator for upconversion of a complex-valued baseband signal *c* to a real-valued passband signal *z*. The downconversion follows the reverse process by reversing the flow of *c* and *z*.

complex quantity, respectively. Traditionally, the IQ modulator can be constructed with a pair of RF mixers and LOs with a 90-degree shift as shown in Figure 2.6. The real-to-complex downconversion of an OFDM signal follows the reverse process of the upconversion by reversing the flow of the baseband signal *c* and RF passband signal *z* in Figure 2.6. The mixer design can be implemented in a mixed-signal integrated circuit (IC) up to 60 GHz with the state-of-the-art silicon technology.[29–31] The IQ modulator/demodulator for optical OFDM up- and downconversion resembles, but is relatively more complicated than, the RF counterpart. We discuss the optical IQ modulator/demodulator in Chapter 8.

Discrete multitone (DMT) is one of the interesting variations of OFDM modulation.[32,33] It uses the real-valued signal even at the baseband. This is achieved by enforcing the following condition to OFDM signal in Eq. (2.9):

$$c_{N-k} = c_k^* \tag{2.18}$$

namely, the information symbols for the subcarriers k and $N - k$ are complex conjugate to each other. The resulting s_m in Eq. (2.9) is thus a real-valued signal. Because both transmit and receive signals are of real value, IQ modulation and demodulation are not necessary. This leads to a much simpler and cost-effective IC design with minimum RF feature.[32,33] For this reason, DMT has been widely incorporated in commercial copper-based digital subscriber line (DSL) systems, such as ADSL and very-high-data-rate DSL.

2.4 Peak-to-Average Power Ratio of OFDM Signals

High PAPR has been cited as one of the drawbacks of the OFDM modulation format. In RF systems, the major problem resides in the power amplifiers at the transmitter end, where the amplifier gain will saturate at high input power. One way to avoid the relatively "peaky" OFDM signal is to operate the power amplifier at the so-called heavy "back-off" regime where the signal power is much lower than the amplifier saturation power. Unfortunately, this requires an excess large saturation power for the power amplifier, which inevitably leads to low power efficiency. In the optical systems, the optical power amplifier (predominately erbium-doped amplifiers are currently in use) is ideally linear regardless of its input signal power due to its slow response time on the order of milliseconds. Nevertheless, PAPR still presents a challenge for optical fiber communications due to the optical fiber nonlinearity.[34–36]

The origin of high PAPR of an OFDM signal can be easily understood from its multicarrier nature. Because cyclic prefix is an advanced time-shifted copy of a part of the OFDM signal in the observation period (see Figure 2.4), we focus on the waveform inside the observation period. The transmitted time domain waveform for one OFDM symbol can be written as

$$s(t) = \sum_{k=1}^{N_{sc}} c_k e^{j2\pi f_k t}, \quad f_k = \frac{k-1}{T_s} \tag{2.19}$$

The PAPR of the OFDM signal is defined as

$$\text{PAPR} = \frac{\max\left\{|s(t)|^2\right\}}{E\left\{|s(t)|^2\right\}}, \quad t \in [0, T_s] \tag{2.20}$$

For simplicity, we assume that an M-PSK encoding is used, where $|c_k| = 1$. The theoretical maximum of PAPR is $10\log_{10}(N_{sc})$ in dB, by setting $c_k = 1$ and $t = 0$ in Eq. (2.19). For OFDM systems with 256 subcarriers, the theoretical maximum PAPR is 24 dB, which obviously is excessively high. Fortunately, such a high PAPR is a rare event, so we do not need to worry about it. A better way to characterize the PAPR is to use the complementary cumulative distribution function (CCDF) of PAPR, P_c, which is expressed as

$$P_c = \Pr\{\text{PAPR} > \zeta_P\} \tag{2.21}$$

Namely, P_c is the probability that PAPR exceeds a particular value of ζ_P.

Figure 2.7 shows CCDF with varying numbers of subcarriers. We have assumed QPSK encoding for each subcarrier. It can be seen that despite the theoretical maximum

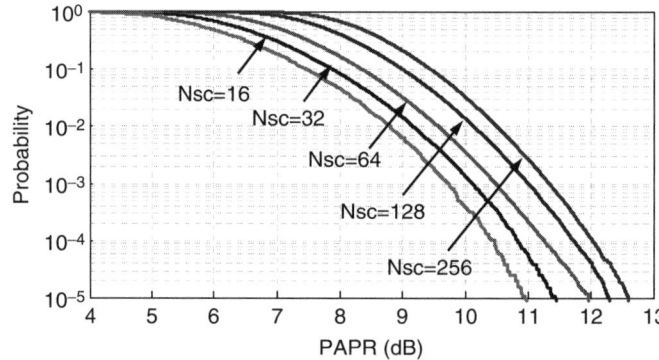

Figure 2.7: Complementary cumulative distribution function, P_c, for the PAPR of OFDM signals with varying numbers of subcarriers. The oversampling factor is fixed at 2.

PAPR of 24 dB for the 256-subcarrier OFDM systems, for more likely probability regimes, such as a CCDF of 10^{-3}, PAPR is approximately 11.3 dB, which is much less than the maximum value of 24 dB. A PAPR of 11.3 dB is still very high because it implies that the peak value is approximately one order of magnitude stronger than the average, and some form of PAPR reduction should be used. It is also interesting to note that the PAPR of an OFDM signal increases slightly as the number of subcarriers increases. For instance, PAPR increases by approximately 1.6 dB when the subcarrier number increases from 32 to 256.

The sampled waveform is used for PAPR evaluation, and subsequently the sampled points may not include the true maximum value of the OFDM signal. Therefore, it is essential to oversample the OFDM signal to obtain an accurate PAPR. Assume that the oversampling factor is h; that is, the number of sampling points increases from N_{sc} to hN_{sc} with each sampling point given by

$$t_l = \frac{(l-1)T_s}{hN_{sc}}, \quad l = 1, 2, \ldots, hN_{sc} \tag{2.22}$$

Substituting Eqs. (2.8) and (2.22) into Eq. (19), the lth sample of $s(t)$ becomes

$$s_l = s(t_l) = \sum_{k=1}^{N_{sc}} c_k e^{j2\pi \frac{(k-1)(l-1)}{hN_{sc}}}, \quad l = 1, 2, \ldots, hN_{sc} \tag{2.23}$$

Expanding the number of subcarriers c_k from N_{sc} to hN_{sc} by appending zeros to the original set, the new subcarrier symbol c'_k after the zero padding is formally given by

$$\begin{aligned} c'_k &= c_k, \quad k = 1, 2, \ldots, N_{sc} \\ c'_k &= 0, \quad k = N_{sc} + 1, N_{sc} + 2, \ldots, hN_{sc} \end{aligned} \tag{2.24}$$

Using the zero-padded new subcarrier set c'_k, Eq. (2.23) is rewritten as

$$s_l = \sum_{k=1}^{hN_{sc}} c'_k e^{j2\pi \frac{(k-1)(l-1)}{hN_{sc}}} = \mathfrak{F}^{-1}(c'_k), \quad l = 1, 2, \ldots, hN_{sc} \tag{2.25}$$

From Eq. (2.25), it follows that the h times oversampling can be achieved by IFFT of a new subcarrier set that zero pads the original subcarrier set to h times the original size.

Figure 2.8 shows the CCDF of PAPR varying oversampling factors from 1 to 8. It can be seen that the difference between the Nyquist sampling ($h = 1$) and 8 times oversampling is approximately 0.4 dB at the probability of 10^{-3}. However, most of the difference takes place below the oversampling factor of 4, and beyond this, PAPR changes very little. Therefore, it seems to be sufficient to use an oversampling factor of 4 for the purpose of PAPR investigation.

It is obvious that the PAPR of an OFDM signal is excessively high for either RF or optical systems. Consequently, PAPR reduction has been an intensely pursued field. Theoretically, for QPSK encoding, a PAPR smaller than 6 dB can be obtained with only a 4%

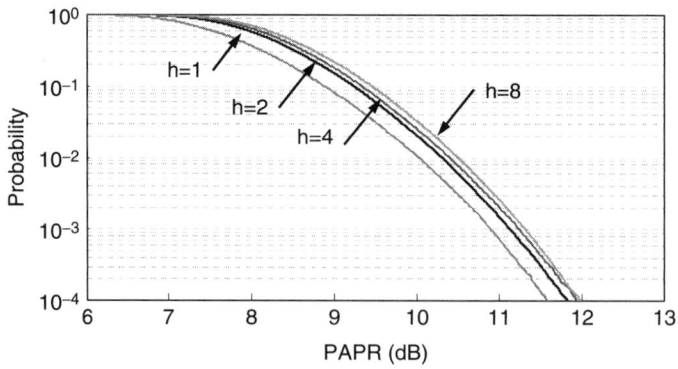

Figure 2.8: Complementary cumulative distribution function for the PAPR of an OFDM signal with varying oversampling factors. The subcarrier number is fixed at 256.

redundancy.[24] Unfortunately, such code has not been identified so far. The PAPR reduction algorithms proposed so far allow for trade-offs among three figures of merits of the OFDM signal: PAPR, bandwidth efficiency, and computational complexity. The following are the most popular PAPR reduction approaches:

PAPR reduction with signal distortion: This is simply done by hard clipping the OFDM signal.[37–39] The consequence of clipping is increased bit error ratio (BER) and out-of-band distortion. The out-of-band distortion can be mitigated through repeated filtering.[39]

PAPR reduction without signal distortion: The idea behind this approach is to map the original waveform to a new set of waveforms that have a PAPR lower than the desirable value, most often with some bandwidth reduction. Distortionless PAPR reduction algorithms include selection mapping,[40,41] optimization approaches such as partial transmit sequence,[42,43] and modified signal constellation or active constellation extension.[44,45]

2.5 Frequency Offset and Phase Noise Sensitivity

Frequency offset and phase noise sensitivity have long been recognized as two major disadvantages of OFDM. Both frequency offset and phase noise lead to ICI. Because of its relatively long symbol length compared to that of the single carrier, OFDM is prone to both frequency offset and phase noise.[46–51] However, it is stressed that these two disadvantages obviously have not prevented OFDM from gaining popularity in RF communications. Frequency offset sensitivity can be mitigated through frequency estimation and compensation,[52–54] and phase noise sensitivity is resolved primarily via careful design of RF local oscillators that satisfy the required phase noise specification. Fortunately, the phase noise specification can be readily met using the state-of-the-art CMOS ASIC chips.[55,56]

Nevertheless, it is important to understand the impairments from both frequency offset and phase noise. Revisit of the phase noise problem is especially important to optical OFDM, in which the laser phase noise is often relatively large, despite the fact that much research effort has focused on designing lasers specifically for the purpose of low laser linewidth.[57,58] The impact of laser phase noise on optical OFDM is thus a critical issue, especially for using higher order constellation to achieve high spectral efficiency modulation.[59]

An OFDM signal with frequency offset and phase noise can be generalized as

$$r(t) = e^{j(2\pi\Delta f \cdot t + \phi(t))} \sum_{k=1}^{N_{sc}} c_k e^{2\pi f_k t} + N(t) \tag{2.26}$$

where Δf is the frequency offset, $\phi(t)$ is the phase noise, and $N(t)$ is the additive white Gaussian noise (AWGN). According to Eq. (2.4), the received information symbol will be

$$c'_{ki} = \frac{1}{T_s} \int_0^{T_s} r(t) e^{-j2\pi f_k t} dt, \quad f_k = \frac{k-1}{T_s} \tag{2.27}$$

Substituting $r(t)$ of Eq. (2.26) into Eq. (2.27), we derive

$$
\begin{aligned}
c'_k &= \frac{1}{T_s} \int_0^{T_s} \left\{ e^{j(2\pi\Delta f t + \phi(t))} \sum_{l=0}^{N_{sc}} c_l e^{2\pi f_l t} + N(t) \right\} e^{-j2\pi f_k t} dt \\
&= \sum_{l=0}^{N_{sc}} c_l \frac{1}{T_s} \int_0^{T_s} e^{j2\pi(f_l - f_k)t + j(2\pi\Delta f t + \phi(t))} dt + n_k = \eta_0 c_k + \sum_{l \neq k} \eta_{k-l} c_l + n_k \\
&= \eta_0 c_k + I_k + n_k
\end{aligned}
\tag{2.28}
$$

We have defined the ICI coefficient η_m as

$$\eta_m = \frac{1}{T_s} \int_0^{T_s} e^{j(2\pi(f'_m + \Delta f)t + \phi(t))} dt, \quad f'_m = \frac{m}{T_s}, \quad m = -(N_{sc} - 1), \dots, 0, 1, \dots, N_{sc} - 1 \tag{2.29}$$

$$n_k = \frac{1}{T_s} \int_0^{T_s} N(t) e^{-j2\pi f_k t} dt \tag{2.30}$$

$$I_k = \sum_{l \neq k} \eta_{k-l} c_l \tag{2.31}$$

In Eq. (2.28), the first, second, and third term respectively correspond to the signal, ICI, and AWGN. The ICI coefficient η_m represents the interference between two subcarriers with the index difference of m. The nonvanishing value of η_m signifies the finite ICI as a result of either frequency offset or phase noise. However, for the sake of simplicity, we investigate frequency offset and phase noise sensitivity separately.

2.5.1 Frequency Offset Effect

When only frequency offset effect is considered, we set $\phi(t)$ to zero and carry out the integration in Eq. (2.29). We have

$$\eta_m = \frac{\sin(\pi(m + \delta))}{\pi(m + \delta)} e^{-j\pi(m+\delta)} \tag{2.32}$$

where $\delta = \Delta f \cdot T_s$ is the normalized frequency offset. Figure 2.9 shows the ICI coefficient η_m when the frequency offset δ is 0 and 0.25. It can be seen that when δ is equal to zero or any integer, the ICI coefficient equals zero or any number of m, which is in essence the orthogonality condition of Eq. (2.5). When δ equals a noninteger value, such as 0.25 as shown in Figure 2.9, there is a residual component of η_m for any number of m, which implies finite interference from one subcarrier to any other subcarrier.

From Eqs. (2.28) and (2.31), the variance of the interference I_m due to frequency offset can be computed as

$$\sigma_{\text{ICI}}^2 = \sigma_c^2 \sum_{m=-N_{sc}/2+1, m\neq 0}^{N_{sc}/2} \eta_m^2 = \chi \cdot \sigma_c^2 \tag{2.33}$$

where χ is the summation of all the ICI terms, given by

$$\chi = \sum_{m=-N_{sc}/2+1, m\neq 0}^{N_{sc}/2} \eta_m^2 \tag{2.34}$$

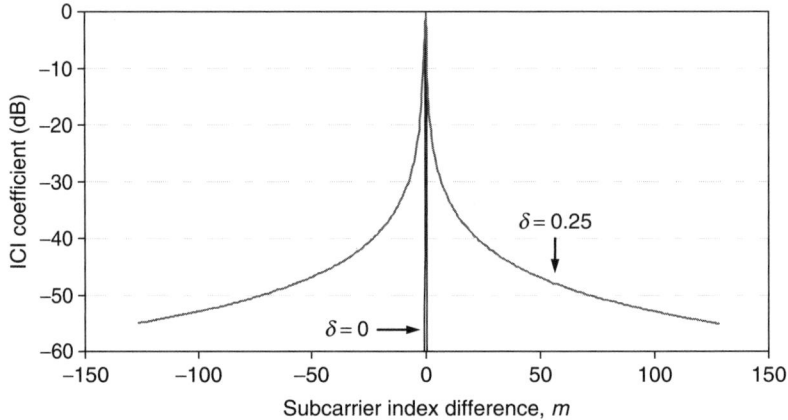

Figure 2.9: The ICI coefficient, η_m, as a function of the subcarrier index difference, m, while varying the frequency offset, δ. The frequency offset is normalized to the subcarrier spacing.

where σ_c^2 is the variance of the transmitted information symbol for each subcarrier. Assume that the interference is also AWGN. The effective SNR γ' is given by

$$\gamma' = \frac{\eta_0^2 \delta_c^2}{\sigma_{ICI}^2 + \sigma_w^2} = \frac{\eta_0^2}{x + \gamma^{-1}} \tag{2.35}$$

where

$$\gamma = \frac{\sigma_c^2}{\sigma_w^2}$$

is the original SNR without the frequency offset. The analytical form of the BER, P_e for a QPSK signal, is given by[60]

$$P_e = \frac{1}{2} erfc\left(\sqrt{\frac{\gamma'}{2}}\right) \tag{2.36}$$

We conduct the simulation for a 256-subcarrier QPSK modulated OFDM system and compare the numerical simulation result with that obtained using the analytical expression of Eq. (2.36). The result is shown in Figure 2.10. It is observed from this figure that the analytical approximation works well for an SNR below 12 dB and an error-floor emerges prematurely for the analytical approximation, implying the AWGN approximation of ICI fails to work well at the regime of high SNR. Figure 2.11 shows the SNR penalty at BER of 10^{-3} as a function of the frequency offset. It can be seen that to maintain the SNR penalty below 1 dB, the frequency offset error δ should be kept below 0.07.

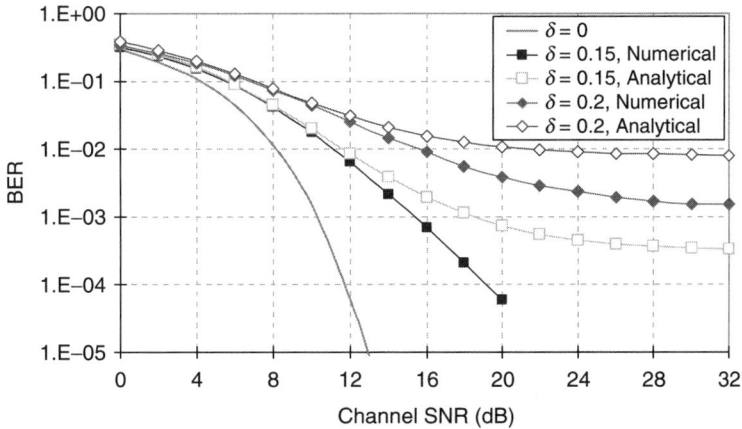

Figure 2.10: BER performance of 256-subcarrier QPSK modulated QPSK systems in the presence of frequency offset. Both the results of numerical simulation and analytical calculation are shown. δ is the frequency offset normalized to the subcarrier spacing.

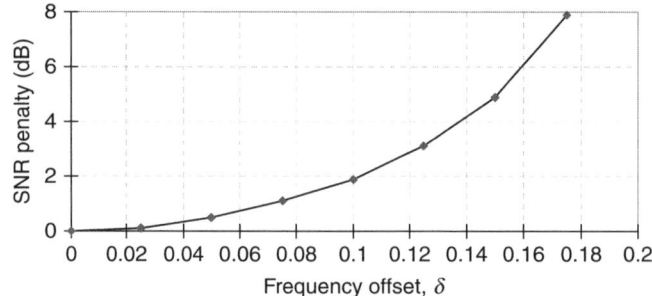

Figure 2.11: SNR penalty as a function of the frequency offset for 256-subcarrier QPSK modulated OFDM systems.

2.5.2 Phase Noise Effect

Phase noise has been extensively studied since OFDM began to be incorporated into the standard. The fundamental work can be found in publications by Pollet et al.,[46] Tomba,[47] and Armada.[48] Assume the frequency offset has been compensated and only phase noise is present. Equation (2.29) becomes

$$\eta_m = \frac{1}{T_s} \int_0^{T_s} e^{j(2\pi f'_m t + \phi(t))} dt, \quad f'_m = \frac{m}{T_s}, m = -(N_{sc} - 1), \ldots, 0, 1, \ldots, N_{sc} - 1 \tag{2.37}$$

Denote $\eta_0 = |\eta_0| e^{j\phi_0}$. ϕ_0 is also called the common phase error (CPE). Substituting Eq. (2.37) into Eq. (2.28) and making some simple rearrangements, we derive

$$\tilde{c}_k = c'_k e^{-j\phi_0} = |\eta_0| c_k + e^{-j\phi_0} \sum_{l \neq k} \eta_{k-l} c_l + e^{-j\phi_0} n_k \tag{2.38}$$

where $\tilde{c}_k$ is the received information symbol after removing the CPE. It can be seen that the phase noise has two major effects. First, CPE, ϕ_0, rotates the entire constellation. This can be estimated and the skewed constellation can be rectified through simple collective rotation as shown in Eq. (2.38). Second, ICI impairment, which is the second term in Eq. (2.38), is manifested by the nonvanishing terms of η_m for nonzero m. To assess the ICI impairment of phase noise, we make a further assumption that the phase noise $\phi(t)$ observes the Wiener process, such that[61]

$$E\left[(\phi(\tau + t_0) - \phi(t_0))^2\right] = 2\pi\beta\tau \tag{2.39}$$

where $E[\ ]$ is the ensemble average, and β is the 3dB laser linewidth or, more precisely, the combined linewidth including both transmit and receive lasers. We denote $|\eta_0| = a + \zeta$, where $a = \langle|\eta_0|\rangle$, and ζ is the residual amplitude noise for each OFDM symbol. From Eq. (2.38), the SNR of $\tilde{c}_k$ can be represented as

$$\gamma' = \frac{a^2 \sigma_k^2}{n^2 + n_P^2} \tag{2.40}$$

where σ_k^2 and n_k^2 are respectively the variance of the signal and AWGN noise for the kth subcarrier, and n_P^2 is the variance for the noise including ICI and the amplitude fluctuation of the $|\eta_0|$. σ_k^2 of 1 is assumed for simplicity. From Eq. (2.38), n_P^2 can be expressed as

$$n_P^2 = E[\zeta^2] + E\left[\sum_{l \neq k} |\eta_{k-l}|^2\right] \tag{2.41}$$

Denoting $a_1 = E\left[|\eta_0|^2\right]$, we have

$$E[\zeta^2] = E\left[|\eta_0|^2\right] - (E[|\eta_0|])^2 = a_1 - a^2 \tag{2.42}$$

It can be also shown that[46]

$$E\left[\sum_{l \neq k} |\eta_{k-l}|^2\right] = 1 - E\left[|\eta_0|^2\right] = 1 - a_1 \tag{2.43}$$

Therefore, the effective SNR impaired by the phase noise becomes

$$\gamma' = \frac{a^2}{\gamma^{-1} + 1 - a^2} = \gamma \frac{a^2}{1 + (1 - a^2)\gamma} \tag{2.44}$$

$$\gamma = \frac{\delta_k^2}{n_k^2} \tag{2.45}$$

where γ is the SNR without phase noise impairment. The SNR penalty or reduction from the phase noise-free system in decibels is thus

$$\Delta\gamma(\text{dB}) = 10 \cdot \log 10(\gamma') - 10 \cdot \log 10(\gamma) = 10 \cdot \log 10\left(\frac{a^2}{1 + (1 - a^2)\gamma}\right) \tag{2.46}$$

$\Delta\gamma(\text{dB})$ can be also interpreted as the additional SNR needed to achieve the same effective SNR or BER. It has been shown[46] for a linewidth of β that $a = \langle|\eta_0|\rangle$ becomes

$$a \approx 1 - \frac{11}{60} 4\pi\beta T_s\gamma \tag{2.47}$$

Keeping only the first-order term proportional to βT_s, we derive

$$\Delta\gamma(\text{dB}) = \frac{10}{\ln(10)} \frac{11}{60} 4\pi\beta T_s\gamma = 10\beta T_s\gamma \tag{2.48}$$

It can be observed from Eq. (2.48) that the phase noise impact on the OFDM signal is enhanced by the relatively long symbol period of T_s compared with the single-carrier signal. Furthermore, the penalty is also proportional to γ, or the SNR where the penalty is referenced, implying that the higher order modulation is more sensitive to the phase noise as it requires higher SNR. Figure 2.12 shows SNR penalty as a function of the linewidth for both 4-QAM and 16-QAM modulations. For 4-QAM modulation, to limit the phase noise impairment below 1 dB, the βT should be kept below 0.01. The higher

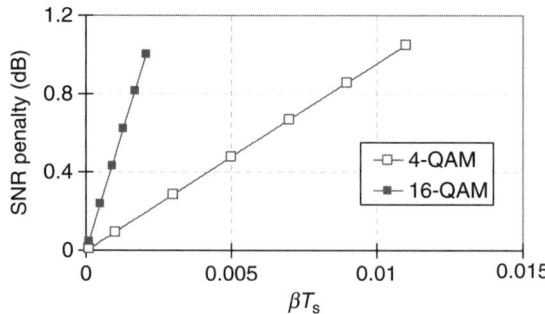

Figure 2.12: SNR penalty against the laser linewidth normalized to $1/T_s$ for both 4-QAM and 16-QAM OFDM systems.

modulation such as 16-QAM tightens the requirement of βT down to 0.002. Therefore, for the CO-OFDM system, the laser linewidth is a critical parameter, especially for the higher order modulation such as 16-QAM.[59]

References

1. Chang RW. Synthesis of band-limited orthogonal signals for multichannel data transmission. *Bell Sys Tech J* 1966;**45**:1775–96.
2. Chang RW. Orthogonal frequency division multiplexing. U.S. Patent no. 3488445, 1970.
3. Toba H, Oda K, Inoue K, Nosu K, Kitoh T. An optical FDM-based self-healing ring network employing arrayed waveguide crating filters and EDFAs with level equalizers. *IEEE J Selected Areas Commun* 1996;**14**:800–13.
4. Kitayama K. Highly spectrum efficient OFDM/PDM wireless networks by using optical SSB modulation. *J Lightwave Technol* 1998;**16**:969–76.
5. Pan Q, Green RJ. Bit-error-rate performance of lightwave hybrid AM/OFDM systems with comparison with AM/QAM systems in the presence of clipping impulse noise. *IEEE Photon Technol Lett* 1996;**8**:278–80.
6. Shi Q. Error performance of OFDM-QAM in subcarrier multiplexed fiber-optic transmission. *IEEE Photon Technol Lett* 1997;**9**:845–7.
7. You R, Kahn JM. Average power reduction techniques for multiple-subcarrier intensity-modulated optical signals. *IEEE Trans Commun* 2001;**49**:2164–71.
8. Ma CP, Kuo JW. Orthogonal frequency division multiplex with multi-level technology in optical storage application. *Jpn J Appl Physics Part 1: Regular Papers Short Notes Rev Papers* 2004;**43**:4876–8.
9. Dixon BJ, Pollard RD, Iezekeil S. Orthogonal frequency-division multiplexing in wireless communication systems with multimode fiber feeds. *IEEE Trans Microwave Theory Techniques* 2001;**49**:1404–9.
10. Jolley NE, Kee H, Rickard R, Tang JM, Cordina K. Generation and propagation of a 1550 nm 10 Gb/s optical orthogonal frequency division multiplexed signal over 1000 m of multimode fibre using a directly modulated DFB. In: *Opt. Fibre Commun. Conf.,* paper no. OFP3. Anaheim, CA; 2005.
11. Tang JM, Lane PM, Shore KA. High-speed transmission of adaptively modulated optical OFDM signals over multimode fibers using directly modulated DFBs. *J Lightwave Technol* 2006;**24**:429–41.
12. Lowery A, Armstrong J. 10Gbit/s multimode fiber link using power-efficient orthogonal-frequency-division multiplexing. *Opt Express* 2005;**13**:10003–9.

13. Lowery AJ, Du L, Armstrong J. Orthogonal frequency division multiplexing for adaptive dispersion compensation in long haul WDM systems. In: *Opt. Fiber Commun. Conf.*, paper no. PDP 39. Anaheim, CA; 2006.

14. Djordjevic I, Vasic B. Orthogonal frequency division multiplexing for high-speed optical transmission. *Opt Express* 2006;**14**:3767–75.

15. Shieh W, Athaudage C. Coherent optical orthogonal frequency division multiplexing. *Electron Lett* 2006;**42**:587–9.

16. Kobayash T, Sano A, Yamada E. Electro-optically subcarrier multiplexed 110 Gb/s OFDM signal transmission over 80 km SMF without dispersion compensation. *IET Electron Lett* 2008;**44**:225–6.

17. Jansen SL, Morita I, Tanaka H. 10 × 121.9-Gb/s PDM-OFDM transmission with 2-b/s/Hz spectral efficiency over 1000 km of SSMF. In: *Opt. Fiber Commun. Conf.*, paper no. PDP2. San Diego; 2008.

18. Shieh W, Yang Q, Ma Y. 107 Gb/s coherent optical OFDM transmission over 1000-km SSMF fiber using orthogonal band multiplexing. *Opt Express* 2008;**16**:6378–86.

19. Mosier RR, Clabaugh RG. Kineplex, a bandwidth-efficient binary transmission system. *AIEE Trans* 1958;**76**:723–8.

20. Zimmerman MS, Kirsch AL. AN/GSC-10 (KATHRYN) variable rate data modem for HF radio. *AIEE Trans* 1960;**79**:248–55.

21. Weinsten SB, Ebert PM. Data transmission by frequency-division multiplexing using the discrete Fourier transform. *IEEE Trans Commun* 1971;**19**:628–34.

22. Duhamel P, Hollmann H. Split-radix FFT algorithm. *IET Elect Lett* 1984;**20**:14–6.

23. Hara S, Prasad R. *Multicarrier Techniques for 4G Mobile Communications*. Boston: Artech House; 2003.

24. Hanzo L, Munster M, Choi BJ, Keller T. *OFDM and MC-CDMA for Broadband Multi-User Communications, WLANs and Broadcasting*. New York: Wiley; 2003.

25. Yi X, Shieh W, Ma Y. Phase noise on coherent optical OFDM systems with 16-QAM and 64-QAM beyond 10 Gb/s. In: *Eur. Conf. Opt. Commun.*, paper no. 5.2.3. Berlin, Germany; 2007.

26. Takahashi H, Amin AA, Jansen SL, Morita I, Tanaka H. 8 × 66.8-Gbit/s coherent PDM-OFDM transmission over 640 km of SSMF at 5.6-bit/s/Hz spectral efficiency. In: *Eur. Conf. Opt. Commun.*, paper no. Th.3.E.4. Brussels, Belgium; 2008.

27. Washburn BR, Diddams SA, Newbury NR, et al. Phase-locked, erbium-fiber-laser-based frequency comb in the near infrared. *Opt Lett* 2004;**29**:250–2.

28. Sano A, Yoshida E, Masuda H, et al. 30 × 100-Gb/s all-optical OFDM transmission over 1300 km SMF with 10 ROADM nodes. In: *Eur. Conf. Opt. Commun.*, paper no. PD 1.7. Berlin, Germany; 2007.

29. Doan C, Emami S, Niknejad A, Brodersen R. A 60-GHz CMOS receiver front-end. *IEEE J Solid-State Circuits* 2005;**40**:144–55.

30. Razavi B. A 60-GHz CMOS receiver front-end. *IEEE J Solid-State Circuits* 2006;**41**:17–22.

31. Zhang F, Skafidas E, Shieh W. 60 GHz double-balanced up-conversion mixer on 130 nm CMOS technology. *IET Electron Lett* 2008;**14**:633–4.

32. Chow PS, Cioffi JM, Bingham JAC. A practical discrete multitone transceiver loading algorithm for data transmission over spectrally shaped channels. *IEEE Trans Commun* 1995;**43**:773–5.

33. Jeffrey Lee SC, Breyer F, Randel S, van den Boom HPA, Koonen AMJ. High-speed transmission over multimode fiber using discrete multitone modulation [Invited]. *J Opt Networking* 2008;**7**:183–96.

34. Lowery AJ, Wang S, Premaratne M. Calculation of power limit due to fiber nonlinearity in optical OFDM systems. *Opt Express* 2007;**15**:13282–7.

35. Dischler R, Buchali F. Measurement of nonlinear thresholds in O-OFDM systems with respect to data pattern and peak power to average ratio. In: *Opt. Fiber Commun. Conf.*, paper no. Mo.3.E.5. San Diego; 2008.

36. Nazarathy M, Khurgin J, Weidenfeld R, et al. Phased-array cancellation of nonlinear FWM in coherent OFDM dispersive multi-span links. *Opt Express* 2008;**16**:15777–810.

37. O'Neil R, Lopes LN. Envelope variations and spectral splatter in clipped multicarrier signals. In: *Proc. IEEE 1995 Int. Symp. Personal Indoor Mobile Radio Commun*; 1995. pp. 71–5.

38. Li X, Cimini Jr LJ. Effects of clipping and filtering on the performance of OFDM. *IEEE Commun Lett* 1998;**2**:131–3.

39. Armstrong J. Peak-to-average power reduction for OFDM by repeated clipping and frequency domain filtering. *IET Elect Lett* 2002;**38**:246–7.

40. Mestdagh DJG, Spruyt PMP. A method to reduce the probability of clipping DMT-based transceivers. *IEEE Trans Commun* 1996;**44**:1234–8.

41. Bauml RW, Fischer RFH, Huber JB. Reducing the peak-to-average power ratio of multicarrier modulation by selected mapping. *IET Electron Lett* 1996;**32**:2056–7.

42. Muller SH, Huber JB. A novel peak power reduction scheme for OFDM. In: *Proc. IEEE 1997 Int. Symp. Personal Indoor Mobile Radio Commun*; 1997. pp. 1090–4.

43. Friese M. OFDM signals with low crest-factor. In: *Proc. 1997 IEEE Global Telecommun. Conf*; 1997. pp. 290–4.

44. Tellado J, Cioffi JM. Peak power reduction for multicarrier transmission. In: *Proc. 1998 IEEE Global Telecommun. Conf.*; 1998. pp. 219–24.

45. Krongold BS, Jones DL. PAR reduction in OFDM via active constellation extension. *IEEE Trans Broadcasting* 2003;**49**:258–68.

46. Pollet T, Van Bladel M, Moeneclaey M. BER sensitivity of OFDM systems to carrier frequency offset and Wiener phase noise. *IEEE Trans Commun* 1995;**43**:191–3.

47. Tomba L. On the effect of Wiener phase noise in OFDM systems. *IEEE Trans Commun* 1998;**46**:580–3.

48. Armada AG. Understanding the effects of phase noise in orthogonal frequency division multiplexing (OFDM). *IEEE Trans Broadcasting* 2001;**47**:153–9.

49. Wu SP, Bar-Ness Y. OFDM systems in the presence of phase noise: Consequences and solutions. *IEEE Trans Commun* 2004;**52**:1988–96.

50. Sathananthan K, Tellambura C. Probability of error calculation of OFDM systems with frequency offset. *IEEE Trans Commun* 2001;**49**:1884–8.

51. Armstrong J. Analysis of new and existing methods of reducing intercarrier interference due to carrier frequency offset in OFDM. *IEEE Trans Commun* 1999;**47**:365–9.

52. Moose PH. A technique for orthogonal frequency division multiplexing frequency offset correction. *IEEE Trans Commun* 1994;**42**:2908–14.

53. Schmidl TM, Cox DC. Robust frequency and timing synchronization for OFDM. *IEEE Trans Commun* 1997;**45**:1613–21.

54. Hsieh M, Wei C. A low-complexity frame synchronization and frequency offset compensation scheme for OFDM systems over fading channels. *IEEE Trans Vehicular Technol* 1999;**48**:1596–609.

55. Maeda T, Matsuno N, Hori S, et al. A low-power dual-band triple-mode WLAN CMOS transceiver. *IEEE J Solid-State Circuits* 2006;**41**:2481–90.

56. Ahola R, Aktas A, Wilson J, et al. A single-chip CMOS transceiver for 802.11a/b/g wireless LANs. *IEEE J Solid-State Circuits* 2004;**39**:2250–8.

57. Wyatt R, Devlin WJ. 10 kHz linewidth 1.5 um InGaAsP external cavity laser with 55 nm tuning range. *IET Elect Lett* 2007;**19**:110–2.

58. Bird DM, Armitage JR, Kashyap R, Fatah RMA, Cameron KH. Narrow line semiconductor laser using fibre grating. *IET Elect Lett* 1991;**27**:1115–6.

59. Yi X, Shieh W, Ma Y. Phase noise effects on high spectral efficiency coherent optical OFDM transmission. *J Lightwave Technol* 2008;**26**:1309–16.

60. Proakis J. *Digital Communications*. 3rd ed. New York: WCB/McGraw-Hill; 1995.

61. Ho K-P. *Phase-Modulated Optical Communication Systems*. New York: Springer; 2005.

CHAPTER 3

Optical Communication Fundamentals

The ultimate goal of the optical signal transmission is to achieve the predetermined bit error ratio (BER) between any two nodes in an optical network. The optical transmission system has to be properly designed to provide reliable operation during its lifetime, which includes the management of key engineering parameters.[1]

This chapter describes the key optical components used in a contemporary optical communication system; basic signal and noise parameters; major channel impairments, including chromatic dispersion, polarization mode dispersion (PMD), and fiber nonlinearities; and the system design process. This chapter is based on references 1–22.

The chapter is organized as follows. After a brief introduction, Section 3.2 identifies and describes the key optical components at a level sufficient to understand other chapters in the book without having any background in optical communications. In Section 3.3, different noise sources are identified and explained. Section 3.4 is devoted to different channel impairments. Section 3.5 deals with different figures of merit, to describe system performance, and basic guidelines for system design.

3.1 Introduction

The optical transmission system design[1–5] involves accounting for different effects that may degrade the signal during modulation, propagation, and detection processes. The transmission quality is assessed by the received *signal-to-noise ratio* (SNR), which is the ratio between signal power and noise power at the decision point. The SNR is related to the receiver *sensitivity*, the minimum received optical power needed to keep SNR at the specified level.

In digital optical communications, BER, defined as the ratio of bits in error to total number of transmitted bits at the decision point, is commonly used as a figure of merit. In that sense, the *receiver sensitivity* is defined as the minimum required received optical power to keep BER below a given value. The three types of parameters that are important from the system engineering standpoint are (1) the optical signal parameters that determine the signal level, (2) the optical noise parameters that determine the BER, and (3) the impairment parameters that determine the power margin to be allocated to compensate for their impact. The optical signal parameters defining the signal level include optical transmitter output power, extinction ratio, optical amplification gain, and photodiode responsivity. The total noise is a stochastic process composed of both additive noise components and multiplicative (nonadditive) noise components. There exist a number of impairments that will deteriorate the signal quality during transmission, such as fiber attenuation, chromatic dispersion, PMD, polarization-dependent loss, fiber nonlinearities, insertion loss, and frequency chirp. A proper design process involves different steps to provide a prespecified transmission system quality and to balance different system parameters. The system parameters can be related to power, time, wavelength, or a combination of these.

Given this general description of different signal and noise parameters, we turn our attention to the key optical components, after which we continue our description of different channel impairments and the transmission system engineering process.

3.2 Key Optical Components

This section describes the basic optical components used in an optical transmission system. An exemplary optical network, identifying the key optical components, is shown in Figure 3.1. The end-to-end optical transmission involves both electrical and optical signal paths. To perform conversion from electrical to optical domain, the optical transmitters are used, whereas to perform conversion in the opposite direction (optical to electrical conversion), the optical receivers are used. The optical fibers serve as the foundation of an

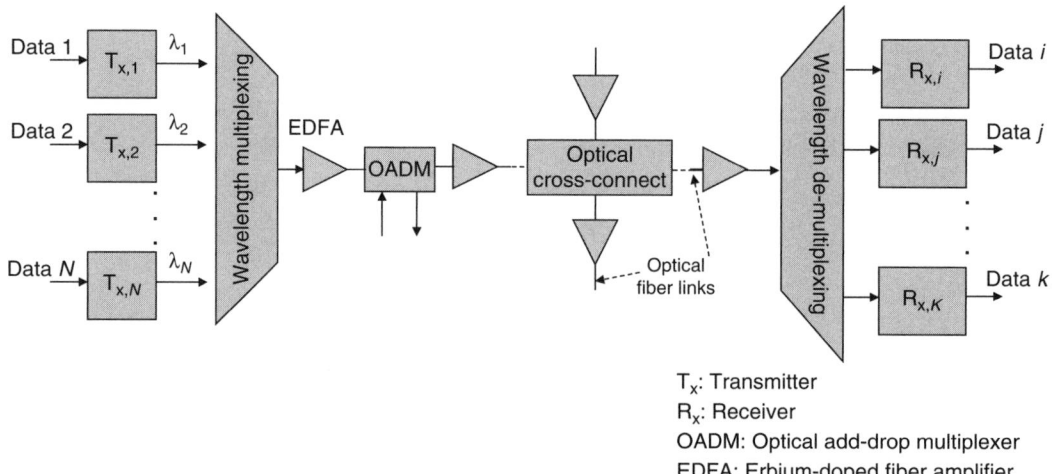

T_x: Transmitter
R_x: Receiver
OADM: Optical add-drop multiplexer
EDFA: Erbium-doped fiber amplifier

Figure 3.1: An exemplary optical network identifying key optical components.

optical transmission system because they are used as a medium to transport the optical signals from source to destination. The optical fibers attenuate the signal during transmission, and optical amplifiers, such as erbium-doped fiber amplifiers (EDFAs), Raman amplifiers, or parametric amplifiers, have to be used to restore the signal quality.[1-5] However, the process of amplification is accompanied by the noise addition. The simplest optical transmission system employs only one wavelength. The wavelength-division multiplexing (WDM) can be considered as an upgrade of the single-wavelength system. WDM corresponds to the scheme in which multiple optical carriers at different wavelengths are modulated by using independent electrical bitstreams, as shown in Figure 3.1, and then transmitted over the same fiber. WDM has the potential of exploiting the enormous bandwidth offered by the optical fiber. During transmission of WDM signals, occasionally one or several wavelengths are added or dropped, which is performed by the optical component known as the optical add–drop multiplexer (OADM), as illustrated in Figure 3.1. The optical networks require the switching of information among different fibers, which is performed in optical cross-connect (OXS). To combine several distinct wavelength channels into a composite channel, the wavelength multiplexers are used. On the other hand, to split the composite WDM channel into distinct wavelength channels, wavelength demultiplexers are used. To impose the information signal, optical modulators are used. The optical modulators are commonly used in combination with semiconductor lasers.

The typical receiver configuration with direct detection is shown in Figure 3.2.[1-5] The main purpose of the optical receiver, terminating the lightwave path, is to convert the signal coming from single-mode fiber from optical to electrical domain and process appropriately such obtained electrical signal to recover the data being transmitted. The incoming optical signal may be preamplified by an optical amplifier and further processed by an optical filter

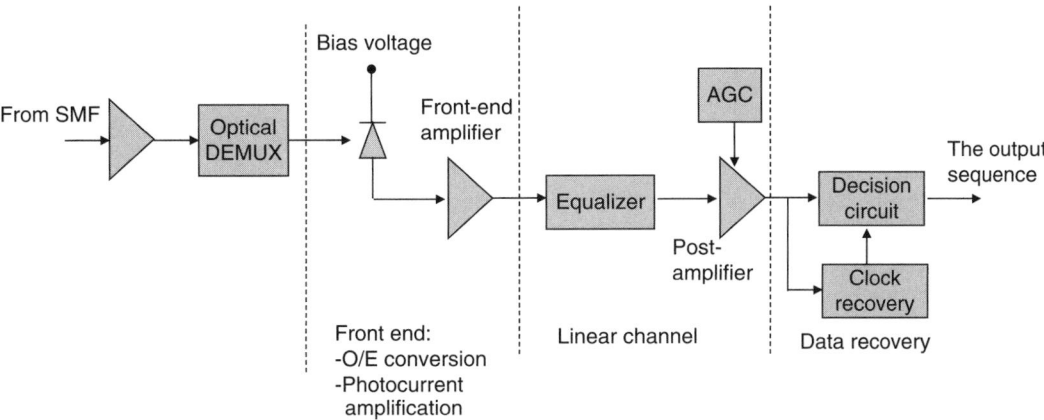

Figure 3.2: A typical direct detection receiver architecture. AGC, automatic gain control; O/E, optical-to-electrical.

to reduce the level of amplified spontaneous emission (ASE) noise or by wavelength demultiplexer to select a desired wavelength channel. The optical signal is converted into electrical domain by using a photodetector, followed by electrical postamplifier. To deal with residual intersymbol interference (ISI), an equalizer may be used. The main purpose of the clock recovery circuit is to provide timing for the decision circuit by extracting the clock from the received signal. The clock recovery circuit is most commonly implemented using the phase-locked loop (PLL). Finally, the purpose of the decision circuit is to provide the binary sequence being transmitted by comparing the sampled signal to a predetermined threshold. Whenever the received sample is larger than the threshold, the decision circuit decides in favor of bit 1; otherwise, it decides in favor of bit 0.

The optical signal generated by the semiconductor laser has to be modulated by the information signal before being transmitted over the optical fiber. This can be achieved by directly modulating the bias current of the semiconductor laser, which can be done even at high speed (up to 40 Gb/s in certain lasers). Unfortunately, although conceptually simple, this concept is rarely used in practice because of the frequency chirp introduced by direct modulation, nonuniform frequency response, and large current swing needed to provide operation. For transmitters operating at 10 Gb/s and above, the semiconductor laser diode is commonly biased at constant current to provide continuous wave (CW) output, and external modulators are used to impose the information signal to be transmitted. The most popular modulators are electro-optic optical modulators, such as Mach–Zehnder modulators, and electroabsorption modulators. The principle of the external modulator is illustrated in Figure 3.3. Through the external modulation process, a certain parameter of the CW signal, used as a signal carrier, is varied in accordance with the information-bearing signal. For example, a monochromatic electromagnetic wave is commonly used as a carrier, and its electrical field $E(t)$ can be represented by

$$E(t) = pA\cos(\omega t + \varphi) \tag{3.1}$$

where A, ω, and φ are amplitude, frequency, and phase, respectively; and p denotes the polarization orientation. Each of these parameters can be used to carry information, and the information-bearing signal can be either CW or discrete. If the information-bearing signal is CW, corresponding modulation formats are amplitude modulation (AM), frequency modulation (FM), phase modulation (PM), and polarization modulation (PolM). On the other

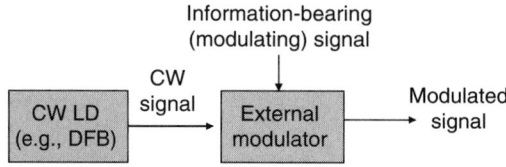

Figure 3.3: An illustration of the external modulation principle.

hand, if the information-bearing signal is digital, the corresponding modulations are amplitude-shift keying (ASK), frequency-shift keying (FSK), phase-shift keying (PSK), and polarization-shift keying (PolSK).

Optical fibers serve as the foundation of an optical transmission system because they transport optical signals from source to destination. The combination of low-loss and large bandwidth allows high-speed signals to be transmitted over long distances before regeneration is needed. A low-loss optical fiber is manufactured from several different materials; the base row material is pure silica, which is mixed with different dopants in order to adjust the refractive index of optical fiber. The optical fiber consists of two waveguide layers, the core and the cladding, protected by buffer coating. The majority of the power is concentrated in the core, although some portion can spread to the cladding. There exists a difference in refractive indices between the core and cladding, which is achieved by mixing dopants, commonly added to the fiber core. There are two types of optical fibers: multimode fiber (MMF) and single-mode fiber (SMF). Multimode optical fibers transfer the light through a collection of spatial transversal modes. Each mode, defined through a specified combination of electrical and magnetic components, occupies a different cross section of the optical fiber core and takes a slightly distinguished path along the optical fiber. The difference in mode path lengths in multimode optical fibers produces a difference in arrival times at the receiving point. This phenomenon is known as *multimode dispersion* (or *intermodal dispersion*), and it causes signal distortion and imposes the limitations in signal bandwidth. The second type of optical fibers, SMFs, effectively eliminate multimode dispersion by limiting the number of propagating modes to a fundamental one. However, SMFs introduce another signal impairment known as chromatic dispersion. *Chromatic dispersion* is caused by the difference in group velocities among different spectral components within the same mode.

The attenuation of signal propagating through optical fiber is low compared to that of other transmission media, such as copper cables or free space. Nevertheless, we occasionally have to amplify the attenuated signal to restore the signal level, without any conversion into the electrical domain. This can be done in optical amplifiers through the process of stimulated emission. The main ingredient of an optical amplifier is the optical gain realized through the amplifier pumping (either electrical or optical) to achieve the so-called population inversion. The common types of optical amplifiers are semiconductor optical amplifiers (SOAs), EDFAs, and Raman amplifiers. The amplification process is commonly followed by the noise process, not related to the signal, that occurs due to spontaneous emission. The amplification process degrades the SNR because of ASE noise added to the signal in every amplifier stage.

Before providing more details about the basic building blocks identified in this section, a more global picture is given by describing a typical optical network shown in Figure 3.4.

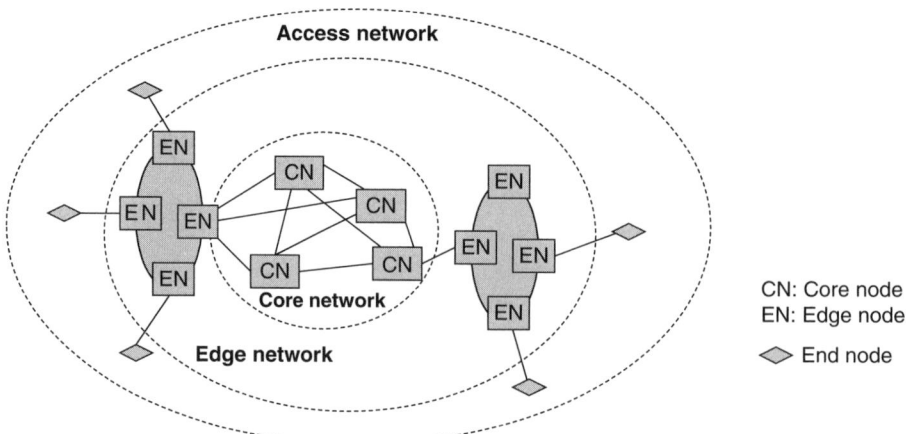

Figure 3.4: A typical optical networking architecture.

We can identify three ellipses representing the core network, the edge network, and the access network.[1] The long-haul core network interconnects large cities, major communications hubs, and even different continents by means of submarine transmission systems. The core networks are often called the wide area networks (WANs) or interchange carrier networks. The edge optical networks are deployed within smaller geographical areas and are commonly recognized as metropolitan area networks (MANs) or local exchange carrier networks. The access networks represent the peripheral part of optical network and provide the last-mile access or the bandwidth distribution to individual end users. The common access networks are local area networks (LANs) and distribution networks. The common physical network topologies are the mesh network (often present in core networks), ring network (in edge networks), and star networks (commonly used in access networks).

Given this general description of key optical components, the remainder of this section provides more details about the basic building blocks: optical transmitters, optical receivers, optical amplifiers, optical fibers, and other optical building blocks, such as multiplexers/demultiplexers, optical filters, OADMs, optical switches, and couplers.

3.2.1 Optical Transmitters

The role of the optical transmitter is to generate the optical signal, impose the information-bearing signal, and launch the modulated signal into the optical fiber. The semiconductor light sources are commonly used in state-of-the-art optical communication systems. The light-generation process occurs in certain semiconductor materials due to recombination of electrons and holes in p-n junctions, under direct biasing. Depending on the nature of the

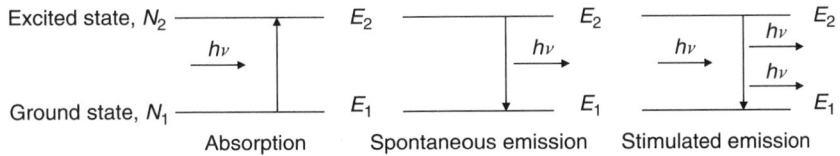

Figure 3.5: Illustrating the interaction of light with matter.

recombination process, we can classify different semiconductor light sources as either light-emitting diodes (LEDs), in which spontaneous recombination dominates, or semiconductor lasers, in which the stimulated emission is a dominating mechanism. Namely, there are three basic processes in semiconductor materials, as illustrated in Figure 3.5, by which light interacts with matter: absorption, spontaneous emission, and stimulated emission. In normal conditions, the number of electrons in ground state (with energy E_1) N_1 is larger than the number of electrons in excited state (with energy E_2) N_2, and in the thermal equilibrium their ratio follows the Boltzmann's statistics[1,3,4]

$$\frac{N_2}{N_1} = \exp\left(-\frac{h\nu}{k_B T}\right) = \exp\left(-\frac{E_2 - E_1}{k_B T}\right) \tag{3.2}$$

where $h\nu$ is a photon energy (h is Planck's constant, and ν is the optical frequency proportional to the energy difference between the energy levels $E_2 - E_1$), k_B is the Boltzmann's constant, and T is the absolute temperature. In the same regime, the spontaneous emission rate $dN_{2,\text{spon}}/dt = A_{21}N_2$ (A_{21} denotes the spontaneous emission coefficient) and the stimulated emission rate $dN_{2,\text{stim}}/dt = B_{21}\rho(\nu)N_2$ (B_{21} denotes the stimulated emission coefficient, and $\rho(\nu)$ denotes the spectral density of electromagnetic energy) are equalized with an absorption rate $dN_{1,\text{abs}}/dt = A_{12}\rho(\nu)N_1$ (A_{12} denotes the absorption coefficient)[1]:

$$A_{21}N_2 + B_{21}\rho(\nu)N_2 = B_{12}\rho(\nu)N_1 \tag{3.3}$$

In the visible or near-infrared region ($h\nu \sim 1$ eV), the spontaneous emission always dominates over stimulated emission in thermal equilibrium at room temperature ($k_B T \approx 25$ meV) (see Eq. 3.2). The stimulated emission rate can exceed the absorption rate only when $N_2 > N_1$; this condition is referred to as *population inversion* and can never be realized for systems in thermal equilibrium. The population inversion is a prerequisite for laser operation, and in an atomic system it is achieved by using three- and four-level pumping schemes (an external energy source raises the atomic population from ground to an excited state). There are three basic components required to sustain stimulated emission and to form useful laser output: the pump source, the active medium, and the feedback mirrors. The active medium can be solid (e.g., in semiconductor lasers), gaseous, or liquid in nature. The pump can be electrical (e.g., semiconductor lasers), optical, or chemical. The purpose of the pump is to achieve the population inversion. The basic structure of the Fabry–Perot

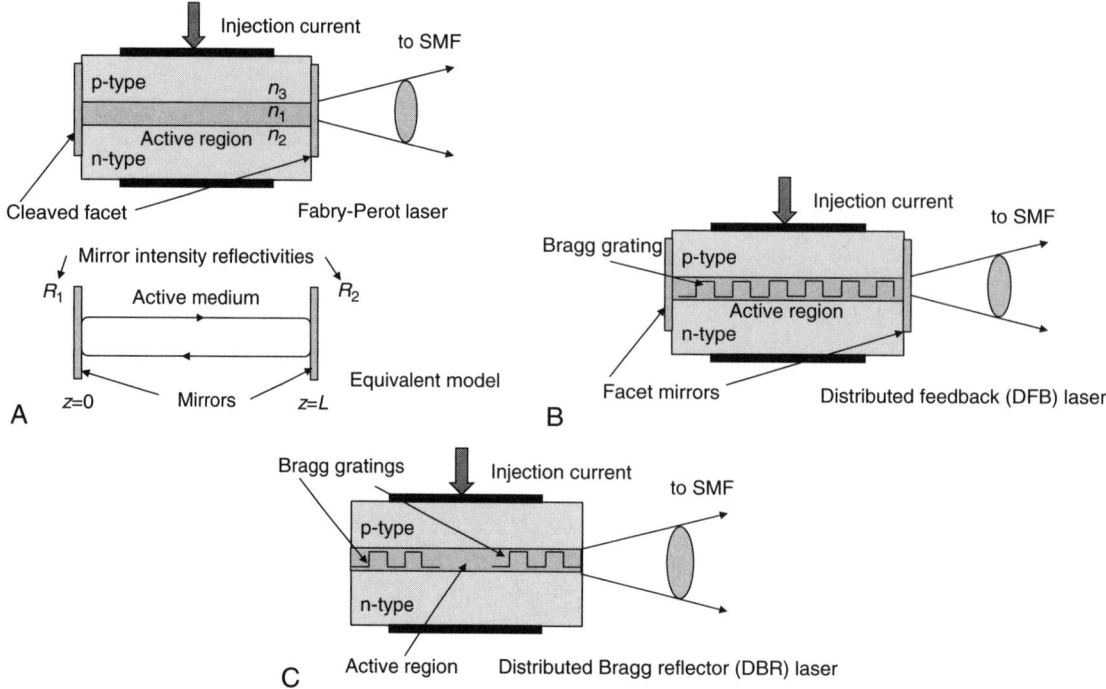

Figure 3.6: Semiconductor lasers: (a) Fabry–Perot semiconductor laser, (b) DFB laser, and (c) DBR laser.

type of semiconductor laser is shown in Figure 3.6a, together with an equivalent model. The injection (bias) current flows through the p-n junction and stimulates the recombination of electron and holes, leading to the generation of photons.

For the lasing action to be sustainable, the gain and phase matching condition should be satisfied. In the active medium, both *gain/absorption*, described by $\gamma(v)$, and *scattering*, described by α_s, are present. The intensity inside the cavity can be described by the following dependence: $I(z) = I_0\exp[(\gamma(v) - \alpha_s)z]$. The lasing is possible when collective gain is larger than loss after a round-trip pass through the cavity:

$$I(2L) = I_0 R_1\, R_2\exp\left[2L(\gamma(v) - \alpha_s)\right] = I_0 \tag{3.4}$$

where R_1 and R_2 are facet reflectivities (see Figure 3.6), L is the length of active medium, and I_0 and $I(2L)$ correspond to initial and round-trip intensities. The gain threshold is obtained by solving Eq. (3.4) per γ.

$$\gamma_{th} = \alpha_s + \frac{1}{2L}\ln\left(\frac{1}{R_1\,R_2}\right) = \alpha_{int} + \alpha_{mir} \tag{3.5}$$

where the internal losses (corresponding to α_s) and mirror losses $((1/2L)\ln(1/R_1R_2))$ are denoted by α_{int} and α_{mir}, respectively. After the round-trip, the resultant phase must be equal to the initial phase, leading to the phase matching condition:

$$\exp[-j2\beta L] = 1 \Rightarrow 2\beta L = q2\pi, \beta = 2\pi n/\lambda \tag{3.6}$$

where β denotes the propagation constant, n is the refractive index of active medium, and λ is the free-space wavelength. The phase matching condition can be satisfied for many different integers q, representing different longitudinal modes of frequency $v_q = qc/(2nL)$. The separation between neighboring longitudinal modes is known as the *free spectral range*:

$$\Delta v = v_q - v_{q-1} = q\frac{c}{2nL} - (q-1)\frac{c}{2nL} = \frac{c}{2nL} \tag{3.7}$$

Because of the presence of longitudinal modes, the Fabry–Perot laser belongs to the class of multimode lasers. To improve the coherence of output light and the laser modulation speed, distributed feedback (DFB) lasers, shown in Figure 3.6b, are used. The key idea of this laser is to effectively select one of the longitudinal modes while suppressing the remaining ones. This is achieved by introducing the Bragg grating inside of the laser cavity. The wavelength of selected longitudinal mode can be determined from the Bragg condition:

$$2\Lambda = m \cdot \frac{\lambda}{n_{av}} \Rightarrow \lambda_B = \frac{2n_{av}\Lambda}{m} \tag{3.8}$$

where Λ is the grating period, n_{av} is the average refractive index of a waveguide mode, and λ/n_{av} is the average wavelength of the light in the waveguide mode. If the grating element is put outside of the active region or used instead of the facet mirrors, the distributed Bragg reflector (DBR) laser is obtained, which is illustrated in Figure 3.6c. Both DFB and DBR lasers belong to the class of single-mode lasers. Different semiconductor lasers shown in Figure 3.6 are edge-emitting lasers.

Another important type of semiconductor laser is the vertical cavity surface emitting laser (VCSEL), which emits the light vertical to the active layer plane.[1-4] VCSELs are usually based on In-GaAs-P layers acting as Bragg reflectors and providing the positive feedback leading to the stimulated emission.

The spectral curve of the single-mode lasers is a result of transition between discrete energy levels and can often be represented using the Lorentzian shape[1-4]:

$$g(v) = \frac{\Delta v}{2\pi\left[(v - v_0)^2 + \left(\frac{\Delta v}{2}\right)^2\right]} \tag{3.9}$$

where v_0 is the central optical frequency, and Δv represents the laser linewidth[1]:

$$\Delta v = \frac{n_{sp}G\left(1 + \alpha_{chirp}^2\right)}{4\pi P} \tag{3.10}$$

where n_{sp} is the spontaneous emission factor, G is the net rate of stimulated emission, P denotes the output power, and α_{chirp} is the chirp factor (representing the amplitude–phase coupling parameter).

The small-signal frequency response of the semiconductor laser is determined by[3,4]

$$H(\omega) = \frac{\Omega_R^2 + \Gamma_R^2}{(\Omega_R + \omega - j\Gamma_R)(\Omega_R - \omega + j\Gamma_R)} \tag{3.11}$$

where Γ_R is the damping factor, and Ω_R is the relaxation frequency

$$\Omega_R^2 \approx \frac{G_N P_b}{\tau_p} \quad (\Gamma_R \ll \Omega_R)$$

with G_N being the net rate of stimulated emission, P_b being the output power corresponding to the bias current, and τ_p being the photon lifetime related to the excited energy level. The modulation bandwidth (defined as 3 dB bandwidth) is therefore determined by the relaxation frequency,

$$\omega_{3dB} = \sqrt{1 + \sqrt{2}\Omega_R}$$

and for fast semiconductor lasers can be 30 GHz. Unfortunately, the direct modulation of semiconductor lasers leads to frequency chirp, which can be described by the instantaneous frequency shift from steady-state frequency v_0 as follows[1]:

$$\delta v(t) = \frac{\alpha_{chirp}}{4\pi}\left[\frac{d}{dt}\ln P(t) + \chi P(t)\right] = C_{dyn}\frac{d}{dt}P(t) + C_{ad}P(t) \tag{3.12}$$

where $P(t)$ is the time variation of the output power, χ is the constant (varying from zero to several tens) related to the material and design parameters, and α_{chirp} is the chirp factor defined as the ratio between the refractive index n change and gain G change with respect to the number of carriers N: $\alpha_{chirp} = (dn/dN)/(dG/dN)$. The first term on the right-hand side of Eq. (3.12) represents dynamic (transient or instantaneous) chirp, and the second term represents the adiabatic (steady-state) frequency chirp. The random fluctuation in carrier density due to spontaneous emission also leads to linewidth enhancement proportional to $(1 + \alpha^2_{chirp})$. To avoid the chirp problem, the external modulation is used, whereas the semiconductor lasers are biased by a dc voltage to produce a continuous wave operation.

Two types of external modulators are commonly used in practice: the Mach–Zehnder modulator (MZM) and the electroabsorption modulator (EAM), whose operational principle is illustrated in Figure 3.7. The MZM is based on electro-optic effect—the effect that in certain materials (e.g., LiNbO$_3$), the refractive index n changes with respect to the voltage V applied across electrodes[4]:

$$\Delta n = -\frac{1}{2}\Gamma n^3 r_{33}(V/d_e) \Rightarrow \Delta\phi = \frac{2\pi}{\lambda}\Delta n L \tag{3.13}$$

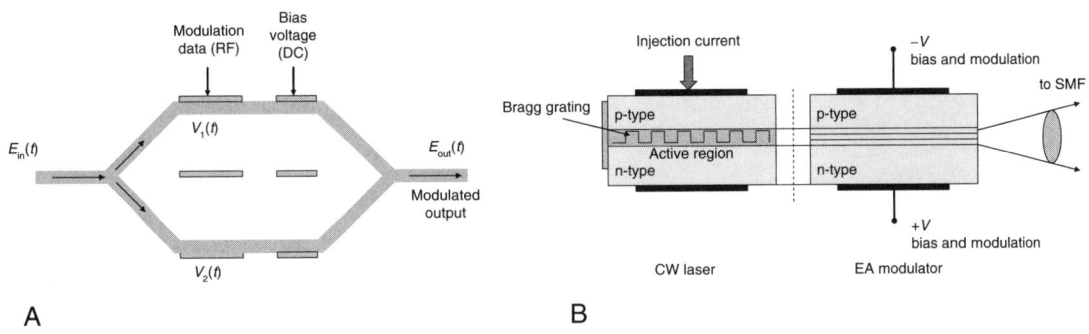

Figure 3.7: External optical modulators: (a) Mach–Zehnder modulator and (b) electroabsorption modulator.

where Δn denotes the refractive index change, $\Delta \phi$ is the corresponding phase change, r_{33} is the electro-optic coefficient ($\sim$30.9 pm/V in LiNbO$_3$), d_e is the separation of electrodes, L is the electrode length, and λ is the wavelength of the light. The MZM (see Figure 3.7a) is a planar waveguide structure deposited on the substrate, with two pairs of electrodes—one pair for high-speed ac voltage representing the modulation data (RF) signal and one pair for dc bias voltage. Let $V_1(t)$ and $V_2(t)$ denote the electrical drive signals on the upper and lower electrodes, respectively. The output electrical field $E_{out}(t)$ of the second Y-branch can be related to the input electrical field E_{in} by

$$E_{out}(t) = \frac{1}{2}\left[\exp\left(j\frac{\pi}{V_\pi}V_1(t)\right) + \exp\left(j\frac{\pi}{V_\pi}V_2(t)\right)\right]E_{in} \quad (3.14)$$

where v_π is differential drive voltage ($V_1 - V_2 = v_\pi$) resulting in differential phase shift of π rad between two waveguides. Possible modulation formats that can be used with this MZM include on–off keying (OOK) with zero/nonzero chirp; binary phase-shift keying (BPSK); differential phase-shift keying; quaternary phase-shift keying (QPSK); differential QPSK; and return-to-zero (RZ) with duty cycle 33%, 50%, or 67%. For example, for zero-chirp OOK or BPSK, the following complementary drive signals should be applied:

$$V_1(t) = V(t) - V_\pi/2, \; V_2(t) = -V(t) + V_\pi/2 \Rightarrow \frac{E_{out}(t)}{E_{in}} = \sin\left(\frac{\pi V(t)}{V_\pi}\right)$$

The EAM is a semiconductor-based planar waveguide composed of multiple p-type and n-type layers that form multiple quantum wells (MQWs). The basic design of EAM is similar to that of semiconductor lasers. The MQW is used to support the quantum-confined Stark effect (the absorption spectrum being a function of applied field) more effectively. Because of similarities between EAM and semiconductor laser designs, it is possible to fabricate them on the same substrate (see Figure 3.7b), provided that the EAM and laser are electrically isolated. Bandgap of quantum wells is larger than photon energy so that the light is completely transmitted in the absence of bias, which corresponds to the ON state. When the reverse bias is applied, the input signal is absorbed, which corresponds to the OFF state.

The modulation speed of EAMs is typically comparable to the modulation speed of MZMs. However, the extinction ratio (the ratio of average powers corresponding to symbol 1 and symbol 0) is lower.

3.2.2 Optical Receivers

The purpose of the optical receiver is to convert the optical signal into the electrical domain and to recover the transmitted data. The typical OOK receiver configuration was already given in Figure 3.2. We can identify three different stages: the front-end stage, the linear channel stage, and the data recovery stage. The front-end stage is composed of a photodetector and a preamplifier. The most commonly used front-end stages are high-impedance front end and transimpedance front end, both of which are shown in Figure 3.8. High-impedance front end (see Figure 3.8a) employs a large value load resistance to reduce the level of thermal noise, and it has good receiver sensitivity. However, the bandwidth of this scheme is low because the RC constant is large. To achieve both the high receiver sensitivity and the large bandwidth, the transimpedance front-end scheme shown in Figure 3.8b is used. Although the load resistance is high, the negative feedback reduces the effective input resistance by a factor of $G - 1$, where G is the front-end amplifier gain. The bandwidth is increased for the same factor compared to the high-impedance front-end scheme.

The photodiode is an integral part of both front-end stage schemes. The key role of the photodiode is to absorb photons in incoming optical signal and convert them back to the electrical level through the process opposite to the one taking place in semiconductor lasers. The common photodiodes are p-n photodiode, p-i-n photodiode, avalanche photodiode (APD), and metal–semiconductor–metal (MSM) photodetectors.[3,4] The p-n photodiode is based on a reverse-biased p-n junction. The thickness of the depletion region is often less than the absorption depth for incident light, and the photons are absorbed outside of the depletion region, leading to the slow response speed. The p-i-n photodiode consists of an

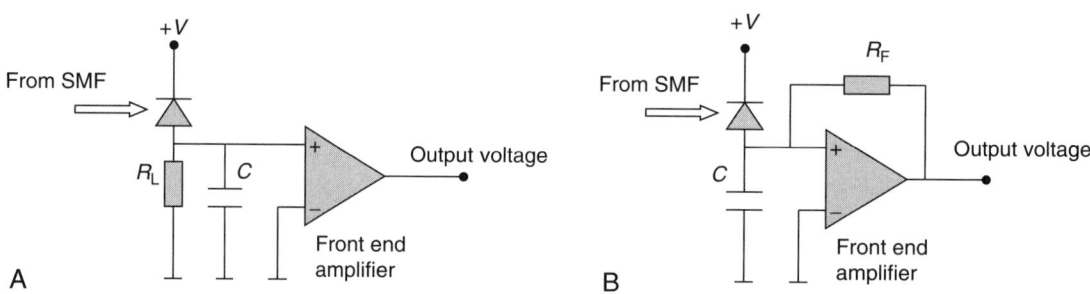

Figure 3.8: Optical receiver front-end stage schemes: (a) high-impedance front end and (b) transimpedance front end.

intrinsic region sandwiched between p- and n-type layers, as shown in Figure 3.9a. Under the reverse bias, the depletion depth can be made sufficiently thick to absorb most of the incident photons. The APD, shown in Figure 3.9b, is a modified p-i-n photodiode that is operated at very high reverse bias. Under high-field conditions, photo-generated carriers induce generation of secondary electron-hole pairs by the process of impact ionization, and this process leads to internal electrical gain. MSM photodetectors employ interdigitated Schottky barrier contacts on one face of the device and are compatible with planar processing and optoelectronic integration. Depending on the device design, the device is illuminated through the p- or n-type contact. In Si, Ge, or GaAs diodes, the substrate is absorbing so that the device has to be illuminated through the top contact, as shown in Figure 3.9a. On the other hand, in InGaAs or InGaAsP, the substrate is transparent, and the device can be designed to be illuminated through either the substrate or the top contact. To increase the depletion region and to minimize the diffusion current component, an intrinsic layer (i-type) is introduced to the p-i-n photodiode structure. The p-i-n photodiode is reverse biased and has very high internal impedance, meaning that it acts as a current source generating the photocurrent proportional to the incoming optical signal power. The equivalent scheme of the p-i-n photodiode is shown in Figure 3.9c. Typically, the internal series resistance R_s is low, whereas the internal shunt resistance is high, so the junction capacitance C_p dominates and can be determined by

$$C_p = \varepsilon_s \frac{A}{w} = \left[\frac{\varepsilon_s N_A N_D}{2(N_A - N_D)(V_0 - V_A)} \right]^{1/2} \tag{3.15}$$

where ε_s is the semiconductor permittivity, A is the area of the space charge region (SCR), w is the width of the SCR, N_A and N_D denote dopant (acceptor and donor) densities, V_0 is the built-in potential across the junction, and V_A is applied negative voltage. The photocurrent $i_{ph}(t)$ is proportional to the power of incident light $P(t)$; that is, $i_{ph}(t) = RP(t)$, where $R[A/W]$

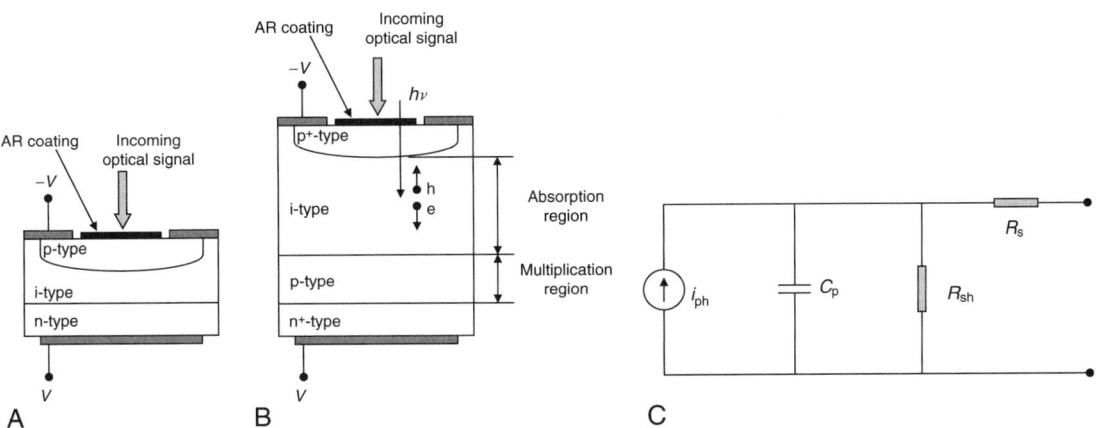

Figure 3.9: The semiconductor photodiodes: (a) p-i-n photodiode and (b) avalanche photodiode. (c) The equivalent p-i-n photodiode model.

is the photodiode responsivity. The photodiode responsivity is related to the quantum efficiency η, defined as the ratio of the number of generated electrons and the number of incident photons, by $R = \eta q/hv$, where q is an electron charge, and hv is a photon energy. Using this model, we can determine the 3 dB bandwidth of the high-impedance front-end scheme as $B_{3dB} = 1/(2\pi R_L C_p)$ and the 3 dB bandwidth of the transimpedance front-end scheme as $B_{3dB} = (G + 1)/(2\pi R_F C_p)$, which is G times higher than the bandwidth of the high-impedance front-end scheme.

3.2.3 Optical Fibers

Optical fibers serve as the foundation of an optical transmission system because they transport optical signals from source to destination. The combination of low-loss and extremely large bandwidth allows high-speed signals to be transmitted over long distances before the regeneration becomes necessary. A low-loss optical fiber is manufactured from several different materials; the base row material is pure silica, which is mixed with different dopants to adjust the refractive index of optical fiber. The optical fiber, shown in Figure 3.10, consists of two waveguide layers—the *core* (of refractive index n_1) and the *cladding* (of refractive index n_2), protected by the jacket (the buffer coating). The majority of the power is

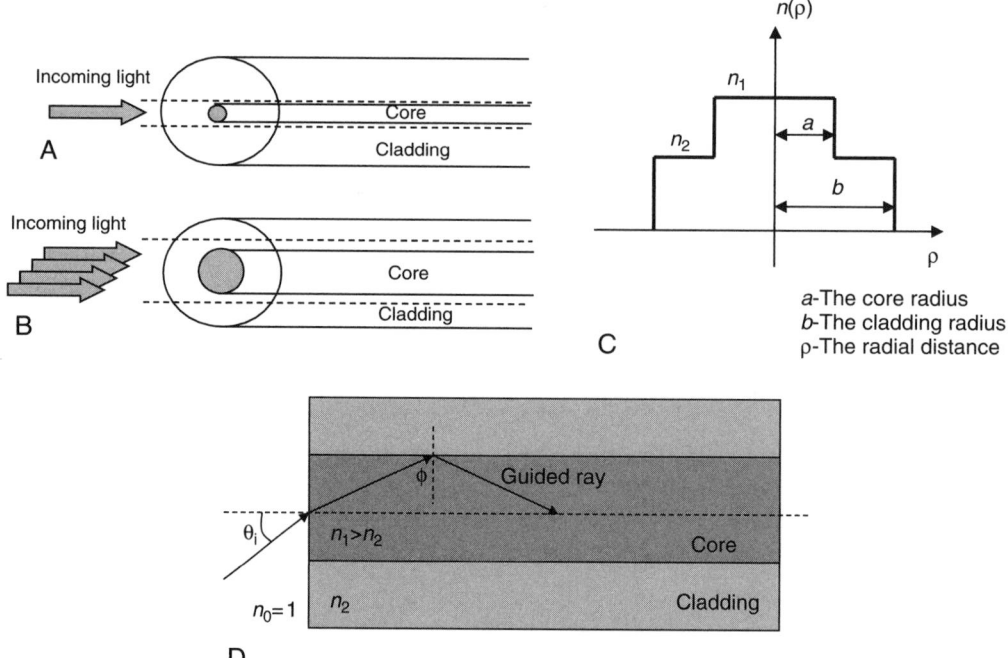

Figure 3.10: Optical fibers: (a) multimode optical fiber, (b) single-mode optical fiber, (c) refractive index profile for step-index fiber, and (d) the light confinement in step-index fibers through the total internal reflection.

concentrated in the core, although some portion can spread to the cladding. There is a difference in refractive indices between the core and cladding ($n_1 > n_2$), which is achieved by a mix of dopants commonly added to the fiber core. The refractive index profile for step-index fiber is shown in Figure 3.10c, whereas light confinement by the total internal reflection is illustrated in Figure 3.10d. The ray will be totally reflected from the core–cladding interface (a guided ray) if the following condition is satisfied:

$$n_0 \sin \theta_i < \sqrt{n_1^2 - n_2^2}$$

where θ_i is the angle of incidence. $\max(n_0 \sin \theta_i)$ defines the light-gathering capacity of an optical fiber and it is called the *numerical aperture* (NA):

$$\text{NA} = \sqrt{n_1^2 - n_2^2} \approx n_1 \sqrt{2\Delta}, \Delta \ll 1$$

where Δ is the *normalized index difference*, defined as $\Delta = (n_1 - n_2)/n_1$. Therefore, from the geometrical optics point of view, light propagates in optical fiber due to a series of total internal reflections that occur at the core–cladding interface. The smallest angle of incidence ϕ (see Figure 3.10d) for which the total internal reflection occurs is called the *critical angle* and equals $\sin^{-1} n_2/n_1$.

There are two types of optical fibers: MMF (shown in Figure 3.10a) and SMF (shown in Figure 3.10b). Multimode optical fibers transfer the light through a collection of spatial transversal modes. Each mode, defined through a specified combination of electrical and magnetic components, occupies a different cross section of the optical fiber core and takes a slightly distinguished path along the optical fiber. The difference in mode path lengths in multimode optical fibers produces a difference in arrival times at the receiving point. This phenomenon is known as *multimode dispersion* (or *intermodal dispersion*) and causes the signal distortion, and it imposes the limitations in signal bandwidth. The second type of optical fibers, single-mode fibers, effectively eliminate multimode dispersion by limiting the number of propagating modes to a fundamental one. The *fundamental mode* occupies the central portion of the optical fiber and has an energy maximum at the axis of the optical fiber core. Its radial distribution can be approximated by a Gaussian curve. The number of modes (M) that can effectively propagate through an optical fiber is determined by the *normalized frequency* (V parameter or V number): $M \approx V^2/2$, when V is large. The normalized frequency is defined by

$$V = \frac{2\pi a}{\lambda} \sqrt{n_1^2 - n_2^2} \tag{3.16}$$

where a is the fiber core radius, λ is the carrier wavelength, and n_1 and n_2 are refractive indices related to the fiber core and the fiber cladding, respectively.

Each mode propagating through the fiber is characterized by its own propagation constant β. The dependence of the electric and magnetic fields on axial coordinate z is expressed through the factor $\exp(-j\beta z)$. The propagation constant must satisfy the following condition:

$$2\pi n_2/\lambda < \beta < 2\pi n_1/\lambda \qquad (3.17)$$

To evaluate the transmission characteristics of the optical fiber, the functional dependence of the mode propagation constant on the optical signal wavelength must be known. The *normalized propagation constant b* is defined for that purpose:

$$b = \frac{\beta^2 - (2\pi n_2/\lambda)^2}{(2\pi n_1/\lambda)^2 - (2\pi n_2/\lambda)^2} \qquad (3.18)$$

The normalized propagation constant is related to the normalized frequency V by[1,3]

$$b(V) \approx (1.1428 - 0.9960/V)^2, \quad 1.5 \le V \le 2.5 \qquad (3.19)$$

The multimode dispersion can effectively be eliminated by limiting the number of propagating modes to a fundamental one: $V \le V_c = 2.405$, with V_c being the cutoff frequency. The cutoff frequency is controlled by keeping the core radius small and the normalized index difference $\Delta = (n_1 - n_2)/n_1$ between 0.2% and 0.3%.

Single-mode optical fibers do, however, introduce another signal impairment known as chromatic dispersion. *Chromatic dispersion* is caused by differences in velocities among different spectral components within the same mode, and it has two components: material dispersion and waveguide dispersion. *Material dispersion* is caused by the fact that the refractive index is a function of wavelength, defined by the Sellmeier equation[1,3]:

$$n(\lambda) = \left[1 + \sum_{i=1}^{M} \frac{B_i \lambda^2}{\lambda^2 - \lambda_i^2}\right] \qquad (3.20)$$

with typical B_i and λ_i parameters for pure silica being

$$B_1 = 0.6961663 \text{ at } \lambda_1 = 0.0684043 \text{ μm}$$
$$B_2 = 0.4079426 \text{ at } \lambda_2 = 0.1162414 \text{ μm}$$
$$B_3 = 0.8974794 \text{ at } \lambda_3 = 9.896161 \text{ μm}$$

The group index n_g can be determined by using the Sellmeier equation and the following definition expression:

$$n_g = n + \omega \mathrm{d}n/\mathrm{d}\omega = n - \lambda \mathrm{d}n/\mathrm{d}\lambda$$

Waveguide dispersion is related to the physical design of the optical fiber. Because the value of Δ is typically small, the refractive indices of the core–cladding are nearly equal, the light is not strictly confined in the fiber core, and the fiber modes are said to be weakly guided. For a given mode, such as a fundamental mode, the portion of light energy that propagates in the core depends on wavelength, giving rise to the pulse spreading phenomenon known as waveguide dispersion. By changing the power distribution across the cross-sectional area, the overall picture related to the chromatic dispersion can be changed. The distribution of the mode power and the total value of waveguide dispersion can be manipulated by multiple cladding layers. This is commonly done in optical fibers for special application purposes, such as dispersion compensation. Different refractive index profiles of single-mode fibers used today are shown in Figure 3.11. The conventional SMF index profile is shown in Figure 3.11a. The nonzero dispersion-shifted fiber (NZDSF), with refractive index profile shown in Figure 3.11b, is suitable for use in WDM systems. Large effective area NZDSFs, with index profile shown in Figure 3.11c, are suitable to reduce the effect of fiber nonlinearities. Dispersion compensating fiber (DCF), with index profile shown in Figure 3.11d, is suitable to compensate for positive chromatic dispersion accumulated along the transmission line.

The *geometrical optics* (or ray theory) approach, used previously to describe the light confinement in step-index fibers through the total internal reflection, is valid when the fiber has a core radius a much larger than the operating wavelength λ. Once the core radius becomes comparable to the operating wavelength, the propagation of light in step-index fiber is governed by Maxwell's equations describing the change of electric (E) and magnetic (H) densities in space and time[1,4,6,8–10]:

$$\nabla \times E = -\partial B / \partial t \qquad (3.21a)$$
$$\nabla \times H = J + \partial D / \partial t \qquad (3.21b)$$
$$\nabla \cdot D = \rho \qquad (3.21c)$$
$$\nabla \cdot B = 0 \qquad (3.21d)$$

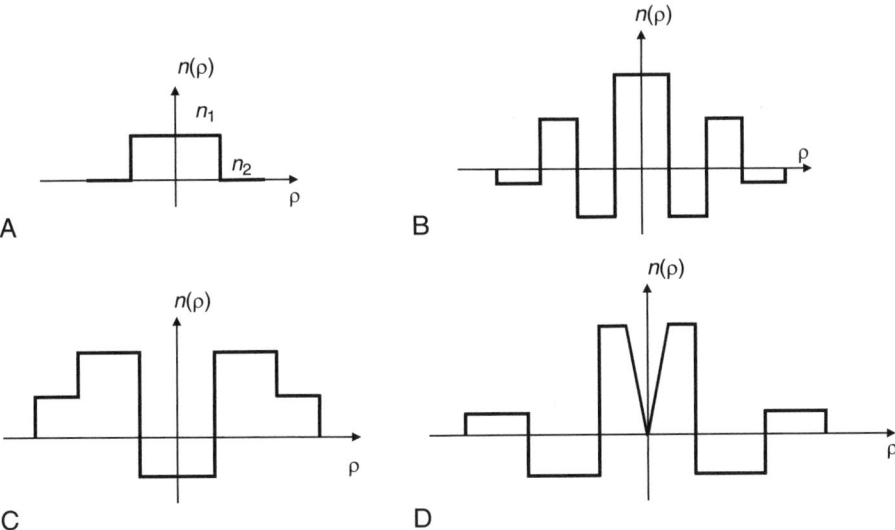

Figure 3.11: Refractive index profiles of different SMFs: (a) standard SMF, (b) NZDSF with reduced dispersion slope, (c) NZDSF with large effective area, and (d) DCF.

where B denotes the magnetic flux density, D is the electric flux density, and $J = \sigma E$ is the current density. Given the fact that there is no free charge in fiber, the charge density $\rho = 0$, and that the conductivity of silica is extremely low ($\sigma \approx 0$), the previous equations can be simplified. The flux densities are related to the field vectors by constitutive relations:

$$D = \varepsilon_0 E + P \tag{3.21e}$$

$$B = \mu_0 H + M \tag{3.21f}$$

where P and M denote the induced electric and magnetic densities, ε_0 is the permeability in the vacuum, and μ_0 is the permittivity in the vacuum. Given the fact that silica is nonmagnetic material, $M = 0$. P and E are mutually connected by

$$P(r,t) = \varepsilon_0 \int_{-\infty}^{\infty} \chi(r, t - t') E(r, t') dt' \tag{3.22}$$

where the linear susceptibility χ is a second-rank tensor that becomes scalar for an isotropic medium. After certain manipulations, the *wave equation* is obtained[1-4,6]:

$$\nabla^2 \tilde{E} + n^2(\omega) k_0^2 \tilde{E} = 0 \tag{3.23}$$

where $k_0 = 2\pi/\lambda$ is the free space wave number, and with $\tilde{E}$ we denoted the Fourier transform of the electric field vector.

An *optical mode* refers to a specific solution of the wave equation, subject to appropriate boundary conditions. The spatial field distributions, $E(r, \omega) = E_0(\rho, \phi) \exp[j\beta(\omega)z - j\omega t]$ and $H(r, \omega) = H_0(\rho, \phi) \exp[j\beta(\omega)z - j\omega t]$, of a mode do not change as the mode propagates along the z-axis except for a multiplicative factor $\exp[j\beta(\omega)z]$, with $\beta(\omega)$ being the *propagation constant* of a mode. The fiber modes may be classified as guided modes, leaky modes, and radiation modes. Different modes in fiber propagate with different values of propagation constant (i.e., with different speeds), causing the pulse spreading. Therefore, it is desirable to design fiber that supports only one mode—the fundamental mode. Such a fiber is called SMF, as mentioned previously. If the weak guidance condition ($\Delta \ll 1$) is not satisfied, the conventional modes TE, TM, EH, and HE can be found. The cylindrical symmetry of fiber suggests the use of cylindrical coordinate system (ρ, ϕ, z). The wave equation is to be solved for six components: E_ρ, E_ϕ, E_z and H_ρ, H_ϕ, H_z. We can solve the wave equation for axial components only (E_z and H_z) and use the system of Maxwell's equations to express the other components as functions of axial ones. The wave equation for the E_z component is as follows[1-4,6]:

$$\frac{\partial^2 E_z}{\partial \rho^2} + \frac{1}{\rho} \frac{\partial E_z}{\partial \rho} + \frac{1}{\rho^2} \frac{\partial^2 E_z}{\partial z^2} + n^2 k_0^2 E_z = 0, \quad n = \begin{cases} n_1, \rho \leq a \\ n_2, \rho > a \end{cases} \tag{3.24}$$

The wave equation for the E_z component can easily be solved using the *method of separation of variables*, leading to the following overall solution[3,4,6]:

$$E_z(\rho,\phi,z) = \begin{cases} A J_m(p\rho)e^{jm\phi}e^{j\beta z}, \; \rho \le a \\ B K_m(q\rho)e^{jm\phi}e^{j\beta z}, \; \rho > a \end{cases} \quad (3.25)$$

where $J_m(x)$ and $K_m(x)$ are corresponding Bessel functions of mth order (A and B are constants to be determined from boundary conditions). A similar equation is valid for H_z.

The other four components can be expressed in terms of axial ones by using Maxwell's equations, and in the core region we obtain[3,4,6]

$$E_\rho = \frac{j}{p^2}\left(\beta\frac{\partial E_z}{\partial\rho} + \mu_0\frac{\omega}{\rho}\frac{\partial H_z}{\partial\phi}\right) \qquad E_\phi = \frac{j}{p^2}\left(\frac{\beta}{\rho}\frac{\partial E_z}{\partial\phi} - \mu_0\omega\frac{\partial H_z}{\partial\rho}\right)$$

$$H_\rho = \frac{j}{p^2}\left(\beta\frac{\partial H_z}{\partial\rho} - \varepsilon_0 n^2\frac{\omega}{\rho}\frac{\partial E_z}{\partial\phi}\right) \qquad H_\phi = \frac{j}{p^2}\left(\frac{\beta}{\rho}\frac{\partial H_z}{\partial\phi} + \varepsilon_0 n^2\omega\frac{\partial H_z}{\partial\rho}\right) \quad (3.26)$$

where $p^2 = n_1^2 k_0^2 - \beta^2$. Similar equations can be obtained for cladding by replacing p^2 with $-q^2$, where $q^2 = \beta^2 - n_2^2 k_0^2$. By satisfying the boundary conditions, the homogeneous system of equations is obtained, which has the nontrivial solution only if the corresponding determinant is zero, leading to the following characteristic (eigenvalue) equation[3,4,6]:

$$\left[\frac{J_m'(pa)}{paJ_m(pa)} + \frac{K_m'(qa)}{qaK_m(qa)}\right] \cdot \left[\frac{J_m'(pa)}{paJ_m(pa)} + \frac{n_2^2}{n_1^2}\frac{K_m'(qa)}{qaK_m(qa)}\right] = m^2\left[\frac{1}{(pa)^2} + \frac{1}{(qa)^2}\right] \cdot \left[\frac{1}{(pa)^2} + \frac{n_2^2}{n_1^2}\frac{1}{(qa)^2}\right]$$

$$(3.27)$$

where pa and qa are related by the following equation:

$$(pa)^2 + (pq)^2 = V^2 \quad (3.28)$$

For given fiber parameters n_1, n_2, a, and the operating wavelength, we can determine the normalized frequency V. For fixed V and a given m there exist multiple solutions $n = 1, 2, 3, \ldots$ that lead to different modes—propagating modes. The case $m = 0$ corresponds to meridional rays (the rays that periodically intersect the center axis of the fiber), has the electric/magnetic components independent of ϕ, and the corresponding modes are classified as TE$_{0n}$ ($E_z = 0$) and TM$_{0n}$ ($H_z = 0$) modes. The case $m \ne 0$ corresponds to skew rays, the electric and magnetic field components are functions of ϕ, and the corresponding modes are classified as HE$_{mn}$ (H_z dominates over E_z) and EH$_{mn}$ (E_z dominates over H_z) modes. Once the mode is identified (m and n are fixed) from Eq. (3.28) and by using the definition expressions for p and q, we can easily determine the propagation constant of the mode β_{mn}.

The propagating light pulse is a composite optical signal containing a number of monochromatic spectral components. Each spectral component behaves differently in a dispersion medium, such as optical fiber, leading to the light pulse distortion. Each axial component of the monochromatic electromagnetic wave can be represented by its complex electric field function[1,3,6]:

$$E(z,t) = E_a(z,t)\exp[j\beta(\omega)z]\exp[-j\omega't] \tag{3.29}$$

The light pulse distortion, which is observed through the pulse broadening along the fiber, can be evaluated by knowing the frequency dependence of the propagation constant $\beta = \beta(\omega)$ at a specific distance z along the fiber. Each spectral component will experience a phase shift proportional to $\beta(\omega)z$. The amplitude spectrum observed at point z along the fiber length in the frequency domain is given by

$$\tilde{E}_a(z,\omega) = \tilde{E}_a(0,\omega)\exp[j\beta(\omega)z] \tag{3.30}$$

The behavior of the pulse envelope during the propagation process can be evaluated through the inverse Fourier transform of the previous equation, which is very complicated to calculate unless additional simplifications are made, such as by expressing the propagation constant in terms of a Taylor series:

$$\beta(\omega) \approx \beta(\omega_c) + \frac{d\beta}{d\omega}\Big|_{\omega=\omega_c}(\omega-\omega_c) + \frac{1}{2}\frac{d^2\beta}{d\omega^2}\Big|_{\omega=\omega_c}(\omega-\omega_c)^2 + \frac{1}{6}\frac{d^3\beta}{d\omega^3}\Big|_{\omega=\omega_c}(\omega-\omega_c)^3 + \ldots \tag{3.31}$$

where $\beta_1 = (d\beta/d\omega)|_{\omega=\omega_c}$ is related to the group velocity v_g by $\beta_1 = 1/v_g$, $\beta_2 = (d^2\beta/d\omega^2)|_{\omega=\omega_c}$ is the group velocity dispersion (GVD) parameter, and $\beta_3 = (d^3\beta/d\omega^3)|_{\omega=\omega_c}$ is the second-order GVD parameter. Introducing the concept of slow varying amplitude $A(z, t)$[1,3,6]

$$E(z,t) = E_a(z,t)\exp[j\beta(\omega)z]\exp(-j\omega_c t) = A(z,t)\exp(j\beta_c z - j\omega_c t) \tag{3.32}$$

we get

$$A(z,t) = \frac{1}{2\pi}\int_{-\infty}^{\infty} \tilde{A}(0,\omega)\exp\left[j\left(\beta_1\Delta\omega + \frac{\beta_2}{2}\Delta\omega^2 + \frac{\beta_3}{6}\Delta\omega^3\right)z\right]\exp(-j(\Delta\omega)t)d(\Delta\omega) \tag{3.33}$$

By taking a partial derivative of $A(z, t)$ with respect to propagation distance z, we obtain

$$\frac{\partial A(z,t)}{\partial z} = \frac{1}{2\pi}\int_{-\infty}^{\infty}\left[j\left(\beta_1\Delta\omega + \frac{\beta_2}{2}\Delta\omega^2 + \frac{\beta_3}{6}\Delta\omega^3\right)\right]$$
$$\tilde{A}(0,\Delta\omega)\exp\left[j\left(\beta_1\Delta\omega + \frac{\beta_2}{2}\Delta\omega^2 + \frac{\beta_3}{6}\Delta\omega^3\right)z\right]\exp(-j(\Delta\omega)t)d(\Delta\omega) \tag{3.34}$$

Finally, by taking the inverse Fourier transform of Eq. (3.34), we derive a basic propagation equation describing the pulse propagation in single-mode optical fibers:

$$\frac{\partial A(z,t)}{\partial z} = -\beta_1\frac{\partial A(z,t)}{\partial t} - j\frac{\beta_2}{2}\frac{\partial^2 A(z,t)}{\partial t^2} + \frac{\beta_3}{6}\frac{\partial^3 A(z,t)}{\partial t^3} \tag{3.35}$$

This equation can be generalized by including the fiber attenuation and the Kerr nonlinear index:

$$\frac{\partial A(z,t)}{\partial z} = -\frac{\alpha}{2}A(z,t) - \beta_1\frac{\partial A(z,t)}{\partial t} - j\frac{\beta_2}{2}\frac{\partial^2 A(z,t)}{\partial t^2} + \frac{\beta_3}{6}\frac{\partial^3 A(z,t)}{\partial t^3} + j\gamma|A(z,t)|^2 A(z,t) \tag{3.36}$$

where α is the attenuation coefficient, and $\gamma = 2\pi n_2/(\lambda A_{\text{eff}})$ is the nonlinear coefficient with n_2 being the nonlinear Kerr coefficient and A_{eff} being the effective cross-sectional area of the fiber core.

After the introduction of a new coordinate system with $T = t - \beta_1 z$, the nonlinear Schrödinger equation (NLSE) is obtained:

$$\frac{\partial A(z,T)}{\partial z} = -\frac{\alpha}{2}A(z,T) - j\frac{\beta_2}{2}\frac{\partial^2 A(z,t)}{\partial T^2} + \frac{\beta_3}{6}\frac{\partial^3 A(z,T)}{\partial T^3} + j\gamma|A(z,T)|^2 A(z,T) \tag{3.37}$$

The NLSE can be solved using the Fourier split-step algorithm.[6] The length of the fiber is divided into small steps of size h, as shown in Figure 3.12. The algorithm proceeds by first applying the linear operator of the equation (L) for one half step length ($h/2$). The nonlinear operator (NL) is applied for one step length h. The linear operator is again applied for the full step size. Therefore, the linear operator is applied at every half step ($h/2$) and the nonlinear one at every full step (h). The linear operator D in the time domain can be described as follows:

$$D = -\frac{\alpha}{2} - j\frac{\beta_2}{2}\frac{\partial^2}{\partial T^2} + \frac{\beta_3}{6}\frac{\partial^3}{\partial t^3} \tag{3.38}$$

and the linear operator in the frequency domain is obtained by applying the Fourier transform (FT) to Eq. (3.38):

$$\tilde{D} = -\frac{\alpha}{2} + j\frac{\beta_2}{2}\omega^2 + j\frac{\beta_3}{6}\omega^3 \tag{3.39}$$

The nonlinear operator is defined as

$$N = j\gamma|A(z,T)|^2 \tag{3.40}$$

so that the Fourier split-step algorithm can be symbolically described by

$$A(z_0 + h,T) = \exp\left(h \cdot \frac{D}{2}\right)\left[\exp(h \cdot N)A(z_0,T)\right]\exp\left(h \cdot \frac{D}{2}\right) \tag{3.41}$$

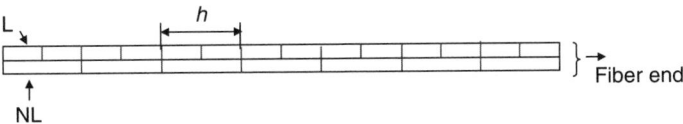

Figure 3.12: Illustration of the Fourier split-step algorithm.

The linear operator is applied in the frequency domain, whereas the nonlinear operator is applied in the time domain. If fiber nonlinearities can be neglected, the study of propagation is greatly simplified. We have to evaluate the FT of pulse $A(0, t)$ to get $\tilde{A}(0,\omega) = \text{FT}[A(0,t)]$. We have to multiply the corresponding FT by dispersion factor to get

$$\tilde{A}(z,\omega) = \tilde{A}(0,\omega)e^{j\left[\frac{\beta_2}{2}\omega^2 + \frac{\beta_3}{6}\omega^3\right]z}e^{-\frac{\alpha}{2}z} \tag{3.42}$$

Finally, to get the pulse shape at distance z we have to evaluate the inverse Fourier transform (FT^{-1}):

$$A(z,t) = \text{FT}^{-1}\left[\tilde{A}(z,\omega)\right]$$

3.2.4 Optical Amplifiers

The purpose of an optical amplifier is to restore the signal power level, reduced due to losses during propagation, without any optical-to-electrical conversion. The general form of an optical amplifier is given in Figure 3.13a. Most optical amplifiers amplify incident light through the stimulated emission, the same mechanism that is used in lasers but without the feedback mechanism. The main ingredient is the optical gain realized through the amplifier pumping (electrical or optical) to achieve the population inversion. The optical gain, in general, is not only a function of frequency but also a function of local beam intensity. To illustrate the basic concepts, we consider the case in which the gain medium is modeled as a two-level system, as shown in Figure 3.13b. The amplification factor G is defined as the ratio of amplifier output P_{out} to input P_{in} powers: $G = P_{out}/P_{in}$. The amplification factor can be determined if the dependence of evolution of power through the gain media is known[3]:

$$\frac{dP}{dz} = gP, \quad g(\omega) = \frac{g_0}{1 + (\omega - \omega_0)^2 T_2^2 + P/P_s} \tag{3.43}$$

where g is the gain coefficient, g_0 is the gain peak value, ω_0 is the atomic transition frequency, T_2 is the dipole relaxation time (<1 ps), ω is the optical frequency of incident signal, P is the incident signal power, and P_s is the saturation power.

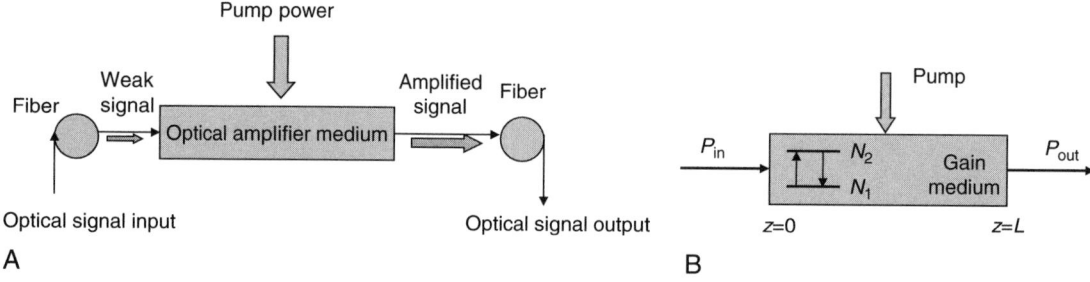

A

B

Figure 3.13: (a) Optical amplifier principle and (b) two-level amplifier system model.

In the unsaturated regime ($P \ll P_s$), the differential equation (3.43) can be solved by separation of variables to get the dependence of power, $P(z) = P(0)\exp(gz)$, so that the amplification factor can be obtained by

$$G(\omega) = \exp[g(\omega)L] \tag{3.44}$$

and corresponding full-width half-maximum (FWHM) bandwidth is determined by

$$\Delta v_A = \Delta v_g \sqrt{\frac{\ln 2}{\ln(G_0/2)}}, \quad G_0 = \exp(g_0 L) \tag{3.45}$$

where Δv_g is the FWHM gain coefficient bandwidth.

The gain saturation comes from the power dependence of the gain coefficient (Eq. 3.43). The coefficient is reduced when the incident power P becomes comparable to the saturation power P_s. Let's assume that the incident frequency is tuned to the peak gain ($\omega = \omega_0$); then from Eq. (3.43) we obtain

$$\frac{dP}{dz} = \frac{g_0 P}{1 + P/P_s} \tag{3.46}$$

By solving the differential equation (3.46) with respect to the boundary conditions, $P(0) = P_{in}$ and $P(L) = P_{out} = GP_{in}$, we get

$$G = G_0 \exp\left[-\frac{G-1}{G}\frac{P_{out}}{P_s}\right] \tag{3.47}$$

From Eq. (3.47), we can determine another important optical amplifier parameter, the output saturation power, as being the optical power at which the gain G is reduced to $G_0/2$ (3 dB down):

$$P_{out}^{sat} = \frac{G_0 \ln 2}{G_0 - 2} P_s \approx (\ln 2) P_s \approx 0.69 P_s (G_0 > 20 \text{ dB}) \tag{3.48}$$

Three common applications of optical amplifiers are as power boosters (of transmitters), in-line amplifiers, and optical preamplifiers (Figure 3.14). The booster (power) amplifiers are placed at the optical transmitter side to enhance the transmitted power level or to compensate for the losses of optical elements between the laser and optical fibers, such as optical couplers, WDM multiplexers, and external optical modulators. The in-line amplifiers are placed along the transmission link to compensate for the losses incurred during propagation of optical signal. The optical preamplifiers are used to increase the signal level before photodetection occurs, improving the receiver sensitivity.

Several types of optical amplifiers have been introduced so far: SOAs, fiber Raman (and Brillouin) amplifiers, rare-earth-doped fiber amplifiers (EDFA operating at 1500 nm and praseodymium-doped operating at 1300 nm), and parametric amplifiers. Semiconductor lasers act as amplifiers before reaching the threshold. To prevent the lasing, antireflection (AR)

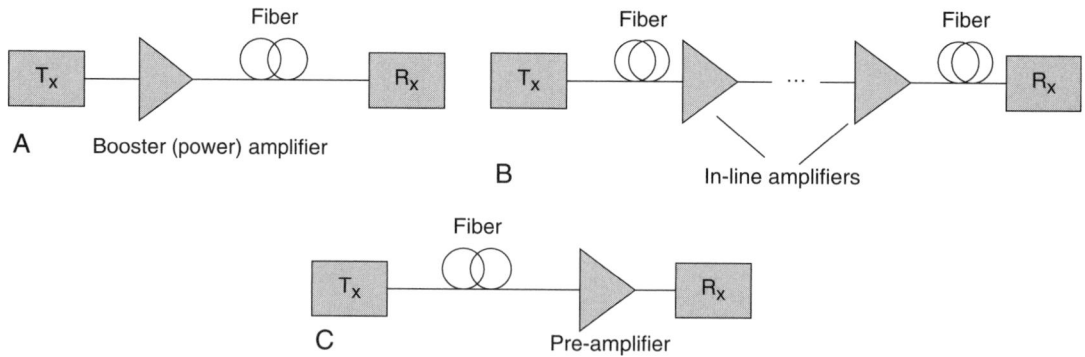

Figure 3.14: Possible applications of optical amplifiers: (a) booster amplifier, (b) in-line amplifiers, and (c) preamplifier.

coatings are used in SOAs, as shown in Figure 3.15. Even with AR coating, the multiple reflections must be included when considering the Fabry–Perot (FP) cavity. The gain of FP amplifiers is given by[3]

$$G_{FP}(v) = \frac{(1 - R_1)(1 - R_2)G(v)}{\left(1 - G(v)\sqrt{R_1 R_2}\right)^2 + 4G\sqrt{R_1 R_2}\sin^2\left[\pi(v - v_m)/\Delta v_L\right]} \tag{3.49}$$

where R_1 and R_2 denote the facet reflectivities, $G(v)$ is the single-pass amplification factor, v_m is the cavity resonance frequency, and Δv_L is the free spectral range.

FP amplifier bandwidth can be determined as follows[3]:

$$\Delta v_A = \frac{2\Delta v_L}{\pi}\sin^{-1}\left[\frac{1 - G\sqrt{R_1 R_2}}{\left(4G\sqrt{R_1 R_2}\right)^{1/2}}\right] \tag{3.50}$$

A fiber-based Raman amplifier employs the stimulated Raman scattering (SRS) occurring in silica fibers when an intense pump propagates through it. SRS is fundamentally different

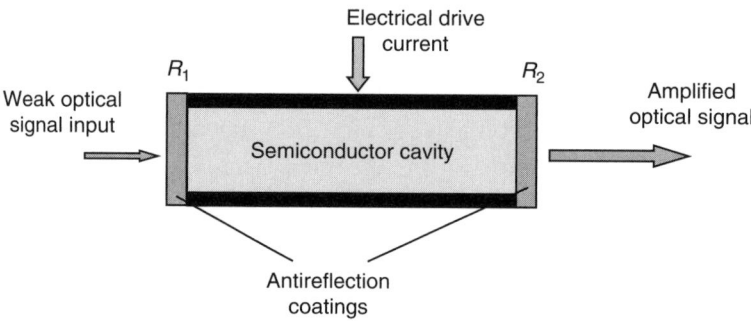

Figure 3.15: The SOA operation principle.

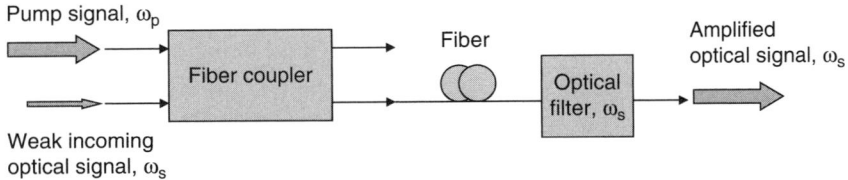

Figure 3.16: The Raman amplifier operation principle in a forward-pumping configuration.

from stimulated emission: In stimulated emission, an incident photon stimulates emission of another identical photon, whereas in SRS the incident pump photon gives up its energy to create another photon of reduced energy at a lower frequency (inelastic scattering); the remaining energy is absorbed by the medium in the form of molecular vibrations (optical phonons). Raman amplifiers must be pumped optically to provide gain, as shown in Figure 3.16.

The Raman gain coefficient g_R is related to the optical gain $g(z)$ as $g(z) = g_R I_p(z)$, where I_p is the pump intensity given by $I_p = P_p/a_p$, with P_p being the pump power and a_p being the pump cross-sectional area. Because the cross-sectional area is different for different types of fibers, the ratio g_R/a_p is the measure of the Raman gain efficiency. The DCF efficiency can be eight times better than that of a standard SMF, as shown in Agrawal.[3] The evolution of the pump P_p and signal P_s powers (in distance z) can be studied by solving the system of coupled differential equations[1,3]:

$$\frac{dP_s}{dz} = -\alpha_s P_s + \frac{g_R}{a_p} P_p P_s \qquad \frac{dP_p}{dz} = -\alpha_p P_p - \frac{\omega_p}{\omega_s} \frac{g_R}{a_p} P_p P_s \tag{3.51}$$

where a_s denotes the signal cross-sectional area, and ω_p and ω_s denote the pump and signal frequency, respectively. The other parameters were previously introduced.

In the small-signal amplification regime (when the pump depletion can be neglected), the pump power evolution is exponential, $P_p(z) = P_p(0)\exp[-\alpha_p z]$, so that the Raman amplifier gain is found to be

$$G_A = \frac{P_s(0)\exp(g_R P_p(0) L_{\text{eff}}/a_p - \alpha_s L)}{P_s(0)\exp(-\alpha_s L)} = \exp(g_0 L), \quad g_0 = g_R \frac{P_p(0)}{a_p} \frac{L_{\text{eff}}}{L} \approx \frac{g_R P_p(0)}{a_p \alpha_p L} (\alpha_p L \gg 1) \tag{3.52}$$

The origin of saturation in Raman amplifiers is pump power depletion, which is quite different from that in SOAs. Saturated amplifier gain G_s can be determined (assuming $\alpha_p = \alpha_s$) by[3]

$$G_s = \frac{1 + r_0}{r_0 + 1/G_A^{1+r_0}}, \quad r_0 = \frac{\omega_p}{\omega_s} \frac{P_s(0)}{P_p(0)} \tag{3.53}$$

The amplifier gain is reduced down by 3 dB when $G_A r_0 \approx 1$, the condition that is satisfied when the amplified signal power becomes comparable to the input pump power $P_s(L) = P_p(0)$. Typically, $P_0 \sim 1$ W, and channel powers in a WDM system are approximately 1 mW, meaning that the Raman amplifier operates in an unsaturated or linear regime.

The rare-earth-doped fiber amplifiers are finding increasing importance in optical communication systems. The most important class is EDFAs due to their ability to amplify in the 1.55 μm wavelength range. The active medium consists of 10–30 m length of optical fiber highly doped with a rare-earth element, such as erbium (Er), ytterbium (Yb), neodymium (Nd), or praseodymium (Pr). The host fiber material can be pure silica, a fluoride-based glass, or a multicomponent glass. General EDFA configuration is shown in Figure 3.17.

The pumping at a suitable wavelength provides gain through population inversion. The gain spectrum depends on the pumping scheme as well as the presence of other dopants, such as Ge or Al, within the core. The amorphous nature of silica broadens the energy levels of Er^{3+} into the bands, as shown in Figure 3.18.

The pumping is primarily done in the optical domain with the primary pump wavelengths at 1.48 and 0.98 μm. The atoms pumped to the $4I_{11/2}$ 0.98 μm decay to the primary emission

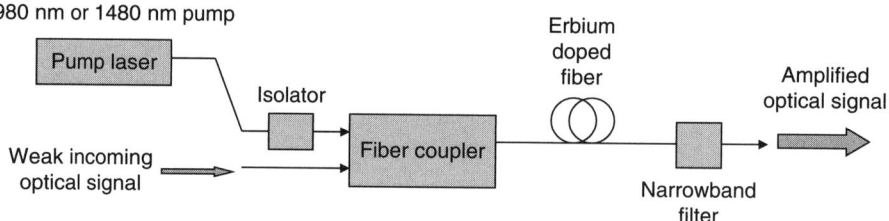

Figure 3.17: The general EDFA configuration.

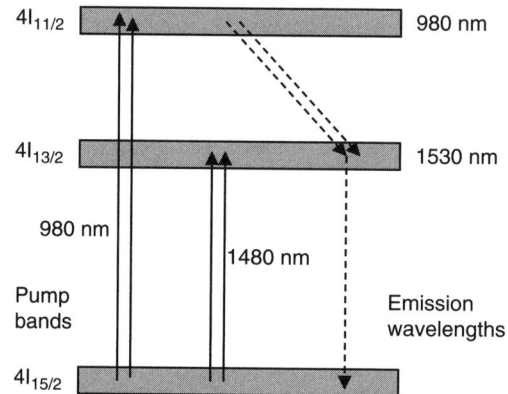

Figure 3.18: The EDFA energy diagram.

transition band. The pumping with 1.48 μm is directly to the upper transition levels of the emission band.

EDFAs can be designed to operate in such a way that the pump and signal travel in opposite directions; this configuration is commonly referred to as *backward pumping*. In *bidirectional pumping*, the amplifier is pumped in both directions simultaneously by using two semiconductor lasers located at both fiber ends.

3.2.5 Other Optical Components

Different optical components can be classified into two broad categories depending on whether or not they can operate without an external electric power source into passive or active components. Important active components are lasers, external modulators, optical amplifiers, photodiodes, optical switches, and wavelength converters. Important passive components are optical couplers, isolators, multiplexers/demultiplexers, and filters. Some components, such as optical filters, can be either passive or active depending on the operational principle. This section briefly explains some important optical components not described in previous sections.

The 2×2 optical coupler is a fundamental device that can be implemented using the fiber fusing or be based on graded-index (GRIN) rods and optical filters, as shown in Figure 3.19. The fused optical couplers (Figure 3.19a) are obtained when the cladding of two optical fibers is removed, the cores are brought together and then heated and stretched. The obtained waveguide structure can exchange energy in the coupling region between the branches. If both inputs are used, a 2×2 coupler is obtained; if only one input is used, a 1×2 coupler is obtained. The optical couplers are recognized either as optical taps (1×2) couplers or directional (2×2) couplers. Depending on purpose, the power coupler splitting ratio can be different, with typical values being 50%/50%, 10%/90%, 5%/95%, and 1%/99%. Directional coupler parameters (defined when only input 1 is active) are splitting ratio $P_{\text{out},1}/(P_{\text{out},1} + P_{\text{out},2})$, excess loss $10\log_{10}[P_{\text{in},1}/(P_{\text{out},1} + P_{\text{out},2})]$, insertion loss $10\log_{10}(P_{\text{in},i}/P_{\text{out},j})$, and crosstalk $10\log_{10}(P_{\text{cross}}/P_{\text{in},1})$. The operation principle of a directional coupler can be

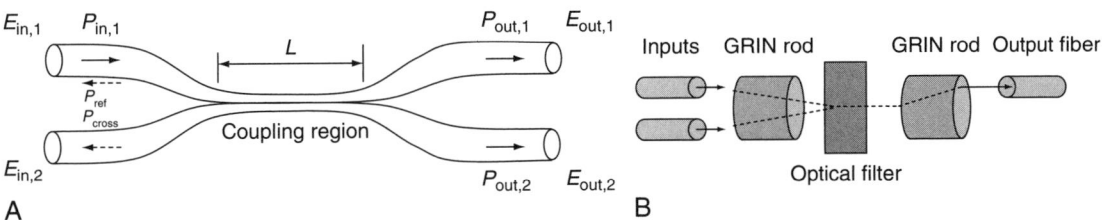

Figure 3.19: Optical couplers: (a) fiber fusing based and (b) GRIN rod based.

explained using the coupled mode theory[15] or the simple scattering (propagation) matrix S approach, assuming that the coupler is a lossless and reciprocal device[2]:

$$\begin{pmatrix} E_{\text{out},1} \\ E_{\text{out},2} \end{pmatrix} = S \begin{pmatrix} E_{\text{in},1} \\ E_{\text{in},2} \end{pmatrix} = \begin{pmatrix} s_{11} & s_{12} \\ s_{21} & s_{22} \end{pmatrix} \begin{pmatrix} E_{\text{in},1} \\ E_{\text{in},2} \end{pmatrix} = e^{-j\beta L} \begin{pmatrix} \cos(kL) & j\sin(kL) \\ j\sin(kL) & \cos(kL) \end{pmatrix} \begin{pmatrix} E_{\text{in},1} \\ E_{\text{in},2} \end{pmatrix} \quad (3.54)$$

where β is the propagation constant, k is the coupling coefficient, L is the coupling region length, $E_{\text{in},1}$ and $E_{\text{in},2}$ are corresponding input electrical fields, and $E_{\text{out},1}$ and $E_{\text{out},2}$ are corresponding output electrical fields. Scattering matrix S elements are denoted with s_{ij}. For example, for a 3 dB coupler we have to select $kL = (2m + 1)\pi/4$ (m is a positive integer) to get

$$\begin{pmatrix} E_{\text{out},1} \\ E_{\text{out},2} \end{pmatrix} = \frac{1}{\sqrt{2}} \begin{pmatrix} 1 & j \\ j & 1 \end{pmatrix} \begin{pmatrix} E_{\text{in},1} \\ E_{\text{in},2} \end{pmatrix} \quad (3.55)$$

The combination of two GRIN rods and an optical filter can effectively be used as an optical coupler, as illustrated in Figure 3.19b. The GRIN rods are used as collimators to collimate the light from two input ports and deliver to the output port, whereas the optical filter is used to select a desired wavelength channel.

The optical couplers can be used to create more complicated optical devices, such as $M \times N$ optical stars, directional optical switches, different optical filters, and multiplexers.

An optical filter modifies the spectrum of incoming light and can mathematically be described by corresponding transfer function $H_{\text{of}}(\omega)$[3]:

$$E_{\text{out}}(t) = \frac{1}{2\pi} \int_{-\infty}^{\infty} \tilde{E}_{\text{in}}(\omega) H_{\text{of}}(\omega) e^{j\omega t} d\omega \quad (3.56)$$

where $E_{\text{in}}(t)$ and $E_{\text{out}}(t)$ denote the input and output electrical field, respectively, and the tilde is used to denote the FT, as before. Depending on the operational principle, the optical filters can be classified into two broad categories as diffraction or interference filters. The important class of optical filters is the tunable optical filters, which are able to dynamically change the operating frequency to the desired wavelength channel. The basic tunable optical filter types include tunable 2×2 directional couplers, FP filters, Mach–Zehnder (MZ) interferometer filters, Michelson filters, and acousto-optical filters. Two basic optical filters, the FP filter and the MZ interferometer filter, are shown in Figure 3.20. An FP filter is in fact a cavity between two high-reflectivity mirrors. It can act as a tunable optical filter if the cavity length is controlled, for example, by using a piezoelectric transducer. Tunable FP filters can also be made by using liquid crystals, dielectric thin films, semiconductor waveguides, etc. A transfer function of an FP filter whose mirrors have the same reflectivity R and cavity length L can be written as[4]

$$H_{\text{FP}}(\omega) = \frac{(1 - R)e^{j\pi}}{1 - Re^{j\omega\tau}}, \quad \tau = 2L/v_{\text{g}} \quad (3.57)$$

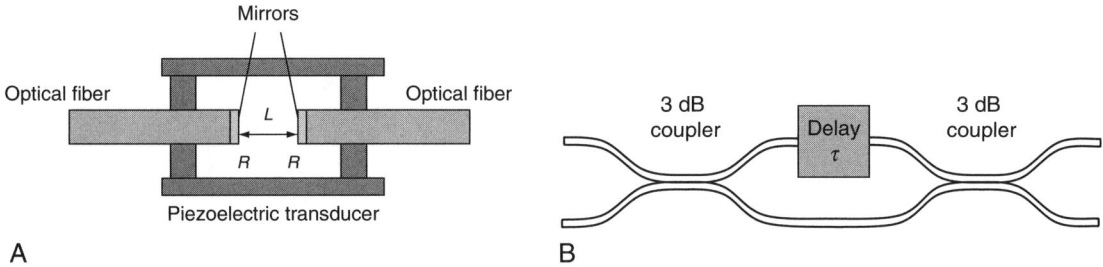

Figure 3.20: Basic optical filters: (a) FP filter and (b) MZ interferometer.

where τ is the round-trip time within the cavity, and v_g is the group velocity. The transfer function of the FP filter is periodic, with the period being the free spectral range $\Delta v = v_g/(2L)$. Another important parameter of the FP filter is the finesse, defined as[1,4]

$$F = \Delta v/\Delta v_{\mathrm{FP}} \cong \frac{\pi\sqrt{R}}{1-R} \tag{3.58}$$

where Δv_{FP} is the FP transmission peak width.

MZ interferometer filters, shown in Figure 3.20b, can also be used as tunable optical filters. The first coupler splits the signal into two equal parts, which acquire different phase shifts before they interfere at the second coupler. Several MZ interferometers can be cascaded to create an optical filter. When cross output of a 3 dB coupler is used, the square magnitude of transfer function is $|H_c(\omega)| = \cos^2(\omega\tau/2)$,[3] so the transfer function of an M-stage MZ filter based on 3 dB couplers can be written as[4]

$$|H_{\mathrm{MZ}}(\omega)|^2 = \prod_{m=1}^{M} \cos^2(\omega\tau_m/2) \tag{3.59}$$

where τ_m is the adjustable delay of the mth ($m = 1, 2, \ldots, M$) cascade.

Multiplexers and demultiplexers are basic devices of a WDM system. Demultiplexers contain a wavelength-selective element to separate the channels of a WDM signal. Based on the underlying physical principle, different demultiplexer devices can be classified as *diffraction-based demultiplexers* (based on a diffraction grating) and *interference-based demultiplexers* (based on optical filters and directional couplers).

Diffraction-based demultiplexers are based on the Bragg diffraction effect and use an angular dispersive element, such as the diffraction grating. The incoming composite light signal is reflected from the grating and dispersed spatially into different wavelength components, as shown in Figure 3.21a. Different wavelength components are focused by lenses and sent to individual optical fibers. The same device can be used as a multiplexer by switching the roles of input and output ports. This device can be implemented using either conventional or GRIN lenses. To simplify design, the concave grating can be used.

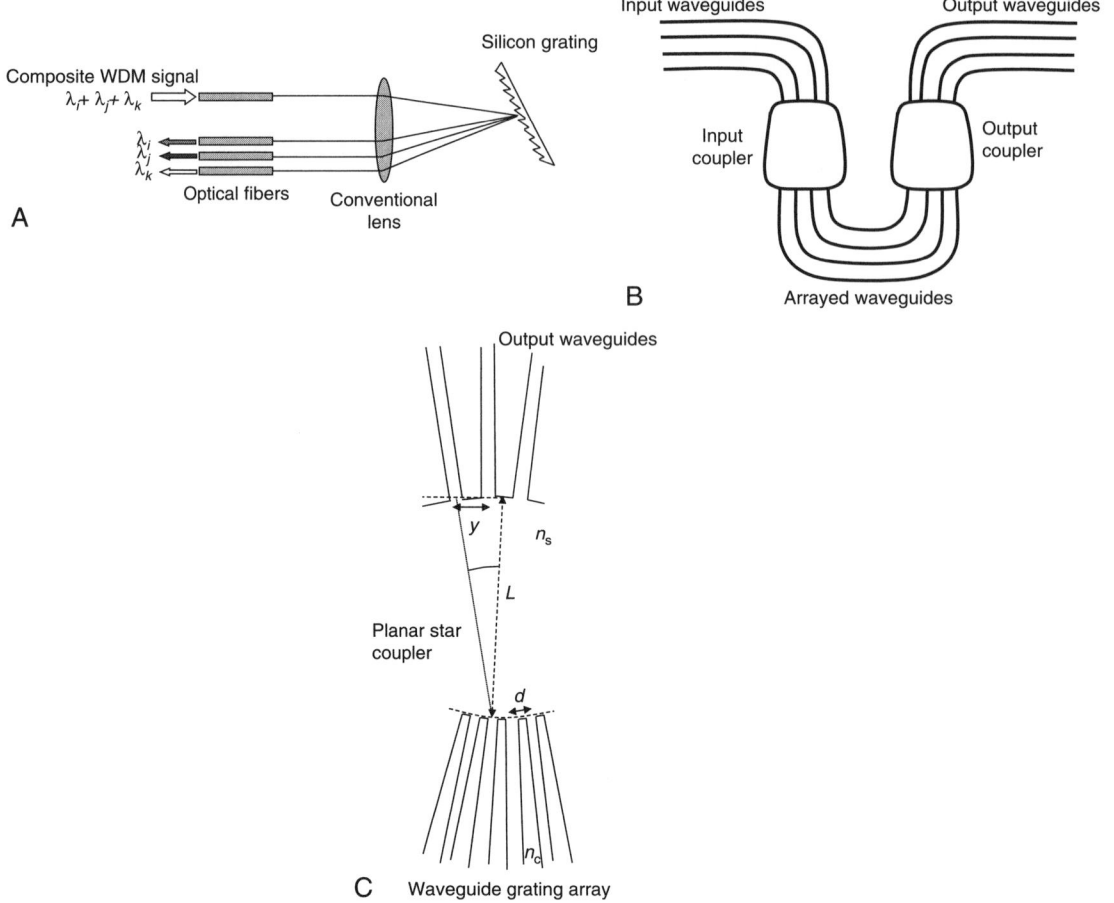

Figure 3.21: Optical multiplexers/demultiplexers: (a) grating-based demultiplexer and (b) array waveguide grating (AWG) demultiplexer. (c) The operating principle of AWG.

The second group of optical multiplexers is based on interference effect, and these employ the optical couplers and filters to combine different wavelength channels into a composite WDM signal. The multiplexers employing the interference effect include thin-film filter multiplexers and the array waveguide grating (AWG),[1–4,8,9,16] which is shown in Figure 3.20b. The AWG is a highly versatile WDM device because it can be used as a multiplexer, a demultiplexer, a drop-and-insert element, or even a wavelength router. It consists of M_{input} and M_{output} slab waveguides and two identical focusing planar star couplers connected by N uncoupled waveguides with a propagation constant β. The lengths of adjacent waveguides in the central region differ by a constant value ΔL, with the corresponding phase difference being $2\pi n_c \Delta L/\lambda$, where n_c is the refractive index of arrayed waveguides. Based on the phase matching condition (see Figure 3.21c) and the knowledge that the focusing is achieved when the path-length difference ΔL between adjacent array waveguides is an integer multiple of the central design wavelength λ_c—that is, $n_c \Delta L = m\lambda_c$—we derive the following expression for the channel spacing[16]:

$$\Delta v = \frac{y}{L} \frac{n_s}{m\lambda^2} \frac{cd}{n_g} \frac{n_c}{} \qquad (3.60)$$

where d is the spacing between the grating array waveguides, y is the spacing between the centers of the output ports, L is the separation between the center of arrayed waveguides and the center of output waveguides, n_c is the refractive index of waveguides in the grating array, n_s is the refractive index of the star coupler, n_g is the group index, and m is the diffraction order. The free spectral range can be obtained by[16]

$$\Delta v_{FSR} = \frac{c}{n_g(\Delta L + d \sin \Theta_{in,i} + d \sin \Theta_{out,j})} \qquad (3.61)$$

where the diffraction angles from the ith input $\Theta_{in,i}$ and jth output $\Theta_{out,j}$ ports (measured from the center of the array) can be found as $\Theta_{in,j} = iy/L$ and $\Theta_{out,j} = jy/L$, respectively.

Given this description of basic building blocks used in the state-of-the-art optical communication systems, we turn our attention to the description of noise sources in Section 3.3 and signal impairments in Section 3.4.

3.3 Noise Sources

The total noise is a stochastic process that has both multiplicative (introduced only when the signal is present) and additive (always present) components. Multiplicative noise components are mode partition noise (MPN), laser intensity noise (RIN), modal noise, quantum shot noise, and avalanche shot noise. Additive noise components are dark current noise, thermal noise, amplified spontaneous emission (ASE) noise, and crosstalk noise.

Originators of the noise in an optical transmission system can be identified as follows. In semiconductor lasers, the laser intensity noise, laser phase noise, and mode partition noise are present. In optical cable, in fiber and splicing, modal noise and reflection-induced noise are present. In optical amplifiers, spontaneous emission and ASE noises are present. At the receiver side, during the photodetection process, thermal noise and quantum noise are generated. The noise components at the receiver side are cumulative. Consider the general receiver structure shown in Figure 3.2. The most relevant optical noise components at the input of the optical preamplifier are intensity noise and spontaneous emission noise, which contains the spontaneous emission noise from the preceding in-line amplifier and accumulated ASE noise from other in-line amplifiers. The optical preamplifier will enhance all optical inputs in proportion to the gain, and it will also generate additional spontaneous emission noise. During the photodetection process, as previously discussed, quantum noise and thermal noise will be generated; in addition, different beating components, such as signal–noise beating and noise–noise beating components, will be created. The most important noise sources coming to the front-end amplifiers are ASE noise and the signal–noise beating component. The front-end amplifier will generate additional thermal noise.

3.3.1 Mode Partition Noise

MPN is the relative fluctuation in power between the main and side modes. The total power can remain unchanged, but the distribution of power among modes changes. MPN is present (up to a certain amount) even in DFBs. MPN can affect RIN significantly by enhancing it. It will occur even when the mode suppression ratio (MSR) is approximately 20 dB. Different modes will travel at different velocities due to dispersion so that MPN can lead to ISI.

3.3.2 Reflection-Induced Noise

Reflection-induced noise is related to the appearance of the back-reflected light due to refractive index discontinuities at optical splices, connectors, and optical fiber ends. The amount of reflected light can be estimated by the *reflection coefficient*[1]:

$$r_{\text{ref}} = \left(\frac{n_a - n_b}{n_a + n_b}\right)^2 \tag{3.62}$$

where n_a and n_b denote the refractive index coefficients of materials facing each other.

The strongest reflection occurs at the glass–air interface, $r_{\text{ref}} \approx [(1.46 - 1)/(1.46 + 1)]^2 \approx 3.5\%$ (-14.56 dB). It can be reduced below 0.1% if index-matching oils or gels are used. The considerable amount of back-reflected light can return and enter the laser cavity, negatively affecting the laser operation and leading to excess intensity noise. RIN can be increased by up to 20 dB if the back-reflected light exceeds -30 dBm. The multiple back-and-forth reflections between splices and connectors can be the source of additional intensity noise.

3.3.3 Relative Intensity Noise and Laser Phase Noise

The operating characteristics of semiconductor lasers are well described by the set of ordinary differential equations—the *rate equations*—which describe the interaction between photons and electrons inside the active region[1,3,4]:

$$\frac{dP}{dt} = GP + R_{\text{sp}} - \frac{P}{\tau_{\text{p}}} + F_{\text{P}}(t) \tag{3.63a}$$

$$\frac{dN}{dt} = \frac{I}{q} - \frac{N}{\tau_{\text{c}}} - GP + F_{\text{N}}(t) \tag{3.63b}$$

$$\frac{d\phi}{dt} = \frac{1}{2}\alpha_{\text{chirp}}\left[G - \frac{1}{\tau_{\text{p}}}\right] + F_{\phi}(t) \tag{3.63c}$$

In Eq. (3.63a), the term GP denotes the increase in photon number (P) due to stimulated emission, R_{sp} denotes the increase in P due to spontaneous emission and can be calculated by $R_{sp} = n_{sp}G$ (n_{sp} is the spontaneous emission factor, typically $n_{sp} \approx 2$), P/τ_p denotes the decrease in P due to emission through the mirrors and scattering/absorption by free carriers, and $F_P(t)$ corresponds to the noise process. In Eq. (3.63b), the term I/q denotes the increase in electron numbers (N) due to injection current I, the term $-N/\tau_c$ denotes the decrease in N due to spontaneous emission and nonradiative recombination, the term $-GP$ denotes the decrease in N due to stimulated emission, and $F_N(t)$ corresponds to the noise process. Finally, in Eq. (3.63c) the first term corresponds to dynamic chirp, the second term to the adiabatic chirp, and $F_\phi(t)$ corresponds to the noise process. (Note that the chirp effect was introduced in Section 3.2.1.) The noise terms $F_P(t)$, $F_N(t)$, and $F_\phi(t)$ are known as *Langevin forces*, which are commonly modeled as zero-mean Gaussian random processes. The photon lifetime τ_p is related to the internal losses (α_{int}) and mirror losses (α_{mir}) by $\tau_p = [v_g(\alpha_{int} + \alpha_{mir})]^{-1}$, with v_g being the group velocity. The carrier lifetime τ_c is related to the spontaneous recombination time (τ_{spon}) and nonradiation recombination time (τ_{nr}) by $\tau_c = \tau_{spon}\tau_{nr}/(\tau_{spon} + \tau_{nr})$. The net rate of stimulate emission G is related to the material gain g_m by $G = \Gamma v_g g_m$, where Γ is the confinement factor.

Noise in semiconductor lasers originates from two sources: spontaneous emission (the dominant noise source) and electron-hole recombination shot noise. As mentioned previously, the noise effects can be modeled by random driving terms in laser rate equations known as *Langevin forces*, which can be described as zero-mean Gaussian random processes with autocorrelation function (Markoffian approximation)[3,4]:

$$\langle F_i(t)F_j(t')\rangle = 2D_{ij}\delta(t - t'); \quad i,j \in \{P,N,\phi\} \tag{3.64}$$

with the dominating factor being $D_{pp} = R_{sp}P$ and $D_{\phi\phi} = R_{sp}/4P$. Fluctuation in intensity can be described by the *intensity autocorrelation function*[3,4]:

$$C_{PP}(\tau) = \langle \delta P(t)\delta P(t + \tau)\rangle/\bar{P}^2, \bar{P} = \langle P\rangle, \delta P = P - \bar{P} \tag{3.65}$$

The RIN spectrum can be found by FT of the autocorrelation function[3,4]:

$$\text{RIN}(\omega) = \int_{-\infty}^{\infty} C_{PP}(\tau)e^{-j\omega t}d\tau \sim \begin{cases} 1/\bar{P}^3 & \text{at low } \bar{P} \\ 1/\bar{P} & \text{at high } \bar{P} \end{cases} \tag{3.66}$$

Finally, the SNR can be estimated by[3,4]

$$\text{SNR} = [C_{PP}(0)]^{-1/2} \sim \left(\frac{\varepsilon_{NL}}{R_{sp}\tau_p}\right)^{1/2}\bar{P} \tag{3.67}$$

where the approximation is valid for SNR above 20 dB.

Noise, especially spontaneous emission, causes phase fluctuations in lasers, leading to a nonzero spectral linewidth Δv. In gas or solid-state lasers, the linewidth Δv typically ranges from the subhertz to the kilohertz range. In semiconductor lasers, the linewidth Δv is often much larger, up to the megahertz range, because of (1) the small number of photons stored in the small cavity and (2) the non-negligible value of the linewidth enhancement factor α_{chirp}. The spectrum of emitted light electric field $E(t) = \sqrt{(P)}\exp(j\phi)$ is related to the field-autocorrelation function $\Gamma_{\text{EE}}(\tau)$ by

$$S(\omega) = \int_{-\infty}^{\infty} \Gamma_{\text{EE}}(t)e^{-j(\omega-\omega_0)\tau}d\tau, \Gamma_{\text{EE}}(t) = \langle E^*(t)E(t+\tau) \rangle = \langle \exp(j\Delta\phi(t)) \rangle = \exp(-\langle\Delta\phi^2(\tau)\rangle/2)$$

(3.68)

where $\Delta\phi(\tau) = \phi(t+\tau) - \phi(t)$. By describing the laser phase noise process as the Wiener–Lévy process[7]—that is, a zero-mean Gaussian process with variance $2\pi\Delta v |t|$ (where Δv is the laser linewidth)—the spectrum of emitted light is found to be Lorentzian:[3]

$$S_{\text{EE}}(v) = \frac{\bar{P}}{2\pi\Delta v}\left[\frac{1}{1+\left(\dfrac{2(v+v_0)}{\Delta v}\right)^2} + \frac{1}{1+\left(\dfrac{2(v-v_0)}{\Delta v}\right)^2}\right]$$

(3.69)

where v_0 is the central frequency, and other parameters were previously introduced. The phase fluctuation variance (neglecting the relaxation oscillation)[3,4] is

$$\langle\Delta\phi^2(\tau)\rangle = \frac{R_{\text{sp}}}{2\bar{P}}(1+\alpha_{\text{chirp}}^2)|\tau| = 2\pi\Delta v|\tau|$$

(3.70a)

The laser linewidth can therefore be evaluated by

$$\Delta v = \frac{R_{\text{sp}}}{4\pi\bar{P}}(1+\alpha_{\text{chirp}}^2)$$

(3.70b)

From Eq. (3.70a), we see that the chirp effect enhances the laser linewidth by factor $(1+\alpha_{\text{chirp}}^2)$.

3.3.4 Modal Noise

Modal noise is related to multimode optical fibers. The optical power is nonuniformly distributed among a number of modes in multimode fibers, causing the so-called speckle pattern at the receiver end containing brighter and darker spots in accordance with the mode distribution. If the speckle pattern is stable, the photodiode effectively eliminates it by registering the total power over the photodiode area. If the speckle pattern changes with time, it will induce the fluctuation in the received optical power, known as the modal noise. The modal noise is inversely proportional to the laser linewidth, so it is a good idea to use LEDs in combination with multimode fibers.

3.3.5 Quantum Shot Noise

The optical signal arriving at the photodetector contains a number of photons generating the electron-hole pairs through the photoelastic effect. The electron-hole pairs are effectively separated by the reversed-bias voltage. However, not every photon is going to generate the electron-hole pair contributing to the total photocurrent. The probability of having n electron-hole pairs during the time interval Δt is governed by the *Poisson probability density function* (PDF):

$$p(n) = \frac{N^n e^{-N}}{n!} \tag{3.71}$$

where N is the average number of generated electron-hole pairs. The Poisson PDF approaches Gaussian for large N. The following is the mean of photocurrent intensity:

$$I = \langle i(t) \rangle = \frac{qN}{\Delta t} = \frac{qN}{T}, \quad \Delta t = T \tag{3.72}$$

For Poisson distribution, the variance equals the mean $\left\langle (n - N)^2 \right\rangle = N$, so the photocurrent can be calculated by $i(t) = qn/T$. The *shot-noise mean square value* can be determined by[1]

$$\langle i^2 \rangle_{sn} = \left\langle [i(t) - I]^2 \right\rangle = \frac{q^2 \left\langle [n - N]^2 \right\rangle}{T^2} = \frac{q^2 N}{T^2} = \frac{qI}{T} = 2qI\Delta f, \quad \Delta f = \frac{R_b}{2} \left(R_b = \frac{1}{T} \right) \tag{3.73}$$

where R_b is the bit rate. Equation (3.73) is derived for NRZ modulation format, but validity is more general. Therefore, the power spectral density of shot noise is determined by $S_{sn}(f) = 2qI$. To determine the APD shot-noise mean square value, the p-i-n shot-noise mean square value has to be multiplied by the excess noise factor $F(M)$ to account for the randomness of impact ionization effect[1,3]:

$$\langle i^2 \rangle_{sn}^{APD} = S_{sn}^{APD}(f)\Delta f = 2qM^2 F(M)I\Delta f \qquad F(M) = k_N M + (1 - k_N)\left[2 - \frac{1}{M}\right] \tag{3.74}$$

where M is the average APD multiplication factor, and k_N is the ratio of impact ionization factors of holes and electrons, respectively. The following is the often used excess noise factor approximation[1]:

$$F(M) \cong M^x, \quad x \in [0,1] \tag{3.75}$$

3.3.6 Dark Current Noise

The dark current I_d flows through the reversed-biased photodiode even in the absence of incoming light. The dark current consists of electron-hole pairs created due to thermal effects in the p-n junction. The dark current noise power can be calculated by modifying Eq. (3.74) as follows:

$$\langle i^2 \rangle_{dcn}^{APD} = S_{dcn}^{APD}(f)\Delta f = 2q\langle M \rangle^2 F(M)I_d\Delta f \tag{3.76}$$

3.3.7 Thermal Noise

The photocurrent generated during the photodetection process is converted to the voltage through the load resistance. The load voltage is further amplified by the front-end amplifier stage. Due to random thermal motion of electrons in load resistance, the already generated photocurrent exhibits an additional thermal noise component, also known as *Johnson noise*. The front-end amplifier enhances the thermal noise generated in the load resistance, which is described by the *amplifier noise figure*, $F_{n,el}$. The *thermal noise PSD* in the load resistance R_L and corresponding *thermal noise power* are defined as[1-4]

$$S_{thermal} = \frac{4k_B\Theta}{R_L} \qquad \langle i^2 \rangle_{thermal} = \frac{4k_B\Theta}{R_L}\Delta f \tag{3.77}$$

where Θ is the absolute temperature, k_B is the Boltzmann's constant, and Δf is the receiver bandwidth.

3.3.8 Spontaneous Emission Noise

The spontaneous emission of light appears during the optical signal amplification, and it is not correlated with signal (it is additive in nature). The noise introduced by spontaneous emission has a flat frequency characterized by Gaussian PDF, and the (double-sided) PSD in one state of polarization is given by[1-10]

$$S_{sp}(v) = (G-1)F_{no}hv/2 \tag{3.78}$$

where G is the amplifier gain, and hv is the photon energy. The *optical amplifier noise figure*, F_{no}, defined as the ratio of SNRs at the input and output of the optical amplifier, is related to the spontaneous emission factor $n_{sp} = N_2/(N_2 - N_1)$ (for the two-level system introduced in Section 3.2.1) by[3]

$$F_{no} = 2n_{sp}\left(1 - \frac{1}{G}\right) \cong 2n_{sp} \geq 2 \tag{3.79}$$

In the most practical cases, the noise figure is between 3 and 7 dB. The effective noise figure of the cascade of K amplifiers with corresponding gains G_i and noise figures $F_{no,i}$ $(I = 1, 2, \ldots, K)$[1] is

$$F_{no} = F_{no,1} + \frac{F_{no,2}}{G_1} + \frac{F_{no,3}}{G_1 G_2} + \ldots + \frac{F_{no,K}}{G_1 G_2 \ldots G_{k-1}} \tag{3.80}$$

The total power of the spontaneous emission noise, for an amplifier followed by an optical filter of bandwidth B_{op}, is determined by

$$P_{sp} = 2|E_{sp}|^2 = 2S_{sp}(v)B_{op} = (G-1)F_{no}hvB_{op} \tag{3.81}$$

where the factor 2 is used to account for both polarizations.

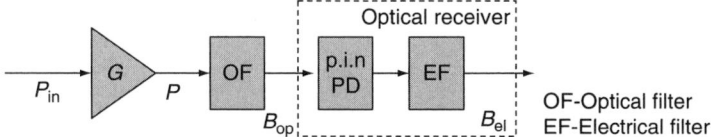

Figure 3.22: The optical receiver with the preamplifier.

3.3.9 Noise Beat Components

Optical preamplifiers are commonly used to improve the receiver sensitivity by preamplifying the signal before it reaches the photodetector, which is illustrated in Figure 3.22. The gain of the optical amplifier is denoted with G, the optical filter bandwidth is denoted with B_{op}, and electrical filter bandwidth is denoted with B_{el}.

An optical amplifier introduces the spontaneous emission noise in addition to the signal amplification:

$$E(t) = \sqrt{2P} \cos(2\pi f_c t + \theta) + n(t) \tag{3.82}$$

where $E(t)$ denotes the electrical field at the receiver input, and $n(t)$ denotes the amplifier spontaneous emission noise, which is the Gaussian process with PSD given by Eq. (3.78). The photocurrent has the following form:

$$i(t) = RP(t) = RE^2(t) = 2RP \cos^2(2\pi f_c t + \theta) + 2R\sqrt{2P}n(t)\cos(2\pi f_c t + \theta) + Rn^2(t) \tag{3.83}$$

Following the procedure described by Ramaswami,[2] we can determine the variance of different noise and beating components as follows[2]:

$$\sigma^2 = \int_{-B_{el}}^{B_{el}} S_i(f)df = \sigma_{sh}^2 + \sigma_{sig-sp}^2 + \sigma_{sp-sp}^2 \tag{3.84}$$

$$\sigma_{sh}^2 = 2qR[GP_{in} + S_{sp}B_{op}]B_{el}, \quad \sigma_{sig-sp}^2 = 4R^2GP_{in}S_{sp}B_{el}, \quad \sigma_{sp-sp}^2 = R^2S_{sp}^2(2B_{op} - B_{el})B_{el}$$

where "sh" denotes the variance of shot noise, "sp–sp" denotes the variance of spontaneous–spontaneous beat noise, and "sig–sp" denotes the variance of signal–spontaneous beating noise. S_{sp} is the PSD of spontaneous emission noise (Eq. 3.78); P_{in} is the average power of incoming signal (see Figure 3.22); and the components of the PSD of photocurrent $i(t)$, denoted by $S_i(f)$, are shown in Figure 3.23.

3.3.10 Crosstalk Components

The crosstalk effect occurs in multichannel (WDM) systems and can be classified as (1) interchannel component (the crosstalk wavelength is sufficiently different from the observed channel wavelength) or (2) intrachannel (the crosstalk signal is at the same wavelength as the observed channel or sufficiently close) component.

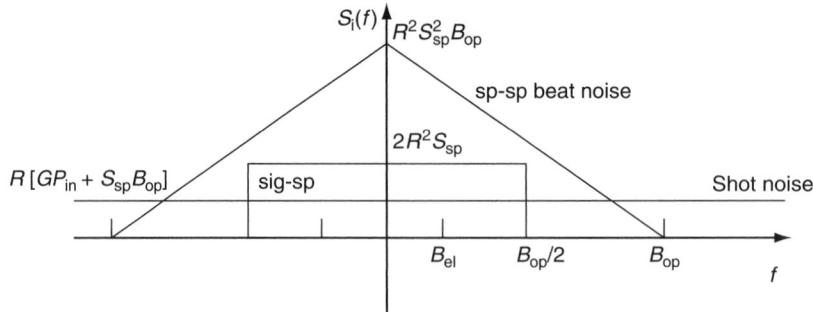

Figure 3.23: The components of the PSD of photocurrent.

Interchannel crosstalk (also known as out-of-band or hetero-wavelength crosstalk) is introduced by either an optical filter or demultiplexer that selects the desired channel and imperfectly rejects the neighboring channels, which is illustrated in Figure 3.24a. Another source of interchannel crosstalk is an optical switch, shown in Figure 3.24b, in which the crosstalk arises because of imperfect isolation among different wavelength ports. The crosstalk signal behaves as noise and it is a source of linear (noncoherent) crosstalk so that the photocurrent at the receiver can be written as[1,3]

$$I = R_m P_m + \sum_{n \neq m} R_n X_{mn} P_n \equiv I_{ch} + I_X, \quad R_m = \eta_m q / h\nu_m, \quad I_X = \sum_{n \neq m} R_n X_{mn} P_n \tag{3.85}$$

where R_m is the photodiode responsivity of the mth WDM channel with average power P_m, $I_{ch} = R_m P_m$ is the photocurrent corresponding to the selected channel, I_X is the crosstalk photocurrent, and X_{mn} is the portion of the nth channel power captured by the optical receiver of the mth optical channel.

The intrachannel crosstalk in transmission systems may occur due to multiple reflections. However, it is much more important in optical networks. In optical networks, it arises from cascading optical (WDM) demux and mux or from an optical switch, which is illustrated in Figure 3.25. Receiver current $I(t) = R|E_m(t)|^2$ contains interference or beat terms, in addition to the desired signal: (1) signal–crosstalk beating terms (e.g., $E_m E_n$) and (2) crosstalk–

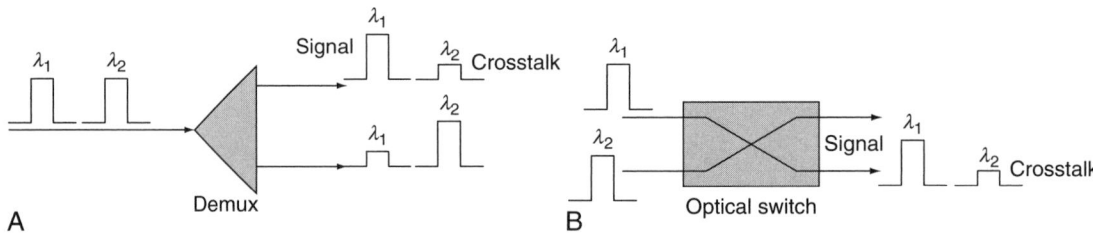

Figure 3.24: Sources of interchannel crosstalk: (a) demux and (b) the optical switch.

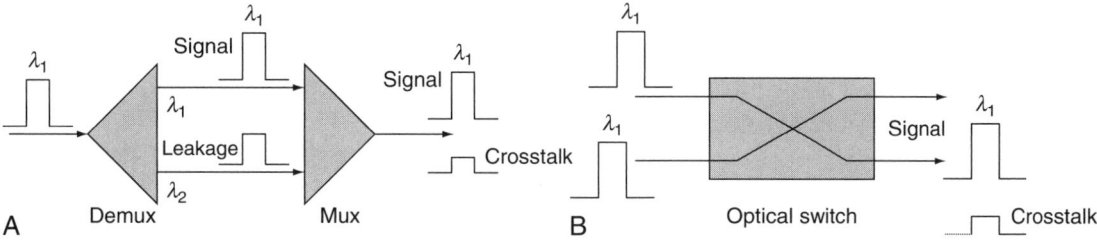

Figure 3.25: Sources of intrachannel crosstalk: (a) a demux–mux pair and (b) the optical switch.

crosstalk beating terms (e.g., $E_n E_k$, $k \neq m$, and $n \neq m$). The total electrical field E_m and receiver current $I(t)$, observing only the in-band crosstalk, can be written as[1–3]

$$E_m(t) = \left(E_m + \sum_{n \neq m} E_n \right) e^{-j\omega_m t}, \quad I(t) = R|E_m(t)|^2 \approx R P_m(t) + 2R \sum_{n \neq m}^{N} \sqrt{P_m P_n} \cos[\phi_m(t) - \phi_n(t)]$$

(3.86)

where $\phi_n(t)$ denotes the phase of the nth channel, and R is the photodiode responsivity. Each term acts as an independent random variable, and the crosstalk effect can be observed as the intensity noise (that is Gaussian for large N):

$$I(t) = R(P_m + \Delta P) \quad (P_n << P_m, n \neq m)$$

where ΔP is the intensity noise due to intrachannel crosstalk.

3.4 Channel Impairments

This section describes different channel impairments, including fiber attenuation, insertion losses, chromatic dispersion, PMD, and fiber nonlinearities. The frequency chirp effect was described in Section 3.2.1.

3.4.1 Fiber Attenuation

Fiber attenuation can be described by the general relation $dP/dz = -\alpha P$, where α is the power attenuation coefficient per unit length. If P_{in} is the power launched into the fiber, the power remaining after propagating a length L within the fiber P_{out} is $P_{\text{out}} = P_{\text{in}} \exp(-\alpha L)$. The absorption coefficient varies with wavelength as many of the absorption and scattering mechanisms vary with λ. For instance, Rayleigh scattering in fiber is due to microscopic variations in the density of glass (density fluctuation sites $< \lambda$) and varies as[3]

$$\alpha_R = C/\lambda^4, C = 0.7 - 0.9 \ (\text{dB/km}) - \mu m^4$$

Mie scattering is caused by imperfections (scattering) in the core–cladding interface that are larger than λ.

Intrinsic absorption can be identified as follows: (1) infrared absorption—in SiO_2 glass, vibrational modes of Si–O bonds cause an absorption peak at $\lambda > 7$ μm, which has a tail extending to the $\lambda = 1.55$ μm range; and (2) ultraviolet absorption is due to the electronic resonances occurring in the ultraviolet region ($\lambda < 0.4$ μm).

Extrinsic absorption results from the presence of impurities: (1) Transition-metal impurities (Fe, Cu, Co, Ni, Mn, and Cr) absorb strongly in 0.6–1.6 μm, and the loss level is reduced below 1 dB/km if their amount is kept below 1 part per billion; and (2) residual OH^- ions cause peaks near $\lambda = 0.95, 1.24, 1.39$ μm.

3.4.2 Insertion Losses

Fiber loss is not the only source of optical signal attenuation along the transmission lines. The fiber splices and fiber connectors also cause signal attenuation. The fiber splices can be fused or joined together by some mechanical means, with typical attenuation being 0.01–0.1 dB per fused splice and slightly above 0.1 dB per mechanical splice.

Optical connectors are removable and allow many repeated connections and disconnections, with typical insertion loss for high-quality SMF not above 0.25 dB. To minimize the connector return loss (a fraction of the optical power reflected back into the fiber at the connector point), the angled fiber-end surfaces are commonly used.

The number of optical splices and connectors depends on transmission length and must be taken into account unless the total attenuation due to fiber joints is distributed and added to the optical fiber attenuation.

3.4.3 Chromatic Dispersion

The dispersion problem can be described as follows. Short optical pulses entering a dispersive channel such as an optical fiber are spread out into a much broader temporal distribution. Both the intermodal dispersion and chromatic dispersion cause the distortion in multimode optical fibers, whereas chromatic dispersion is the only cause of the signal distortion in SMF. The intermodal dispersion can be specified through the optical fiber bandwidth B_{fib} that is related to a 1 km long optical fiber, and this parameter is specified by the manufacturers and commonly measured at a wavelength of approximately 1310 nm (chromatic dispersion is negligible in this region compared to intermodal dispersion)[1]:

$$B_{fib,L} = \frac{B_{fib}}{L^\mu}, \quad \mu = 0.5 - 1 \tag{3.87}$$

SMFs effectively eliminate intermodal dispersion by limiting the number of modes to just one through a much smaller core diameter. Chromatic dispersion (intramodal dispersion) is related to each individual optical frequency and does not exist for a monochromatic wave.

Because the source is nonmonochromatic, different spectral components within a pulse will travel at different velocities, inducing the pulse broadening. When the neighboring pulses cross their allocated time slots, ISI occurs, and the signal bit rates that can be effectively transmitted can be severely limited. A specific spectral component, characterized by the angular optical frequency ω, will arrive at the output end of the fiber of length L after some delay τ_g, known as the group delay:

$$\tau_g = \frac{L}{v_g} = L\frac{d\beta}{d\omega} = \frac{L}{c}\frac{d\beta}{dk} = -L\frac{\lambda^2}{2\pi c}\frac{d\beta}{d\lambda} \tag{3.88}$$

where β is the propagation constant introduced previously, and $v_g = [d\beta/d\omega]^{-1}$ is the group velocity, the speed at which the energy of an optical pulse travels. As a result of the difference in time delays, the optical pulse disperses after traveling a certain distance, and the pulse broadening ($\Delta\tau_g$) can be characterized by

$$\Delta\tau_g = \frac{d\tau_g}{d\omega}\Delta\omega = \frac{d\tau_g}{d\lambda}\Delta\lambda \tag{3.89a}$$

where $\Delta\omega$ represents the frequency bandwidth of the source, and $\Delta\lambda$ represents the wavelength bandwidth of the source. By substituting Eq. (3.88) into Eq. (3.89a), we get

$$\Delta\tau_g = \frac{d\tau_g}{d\omega}\Delta\omega = \frac{d\left(L\dfrac{d\beta}{d\omega}\right)}{d\omega}\Delta\omega = L\frac{d^2\beta}{d\omega^2}\Delta\omega = L\frac{d^2\beta}{d\omega^2}\left(-\frac{2\pi c}{\lambda^2}\Delta\lambda\right) = LD\Delta\lambda \tag{3.89b}$$

where D [ps/nm-km] represents the chromatic dispersion parameter defined by

$$D = -\frac{2\pi c}{\lambda^2}\frac{d^2\beta}{d\omega^2} = -\frac{2\pi c}{\lambda^2}\beta_2, \quad \beta_2 = \frac{d^2\beta}{d\omega^2} \tag{3.90}$$

where β_2 denotes the previously introduced GVD parameter. The chromatic dispersion can be expressed as the sum of two contributing factors[1,7]:

$$D = \frac{d\left(-\dfrac{\lambda^2}{2\pi c}\dfrac{d\beta}{d\lambda}\right)}{d\lambda} = -\frac{1}{2\pi c}\left(2\lambda\frac{d\beta}{d\lambda} + \lambda^2\frac{d^2\beta}{d\lambda^2}\right) = D_M + D_W, D_M = -\frac{\lambda^2}{2\pi c}\frac{d^2\beta}{d\lambda^2} \quad D_W = -\frac{\lambda}{\pi c}\frac{d\beta}{d\lambda} \tag{3.91}$$

where D_M represents the material dispersion, and D_W represents the waveguide dispersion.

The material dispersion arises due to wavelength dependence of the refractive index on the fiber core material, which causes the wavelength dependence of the group delay. The wavelength dependence of the refractive index $n(\lambda)$ is well approximated by the Sellmeier equation, introduced previously. The material dispersion is related to the slope of $n_g = n - \lambda dn(\lambda)/d\lambda$ by $D_M = (1/c)dn_g/d\lambda$. For pure silica fiber, the following approximation is valid[3]:

$$D_M \approx 122(1 - \lambda_{ZD}/\lambda), \quad 1.25\ \mu m < \lambda < 1.66\ \mu m \tag{3.92}$$

where $\lambda_{ZD} = 1.276$ μm is the zero dispersion wavelength. It can be seen that for wavelengths larger than zero dispersion wavelength, the material dispersion coefficient is positive, whereas the GVD is negative, and this regime is known as the *anomalous dispersion region.*

The waveguide dispersion occurs because the propagation constant is a function of the fiber parameters (core radius and difference between refractive indices in fiber core and cladding) and at the same time is a function of wavelength. Therefore, the propagation paths for a mode due to different boundary matching conditions are slightly different at different wavelengths. Waveguide dispersion is related to the physical design of the optical fiber. Because the value of Δ is typically small, the refractive indices of the core–cladding are nearly equal, and the light is not strictly confined in the fiber core. For a given mode, such as the fundamental mode, the portion of light energy that propagates in the core depends on the wavelength: The longer the wavelength, the more power in the cladding. The effect of waveguide dispersion on pulse spreading can be approximated by assuming that the refractive index of material is independent of wavelength. To make the result independent of fiber configuration, we should express the group delay in terms of the normalized propagation constant b, defined as[8]

$$\beta \approx n_2 k (b\Delta + 1), \quad \Delta \ll 1 \tag{3.93}$$

The group delay due to waveguide dispersion can be found as[8]

$$\tau_W = \frac{L}{c}\frac{d\beta}{dk} \simeq \frac{L}{c}\left(n_2 + n_2\Delta\frac{d(Vb)}{dV}\right), \quad \frac{d(Vb)}{dV} = b\left[1 - \frac{2J_m^2(pa)}{J_{m+1}(pa)J_{m-1}(pa)}\right] \tag{3.94}$$

where p is the core parameter introduced previously ($p^2 = n_1^2 k^2 - \beta^2$). The pulse broadening due to waveguide dispersion can be determined using Eq. (3.89) by

$$\Delta\tau_W = \frac{d\tau_W}{d\lambda}\Delta\lambda = \frac{d\tau_W}{dV}\frac{dV}{dk}\frac{dk}{d\lambda}\Delta\lambda = L\frac{n_2\Delta}{c}\frac{d^2(Vb)}{dV^2}\frac{V}{k}\left(-\frac{k}{\lambda}\right)\Delta\lambda = LD_W\Delta\lambda \tag{3.95}$$

where the waveguide dispersion parameter D_W is defined by

$$D_W \simeq -\frac{n_2\Delta}{c}\frac{1}{\lambda}\left[V\frac{d^2(Vb)}{dV^2}\right] \tag{3.96}$$

and it is negative for the normalized frequencies between 0 and 3.

The total dispersion can be written as the sum of two contributions, $D = D_M + D_W$. The waveguide dispersion (that is negative) shifts the zero dispersion wavelength to approximately 1.308 μm. $D \sim 15$–18 ps/(km-nm) near 1.55 μm. This is the low-loss region for fused silica optical fibers. This approach for mutual cancellation was used to produce several types of single-mode optical fibers that are different in design compared to standard SMF, standardized by ITU-T. In addition to standard SMF, there are two major fiber types: (1) dispersion shifted fibers (DSFs; described in ITU-T recommendation G.653) with dispersion minimum shifted from the 1310 nm wavelength region to the 1550 nm region, and

(2) NZDSFs (described in G.655 recommendation), with dispersion minimum shifted from the 1310 nm window to anywhere within C or L bands (commercial examples are TrueWave fiber and LEAF).

Dispersion effects do not disappear completely at zero dispersion wavelength. Residual dispersion due to higher order dispersive effects still exists. Higher order dispersive effects are governed by the dispersion slope parameter $S = dD/d\lambda$. Parameter S is also known as the differential dispersion parameter or second-order dispersion parameter, and it is related to the GVD parameter and second-order GVD parameter $\beta_3 = d\beta_2/d\omega$ by

$$S = \frac{dD}{d\lambda} = \frac{d}{d\lambda}\left(-\frac{2\pi c}{\lambda^2}\beta_2\right) = \frac{4\pi c}{\lambda^3}\beta_2 + \left(\frac{2\pi c}{\lambda^2}\right)^2 \beta_3 \tag{3.97}$$

3.4.4 Polarization Mode Dispersion

The polarization unit vector, representing the state of polarization (SOP) of the electric field vector, does not remain constant in practical optical fibers; rather, it changes in a random manner along the fiber because of its fluctuating birefringence. Two common birefringence sources are geometric birefringence (related to small departures from perfect cylindrical symmetry) and anisotropic stress (produced on the fiber core during manufacturing or cabling of the fiber). The degree of birefringence is described by the difference in refractive indices of orthogonally polarized modes $B_m = |n_x - n_y| = \Delta n$. The corresponding difference in propagation constants of two orthogonally polarized modes is $\Delta\beta = |\beta_x - \beta_y| = (\omega/c)\Delta n$. Birefringence leads to a periodic power exchange between the two polarization components, described by beat length $L_B = 2\pi/\Delta\beta = \lambda/\Delta n$. Typically, $B_m \sim 10^{-7}$, and therefore $L_B \sim 10$ m for $\lambda \sim 1$ μm. Linearly polarized light remains linearly polarized only when it is polarized along one of the principal axes; otherwise, its state of polarization changes along the fiber length from linear to elliptical and then back to linear in a periodic manner over the length L_B.

In certain applications, it is necessary to transmit a signal through a fiber that maintains its SOP. Such a fiber is called a polarization-maintaining fiber (PMF). (PANDA fiber is a well-known PMF.) Approaches to design PMF having high but constant birefringence are (1) the shape birefringence (the fiber has an elliptical core) and (2) stress birefringence (stress-inducing mechanism is incorporated in fiber). Typical PMF has a birefringence $B_m \sim 10^{-3}$ and a beat length $L_B \sim 1$ mm, resulting in a polarization crosstalk smaller than -30 dB/km. Unfortunately, the loss of PMFs is high ($\alpha \sim 2$ dB/km).

The modal group indices and modal group velocities are related by

$$\bar{n}_{gx,y} = \frac{c}{v_{gx,y}} = \frac{c}{1/\beta_{1x,y}}, \quad \beta_{1x,y} = \frac{d\beta x,y}{d\omega} \tag{3.98}$$

so that the difference in time arrivals (at the end of fiber of length L) for two orthogonal polarization modes, known as the *differential group delay* (DGD), can be calculated by[3]

$$\Delta\tau = \left| \frac{L}{v_{gx}} - \frac{L}{v_{gy}} \right| = L|\beta_{1x} - \beta_{1y}| = L\Delta\beta_1 \tag{3.99}$$

DGD can be quite large in PMFs ($\sim$1 ns/km) due to their large birefringence. Conventional fibers exhibit much smaller birefringence, but its magnitude and polarization orientation change randomly at a scale known as the correlation length l_c (with typical values of 10–100 m). The analytical treatment of PMD is quite complex due to its statistical nature. A simple model divides the fiber into a large number of segments (tens to thousands), with both the degree of birefringence and the orientation of the principal states being constant in each segment but changing randomly from section to section, as shown in Figure 3.26. Although the fiber is composed of many segments having different birefringence, there exist principal states of polarization (PSPs). If we launch a pulse light into one PSP, it will propagate to the corresponding output PSP without spreading due to PMD. Output PSPs are different from input PSPs. Output PSPs are frequency independent up to the first-order approximation. A pulse launched into the fiber with arbitrary polarization can be represented as a superposition of two input PSPs. The fiber is subject to time-varying perturbations (e.g., temperature and vibrations) so that PSPs and DGD vary randomly over time.

DGD is a Maxwellian distributed random variable with the mean square DGD value[3]:

$$\left\langle (\Delta T)^2 \right\rangle = 2(\Delta\beta_1)^2 l_c^2 \left[\exp\left(-\frac{L}{l_c} \right) + \frac{L}{l_c} - 1 \right] \tag{3.100}$$

where the parameters in Eq. (3.100) were previously introduced. For long fibers, $L \gg l_c$, the root mean square DGD follows:

$$\left\langle (\Delta T)^2 \right\rangle^{1/2} = \Delta\beta_1 \sqrt{2l_c L} = D_p \sqrt{L} \tag{3.101}$$

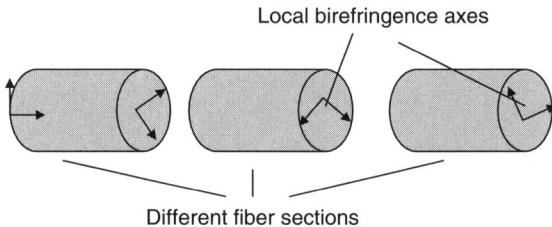

Local birefringence axes

Different fiber sections

Figure 3.26: A simple PMD model.

where D_p [ps/$\sqrt{\text{km}}$] is the PMD parameter. Typical values for the PMD parameter are 0.01–10 ps/(km)$^{1/2}$; in new fibers $D_p < 0.1$ ps/(km)$^{1/2}$. The first-order PMD coefficient, D_p, can be characterized again by Maxwellian PDF[1]:

$$P(D_p) = \sqrt{\frac{2}{\pi}} \frac{D_p^2}{\alpha^3} \exp\left(-\frac{D_p^2}{2\alpha^2}\right)$$

(3.102)

where α is the coefficient with a typical value of approximately 30 ps. The mean value $<D_p>$ determined from Maxwellian PDF is $<D_p> = (8/\pi)^{1/2}\alpha$. The overall probability $P(D_p)$ that coefficient D_p will be larger than the prespecified value can be determined by[1]

$$P(D_p) = \int_0^{D_p} p(D_p) dD_p$$

(3.103)

$P(3<D_p>)$ is approximately 4×10^{-5}, so the practical expression to characterize the first-order PMD is[1]

$$\Delta T = 3\langle D_p \rangle \sqrt{L}$$

(3.104)

The second-order PMD occurs due to frequency dependence of both DGD and PSP, and it can be characterized by coefficient D_{P2}[1,17]:

$$D_{P2} = \left[\left(\frac{1}{2}\frac{\partial D_{P1}}{\partial \omega}\right)^2 + \left(\frac{D_{P1}}{2}\frac{\partial|s|}{\partial \omega}\right)^2\right]^{1/2}$$

(3.105)

where the first term describes the frequency dependence of DGD, and the second term describes the frequency dependence of Stokes vector s (describing the position of PSP).[18] The statistical nature of the second-order PMD coefficient in real fiber is characterized by PDF of D_{P2}[1,19]:

$$P(D_{P2}) = \frac{2\sigma^2 D_{P2}}{\pi} \frac{\tanh(\sigma D_{P2})}{\cosh(\sigma D_{P2})}$$

(3.106)

The probability that the second-order PMD coefficient is larger than $3<D_{P2}>$ is still not negligible. However, it becomes negligible for larger values of $5<D_{P2}>$ so that the total pulse spreading due to the second-order PMD effect can be expressed as[1]

$$\Delta \tau_{P2} = 5\langle D_{P2} \rangle L$$

(3.107)

3.4.5 Fiber Nonlinearities

The basic operational principles of optical transmission can be explained assuming that optical fiber medium is linear. The linearity assumption is valid if the launched power does not exceed several milliwatts in a single channel system. In modern WDM technology, high-power semiconductor lasers and optical amplifiers are employed, and the influence of

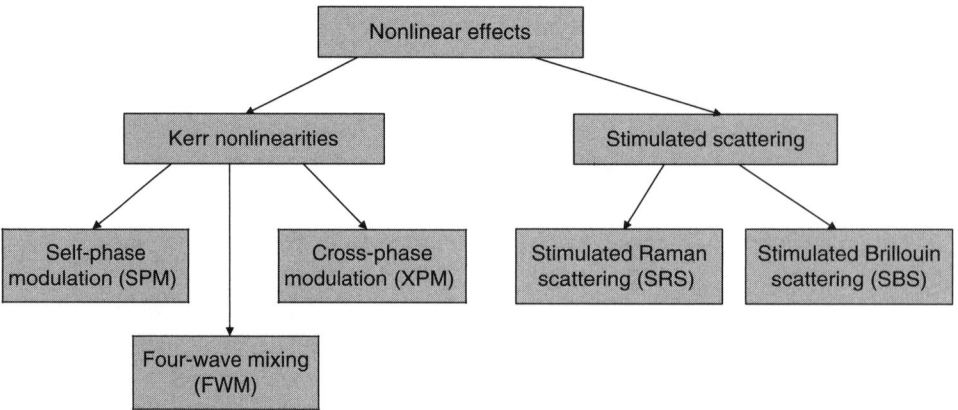

Figure 3.27: Classification of fiber nonlinearities.

fiber nonlinearities becomes important. Moreover, in some special cases, the fiber nonlinearities may be used to enhance the fiber transmission capabilities (e.g., soliton transmission). There are two major groups of fiber nonlinearities related either to nonlinear refractive index (Kerr effect) or to nonlinear optical scattering, which is illustrated in Figure 3.27.

The Kerr effect occurs due to the dependence of the index of refraction on light intensity, and fiber nonlinearities belonging to this category are (1) self-phase modulation (SPM), (2) cross-phase modulation (XPM), and (3) four-wave mixing (FWM). The SPM is related to the single optical channel. The variation of power within the channel causes changes in the refractive index, which leads to pulse distortion. In XPM, the refractive index changes due to variations in power not only in the observed channel but also due to variation in powers of other wavelength channels leading to the pulse distortion. In FWM, several wavelength channels interact to create new channels. This effect is dependent on both the powers of the interacting channels and chromatic dispersion.

The stimulated scattering effects are caused by parametric interaction between the light and materials. There are two types of stimulated scattering effects: SRS and stimulated Brillouin scattering (SBS). In SRS, the light interacts with material through the vibrations and causes the energy transfer from short wavelength channels to longer wavelength channels, leading to the interchannel crosstalk. In SBS, the light interacts with the matter through acoustic waves, leading to the coupling with backward propagating waves, thus limiting the available power per channel.

The nonlinear effects introduced previously are a function of the transmission length and cross-sectional area of the optical fiber. The nonlinear interaction is stronger for longer fibers and smaller cross-sectional areas. On the other hand, the nonlinear interaction decreases along the transmission line because the signal is attenuated as it propagates.

Therefore, the strength of the nonlinear effects may be characterized by introducing the concept of *effective length* L_{eff} as an equivalent length so that the product of launched power P_0 and the equivalent length is the same as the area below the power evolution curve $P(z) = P_0 \exp(-\alpha z)$:

$$P_0 L_{eff} = \int_{z=0}^{L} P(z)dz = \int_{z=0}^{L} P_0 e^{-\alpha z}dz \Rightarrow L_{eff} = \frac{1 - e^{-\alpha L}}{\alpha} \approx 1/\alpha \qquad (3.108)$$

where α denotes the attenuation coefficient, and the approximation is valid for fibers that are at least several tens of kilometers in length. The affective length after M amplifiers is

$$L_{eff,total} = \frac{1 - e^{-\alpha L}}{\alpha} M = \frac{1 - e^{-\alpha L}}{\alpha} \frac{L}{l} \qquad (3.109)$$

where L is the total transmission length, l is the amplifier spacing, and M is the number of amplifiers.

If the amplifier spacing increases, the optical amplifier gain increases in proportion to $\exp(\alpha l)$ to compensate for the fiber losses. On the other hand, the increase in power enhances the nonlinear effects. Therefore, what is important is the product of launched power P_0 and the total effective length $L_{eff,total}$, as follows: $P_0 L_{eff,total}$.

The nonlinear effects are inversely proportional to the area of the fiber core because the concentration of the optical power per unit cross-sectional area (power density) is higher for smaller cross-sectional areas and vice versa. Optical power distribution across the cross-sectional area is closely related to the overall refractive index, and to characterize it, the concept of effective cross-sectional area (A_{eff}) is introduced[1,2]:

$$A_{eff} = \frac{\left[\iint_{r\ \theta} rdrd\theta |E(r,\theta)|^2\right]^2}{\iint_{r\ \theta} rdrd\theta |E(r,\theta)|^4} \qquad (3.110)$$

where $E(r, \theta)$ is the distribution of the electrical field in cross-section, and (r, θ) denote the polar coordinates. For the Guassian approximation $E(r, \theta) = E_0 \exp(-r^2/w^2)$, the effective cross-sectional area is simply $A_{eff} = \pi w^2$, where w is the mode radius. The Gaussian approximation is not applicable to optical fibers with a more complicated index profile, such as DSFs, NZDFs, and DCFs. In these cases, the effective cross-sectional area can be found by[1]

$$A_{eff} = k\pi w^2 \qquad (3.111)$$

where k is smaller than 1 for DCFs and some NZDSFs, and it is larger than 1 for NZDSF with a larger effective cross-section area.

3.4.5.1 Self-Phase Modulation

For relatively high power density, the index of refraction acts as the function density; the effect is known as the Kerr effect:

$$n(P) = n_0 + n_2 \frac{P}{A_{\text{eff}}} \tag{3.112}$$

where n_2 is the Kerr coefficient (second-order refractive index coefficient), with typical values of 2.2–3.4 $\cdot$ 10^{-20} m^2/W. The variations in the refractive index due to the Kerr effect yield to the propagation constant β variations because $\beta = 2\pi n/\lambda$ (the propagation function is a function of refractive index):

$$\beta(P) = \beta_0 + \gamma P, \; \beta_0 = 2\pi n_0/\lambda, \quad \gamma = \frac{2\pi n_2}{\lambda A_{\text{eff}}} \tag{3.113}$$

where γ is the nonlinear coefficient, with typical values of 0.9–2.75 W^{-1}km^{-1} at 1550 nm. The propagation constant β will vary along the duration of the optical pulse because the different points along the pulse will "see" different optical powers, and the frequency chirping is introduced. The propagation constant associated with the leading edge of the pulse will be lower than that related to the central part of the pulse. The difference in propagation constants will cause the difference in phases associated with different portions of the pulse. The central part of the pulse will acquire phase more rapidly than the leading and trailing edges. The total nonlinear phase shift after some length L can be found by

$$\Delta\Phi[P(t)] = \int_0^L \left[\beta(P_0) - \beta\right]dz = \int_0^L \gamma P(z)dz = \frac{\gamma P_0(t)\left[1 - e^{-\alpha L}\right]}{\alpha} = \gamma P_0(t)L_{\text{eff}} = \frac{L_{\text{eff}}}{L_{\text{NL}}}, \; L_{\text{NL}} = \frac{1}{\gamma P_0} \tag{3.114}$$

where L_{NL} is the nonlinear length.

The frequency variation (chirp) can be found as the first derivative of the nonlinear phase shift with respect to time:

$$\delta\omega_{\text{SPM}}(t) = \frac{d[\Delta\Phi(t)]}{dt} = \frac{2\pi}{\lambda}\frac{L_{\text{eff}}n_2}{A_{\text{eff}}}\frac{dP_0(t)}{dt} \tag{3.115}$$

In the general case, SPM causes the pulse broadening. However, in some situations, the frequency chirp due to SPM may be opposite to that due to chromatic dispersion (e.g., the anomalous dispersion region). In such a case, SPM helps to reduce the impact of chromatic dispersion. To understand the relative effects of chromatic dispersion and SPM, let's observe the NLSE (Eq. 3.37), assuming that the fiber loss and the second-order GVD can be neglected:

$$\frac{\partial A(z,t)}{\partial z} = -\beta_1 \frac{\partial A(z,t)}{\partial t} - j\frac{\beta_2}{2}\frac{\partial^2 A(z,t)}{\partial t^2} + j\gamma|A(z,t)|^2 A(z,t) \tag{3.116}$$

Substituting $\tau = (t - \beta_1 z)/\tau_0$, $\xi = z/L_D (L_D = \tau^2_0/|\beta_2|)$, and $U = A/\sqrt{P_0}(P_0 = A^2_0)$, we get[2]

$$j\frac{\partial U}{\partial \xi} - \frac{\text{sgn}(\beta_2)}{2}\frac{\partial^2 U}{\partial \tau^2} + N^2|U|^2 U = 0, \quad N^2 = L_D/L_{NL} \tag{3.117}$$

When $N \ll 1$, the nonlinear length is much larger than the dispersion length so that the nonlinear effects can be neglected compared to those of chromatic dispersion. If chromatic dispersion can be neglected ($L_D \to \infty$), the NLSE can be solved analytically[2]:

$$U(z,\tau) = U(0,\tau)e^{jz|U(0,\tau)|^2/L_{NL}} \tag{3.118}$$

Interestingly, the SPM modulation causes a phase change but no change in the envelope of the pulse. Thus, SPM by itself leads only to chirping, regardless of the pulse shape. The SPM-induced chirp, however, modifies the pulse-broadening effects of chromatic dispersion.

3.4.5.2 Cross-Phase Modulation

XPM is another effect caused by the intensity dependence of the refractive index, but it is related to the multichannel transmission and occurs during propagation of a composite signal. The nonlinear phase shift of a specific optical channel is affected not only by the power of that channel but also by the optical power of the other channels[1]:

$$\Delta\Phi_m(t) = \gamma L_{\text{eff}}P_{0m}(t) + \sum_{i=1,i\neq m}^{M}P_{0i}(t) = \frac{2\pi}{\lambda}\frac{L_{\text{eff}}n_2}{A_{\text{eff}}}\left[P_{0m}(t) + \sum_{i=1,i\neq m}^{M}P_{0i}(t)\right] \tag{3.119}$$

where $P_{0m}(t)$ denotes the pulse shape in the mth channel, and the other parameters are the same as described previously. The nonlinear phase shift is bit-pattern dependent. In the worst-case scenario (assuming that all channels are transmitting ones and are loaded with the same power P_{0m}),

$$\Delta\Phi_m(t) = \frac{2\pi}{\lambda}\frac{L_{\text{eff}}n_2}{A_{\text{eff}}}(2M - 1)P_{0m}(t) \tag{3.120}$$

In practice, the optical pulses from different optical channels propagate at different speeds (they have different group velocities). The phase shift given previously can occur only during the overlapping time. The overlapping among neighboring channels is longer than the overlapping of channels spaced apart, and it will produce the most significant impact on the phase shift. If pulses walk through one another quickly (due to significant chromatic dispersion or when the channels are widely separated), the described effects on both pulses are diminished because the distortion done by the trailing edge is undone by the leading edge.

3.4.5.3 Four-Wave Mixing

FWM is another effect that occurs in optical fibers during the propagation of a composite optical signal, such as the WDM signal. It gives rise to the optical signals. The three optical signals with different carrier frequencies f_i, f_j, and f_k ($i, j, k = 1, \ldots, M$) interact to generate the new optical signal at frequency $f_{ijk} = f_i + f_j - f_k$, providing that the phase matching condition is satisfied[1-3]:

$$\beta_{ijk} = \beta_i + \beta_j - \beta_k \tag{3.121}$$

where βs are corresponding propagation constants. The measure of the phase-matching condition of wavelength channels involved in the interaction is defined by

$$\Delta\beta = \beta_i + \beta_j - \beta_k - \beta_{ijk} \tag{3.122}$$

The FWM is effective only if $\Delta\beta$ approaches zero. Therefore, the phase matching condition is a requirement for the momentum conservation. The FWM process can also be considered from the quantum-mechanic standpoint as the annihilation of two photons with energies hf_i and hf_j and the generation of two new photons with energies hf_k and hf_{ijk}. In the FWM process, the indices i and j do not necessarily need to be distinct, meaning that only two channels may interact to create the new one; this case is known as the degenerate case.

The power of the newly generated optical frequency is a function of the powers of optical signals involved in the process, the Kerr coefficient, and the degree of satisfaction of the phase matching condition, so we can write[1,3,5]

$$P_{ijk} \sim \left(\frac{2\pi f_{ijk} n_2 d_{ijk}}{3cA_{\text{eff}}}\right)^2 P_i P_j P_k L_{\text{eff}}^2 \tag{3.123}$$

where d_{ijk} is the measure of degeneracy, which takes value 3 in the degenerate case or value 6 in the nondegenerate case. FWM can produce significant signal degradation in WDM systems because several newly generated frequencies can coincide with any specific channel. Namely, the total number of newly generated frequencies N through the FWM process is $N = M^2(M - 1)/2$. Fortunately, some of the newly generated frequencies have negligible impact. However, some of them will have significant impact, especially those coinciding with already existing WDM channels.

FWM can be reduced either by reducing the power per channel or by preventing the perfect phase matching by increasing the chromatic dispersion or increasing the channel spacing. One option would be to use nonuniform channel spacing. On the other hand, the FWM process can be used to perform some useful functions, such as wavelength conversion through the process of phase conjugation.[3]

In the rest of this section, we provide another interpretation of Kerr nonlinearities.[2] Under the isotropy assumption, in the presence of nonlinearities, induced polarization $P(r, t)$ is along the same direction as the electric field $E(r, t)$ so that corresponding vector notation can be substituted by the scalar functions, and the induced polarization can be written in the form[2]

$$P(r,t) = \varepsilon_0 \int_{-\infty}^{t} \chi^{(1)}(t - t')E(r,t')dt' + \varepsilon_0 \chi^{(3)} E^3(r,t) \tag{3.124}$$

where the first term corresponds to the linear part, and the second term, $P_{NL}(r, t)$, corresponds to the nonlinear part. $\chi^{(1)}$ denotes the linear susceptibility, $\chi^{(3)}$ denotes the third-order susceptibility, and ε_0 is the permittivity of vacuum. (Because of the symmetry of silica molecules, only odd terms are present in Taylor expansion of $P(r, t)$.) A WDM signal can be represented as the sum of n monochromatic plane waves of angular frequency ω_i (the modulation process is ignored in this simplified analysis):

$$E(r,t) = \sum_{i=1}^{M} E_i \cos(\omega_i t - \beta_i z) \tag{3.125}$$

The nonlinear polarization can be written after substitution of Eq. (3.125) into the second term of Eq. (3.124)[2]:

$$
\begin{aligned}
P_{NL}(r,t) &= \varepsilon_0 \chi^{(3)} \sum_{i=1}^{M} \sum_{j=1}^{M} \sum_{k=1}^{M} E_i \cos(\omega_i t - \beta_i z) E_j \cos(\omega_j t - \beta_j z) E_k \cos(\omega_k t - \beta_k z) \\
&= \frac{3}{4} \varepsilon_0 \chi^{(3)} \sum_{i=1}^{M} E_i^2 E_i \cos(\omega_i t - \beta_i z) + \frac{3}{4} \varepsilon_0 \chi^{(3)} \sum_{i=1}^{M} \sum_{j \neq i} 2 E_i E_j E_i \cos(\omega_i t - \beta_k z) \\
&\quad + \frac{3}{4} \varepsilon_0 \chi^{(3)} \sum_{i=1}^{M} \sum_{j \neq i} E_i^2 E_j \cos\left[(2\omega_i - \omega_j)t - (2\beta_i - \beta_j)z\right] \\
&\quad + \frac{6}{4} \varepsilon_0 \chi^{(3)} \sum_{i=1}^{M} \sum_{j > i} \sum_{k > j} E_i E_j E_k \cos\left[(\omega_i + \omega_j - \omega_k)t - (\beta_i + \beta_j - \beta_k)z\right] + \ldots
\end{aligned}
\tag{3.126}
$$

The first term in the second line of Eq. (3.126) corresponds to the SPM, the second term in the same line corresponds to the XPM, the first term the in third line corresponds to the degenerate FWM case, and the last term corresponds to the nondegenerate FWM.

3.4.5.4 Stimulated Raman Scattering

Raman scattering is a nonlinear effect that occurs when a propagating optical signal interacts with glass molecules in the fiber undergoing a wavelength shift[1–3]. The result of interaction is a transfer of energy from some photons of the input optical signal to vibrating silica molecules and creation of new photons with lower energy than the energy of incident photons. The incident optical signal is commonly referred to as a pump, and the generated signal is called the *Stokes signal*. The difference in frequencies between the pump (ω_p) and the Stokes signal (ω_s) is known as the *Raman frequency shift* $\omega_R = \omega_p - \omega_s$. Scattered photons are not in phase with each other and do not follow the same scattering pattern, meaning that energy transfer from the pump to Stokes photons is not a uniform process. As a result, there exists the frequency band that includes frequencies of all scattered Stokes photons. Scattered Stokes photons can take any direction that can be either forward or backward with respect to the direction of the pump, meaning that Raman scattering is an isotropic process. If the pump power is lower than a certain threshold value, the Raman scattering process will have a spontaneous quality, characterized by the small number of pump photons that will be scattered and converted to the Stokes photons. However, when the pump power exceeds the threshold value, Raman scattering becomes a stimulated process. SRS can be explained as a positive feedback process in the pump signal that interacts with the Stokes signal and creates the beat frequency $\omega_R = \omega_p - \omega_s$. The beat frequency then serves as a stimulator of molecular oscillations, and the process is enhanced or amplified. Assuming that the Stokes signal propagates in the same direction as the pump signal, the intensity of the pump signal $I_p = P_p/A_{eff}$ and the intensity of the Stokes signal $I_s = P_s/A_{eff}$ are related by the following systems of coupled differential equations[1,3]:

$$\frac{dI_p}{dz} = -g_R\left(\frac{\omega_p}{\omega_s}\right)I_pI_s - \alpha_pI_p \qquad \frac{dI_s}{dz} = g_RI_pI_s - \alpha_sI_s \qquad (3.127)$$

where α_p (α_s) is the fiber attenuation coefficient of the pump (Stokes) signal, and g_R is the Raman gain coefficient. Because the SRS process is not uniform, the scattered Stokes photons will occupy a certain frequency band, and the Raman gain is not constant but, rather, the function of frequency. It can roughly be approximated by the Lorentzian spectral profile[1]:

$$g_R(\omega_R) = \frac{g_R(\Omega_R)}{1 + (\omega_R - \Omega_R)^2 T_R^2} \qquad (3.128)$$

where T_R is vibration states decay time (on the order of 0.1 ps for silica-based materials), and Ω_R is the Raman frequency shift corresponding to the Raman gain peak. The actual gain profile extends over 40 THz ($\sim$320 nm), with a peak at approximately 13.2 THz. The Raman gain peak $g_R(\Omega_R) = g_{R,max}$ is between 10^{-12} and 10^{-13} m/W for wavelengths above 1300 nm. The gain profile can also be approximated by a triangle function, which is commonly used in analytical studies of SRS[1,3]:

$$g_R(\omega_R) = \frac{g_R(\Omega_R)\omega_R}{\Omega_R} \qquad (3.129)$$

The *Raman threshold*, defined as the incident power at which half of the pump power is converted to the Stokes signal, can be estimated using the system of differential equations by replacing g_R with $g_{R,max}$. The amplification of Stokes power along distance L can be determined by

$$P_s(L) = P_{s,0} \exp\left(\frac{g_{R,max}P_{s,0}L}{2A_{eff}}\right) \qquad (3.130)$$

where $P_{s,0} = P_s(0)$. The Raman threshold can be determined by[1-3]

$$P_{R,threshold} = P_{s,0} \approx \frac{16A_{eff}}{g_{R,max}L_{eff}} \qquad (3.131)$$

and for typical SMF parameters ($A_{eff} = 50 \ \mu m^2$, $L_{eff} = 20$ km, and $g_{R,max} = 7 \times 10^{-13}$ m/W), it is approximately 500 mW.

The SRS can effectively be used for optical signal amplification (Raman amplifiers), as previously explained. In WDM systems, however, the SRS effect can be quite detrimental because the Raman gain spectrum is very broad, which enables the energy transfer from lower to higher wavelength channels. The shortest wavelength channel within the WDM system acts as a pump for several long wavelength channels while undergoing the most intense depletion. The fraction of power coupled out of channel 0 to all the other channels $(1, 2, \ldots, M - 1$, where M is the number of wavelengths), when the Raman gain shape is approximated as a triangle, is[2]

$$P_0 = \sum_{i=1}^{M-1} P_0(i) = \frac{g_{R,\max} \Delta \lambda P L_{\text{eff}}}{2 \Delta \lambda_c A_{\text{eff}}} \frac{M(M - 1)}{2} \qquad (3.132)$$

where $\Delta \lambda$ is the channel spacing, and $\Delta \lambda_c \cong 125$ nm. To keep the power penalty, $-10\log(1 - P_0)$, below 0.5 dB, the following inequality must be satisfied[2]:

$$PM(M - 1)\Delta \lambda L_{\text{eff}} < 40{,}000 \ \text{mW-nm-km} \qquad (3.133)$$

With chromatic dispersion present, the previous inequality can be relaxed to 80,000 mW-nm-km. For example, in 32-wavelength channels spaced 0.8 nm apart and $L_{\text{eff}} = 20$ km, $P \leq 2.5$ mW. The energy transfer between two channels is bit pattern dependent and occurs only if both wavelengths are synchronously loaded with 1 bit, meaning that the energy transfer will be reduced if dispersion is higher due to the walk-off effect (the difference in velocities of different wavelength channels will reduce the time of overlapping).

3.4.5.5 Stimulated Brillouin Scattering

Brillouin scattering is a physical process that occurs when an optical signal interacts with acoustical phonons rather than the glass molecules. During this process, an incident optical signal reflects backward from the grating formed by acoustic vibrations and downshifts in frequency. The acoustic vibrations originate from the thermal effect if the power of an incident optical signal is relatively small. If the power of incident light increases, it increases the material density through the electrostrictive effect. The change in density enhances acoustic vibrations and forces Brillouin scattering to become stimulated. The SBS process can also be explained as a positive feedback mechanism in which the incident light (the pump) interacts with the Stokes signal and creates the beat frequency $\omega_B = \omega_p - \omega_s$. The parametric interaction between pump, Stokes signal, and acoustical waves requires both the energy and momentum conservation: The energy is effectively preserved through the downshift in frequency, whereas momentum conservation occurs through the backward direction of the Stokes signal. The Brillouin shift (the frequency downshift) can be determined by[1]

$$f_B = \frac{2n v_A}{\lambda_p} \qquad (3.134)$$

where v_A is the velocity of the acoustic wave. The frequency shift is fiber material dependent and can vary from 10.5 to 12 GHz. The SBS is governed by the following set of coupled equations, similarly as in SRS[1,3]:

$$\frac{dI_p}{dz} = -g_B I_p I_s - \alpha_p I_p \qquad -\frac{dI_s}{dz} = g_B I_p I_s - \alpha_s I_s \qquad (3.135)$$

where I_p and I_s are the intensities of the pump and Stokes waves, respectively, g_B is the Brillouin gain coefficient, and $\alpha_p(\alpha_s)$ is the attenuation coefficient of the pump (Stokes) signal. The scattered Stokes photons will not have equal frequencies but will be dispersed within a certain frequency band. The spectrum of Brillouin gain is related to the acoustic phonons lifetime (characterized by the time constant T_B) and can be approximated by a Lorentzian spectral profile[1]:

$$g_B(\omega_B) = \frac{g_B(\Omega_B)}{1 + (\omega_B - \Omega_B)^2 T_B^2} \qquad (3.136)$$

where $\Omega_B = 2\pi f_B$ is the Brillouin shift. The Brillouin gain is also dependent on the fiber waveguide characteristics. The SBS gain bandwidth Δf_B is approximately 17 MHz at $\lambda_p = 1520$ nm for pure silica, but it is almost 100 MHz in doped silica fibers, and the typical bandwidth is approximately 50 MHz. The maximum value of the gain $g_{B,max}$ is also dependent on the type of fiber material, and it is between 10^{-10} and 10^{-11} m/W for silica-based optical fiber above 1 μm. The threshold pump power at which the Brillouin scattering becomes stimulated can be calculated in a similar manner as done for SRS[1]:

$$P_{B,threshold} \approx \frac{21 A_{eff}}{g_{B,max} L_{eff}} \qquad (3.137)$$

and the estimated Brillouin threshold is approximately 1.05 mW for typical values of fiber parameters ($A_{eff} = 50$ μm^2, $L_{eff} = 20$ km, and $g_{B,max} = 5 \times 10^{-11}$ m/W). The preceding expression assumes that the pump signal has a negligible linewidth and lies within the gain bandwidth of SBS. The threshold power is considerably increased if the signal has a broad linewidth, and thus much of the pump power lies outside of the gain bandwidth of SBS[2]:

$$P_{B,threshold} \approx \frac{21 A_{eff}}{g_{B,max} L_{eff}} \left(1 + \frac{\Delta f_s}{\Delta f_B}\right) \qquad (3.138)$$

where Δf_s is the spectral width of the source. For example, when the source bandwidth is 10 times larger than the SBS gain bandwidth, the SBS threshold increases to 11.55 mW. The SBS suppression can be achieved by dithering the source (directly modulating the laser with a sinusoid at a frequency much lower than the receiver low-frequency cutoff).

3.5 Transmission System Performance Assessment and System Design

The transmission quality can be estimated by using different figures of merit, such as SNR, optical SNR (OSNR), Q-factor, eye opening penalty, and BER. The SNR at decision point is defined as the ratio between the signal power and the noise power. The optical amplification process is accompanied by ASE noise that accumulates along the transmission line and degrades the OSNR. Therefore, the OSNR can be defined as the ratio of the received optical power and accumulated noise power. It is a common practice in optical communications to use the Q-factor as a figure of merit instead of SNR. It is defined as

$$Q = \frac{I_1 - I_0}{\sigma_1 + \sigma_0} \tag{3.139}$$

where I_1 and I_0 represent the average photocurrents corresponding to one- and zero-symbol levels, and σ_1 and σ_0 are corresponding standard deviations. The total variance in the photocurrents corresponding to symbol 1/0 is[1-3]

$$\sigma_{1,0}^2 = \left\langle i_{sn,1,0}^2 \right\rangle + \left\langle i_{thermal,1,0}^2 \right\rangle \tag{3.140}$$

where $\left\langle i_{sn,1,0}^2 \right\rangle$ is the variance corresponding to the shot noise, and $\left\langle i_{thermal,1,0}^2 \right\rangle$ is the variance corresponding to the thermal noise.

BER is an important parameter in digital transmission quality assessment. It defines the probability that the bit being transmitted will be mistaken by the decision circuit. The fluctuating signal levels corresponding to 0 and 1 bits can be characterized by corresponding PDFs. If at decision point the received sample is larger than the threshold ($I > I_{tsh}$), we decide in favor of bit 1; otherwise we decide in favor of bit 0. An error occurs if a bit 1 was transmitted but we decided in favor of 0 or a bit 0 was transmitted but we decided in favor of 1.

The BER is defined by

$$\text{BER} = \Pr(0|1)\Pr(1) + \Pr(1|0)\Pr(0) \tag{3.141}$$

where $\Pr(0|1)$ is the conditional probability that symbol 1 was transmitted by the decision circuit decided in favor of 1, $\Pr(1|0)$ is the conditional probability that symbol 0 was transmitted by the decision circuit decided in favor of 1, and $\Pr(0)$ and $\Pr(1)$ are *a priori* probabilities of symbols 0 and 1, respectively. Assuming an equiprobable transmission, $\Pr(0) = \Pr(1) = 1/2$, we obtain $\text{BER} = 0.5(\Pr(0|1) + \Pr(1|0))$. The PDFs for symbols 0 and 1 are commonly considered as Gaussian,[1,3,5,7-9] so conditional error probabilities can be determined by

$$\text{Pr}(0|1) = \frac{1}{\sigma 1 \sqrt{2\pi}} \int_{-\infty}^{I_{\text{tsh}}} \exp\left[-\frac{(I-I_1)^2}{2\sigma_1^2}\right] dI = \frac{1}{2} \text{erfc}\left(\frac{I_1 - I_{\text{tsh}}}{\sigma_1 \sqrt{2}}\right)$$

$$\text{Pr}(1|0) = \frac{1}{\sigma_0 \sqrt{2\pi}} \int_{I_{\text{tsh}}}^{\infty} \exp\left[-\frac{(I-I_0)^2}{2\sigma_0^2}\right] dI = \frac{1}{2} \text{erfc}\left(\frac{I_{\text{tsh}} - I_0}{\sigma_0 \sqrt{2}}\right)$$

(3.142)

where the complementary error function erfc(x) is defined by

$$\text{erfc}(x) = \frac{2}{\sqrt{\pi}} \int_x^{+\infty} e^{-z^2} dz$$

The optimum decision, minimizing the BER, is the maximum *a posteriori* probability (MAP) decision rule, which can be formulated as follows:

$$p(I|0)\text{Pr}(0) \underset{H_1}{\overset{H_0}{\gtrless}} p(I|1)\text{Pr}(1), \quad p(I|i) = \frac{1}{\sigma_i \sqrt{2\pi}} \exp\left[-\frac{(I-I_i)^2}{2\sigma_i^2}\right], \quad i \in \{0,1\}$$

(3.143)

where we used $H_i(I = 0, 1)$ to denote the hypothesis that the symbol i was transmitted, and $p(I|i)(i = 0, 1)$ is the corresponding PDF. For the equiprobable transmission ($\text{Pr}(0) = \text{Pr}(1) = 1/2$), the MAP rule becomes the maximum likelihood (ML) decision rule ($\text{Pr}(0) = \text{Pr}(1) = 1/2$):

$$p(I|0) \underset{H_1}{\overset{H_0}{\gtrless}} p(I|1)$$

(3.144)

The resulting BER can be obtained by substituting Eq. (3.142) into Eq. (3.141):

$$\text{BER} = \frac{1}{4}\left[\text{erfc}\left(\frac{I_1 - I_{\text{tsh}}}{\sigma_1 \sqrt{2}}\right) + \text{erfc}\left(\frac{I_{\text{tsh}} - I_0}{\sigma_0 \sqrt{2}}\right)\right]$$

(3.145)

The optimum threshold can be obtained by minimizing the BER, which yields the following equation to be solved numerically:

$$\ln\left[\frac{\text{Pr}(0)}{\text{Pr}(1)}\right] + \ln\left(\frac{\sigma_1}{\sigma_0}\right) = \frac{(I_1 - I_{\text{tsh}})^2}{2\sigma_1^2} - \frac{(I_{\text{tsh}} - I_0)^2}{2\sigma_0^2}$$

(3.146)

When the standard deviations are close to each other ($\sigma_1 \approx \sigma_0$), the following approximation is valid:

$$I_{\text{tsh}} \approx \frac{\sigma_0 I_1 + \sigma_1 I_0}{\sigma_0 + \sigma_1}$$

(3.147)

By substituting Eq. (3.147) into Eq. (3.145), we obtain the following relation:

$$\text{BER} = \frac{1}{2} \text{erfc}\left(\frac{Q}{\sqrt{2}}\right), \quad Q = \frac{I_1 - I_0}{\sigma_1 + \sigma_0}$$

(3.148)

establishing the connection between the BER and Q-factor introduced in Eq. (3.139). For example, $Q = 7$ (16.9 dB) corresponds to a BER of 10^{-12}, and $Q = 8$ (18.06 dB) corresponds to a BER of 10^{-15}.

Another important receiver parameter is receiver sensitivity. *Receiver sensitivity* is defined as the minimum average received optical power $P_R = (P_0 + P_1)/2$ (P_i is the average power corresponding to symbol i) required to operate at a given BER (e.g., 10^{-12} or 10^{-15}). By including the impact of the basic noise components (shot, thermal, and beat noise components) into three basic real case detection scenarios, we can establish the new reference point. All other impairments degrade the receiver sensitivity by increasing the value of the optical power P_R required to achieve a desired BER. Accordingly, each individual impairment causes the receiver sensitivity degradation, which is evaluated by the power penalty. The power penalty, denoted as ΔP, is defined as follows:

$$\Delta P[\text{dB}] = 10\log_{10} \frac{P_R(\text{in the presence of impairment, at given BER})}{P_R(\text{in the absence of impairment, at given BER})} \qquad (3.149)$$

3.5.1 Quantum Limit for Photodetection

The quantum limit is defined for an ideal photodiode (no thermal noise, no dark current, no shot noise, and 100% quantum efficiency) and can be used as the reference point to compare different receiver architectures. In the quantum limit case, the Gaussian assumption is no longer valid, and Poisson statistics should be used instead. Assuming that N_p is the average number of photons in each 1 bit (the number of photons in a 0 bit is zero), the probability for the formation of m e-h pairs is given by Poisson distribution:

$$P_m = \exp(-N_p)N_p^m/m!$$

Using our previous definitions, $\Pr(1|0) = 0$ because no e-h pairs are formed when $N_p = 0$, whereas $\Pr(0|1) = P_0 = \exp(-N_p)$. Therefore, BER for the quantum limit is as follows:

$$\text{BER} = \frac{1}{2}\exp(-N_p) \qquad (3.150)$$

The receiver sensitivity is related to the average number of photons extended over the stream of 0's and 1's $<N_p>$ by

$$P_R = \frac{P_0 + P_1}{2} = \frac{N_p h\nu}{2T} = \langle N_p \rangle R_b h\nu \qquad (3.151)$$

where $h\nu$ is the photon energy, T is the bit duration, and R_b is the bit rate. For example, $N_p = 34$ ($<N_p> = 17$) for a BER of 10^{-15}, whereas the corresponding receiver sensitivity is $P_R = -46.61$ dBm at $R_b = 10$ Gb/s and $P_R = -40.59$ dBm at $R_b = 40$ Gb/s.

3.5.2 Shot Noise and Thermal Noise Limit

The impact of shot noise and thermal noise is unavoidable in any optical receiver, and as such it can be used as the referent case. Let's assume that $I_0 = 0(P_0 = 0$, the infinite extinction ratio); then $P_1 = 2P_R$ and $I_1 = 2\langle M \rangle RP_R$ for APDs. For bit 1, the variance of noise includes both thermal and shot noise terms, whereas for bit 0 only the thermal noise term is included. In the case of APDs, the corresponding variances can be written as[1–3]

$$\sigma_1^2 = \sigma_{sn}^2 + \sigma_{thermal}^2 \quad \sigma_0^2 = \sigma_{thermal}^2, \quad \sigma_{sn}^2 = \langle i_{sn}^2 \rangle = 4qM^2F(M)RP_RB_{el} \quad \sigma_{thermal}^2 = \langle i_{thermal}^2 \rangle = \frac{4k_B\Theta F_{ne}}{R_L}B_{el}$$

$$(3.152)$$

where $F(M)$ is the excess noise factor introduced by Eq. (3.74), M is the APD multiplication factor, R is the photodiode responsivity, k_B is the Boltzmann's constant, Θ is the absolute temperature, F_{ne} is the noise figure of postamplifier (see Figure 3.2), and B_{el} is the electrical filter bandwidth. The corresponding expressions for PIN photodetectors can be obtained by setting $M = 1$ and $F(M) = 1$. The receiver sensitivity can be obtained from the Q-factor

$$Q = \frac{I_1}{\sigma_1 + \sigma_0} = \frac{MR(2P_R)}{\left(2qM^2F(M)R(2P_R)B_{el} + \sigma_{thermal}^2\right)^{1/2} + \sigma_{thermal}} \tag{3.153}$$

as follows

$$P_R = \frac{Q_{req}}{R}\left(qF(M)Q_{req}B_{el} + \frac{\sigma_{thermal}}{M}\right) \tag{3.154}$$

The optimum multiplication factor is obtained by solving for $M = M_{opt}$ from $dP_R/dM = 0$ by

$$M_{opt} = \sqrt{\frac{\frac{\sigma_{thermal}}{qQB_{el}} + k_N - 1}{k_N}} \approx \sqrt{\frac{\sigma_{thermal}}{k_N qQB_{el}}} \tag{3.155}$$

where k_N is the ratio of impact ionization factors of holes and electrons, respectively. By substituting Eq. (3.155) into Eq. (3.154), we obtain

$$P_{R,APD} = \frac{2qB_{el}}{R}Q^2(k_N M_{opt} + 1 - k_N) \tag{3.156}$$

The receiver sensitivity for the thermal noise limit is obtained by setting the shot noise contributor in Eq. (3.154) to zero, and this case corresponds to the PIN photodiode:

$$P_{R,PIN} \approx \frac{\sigma_{thermal}Q}{R} = \frac{Q\sqrt{4k\Theta F_{ne}B_{el}}}{R\sqrt{R_L}} \tag{3.157}$$

3.5.3 Receiver Sensitivity for Receivers with an Optical Preamplifier

The receiver sensitivity can be greatly improved by using the optical preamplifier in front of the PIN photodiode. Following a similar procedure to that in Section 3.5.2 and using Eq. (3.84) from Section 3.3.9, the following expression for the receiver sensitivity is obtained[1,3] (assuming again that $I_0 = 0$):

$$P_R = \frac{2S_{sp}B_{el}\left[Q^2 - Q\sqrt{B_{op}/B_{el}}\right]}{G-1} = F_{no}hfB_{el}\left[Q^2 - Q\sqrt{B_{op}/B_{el}}\right], \quad S_{sp} = F_{no}hf(G-1)/2 \quad (3.158)$$

where S_{sp} is the PSD of spontaneous emission noise, G is the gain of the optical preamplifier, F_{no} is the corresponding noise figure, hf is photon energy, B_{op} is the optical filter bandwidth, and B_{el} is the electrical filter bandwidth.

3.5.4 Optical Signal-to-Noise Ratio

The optical amplification process is accompanied by ASE noise that accumulates along the transmission line, degrading the OSNR, defined by

$$OSNR = \frac{P_s}{P_{ASE}} = \frac{P_s}{S_{sp}B_{op}} = \frac{P_s}{2n_{sp}hf(G-1)B_{op}} \quad (3.159)$$

where P_s is the signal power, P_{ASE} is the noise power, and the other parameters are as previously defined. The OSNR can be measured at the receiver entrance point (just before the photodetector) and as such can be related to the Q-factor as follows[1,2]:

$$Q = \frac{2OSNR\sqrt{B_{op}/B_{el}}}{1 + \sqrt{1 + 4OSNR}} \quad (3.160)$$

where B_{op} and B_{el} are optical and electrical filter bandwidths, respectively. Notice that this expression is valid only for the NRZ modulation format.

3.5.5 Power Penalty Due to Extinction Ratio

Previously, we considered different detection scenarios by including the basic receiver noise components. All other factors degrading the system performance, both noise components and signal impairments, can be characterized by introducing the power penalty (Eq. 3.149), which is the increase in receiver sensitivity required to keep the same quality of transmission.

The power penalty (PP) due to the extinction ratio (the extinction ratio is defined as $r_{ex} = P_0/P_1$), denoted as ΔP_{ex}, for the thermal noise-dominated scenario can be obtained by using the PP definition expression (Eq. 3.149) and receiver sensitivity expression as follows:

$$\Delta P_{ex}[dB] = 10\log_{10}\frac{P_R(r_{ex} \neq 0)}{P_R(r_{ex} = 0)} = 10\log_{10}\frac{1 + r_{ex}}{1 - r_{ex}} \quad (3.161)$$

The power penalty due to the extinction ratio for the shot noise-dominated scenario can be obtained by repeating a similar procedure as given previously but employing Eq. (3.156) instead to get

$$\Delta P_{ex}[dB] = 10\log_{10}\frac{1+r_{ex}}{\left(1-\sqrt{r_{ex}}\right)^2} \tag{3.162}$$

3.5.6 Power Penalty Due to Intensity Noise

The intensity noise is related to the fluctuations of the incoming optical power, mostly due to the light source. The power penalty due to intensity noise can be obtained by first calculating the Q-factors in the presence and absence of intensity noise, then calculating the corresponding receiver sensitivities, and, finally, calculating the power penalty as the ratio of receiver sensitivities in the presence and absence of intensity noise to get

$$\Delta P_{int}[dB] = 10\log_{10}\frac{P_R(r_{int} \neq 0)}{P_R(r_{int} = 0)} = -10\log_{10}(1 - r_{int}^2 Q^2) \tag{3.163}$$

where r_{int} is the intensity noise parameter defined as $\sqrt{(2RIN_{laser}\Delta f)}$, where RIN_{laser} is the average value of RIN spectrum (see Eq. 3.66), which is typically below -160 dBm/Hz ($r_{int} \sim 0.0004$) for high-quality lasers; and Δf is the receiver bandwidth. The other sources of intensity noise include reflections (characterized by intensity reflection noise parameter r_{ref}), mode partition noise (MPN) (characterized by r_{MPN}), and phase noise-to-intensity noise conversion (characterized by r_{phase}), so the effective intensity noise parameter can be determined by[1]

$$r_{eff}^2 = r_{int}^2 + r_{ref}^2 + r_{MPN}^2 + r_{phase}^2, \quad r_{ref} \approx \frac{(r_1 r_2)^{1/2}}{\alpha_{iso}}, \quad r_{MPN} \approx (k/\sqrt{2})\left\{1 - \exp\left[-(\pi R_b DL\sigma_\lambda)^2\right]\right\}$$
$$\tag{3.164}$$

where r_1 and r_2 are reflection coefficients of two disjoints, α_{iso} is the attenuation of the optical isolator, R_b is the bit rate, D is the chromatic dispersion coefficient, L is the transmission distance, σ_λ is the light source spectral linewidth, and k is the MPN coefficient ($k = 0.6$–0.8).

3.5.7 Power Penalty Due to Timing Jitter

Timing is typically determined by the clock recovery circuit. Because the input to this circuit is noisy, the sampling time fluctuates from bit to bit, and this fluctuation is commonly referred to as *timing skew* or *timing jitter*. Timing skew occurs when the sampling time error is fixed (e.g., offsets in circuits or insufficient dc gain in PLL). Timing jitter occurs when the sampling time error is dynamic (e.g., interplay of optical amplifier noise, GVD, and fiber nonlinearities, or due to clock recovery circuit). By assuming that the PIN photodiode is used, the thermal noise-dominated scenario is being observed, and the extinction ratio is ideal, the following power penalty expression is obtained:

$$\Delta P_{\text{jitter}}[\text{dB}] = 10\log_{10}\frac{P_{\text{R}}(b \neq 0)}{P_{\text{R}}(b = 0)} = -5\log_{10}\left[(1-b)^2 - 2b^2 Q^2\right], \quad b = \left(\frac{\pi R_{\text{b}}\langle \Delta t^2\rangle^{1/2}}{2}\right)^2 \quad (3.165)$$

where R_{b} is the bit rate, and Δt is the timing error.

3.5.8 Power Penalty Due to GVD

The impact of GVD is difficult to evaluate because of the nonlinear relationship between transmitted and received intensities, and because of the presence of fiber nonlinearities. Under assumptions that the noise and variation in intensity due to ISI can be modeled using Gaussian approximation (and neglecting the extinction ratio), the power penalty due to GVD for the thermal noise-dominated scenario can be evaluated as follows[12,13]:

$$\Delta P_{\text{GVD}}[\text{dB}] = 10\log_{10}\frac{1 + \langle \Delta I_1\rangle - \langle \Delta I_0\rangle}{\sqrt{\left[(1 + \langle \Delta I_1\rangle - \langle \Delta I_0\rangle)^2 - Q^2\left(\sigma_{I_0}^2 + \sigma_{I_1}^2\right)\right]^2 - 4Q^2\sigma_{I_0}^2\sigma_{I_1}^2}} \quad (3.166)$$

where $\langle \Delta I_1\rangle$ and $\langle \Delta I_0\rangle$ are the average values of variations in intensities for bit 1 and bit 0 (normalized with I_1, the intensity of bit 1 in the absence of GVD), respectively, whereas σ_{I_1} and σ_{I_0} denote corresponding standard deviations.

3.5.9 Power Penalty Due to Signal Crosstalk

The crosstalk noise, introduced in Section 3.3.10, is related to multichannel systems (WDM systems) and can be either out-of-band or in-band in nature. Following a similar procedure as used previously, the power penalty expression due to signal crosstalk can be found as follows[1]:

$$\Delta P_{\text{cross}} = -10\log_{10}\left(1 - r_{\text{cross}}^2 Q^2\right), \quad r_{\text{cross}} = \sqrt{r_{\text{cross,out}}^2 + r_{\text{cross,in}}^2} \quad (3.167)$$

where

$$r_{\text{cross,out}}^2 = \left(\sum_{n=1,n\neq m}^{M} X_{n,\text{out}}\right)^2$$

and

$$r_{\text{cross,in}}^2 = 2\left(\sum_{n=1,n\neq m}^{M} \sqrt{X_{n,\text{in}}}\right)^2$$

$X_{n,\text{out}}$ denotes the out-of-band crosstalk captured by the observed channel, $X_{n,\text{in}}$ denotes the in-band crosstalk originating from the nth channel, and M denotes the number of wavelength channels.

3.5.10 Accumulation Effects

A typical optical transmission system contains the amplifiers that are spaced l km apart, and the span loss between two amplifiers is $\exp(-\alpha l)$, where α is the fiber attenuation coefficient. Each amplifier generates spontaneous emission noise in addition to providing the signal amplification. Both the signal and the noise propagate together toward the next amplifier stage to be amplified. When dealing with long distances, the buildup of amplifier noise is the most critical factor in achieving prespecified performance. The total ASE noise will be accumulated and increased after each amplifier stage, thus contributing to SNR degradation. Moreover, the ASE noise power accumulation will contribute to the saturation of optical amplifiers, thereby reducing the amplifier gain. The optical amplifier gain is usually adjusted to compensate for the span loss. The spatial steady-state condition is achieved when the amplifier output power and gain remain the same from stage to stage[2]:

$$P_{o,\text{sat}}e^{-\alpha l}G_{\text{sat}} + P_{\text{sp}} = P_{\text{out}}, \quad P_{\text{sp}} = 2S_{\text{sp}}B_{\text{op}} = F_{\text{no}}hv(G_{\text{sat}} - 1)B_{\text{op}} \tag{3.168}$$

where $P_{o,\text{sat}}$ is the optical amplifier saturation power, P_{sp} is the spontaneous emission factor, B_{op} is the optical filter bandwidth, F_{no} is the optical amplifier noise figure, hv is the photon energy, and G_{sat} is the saturation gain determined by $G_{\text{sat}} = 1 + (P_{\text{sat}}/P_{\text{in}})\ln(G_{\text{max}}/G_{\text{sat}})$ (G_{max} is the maximum gain, and P_{in} is the amplifier input). The total noise power at the end of the transmission line is related to the steady-state condition[2]:

$$NP_{\text{sp}} = NF_{\text{no}}hv(G_{\text{sat}} - 1)B_{\text{op}} = (L/l)F_{\text{no}}hv(e^{\alpha l} - 1)B_{\text{op}} \tag{3.169}$$

where N is the number of optical amplifiers, and L is the total transmission length.

The OSNR calculated per channel basis, at the end of the amplifier chain, can be obtained as follows[1]:

$$\text{OSNR} = \frac{P_{s,\text{sat}}/M - F_{\text{no}}hv(e^{\alpha l} - 1)B_{\text{op}}L/l}{F_{\text{no}}hv(e^{\alpha l} - 1)B_{\text{op}}L/l} = \frac{P_{\text{ch}} - F_{\text{no}}hv(e^{\alpha l} - 1)B_{\text{op}}N}{F_{\text{no}}hv(e^{\alpha l} - 1)B_{\text{op}}N} \tag{3.170}$$

where P_{ch} is the required launch power to obtain a given OSNR. Equation (3.170) can be used to design a system satisfying a given BER (or OSNR). The minimum launched power to achieve the prespecified OSNR is as follows:

$$P_{\text{ch}} \geq (1 + \text{OSNR})\left[F_{\text{no}}hv(e^{\alpha l} - 1)B_{\text{op}}N\right] \approx \text{OSNR}\left[F_{\text{no}}hv(e^{\alpha l} - 1)B_{\text{op}}N\right] \tag{3.171}$$

The minimum launched power expressed in decibel scale obtained from Eq. (3.171) is:

$$P_{\text{ch}}[\text{dB}] \geq \text{OSNR}[\text{dB}] + F_{\text{no}}[\text{dB}] + 10\log_{10}N + 10\log_{10}(hvB_{\text{op}}) \tag{3.172}$$

Using Eq. (3.160), Eq. (3.172) can be written in the following form[1]:

$$\text{OSNR}[\text{dB}] \approx 10\log_{10}\left(\frac{Q^2 B_{\text{el}}}{B_{\text{op}}}\right) \geq P_{\text{ch}}[\text{dB}] - F_{\text{no}}[\text{dB}] - \alpha l - 10\log_{10}N - 10\log_{10}(hvB_{\text{op}}) \tag{3.173}$$

OSNR can be increased by increasing the output optical power, decreasing the amplifier noise figure, and decreasing the optical loss per fiber. However, the number of amplifiers can be decreased only when the span length is increased (meaning that the total signal attenuation will be increased). The transmission system design is based on prespecified transmission system quality. The following are the basic steps involved in the design process[1]: (1) Identify system requirements in terms of transmission distance, bit rate, and BER; (2) identify system elements; (3) set up major system parameters (output signal power, amplifier spacing, etc.) that satisfy initial requirements; (4) allocate the margins to account for penalties; and (5) repeat the design process in order to perform fine-tuning of parameters and identify possible trade-offs. In the following section, the design process is described for several systems of interest: (1) power budget-limited point-to-point lightwave system, (2) bandwidth-limited system, and (3) high-speed optical transmission system.

3.5.11 Systems Design

When the optical amplifiers are not employed, the power budget is expressed as follows:

$$P_{out} - \alpha L - \alpha_c - \Delta P_M \geq P_R(Q, B_{el}) \tag{3.174}$$

where P_{out} is the output power from the light source pigtail, α is the fiber attenuation coefficient, ΔP_M is the system margin to be allocated for different system imperfections, and P_R is the receiver sensitivity that is the function of required Q-factor and electrical filter bandwidth. All parameters from Eq. (3.174) are expressed in decibel scale.

The performance of an optical transmission system can be limited due to limited frequency bandwidth of some of the key components that are used, such as light source, photodetector, or optical fiber. Available fiber bandwidth at length L, for different fiber types and bit rate R_b, can be summarized as follows[1]:

$$\begin{aligned}
R_b L^\mu &\leq B_{fiber}, & &\text{for MMFs} \\
R_b L &\leq (4D\sigma_\lambda)^{-1}, & &\text{for SMFs and large source linewidth} \\
R_b^2 L &\leq (16|\beta_2|)^{-1}, & &\text{for SMFs and narrow source linewidth}
\end{aligned} \tag{3.175}$$

The maximum transmission system length to the power budget can be expressed as follows[1]:

$$L(\lambda, B) \leq \frac{P_{out}(\lambda, B) - [P_R(\lambda, B) + \alpha_c + \Delta P_M]}{\alpha(\lambda)} \tag{3.176}$$

The bandwidth limitation is related to the pulse rise times occurring in individual modules that can be expressed using the following expression[1,3]:

$$T_r^2 \geq T_{T_x}^2 + T_{fiber}^2 + T_{R_x}^2 \tag{3.177}$$

where T_r is the overall response time of the system, T_{Tx} is the rise time of the transmitter, T_{fiber} is the rise time of the fiber, and T_{Rx} is the rise time of the receiver. The response time and 3 dB system bandwidth are related by[3] $T_r = 0.35/B_{el}$, where $B_{el} = R_b$ for RZ and

$B_{el} = 0.5R_b$ for NRZ, with R_b being the bit rate. Based on this relationship and the previous expression, we can impose the following design criteria[1]:

$$\frac{a^2}{R_b^2} \geq \frac{1}{B_{el,Tx}^2} + \frac{1}{B_{fiber,L}^2} + \frac{1}{B_{el,Rx}^2} \tag{3.178}$$

where $a = 1$ for RZ and $a = 2$ for NRZ. In Eq. (3.178), the optical fiber bandwidth is determined from Eq. (3.175), $B_{el,Tx}$ denotes the transmitter bandwidth, and $B_{el,Rx}$ denotes the receiver bandwidth.

The results related to point-to-point transmission can be generalized for an optically amplified system, with certain modifications. For example, the power budget equation should be replaced by the OSNR equation (Eq. 3.173):

$$\text{OSNR[dB]} \approx 10\log_{10}\left(\frac{Q^2 B_{el}}{B_{op}}\right) \geq P_{ch}[\text{dB}] - F_{no}[\text{dB}] - \alpha l - 10\log_{10}N - 10\log_{10}\left(hvB_{op}\right)$$

The in-line optical amplifiers are employed to compensate not only for attenuation loss but also for chromatic dispersion. Such compensation is not perfect in multichannel systems, and some residual penalty remains, thus causing the power penalty so that required OSNR expression should be modified to account for this imperfection as follows:

$$\text{OSNR}_{req}[\text{dB}] = \text{OSNR} + \Delta P \approx 10\log_{10}\left(\frac{Q^2 B_{el}}{B_{op}}\right)$$
$$+ \Delta P \geq P_{ch}[\text{dB}] - F_{no}[\text{dB}] - \alpha l - 10\log_{10} N - 10\log_{10}\left(hvB_{op}\right) \tag{3.179}$$

where OSNR_{req} is the OSNR needed to account for different system imperfections, and ΔP is the system margin.

The transmission system design process previously presented represents a conservative scenario and has the purpose of providing an understanding of design parameters and design processes.[1] This conservative scenario can be used as a reference case and feasibility check study before we continue with further considerations. The most complex case from the system design standpoint is the high-speed long-haul transmission with many WDM channels and possible optical routing. In this case, the conservative scenario can be used only as the guidance, whereas the more precise design should be performed by using the computer-aided deign approach. Commercially available software packages include Virtual Photonics, OptiWave, and BroadNeD. Simulation software can be written using different tools, including C/C++, Matlab, Mathcad, and Fortran.

3.5.12 Optical Performance Monitoring

Performance monitoring is very important for the network operator to control the overall status of transmission lines and the status of the network and also to deliver desired quality of service to the end user. Performance monitoring techniques are either analog or digital. Commonly used analog techniques for performance monitoring include (1) optical spectrum analysis (OSNR, optical power, and optical frequency monitoring), (2) detection of a special pilot tone (the amplitude of the pilot tone is related to the signal power and can be used to extract the OSNR), and (3) the histogram method (suitable to identify dispersion and nonlinear distortions). The OSNR measured by the OSA or pilot method is related to the Q-factor by[1]

$$Q = \frac{2\text{OSNR}_{\text{reference band}}\sqrt{B_{\text{op}}/B_{\text{el}}}}{1 + \sqrt{1 + 4\text{OSNR}_{\text{reference band}}}}, \quad \text{OSNR}_{\text{reference band}} = \left(B_{\text{reference band}}/B_{\text{op}}\right)\text{OSNR} \tag{3.180}$$

where $\text{OSNR}_{\text{reference band}}$ is the OSNR defined in the referent bandwidth (0.1 nm is commonly used), and B_{op} and B_{el} are the optical and electrical filter bandwidths, respectively. The Q-factor penalty can be expressed by

$$\Delta Q = 10\log_{10}\frac{Q}{Q_{\text{ref}}} \tag{3.181}$$

where Q_{ref} is the referent Q-factor. The histogram method provides more details about signal degradation[1,20–22]:

$$\text{BER} = \frac{1}{2p(1)}\sum_i H(I_{1,i})\text{erfc}\left(\frac{I_{1,i} - I_{\text{tsh}}}{\sigma_{1,i}\sqrt{2}}\right) + \frac{1}{2p(0)}\sum_i H(I_{0,i})\text{erfc}\left(\frac{I_{\text{tsh}} - I_{0,i}}{\sigma_{1,i}\sqrt{2}}\right) \tag{3.182}$$

where $H(I_{0,i})$ and $H(I_{1,i})$ represent the occurrences associated with bits 0 and 1, respectively, and $p(i)$ is the *a priori* probability of bit i ($I = 0, 1$).

The digital methods of performance monitoring are based on error detection codes, such as CRC codes and bit interleaved parity codes. The information is processed by receiving data in blocks and performing the parity checks. These methods are suitable for in-service monitoring but not for fault localization.

3.6 Summary

This chapter described the basic concepts of optical transmission systems based on intensity modulation with direct detection. In Section 3.2, basic principles of optical transmission were provided, the basic building blocks were identified, and fundamental principles of those building blocks were provided. In Section 3.3, different noise sources, both additive and multiplicative, were described. Section 3.4 was devoted to the description of basic channel impairments, including fiber attenuation, chromatic dispersion, PMD, and fiber nonlinearities. In Section 3.5, the basic figures of merit of an optical transmission system were introduced, including SNR, optical SNR, receiver sensitivity, and BER. Furthermore, the receiver sensitivity was determined in the presence of shot, thermal, and beat noise; all other noise sources and channel impairments were described through the power penalty—the increase in receiver sensitivity needed to preserve the desired BER. This approach leads to the worst-case scenario but helps to verify the feasibility of a particular system design before further considerations. The guidelines for the system design were provided as well.

References

1. Cvijetic M. *Optical Transmission Systems Engineering*. Boston: Artech House; 2004.
2. Ramaswami R, Sivarajan K. *Optical Networks: A Practical Perspective*. 2nd ed. Boston: Morgan Kaufmann; 2002.
3. Agrawal GP. *Fiber-Optic Communication Systems*. 3rd ed. New York: Wiley; 2002.
4. Agrawal GP. *Lightwave Technology: Components and Devices*. New York: Wiley-Interscience; 2004.
5. Agrawal GP. *Lightwave Technology: Telecommunication Systems*. New York: Wiley-Interscience; 2005.
6. Agrawal GP. *Nonlinear Fiber Optics*. 4th ed. Boston: Academic Press; 2007.
7. Cvijetic M. *Coherent and Nonlinear Lightwave Communications*. Boston: Artech House; 1996.
8. Keiser G. *Optical Fiber Communications*. 3rd ed. New York: McGraw-Hill; 2000.
9. Kazovsky L, Benedetto S, Willner A. *Optical Fiber Communication Systems*. Boston: Artech House; 1996.
10. Palais JC. *Fiber Optic Communications*. 5th ed. Upper Saddle River, NJ: Pearson Prentice Hall; 2005.
11. Ramamurthy B. Switches, wavelength routers, and wavelength converters. In: Sivalingam KM, Subramaniam S, editors. *Optical WDM Networks: Principles and Practice*. Norwell, MA: Kluwer Academic; 2000.
12. Djordjevic IB. *ECE 430/530: Optical communication systems* [lecture notes]. University of Arizona; 2006/2007.
13. Kostuk R. *ECE 430/530: Optical communication systems* [lecture notes]. University of Arizona; 2002/2004.
14. Djordjevic IB. *ECE 632: Advanced optical communication systems* [lecture notes]. University of Arizona; 2007.
15. Yariv A. *Optical Electronics in Modern Communications*. New York: Oxford University Press; 1997.
16. Amersfoort M. Arrayed waveguide grating. Application note A1998003. Available at www.c2v.nl; 1998.
17. Staif M, Mecozzi A, Nagel J. Mean square magnitude of all orders PMD and the relation with the bandwidth of principal states. *IEEE Photon Technol Lett* 2000;**12**:53–5.
18. Kogelnik H, Nelson LE, Jobson RM. Polarization mode dispersion. In: Kaminow IP, Li T, editors. *Optical Fiber Telecommunications*. San Diego: Academic Press; 2002.
19. Foschini GJ, Pole CD. Statistical theory of polarization mode dispersion in single-mode fibers. *IEEE/OSA J Lightwave Technol* 1991;**LT-9**:1439–56.
20. Bendelli G, et al. Optical performance monitoring techniques. In: *Proc. ECOC 2000*, vol. 4. Munich; 2000. pp. 213–6.
21. Djordjevic IB, Vasic B. An advanced direct detection receiver model. *J Opt Commun* 2004;**25**(1):6–9.
22. Winzer PJ, Pfennigbauer M, Strasser MM, Leeb WR. Optimum filter bandwidths for optically preamplified NRZ receivers. *J Lightwave Technol* 2001;**19**:1263–72.

CHAPTER 4

Signal Processing for Optical OFDM

4.1 Introduction

One of the central features that sets orthogonal frequency-division multiplexing (OFDM) apart from single-carrier modulation is its uniqueness of signal processing. The single-carrier technique has been employed in optical communication systems for the past three decades. It is no surprise that OFDM signal processing seems to be quite strange to an optical engineer at first glance. However, as alluded to in Chapter 1, OFDM technology is exceptionally scalable for migration to higher data rate. Once the algorithms and hardware designs are developed for the current-generation product, it is very likely that these skill sets can be incorporated for the next-generation product. In this respect, OFDM is a future-proof technology, and subsequently various aspects of OFDM signal processing deserve careful perusal.

For conventional optical single-carrier systems, as the transmission speed increases, the requirement for optimal timing sampling precision becomes critical. Excessive timing jitter would place the sampling point away from the optimal, incurring severe penalty. On the other hand, for optical OFDM systems, a precise time sampling is not necessary. As long as an appropriate "window" of sampling points is selected that contains an uncontaminated OFDM symbol, it is sufficient. However, this tolerance to sampling point imprecision is traded off against the stringent requirement of frequency offset and phase noise in OFDM systems, as discussed in Chapter 2.

In this chapter, we lay out various aspects of OFDM signal processing associated with (1) three levels of synchronization, including window synchronization, frequency synchronization, and channel estimation; (2) the analog-to-digital converter (ADC)/digital-to-analog converter (DAC) impact on system performance; and (3) multiple-input multiple-output (MIMO)-OFDM systems.

4.2 End-to-End OFDM Signal Processing

In general, the OFDM signal transmission model that describes signal evolution across the transmitter, the transmission channel, and the receiver depends on the specific application.

Here, we use the coherent optical (CO)-OFDM system as an example to illustrate the essential elements of OFDM signal processing.

Figure 4.1 shows the conceptual diagram of a generic CO-OFDM system, including five basic functional blocks[1]: RF OFDM transmitter, RF-to-optical (RTO) up-converter, optical link, optical-to-RF (OTR) down-converter, and RF OFDM receiver. In this section, the term "RF" is used interchangeably with "electrical" to signify the physical interface in contrast with that in the "optical" domain. Transmission channel linearity is the fundamental assumption for OFDM. Therefore, study of the nonlinearity in each functional block is of critical importance. The RF OFDM transmitter and receiver have been studied in RF systems[2,3] and thus will hold the same importance as CO-OFDM systems. One unique aspect of CO-OFDM transmission is that it brings new sources of nonlinearity in RTO up-conversion,[1] OTR down-conversion, and fiber link transmission.[4–6] In this section, however, we concentrate on the signal processing aspect of CO-OFDM and therefore assume perfect linearity in each functional block.

We will trace the signal flow end-to-end and illustrate the signal processing that arises in each functional block. In the RF OFDM transmitter, the input digital data are first converted from serial to parallel into a "block" of bits consisting of N_{sc} "information symbol," each of which may comprise multiple bits for m-ary coding. This information symbol is mapped

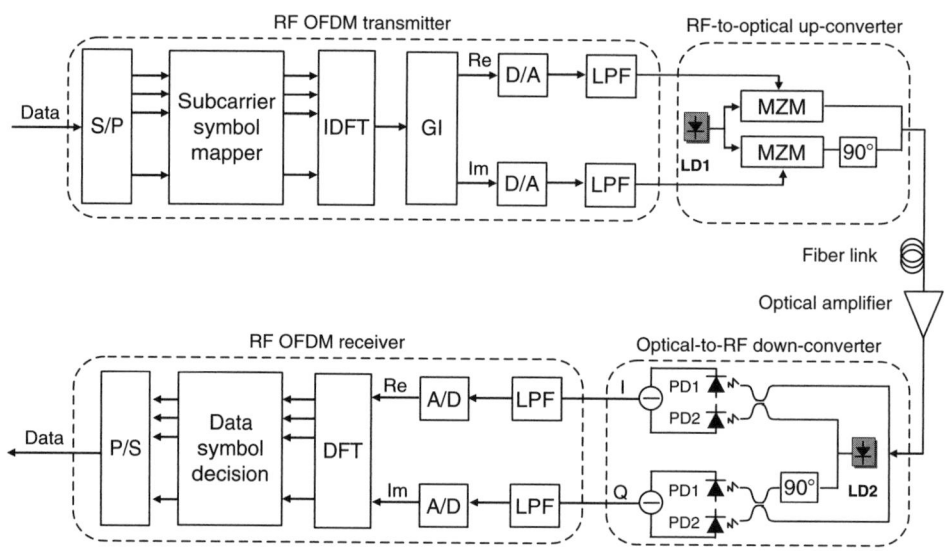

S/P: Serial-to-parallel GI: Guard time insertion (I)DFT: (inverse) Discrete Fourier transform
D/A: Digital-to-analog LPF: Low pass filter. MZM: Mach-Zehnder modulator PD: Photodiode
LD: Laser diode

Figure 4.1: Conceptual diagram for a generic CO-OFDM system with a direct up-/down-conversion architecture.

into a two-dimensional complex signal c_{ki}, for instance, using Gray coding, where c_{ki} stands for the mapped complex information symbol. The subscripts of c_{ki} correspond to the sequence of the subcarriers and OFDM blocks, which are further defined later. The time domain OFDM signal is obtained through inverse discrete Fourier transform (IDFT) of c_{ki}, and a guard interval is inserted to avoid channel dispersion.[2,3] The resultant baseband time domain signal can be described as

$$s_B(t) = \sum_{i=-\infty}^{+\infty} \sum_{k=-N_{sc}/2+1}^{k=N_{sc}/2} c_{ki} \Pi(t - iT_s) e^{j2\pi f_k(t-iT_s)} \tag{4.1}$$

$$f_k = \frac{k-1}{t_s} \tag{4.2}$$

$$\Pi(t) = \begin{cases} 1, (-\Delta_G < t \le t_s) \\ 0, (t \le -\Delta_G, t > t_s) \end{cases} \tag{4.3}$$

where c_{ki} is the ith information symbol at the kth subcarrier; f_k is the frequency of the kth subcarrier; N_{sc} is the number of OFDM subcarriers; T_s, Δ_G, and t_s are the OFDM symbol period, guard interval length, and observation period, respectively; and $\Pi(t)$ is the rectangular pulse waveform of the OFDM symbol. The extension of the waveform in the time frame of $[-\Delta_G, 0]$ in Eq. (4.3) represents the insertion of the cyclic prefix, or guard interval, discussed in Chapter 2. The digital signal is then converted to an analog form through a DAC and filtered with a low-pass filter to remove the alias signal. The baseband OFDM signal can be further converted to an RF passband through an RF IQ mixer (not shown in Figure 4.1). Figure 4.1 shows the direct up-conversion architecture, where the RF OFDM transmitter outputs a baseband OFDM signal. The subsequent RTO up-converter transforms the baseband signal to the optical domain using an optical IQ modulator comprising a pair of Mach–Zehnder modulators (MZMs) with a 90 degree phase offset. The baseband OFDM signal is directly up-converted to the optical domain given by

$$E(t) = e^{j(\omega_{LD1}t + \phi_{LD1})} \cdot s_B(t) \tag{4.4}$$

where ω_{LD1} and ϕ_{LD1}, respectively, are the angular frequency and phase of the transmitter laser. The up-converted signal $E(t)$ traverses the optical medium with an impulse response of $h(t)$, and the received optical signal becomes

$$E'(t) = e^{j(\omega_{LD1}t + \phi_{LD1})} s_B(t) \otimes h(t) \tag{4.5}$$

where $\otimes$ stands for convolution. The optical OFDM signal is then fed into the OTR down-converter, where the optical OFDM signal is converted to an RF OFDM signal. Figure 4.1 shows the direct down-conversion architecture in which the intermediate frequency (IF) is near-DC. The directly down-converted near-DC signal can be expressed as

$$r(t) = e^{j(\omega_{off}t + \Delta\phi)} r_0(t), \quad r_0(t) = s_B(t) \otimes h(t) \tag{4.6}$$

$$\omega_{off} = \omega_{LD1} - \omega_{LD2}, \quad \Delta\phi = \phi_{LD1} - \phi_{LD2} \tag{4.7}$$

where $\Delta\omega_{\text{off}}$ and $\Delta\phi$ are respectively the angular frequency offset and phase offset between the transmit and receive lasers.

In the RF OFDM receiver, the down-converted near-DC OFDM signal is first sampled with an ADC. Then the signal needs to go through the following three levels of sophisticated synchronizations before the symbol decision can be made:

1. DFT window synchronization in which OFDM symbols are properly delineated to avoid intersymbol interference

2. Frequency synchronization, namely frequency offset ω_{off} being estimated, compensated, and, preferably, adjusted to a small value at the start

3. The subcarrier recovery, where each subcarrier channel is estimated and compensated

Detailed discussion of these synchronizations is presented in Sections 5.3–5.5. Assuming successful completion of DFT window synchronization and frequency synchronization, the RF OFDM signal through the DFT of the sampled value of Eq. (4.6) becomes

$$r_{ki} = e^{\phi_i} H_{ki} c_{ki} + n_{ki} \tag{4.8}$$

where r_{ki} is the received information symbol, ϕ_i is the OFDM symbol phase or common phase error (CPE), H_{ki} is the frequency domain channel transfer function, and n_{ki} is the random noise. The third-level synchronization of subcarrier recovery involves estimation of CPE noise, ϕ_i, and the channel transfer function, H_{ki}. Once they are known, an estimated value of c_{ki}, $\hat{c}_{ki}$, is given by the zero-forcing method as

$$\hat{c}_{ki} = \frac{H_{ki}^*}{|H_{ki}|^2} e^{-i\phi_i} r_{ki} \tag{4.9}$$

where $\hat{c}_{ki}$ is used for symbol decision, which is subsequently mapped to the closest constellation point to recover the original transmitted digital bits.

The previous brief description of CO-OFDM processing left out the pilot subcarrier or pilot symbol insertion, where a proportion of the subcarriers or all the subcarriers in one OFDM symbol are known values to the receiver. The purpose of these pilot subcarriers or symbols is to assist the aforementioned three-level synchronization. Another important aspect of CO-OFDM signal processing that has not been discussed is the error-correction coding involving the error-correction encoder/decoder and interleaver/de-interleaver,[2,7] which are discussed in detail in Chapter 6.

4.3 DFT Window Synchronization

Synchronization is one of the most critical functionalities for a CO-OFDM receiver. As discussed in the previous section, it can be divided into three levels of synchronization: DFT window timing synchronization, carrier frequency offset synchronization, and subcarrier

Figure 4.2: Time domain structure of an OFDM signal.

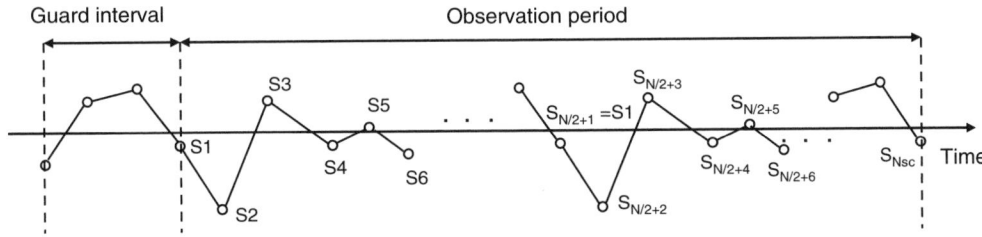

Figure 4.3: Schematic of the Schmidl synchronization format.

recovery. Figure 4.2 shows the time domain structure of an OFDM signal consisting of many OFDM symbols. Each OFDM symbol comprises a guard interval and an observation period. It is imperative that the start of the DFT window (i.e., the observation period) be determined properly because an improper DFT window will result in intersymbol interference (ISI) and intercarrier interference (ICI).[3]

One of the popular methods for window synchronization was proposed by Schmidl and Cox.[8] In such a method, a pilot symbol or preamble is transmitted that consists of two identical segments, as shown in Figure 4.3, that can be expressed as

$$s_m = s_{m-N_{sc}/2}, \quad m \in \left[N_{sc}/2 + 1, \, N_{sc} \right] \tag{4.10}$$

where s_m is the mth sample with a random value when m is from 1 to $N_{sc}/2$. Assuming a time-invariant channel impulse response function $h(t)$, from Eq. (4.6) the sampled received signal has the following form:

$$r_m = r(mt_s/N_{sc}) = r_m^0 e^{j\omega_{off}mt_s/N_{sc}} + n_m \tag{4.11}$$

We have assumed that the constant phase across the entire OFDM symbol, or $\Delta\phi$, equals zero in Eq. (4.6). The delineation can be identified by studying the following correlation function,[8] defined as

$$R_d = \sum_{m=1}^{N_{sc}/2} r_{m+d}^* r_{m+d+N_{sc}/2} \tag{4.12}$$

The principle is based on the fact that the second half of r_m is identical to the first half except for a phase shift. Assuming the frequency offset ω_{off} is small to start with, we anticipate that when $d = 0$, the correlation function R_d reaches its maximum value. The correlation function can be normalized to its maximum value given by

$$M(d) = |R_d/S_d|^2, \qquad S_d = \sqrt{\left(\sum_{m=1}^{N_{sc}/2} |r_{m+d}^2|\right)\left(\sum_{m=1}^{N_{sc}/2} |r_{m+d+N_{sc}/2}^2|\right)} \qquad (4.13)$$

where $M(d)$ is defined as the DFT window synchronization timing metric. The optimal timing metric has its peak at the correct starting point of the OFDM symbol—that is,

$$\hat{d} = \arg\{\max[M(d)]\} \qquad (4.14)$$

where $\arg\{\max[M(d)]\}$ in general stands for searching the optimal argument of d that maximizes the objective function of $M(d)$, and $\hat{d}$ stands for the optimal timing point.

We have conducted a Monte Carlo simulation to confirm the DFT window synchronization using the Schmidl format for a CO-OFDM system at 10 Gb/s under the influence of chromatic dispersion, linewidth, and optical-to-signal noise ratio (OSNR). The OFDM system parameters used for the simulation are a symbol period of 25.6 ns, guard time of 3.2 ns, and number of subcarriers of 256. Binary phase-shift keying encoding is used for each subcarrier, resulting in a total bit rate of 10 Gb/s. The linewidth of the transmitter and receiver lasers is assumed to be 100 kHz each, which is close to the value achieved with commercially available external-cavity semiconductor lasers.[9] The optical link noise from the optical amplifiers is assumed to be additive white Gaussian noise (AWGN), and the phase noise of the laser is modeled as white frequency noise characterized by its linewidth.

Figure 4.4a shows that the peak of the timing metric decreases from an ideal value of 1 to approximately 0.7 when the OSNR is 6 dB. For reference, to achieve a bit error ratio (BER) of 10^{-3}, an OSNR of 3.5 dB is needed. Both curves at the OSNR of infinity and 6 dB show the flat platform of 32 samples corresponding to 3.2 ns of guard interval under no chromatic dispersion. However, at a chromatic dispersion of 34,000 ps/nm (Figure 4.4b),

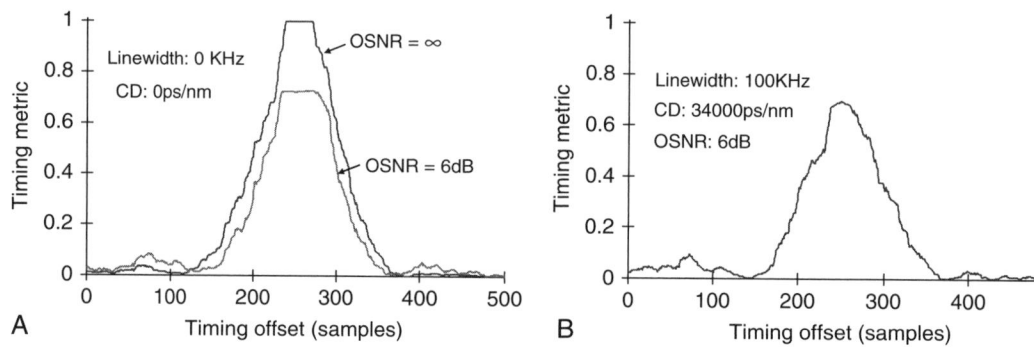

Figure 4.4: Timing metric for CO-OFDM systems at a chromatic dispersion of (a) 0 ps/nm and (b) 34,000 ps/nm.

the flat platform has almost disappeared, indicating that most of the guard interval has been affected by the ISI from the neighbor OFDM symbol. Another consequence is that the timing metric peak is not obvious anymore, and multiple peaks may coexist in the timing metric diagram. Other OFDM window synchronization approaches, such as Minn's approach,[10] in which the pilot symbol is partitioned into more than two identical or sign-inverted segments may be more robust to chromatic dispersion and thus more suitable for optical communications, which is left for future research.

4.4 Frequency Offset Synchronization

As discussed in Chapter 2, frequency offset in an OFDM system breaks the orthogonality among the subcarriers, incurring ICI.[11,12] We envisage that for CO-OFDM systems, the frequency synchronization process is divided into two phases—a frequency acquisition phase and a tracking phase.

4.4.1 Frequency Acquisition

The telecommunication lasers are usually locked to an ITU frequency standard through a wavelength locker, but only with an accuracy of approximately 2.5 GHz. This implies that the frequency offset could be anywhere from -5 to 5 GHz. Figure 4.5 shows the spectrum of a received CO-OFDM signal in relation to the LO laser frequency. The excessive frequency offset f_{off} creates two problems for the CO-OFDM system: (1) The highest RF frequency (f_{max} in Figure 4.5) of the RF OFDM signal after "direct" down-conversion is increased by the amount of frequency offset f_{off}, for instance, by as much as 5 GHz, and (2) the two signal subcarriers that happen to overlap with the receiver LO will have degraded performance due

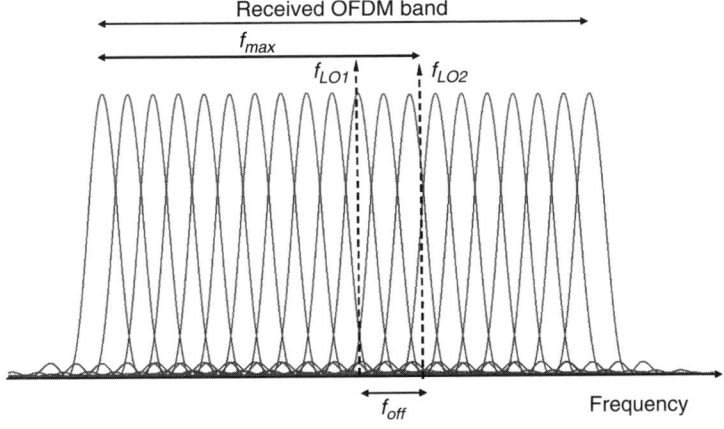

Figure 4.5: Frequency spectrum of a received OFDM signal in relation to the local oscillator laser. f_{LO1} and f_{LO2} are, respectively, the transmitter and receiver laser frequencies.

to direct down-conversion DC leakage. The former problem leads to excessive RF bandwidth expansion, thus increasing the receiver cost. The latter problem may not be a severe issue for point-to-point networks in which a feedback to the transmitter is allowed such that the subcarriers with degraded performance will not be filled. However, it will be quite challenging for an optical broadcast network in which one CO-OFDM signal stream will be dropped at multiple points. The local receiver laser frequencies of multiple users are different if they are not frequency tracked to the transmit laser, and subsequently there are no common subcarriers that can be unfilled to accommodate the DC leakage problem. The purpose of the frequency acquisition is to first coarse estimate the frequency offset and bring the receiver laser within approximately one or two times subcarrier frequency spacing, typically approximately 100–200 MHz from that of the transmit laser. Therefore, we believe that a coarse frequency (100–200 MHz accuracy) control is preferred in CO-OFDM systems to simplify the receiver signal processing. Many frequency acquisition algorithms have been proposed in the RF domain, such as the pilot tone approach[13] or the repeated DFT approach.[14]

In CO-OFDM systems, because the initial frequency offset could be as much as several gigahertz, the initial acquisition can also be obtained by sending a long-stream CW signal (tens or hundreds of OFDM symbols) and measuring the frequency of the offset tone. The error signal is used to control the local laser to bring its frequency close to that of the transmit laser. Then, the second phase of frequency tracking can be initiated.

4.4.2 Frequency Tracking

The purpose of the frequency tracking is to identify the magnitude of the frequency offset and compensate for it. Again, various approaches can be used for this purpose. We use the Schmidl approach.[8] From Eq. (4.12), we derive

$$R_{\hat{d}} = \sum_{m=1}^{N_{sc}/2} \left| r_{m+\hat{d}}^{0} \right|^{2} e^{j\frac{\pi f_{off}}{\Delta f}} + o(n) \tag{4.15}$$

where $o(n)$ is the residual term with the magnitude in the order of noise component. It follows from Eq. (4.15) that the offset frequency can be estimated as

$$\hat{f}_{off} = \frac{\Delta f}{\pi} \cdot \angle\left(R_{\hat{d}}\right) \tag{4.16}$$

where $\hat{d}$ is the estimated DFT window timing from Eq. (4.14), and $\angle\left(R_{\hat{d}}\right)$ is the angle of the complex value of the correlation function of $R_{\hat{d}}$ in radian. The accuracy of Eq. (4.16) can be improved by sending multiple Schmidl pilot symbols.

Once the frequency offset is estimated, the received sampled signal can be compensated as

$$r_c(t) = \exp\left(-j2\pi \hat{f}_{off} t\right) r(t) \tag{4.17}$$

The frequency offset compensated signal $r_c(t)$ can be used for DFT to obtain the received information symbol r_{ki} in Eq. (4.8) and subsequent subcarrier recovery for symbol decision discussed in the following section.

4.5 Subcarrier Recovery: Channel Estimation and Phase Estimation

From the channel model of Eq. (4.8), there are three factors that lead to the rotation of the receiver information symbol constellation for r_{ki}: (1) the channel dispersion H_{ki}, which gives frequency dependence across the OFDM spectrum; (2) the DFT sampling timing offset that generates a phase term linear with the subcarrier frequency; and (3) the phase noises from the transmit and receive lasers. The time constants for the three factors are different. The first one changes on the timescale of milliseconds due to mechanical movement of fiber link. In particular, the chromatic dispersion varies in response to the diurnal temperature fluctuation. The polarization mode dispersion varies due to the mechanical and temperature fluctuation on the timescale of millisecond. The second one is caused by the sampling clock rate offset, and it may need to be reset every microsecond to tens of microseconds. The third one comes from the laser phase noise with linewidth ranging from 100 KHz to several megahertz, which needs to be tracked on a per-symbol basis. The first two factors are dealt with through channel estimation. The third factor is treated through phase estimation and compensation.

We further assume that the signal processing is performed in blocks, with each containing a large number of OFDM symbols. Within each block, the optical channel is assumed to be invariant, whereas the OFDM CPE varies on the per-symbol basis. Subsequently, the subcarrier recovery includes two baseband signal processings—channel estimation and phase estimation. There are various methods of channel estimation, such as the time domain pilot-assisted approach and frequency domain-assisted approach.[3,15,16] We focus on the carrier recovery based on frequency domain pilot carriers or pilot symbols. Figure 4.6 shows the two-dimensional time/frequency structure for one OFDM block, which includes N_{sc} subcarriers in frequency and N_f OFDM symbols in time. The preamble is added at the beginning to realize DFT window synchronization and channel estimation. The channel transfer function can be estimated as

$$\hat{H}_k = \sum_{i=1}^{p} e^{-j\angle r_{k_1 i}} r_{ki}/c_{ki} \tag{4.18}$$

where c_{ki} and r_{ki} are, respectively, transmitted and received pilot subcarriers, $\angle r_{k_1 i}$ is the angle for the k_1th carrier (an arbitrary reference carrier) in the ith OFDM symbol, and p is the number of pilot symbols. The additional phase compensation of $-\angle r_{k_1 i}$ is needed to remove the influence of the CPE. The accuracy of $\hat{H}_k$ can be further improved if the functional dependence of $\hat{H}_k$ on the subcarrier frequency is known. For instance, one pilot symbol could be sufficient for chromatic dispersion estimation if a quadratic function is used to interpolate

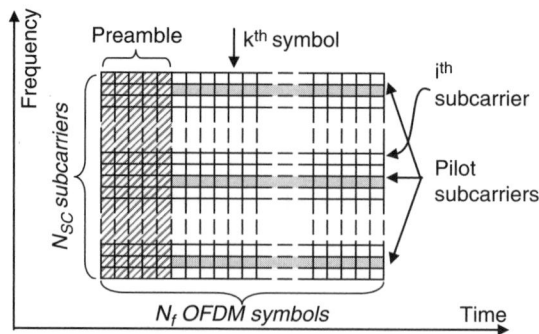

Figure 4.6: Time and frequency representation of a block of CO-OFDM signal.

the channel transfer function. The estimated channel transfer function has uncertainty of a common phase, which can be included in the CPE ϕ_i. The phase estimation is to estimate the CPE ϕ_i due to the laser phase noise. We assume the N_p pilot subcarriers are used for phase estimation, and the maximum likelihood CPE can be estimated as[17]

$$\phi_i = \angle\left(\sum_{k=1}^{N_p} C'_{ki} H_k^* C_{ki}^* / \delta_k^2\right) \tag{4.19}$$

where δ_k is the standard deviation of the constellation spread for the kth subcarrier. When δ_k is assumed to be constant across all subcarriers, Eq. (4.19) will reduce to the least squares method.[18]

After the channel estimation of Eq. (4.18) and the phase estimation of Eq. (4.19), the subcarrier recovery can be completed using Eq. (4.9), in essence, to reconstruct the constellation for each subcarrier via the so-called "one-tap" OFDM equalizer.

4.6 Channel Estimation

Channel estimation is a part of the subcarrier recovery described briefly in the previous section on the complete process of three-level OFDM synchronization. However, due to its critical importance to the overall system performance and its intimate connection to other issues, such as tracking speed and error correction, it is one of the most extensively researched topics in the field of OFDM.[19–23] Therefore, it is worthwhile to dedicate a separate section to this important topic.

4.6.1 Why Is Channel Estimation Needed?

Although we have previously shown that channel estimation is needed to de-rotate the constellation at the receiver, it may not always be required for incoherent modulation such as differential phase-shift keying. Differential coding can be realized through two approaches:

(1) using two consecutive OFDM symbols and (2) using two consecutive subcarriers. The differential encoding eliminates the need for either the absolute reference of the channel characteristics or the absolute symbol phase, and therefore channel estimation is not necessary. However, there is an SNR sensitivity penalty associated with the differential encoding. Assuming an AWGN channel, the BER for differential quaternary phase-shift keying (DQPSK) modulation is given by[24]

$$P_e(\gamma) = Q_1(a,b) - \frac{1}{2}I_0(ab)\exp\left[-\frac{1}{2}(a^2+b^2)\right] \tag{4.20}$$

where $Q_1(a,b)$ is the Markum Q function, $I_0(x)$ is the modified Bessel function of zeroth order, and the parameters a and b are defined as

$$a = \sqrt{2\gamma_b\left(1-\sqrt{\frac{1}{2}}\right)} \text{ and } b = \sqrt{2\gamma_b\left(1+\sqrt{\frac{1}{2}}\right)} \tag{4.21}$$

whereas the BER for the QPSK is[24]

$$P_e(\gamma) = Q\left(\sqrt{2\gamma_b}\right) \tag{4.22}$$

where $Q(x)$ is the Q function defined as

$$Q(x) = \frac{1}{\sqrt{2\pi}}\int_x^\infty e^{-t^2/2}dt \tag{4.23}$$

γ_b is the SNR per bit. Figure 4.7 shows the BER performance using the formula in Eq. (4.20) and Eq. (4.22). It can be seen that QPSK has an approximately 2.3 dB advantage over DQPSK at a BER of 10^{-3}. This advantage is significant, and most often it is worthwhile to

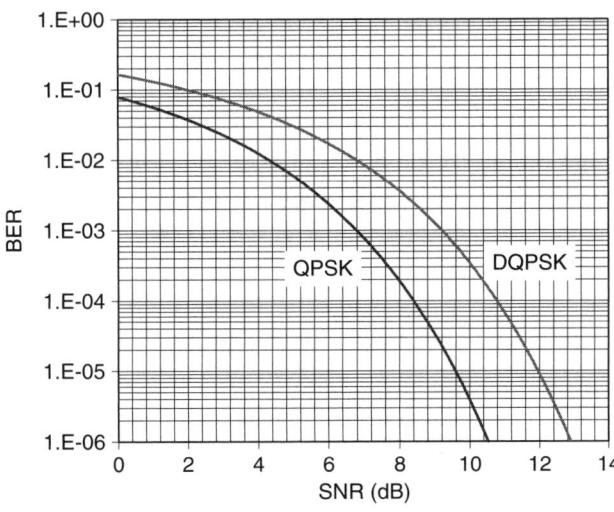

Figure 4.7: BER performance comparison between QPSK and DQPSK signals.

add additional complexity of channel and phase estimation and use PSK to gain that advantage. Moreover, channel estimation is needed for MIMO systems, which are discussed later. In such systems, channel estimation is mandatory to perform the coherent combining at the output to gain the MIMO advantage.

4.6.2 Channel Estimation Algorithms

There are a wide variety of algorithms for channel estimation. They can all be derived from two basic principles: maximum likelihood (ML) estimation and minimum mean square error (MMSE) estimation. From these two fundamental algorithms, there exist rich variations of channel estimation algorithms that are adapted to different applications. In the following two sections, we discuss ML and MMSE channel estimation in detail.

4.6.2.1 Maximum Likelihood Channel Estimation

For simplicity, we omit the procedure of the phase noise estimation and rewrite Eq. (4.8) in a more concise form as

$$r_k = H_k c_k + n_k \tag{4.24}$$

For explanatory simplicity, we ignore the index of OFDM symbol sequence "i" in Eq. (4.8), namely assuming the time-invariant channel. The channel estimation over the time-varying channel can also be treated in a similar manner[19] but is outside the scope of this chapter. Assuming n_k is an AWGN noise, or Eq. (4.24) represents an AWGN channel, and each subcarrier is independent, the joint probability density function (PDF) of the received signal r at a deterministic but unknown function H_k can be expressed as

$$p(r_1, r_2, \ldots, r_{N_{sc}} | H_1, H_2, \ldots, H_{N_{sc}}) \propto \exp\left(\sum_{k=1}^{N_{sc}} \frac{(r_k - H_k c_k)^* (r_k - H_k c_k)}{2\sigma^2} \right) \tag{4.25}$$

where σ is the standard deviation of the noise for each subcarrier, which is assumed to have the same magnitude, and the asterisk stands for complex conjugate. The purpose of the channel estimation is to find the matrix H_k that maximizes the joint PDF function of $p(r_1, r_2, \ldots, r_{N_{sc}} | H_1, H_2, \ldots, H_{N_{sc}})$. Equivalently, we can search for H_k that minimizes the likelihood function

$$\Lambda(H_1, H_2, \ldots, H_{N_{sc}}) = \sum_{k=1}^{N_{sc}} (r_k - H_k c_k)^* (r_k - H_k c_k) \tag{4.26}$$

Because H_k is the complex variable, we treat H_k and its conjugate, H_k^*, as independent variables and differentiate Eq. (4.26) against H_k^*; we have

$$\frac{\partial \Lambda(H_1, H_2, \ldots, H_{N_{sc}})}{\partial h_k} = c_k^* (r_k - H_k c_k) \tag{4.27}$$

We arrive at H_k^{ML}, or the ML channel estimation of H_k, by setting Eq. (4.27) equal to zero, and we have

$$H_k^{\text{ML}} = \frac{r_k}{c_k} = H_k + \frac{n_k}{c_k} = H_k + \tilde{n}_k \tag{4.28}$$

where

$$\tilde{n}_k = \frac{n_k}{c_k},$$

$$H_k^{\text{ML}}$$

is the ML estimation for the kth subcarrier. From Eq. (4.28), it follows that the estimated channel transfer function is contaminated by the noise. We use two means to characterize the errors as a result of this noise. The first one is the mean square error (MSE), defined as

$$\text{MSE} = \left\langle \left| \frac{H_k - \hat{H}_k}{H_k} \right|^2 \right\rangle = \left\langle \left| \frac{n_k}{H_k c_k} \right|^2 \right\rangle = \frac{\sigma^2}{S^2} = \frac{1}{\text{SNR}} \tag{4.29}$$

where $\hat{H}_k$ is the channel estimation for the kth subcarrier, and in this case is H_k^{ML} in Eq. (4.28). The second means is to compute the SNR degradation due to the channel estimation error, which we will derive later.

The estimated information symbol is given by

$$\hat{c}_k = \frac{r_k}{\hat{H}_k} = \frac{H_k c_k + n_k}{H_k + n_{jk}/c_{jk}} = c_k + \frac{n_k + n_{jk}\frac{c_k}{c_{jk}}}{H_k} = c_k + n_k' \tag{4.30}$$

$$n_k' = \frac{n_k + n_{jk}\frac{c_k}{c_{jk}}}{H_k} \tag{4.31}$$

where c_{jk} and n_{jk} with the index j retained are, respectively, the transmitted symbol and noise when the channel estimation is performed. The first and second terms on the right side of Eq. (4.31), respectively, stand for the signal and noise of the estimated information symbol. Therefore, the effective SNR of the detected signal is

$$\text{SNR}' = \frac{\left\langle |c_k|^2 \right\rangle}{\left\langle |n_k'|^2 \right\rangle} = \frac{\left\langle |c_k|^2 \right\rangle}{\left\langle 2|n_k/H_k|^2 \right\rangle} = \frac{|H_k|^2 \left\langle |c_k|^2 \right\rangle}{2\sigma^2} = \frac{\text{SNR}}{2} \tag{4.32}$$

It can be seen that the effective SNR is decreased by half. To maintain the effective SNR equal to that without channel estimation noise, the SNR has to be increased by 3 dB. This is obviously not acceptable because this defeats the purpose of the channel estimation to maintain the SNR advantage of 2.3 dB over differential encoding.

On the other hand, in the previous derivation, we assumed that we only use one pilot symbol to perform one-time channel estimation. The channel estimation can be improved by sending the pilot symbol multiple times. Suppose that we can send p pilot symbols or pilot subcarriers. It is easy to show the following equations:

$$MSE = \frac{1}{p \cdot SNR} \tag{4.33}$$

and

$$SNR' = \frac{1}{1 + p^{-1}} SNR \tag{4.34}$$

From Eq. (4.34), it is observed that to limit the SNR degradation to 0.5 dB, the pilot symbol has to be resent nine times.

From the previous exercise, it is quite clear that the ML channel estimation is simple to perform. However, the disadvantage is that MSE and SNR degradation is significant and unacceptable. Sending pilots more than nine times can reduce the penalty to 0.5 dB, which is acceptable for a slow-varying channel. However, for a fast-varying channel, it may not be possible to send a pilot symbol an excessive number of times and average down the noise. A different means must be explored to improve the channel estimation algorithm.

4.6.2.2 MMSE Channel Estimation

As discussed in the previous section, some kind of averaging is needed to improve ML channel estimation performance. This kind of averaging can be performed by taking advantage of the frequency domain correlation. Take a simple extreme case for which the channel is flat across all the subcarriers by sending one set of pilot subcarriers $c = (c_1, c_2, \ldots, c_M)^T$ and obtaining one set of ML channel estimation $H_{ML} = (H_1, H_2, \ldots, H_M)^T$, where M is the number of pilot subcarriers used, the bold font convention is used to indicate a nonscalar quantity such as a matrix or a vector, and T is the matrix transpose operator. Because we have the prior knowledge that the channel is flat, then we can perform further averaging across all the subcarriers to improve the MSR by M-fold. In reality, the channel usually is not flat across the entire OFDM spectrum but does look "flat" in a finite narrow spectrum band. The range of this strong correlation is called correlation bandwidth.[2] It is determined by the maximum delay of the channel. The fundamental work of improvement of channel estimation using frequency domain correlation was first performed by Edfors et al.[20] and has been broadened and extended by other researchers.[19,21–23]

MMSE is a very popular signal processing algorithm with a wide range of applications.[25] The adaptive filter via MMSE is also called the Wiener filter, representing the ultimate performance for filtering or estimation. Many forms of filtering algorithms can trace their origin to the Wiener filter, with a goal to obtain the performance of the Wiener filter but with reduced complexity.

We start the treatment of MMSE channel estimation with an assumption that a set of channel estimation has been derived using ML estimation described in the previous section. The task here is further improvement of the channel estimation of $\boldsymbol{H} = (H_1, H_2, \ldots, H_M)^T$ derived from the set of ML estimation $\boldsymbol{H}_{\text{ML}} = (H_1, H_2, \ldots, H_M)^T$. We further assume that transformation from $\boldsymbol{H}_{\text{ML}}$ to a new set of coefficient $\boldsymbol{H}'$ is linear given by

$$\boldsymbol{H}' = \boldsymbol{W}\boldsymbol{H}_{\text{ML}} \tag{4.35}$$

where $\boldsymbol{W} = [W_{ij}]$ is an $M \times M$ matrix. The problem of MMSE estimation becomes one of searching for each matrix element W_{ij}, or the so-called Wiener filter coefficient, that will minimize the MSE given by

$$\text{MSE} = \left\langle (\boldsymbol{H}' - \boldsymbol{H})^+ (\boldsymbol{H}' - \boldsymbol{H}) \right\rangle = \left\langle (\boldsymbol{W}\boldsymbol{H}_{\text{ML}} - \boldsymbol{H})^+ (\boldsymbol{W}\boldsymbol{H}_{\text{ML}} - \boldsymbol{H}) \right\rangle$$
$$= \left\langle \sum_{i=1}^{M} \left(\sum_{j=1}^{M} W_{ij}^* \left(H_j^{\text{ML}} \right)^* - H_i^* \right) \left(\sum_{k=1}^{M} W_{ik} H_k^{\text{ML}} - H_i \right) \right\rangle \tag{4.36}$$

where the superscript "+" is the Hermitian conjugate performing transpose and complex conjugate over a vector or matrix, and $\langle \, \rangle$ is the ensemble average. Taking the derivative of Eq. (4.36) with respect to W_{ij}^* and equaling it to zero, we arrive at the MMSE condition as

$$\frac{\partial \text{MSE}}{\partial W_{ij}^*} = \left\langle \left(H_j^{\text{ML}} \right)^* \left(\sum_{k=1}^{M} W_{ik}^{MMSE} H_k^{\text{ML}} - H_i \right) \right\rangle$$
$$= \sum_{k=1}^{M} W_{ik}^{MMSE} \left\langle H_j^{\text{ML}} \left(H_k^{\text{ML}} \right)^* \right\rangle - \left\langle H_i \left(H_j^{\text{ML}} \right)^* \right\rangle = 0 \tag{4.37}$$

where W_{ik}^{MMSE} is the matrix element in the ith row and kth column of the MMSE filter matrix $\boldsymbol{W_{MMSE}}$. We also denote

$$V = \left\langle \boldsymbol{H}_{\text{ML}} \boldsymbol{H}_{\text{ML}}^+ \right\rangle \tag{4.38}$$

$$L = \left\langle \boldsymbol{H} \boldsymbol{H}_{\text{ML}}^+ \right\rangle \tag{4.39}$$

Using the expression of $\boldsymbol{R}$ and $\boldsymbol{L}$ in Eqs. (4.38) and (4.39), Eq. (4.37) is transformed into a more compact form expressed as

$$\boldsymbol{W_{MMSE}} V = L \tag{4.40}$$

Thus, the optimal Wiener coefficient is given by

$$\boldsymbol{W_{MMSE}} = L V^{-1} \tag{4.41}$$

We can further simplify V and L by substituting Eq. (4.28) into Eqs. (4.38) and (4.39); then we obtain

$$V_{kl} = \left\langle (H_k + \tilde{n}_k)(H_l + \tilde{n}_l)^* \right\rangle = \left\langle H_k H_l^* \right\rangle + \rho_i \delta_{ij} = R_{ij} + \rho \delta_{ij} \tag{4.42}$$

$$L_{ij} = \left\langle H_k (H_l + \tilde{n}_l)^* \right\rangle = R_{ij} \tag{4.43}$$

where $R_{kl} = \langle H_k H_l^* \rangle$ is the channel correlation function, and $\rho_i = \langle \tilde{n}_i \tilde{n}_i^* \rangle = \sigma_i^2 / s_i^2$ is the inverse of the SNR for the ith subcarrier, which is assumed to be equal for all subcarriers. The MMSE can now be expressed in terms of the channel correction matrix R as

$$W_{\text{MMSE}} = R(R + \rho I)^{-1} \tag{4.44}$$

The MMSE channel estimation can be obtained by substituting Eq. (4.44) into Eq. (4.35) and is given by

$$H_{\text{MMSE}} = R(R + \rho I)^{-1} H_{\text{ML}} \tag{4.45}$$

In theory, if the channel correlation function R and the SNR for each subcarrier are known, the MMSE estimate H_{MMSE} can be obtained. In practice, especially in relation to implementing the algorithm in real time, it is highly challenging, if not impossible. Because the complexity of the inverse function in Eq. (4.45) is proportional to the third power of the dimension of the matrix, the three matrix multiplications in Eq. (4.45) are each scaled with the square of the dimension of the matrix. Therefore, the computation complexity of the MMSE estimate of Eq. (4.45) is extremely high. Some approximation is mandatory.

The complexity can be reduced through singular value decomposition (SVD), which was first discussed by Edfors et al.[20] and Hsieh and Wei.[26] The channel correlation function through SVD can be expressed as[27]

$$R = U \Lambda U^+ \tag{4.46}$$

where U is a unitary matrix containing the singular eigenvectors, and Λ is a diagonal matrix containing the singular eigenvalues $\lambda_k (\lambda_1 \geq \lambda_1 \geq \ldots \geq \lambda_M)$. It is worth noting that the summation of eigenvalues observes that

$$\sum_{j=1}^{M} \lambda_j = \text{Tr}(R) = \sum_{j=1}^{M} \langle |H_j|^2 \rangle \tag{4.47}$$

which is proportional to the total received signal power.

Using SVD, the MMSE estimation can be written as

$$H_{MMSE} = U \Gamma U^+ H_{\text{ML}} \tag{4.48}$$

where Γ is a diagonal matrix expressed as

$$\Gamma = \text{diag}\{\Gamma_1, \Gamma_2, \ldots, \Gamma_M\} \tag{4.49}$$

and Γ_j is defined as

$$\Gamma_j = \frac{\lambda_k}{\lambda_k + \sigma^2} \tag{4.50}$$

Edfors et al.[20] showed that the unitary matrix U can be approximated as a DFT matrix given by

$$U_{lk} = e^{-j \frac{2\pi(l-1)(k-1)}{M}} \tag{4.51}$$

The complexity of the matrix U is reduced because it is no longer a time-varying matrix. Second, because the channel frequency response is usually highly correlated, this implies that only a few eigenvalues λ_k are significant. Further simplification can be made by assuming that

$$\Gamma_j = \left\{ \begin{array}{l} \dfrac{\lambda_k}{\lambda_k + \sigma^2}, 1 \leq k \leq k_0 \\ 0, \quad k_0 + 1 \leq k \leq M \end{array} \right\} \tag{4.52}$$

namely, some of the diagonal elements of Γ are approximated as zeros when the index is beyond a certain number k_0. This is the so-called reduced rank estimator, and its conceptual diagram is depicted in Figure 4.8, which is essentially a symbolic interpretation of Eq. (4.48) using SVD with a reduced rank k_0 approximation.

In this section, we focused on the MMSE estimation using only frequency domain correlation. Note that a similar approach can be performed by taking advantage of the time domain correlation or both frequency domain and time domain correlations through a two-dimensional Wiener filter.[19]

4.6.2.3 Channel Estimation through Pilot Subcarriers

For either ML or MMSE channel estimation, we have assumed that the transmitted information symbols are known. In general, two approaches can be applied to these two estimation methods: (1) using decision feedback, namely the decision is made with some prior knowledge of the rough channel estimation, and then these sliced data are used for the channel estimation for improved accuracy; and (2) using the pilot subcarriers, where only a certain proportion of the subcarriers are allocated for the channel estimation. The former approach does not require overhead for pilot subcarriers but is always susceptible to error propagation. Therefore, the pilot-based estimation is more robust and is thus the most popular and widely studied approach. In this section, we focus on some fundamental issues associated with pilot-assisted channel estimation.

The straightforward way to perform the channel estimation is to populate the entire OFDM block with pilot subcarriers and obtain the frequency channel transfer function using either

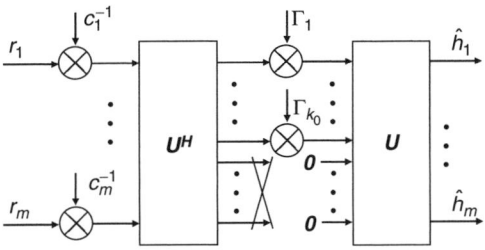

Figure 4.8: Conceptual diagram of the reduced rank k_0 channel estimator using SVD.

ML or MMSE channel estimation as shown previously. There are two disadvantages of doing so. First, this causes a large loss of the throughput because of the huge overhead of the pilot subcarriers. Second, in either a wireless or an optical channel, the dynamic elements that have to be updated frequently usually have a short delay duration, much shorter than the OFDM symbol period. Therefore, it does not make sense to sample much more often than its correlation function of the channel response. For instance, in wireless communications, the maximum delay (normalized to the sampling period) of the three-path indoor wireless asynchronous transfer mode channel is much smaller than the number of the subcarriers.[3] In optical communications, polarization mode dispersion can be fast varying, but it is usually less than 100 ps, which is much shorter than the symbol period of approximately a few nanoseconds. As such, there are two important questions to be answered: How many subcarriers are necessary for accurate channel estimation? and What is the optimal way to distribute the subcarriers—distributively within each OFDM symbol or lumped together in one OFDM subcarrier? These two questions have been explored in Negi and Cioffi[23] and are discussed in detail here.

The channel is assumed to have L taps characterized by a time domain impulse response function $h(t)$, given by

$$h(t) = \sum_{i=1}^{L} h_i \delta(t - iT_s) \tag{4.53}$$

where T_s is the sampling period of the OFDM signal. It can be easily shown that once the OFDM symbol period is chosen long enough, the channel transfer function in Eq. (4.53) can be expressed in the frequency domain as

$$H_{ki} = \mathfrak{F}[h(t)] = \sum_{j=1}^{N_{sc}} h_{ji} e^{-j\frac{2\pi}{N_{sc}}(j-1)(i-1)} = \sum_{j=1}^{L} h_{ji} e^{-j\frac{2\pi}{N_{sc}}(j-1)(i-1)} \tag{4.54}$$

where h_{ji} is the jth tap in the ith OFDM frame, to account for the timing-varying nature of the transfer function. In this section, as can be seen from Eqs. (4.53) and (4.54), we use h_{ji} and H_{ki} to represent the time and frequency domain response function, respectively.

We first assume that in the absence of the noise, arbitrary L subcarriers are used for the channel estimation. Using the ML estimation as shown in Eq. (4.28), we obtain the frequency domain channel response of $\boldsymbol{H}_{pl} = (H_{k_1}, H_{k_2}, \ldots, H_{k_L})^T$, where k_n is the OFDM subcarrier sequence number for the nth pilot subcarrier, and "pl" is the quantity associated with the pilot tones. From Eq. (4.54), it follows that the time domain response function $\boldsymbol{h}_{pl}$ can be obtained with IDFT, given by

$$\boldsymbol{h}_{pl} = W_{pl} \boldsymbol{H}_{pl} \tag{4.55}$$

where

$$
W_{pl} = \begin{pmatrix}
1 & w_{k_1} & w_{k_1}^2 & \cdots & w_{k_1}^{L-1} \\
1 & w_{k_2} & w_{k_2}^2 & \cdots & w_{k_2}^{L-1} \\
1 & w_{k_3} & w_{k_3}^2 & \cdots & w_{k_3}^{L-1} \\
\vdots & \vdots & \vdots & \vdots & \vdots \\
1 & w_{k_L} & w_{k_L}^2 & \cdots & w_{k_L}^{L-1}
\end{pmatrix}, w_{k_n} = e^{j\frac{2\pi}{N_{sc}}k_n}
\tag{4.56}
$$

It can be seen that W_{pl} is a Vandermond matrix with each L parameter being distinct, and therefore W_{pl} is nonsingular. We can then derive the time domain response of h_{pl} as

$$
h_{pl} = W_{pl}^{-1} H_{pl}
\tag{4.57}
$$

Once the time domain response of h_{pl} is known, the frequency response for all the other subcarriers can be interpolated with Eq. (4.54).

From the previous exercise, we arrive at the important conclusion that any L subcarrier tones can be used as the pilot tones to derive the complete frequency response for the OFDM signal. In addition, if less than L subcarriers are used, then W_{pl} is noninvertible and L coefficients for the time response taps cannot be determined uniquely.

Although any L subcarriers can be used as pilot subcarriers, the important question remains regarding what is the optimum set of subcarriers for the channel estimation. The answer can be found by minimizing the MSE of the estimated function.[23] Under the influence of the noise, substituting the ML estimation of Eq. (4.28) for H_{pl}, Eq. (4.57) is modified as

$$
h_{pl} = W_{pl}^{-1}(H_{pl} + N_{pl}) = h + W_{pl}^{-1}N_{pl}
\tag{4.58}
$$

$$
N_{pl} = \left(\frac{n_{k1}}{c_{k_1}}, \frac{n_{k2}}{c_{k_2}}, \ldots, \frac{n_{kL}}{c_{k_L}}\right)^T
\tag{4.59}
$$

The MSE of the estimation is thus

$$
\begin{aligned}
\left\langle (h_{pl} - h)^+ (h_{pl} - h) \right\rangle &= \left\langle \left(W_{pl}^{-1}N_{pl}\right)^+ W_{pl}^{-1}N_{pl} \right\rangle \\
&= \left\langle \mathrm{Tr}\left[W_{pl}^{-1}N_{pl}\left(W_{pl}^{-1}N_{pl}\right)^+\right] \right\rangle = \mathrm{Tr}\left[W_{pl}^{-1}\left\langle N_{pl}N_{pl}^+\right\rangle\left(W_{pl}^{-1}\right)^+\right] \\
&= \frac{\sigma^2}{s^2}\mathrm{Tr}\left\{W_{pl}^{-1}I_L\left(W_{pl}^{-1}\right)^+\right\} = \frac{\sigma^2}{s^2}\mathrm{Tr}\left\{W_{pl}^+\left(W_{pl}\right)^{-1}\right\}
\end{aligned}
\tag{4.60}
$$

where Tr is the trace of a matrix. We denote

$$
Q \equiv W_{pl}^+ W_{pl}
\tag{4.61}
$$

It can be shown that Q can be expressed as

$$Q \equiv L \begin{pmatrix} 1 & x & x & x \\ x & 1 & x & x \\ \vdots & \vdots & \vdots & \vdots \\ x & x & x & 1 \end{pmatrix} \tag{4.62}$$

The eigenvalues for Q are assumed to be $\lambda_1, \lambda_2, \ldots, \lambda_L$, and from Eq. (4.62) their summation should be equal to the trace of Q, namely

$$\sum_{k=1}^{L} \lambda_k = \text{Tr}\{Q\} = L^2 \tag{4.63}$$

The eigenvalues for Q^{-1} are thus $\lambda_1^{-1}, \lambda_2^{-1}, \ldots, \lambda_L^{-1}$, and the MSE of the channel estimation in Eq. (4.60) is thus given by

$$\text{MSE}_{h_{\text{pl}}} = \frac{\sigma^2}{s^2} \sum_{k=1}^{N} \frac{1}{\lambda_k} \tag{4.64}$$

Because $1/\lambda$ is the convex function for positive λ, we use Jensen's inequality that

$$\text{MSE}_{h_{\text{pl}}} = \frac{\sigma^2}{s^2} \sum_{k=1}^{L} \frac{1}{\lambda_k} \leq \frac{\sigma^2}{s^2} \frac{L}{\frac{1}{L} \sum_{k=1}^{L} \lambda_k} = \frac{\sigma^2}{s^2} \frac{L^2}{L^2} = \frac{\sigma^2}{s^2} \tag{4.65}$$

Obviously the equality takes place if and only if all λ_k's are equal, implying that Q is proportional to the identity matrix. From Eq. (4.61), it follows that W_{pl} is proportional to a unitary matrix. This can happen if and only if for the pilot subcarriers sets

$$\left\{ i, i + \frac{N\text{sc}}{L}, i + \frac{2N\text{sc}}{L}, \quad i + \frac{(L-1)N\text{sc}}{L} \right\}, \quad \text{where} \quad i \in \left\{ 1, 2, \ldots, \frac{N\text{sc}}{L} \right\} \tag{4.66}$$

Namely, the evenly distributed L-subcarrier set is the optimum selection for a channel with the dispersion length of L. This evenly distributed pilot subcarrier set is used in practice. For instance, in IEEE standard 802.11a,[28] four subcarriers are allocated as pilot subcarriers at the positions of $-21, -7, 7$, and 21 with overall subcarrier index from -26 to 26, including 48 data subcarriers and 1 null carrier in the middle. The application of the pilot subcarriers in optical systems is discussed in Chapters 7 and 8.

4.7 ADC/DAC Impact

OFDM has the advantages of resilience to channel dispersion, ease of channel and phase estimation, and capability of dynamic bit and power loading, but it has the disadvantages of high peak-to-average power ratio (PAPR) and high sensitivity to phase and frequency noise. In particular, the high PAPR potentially poses a challenge for signal processing elements such as ADC/DACs that have limited resolution. In this section, we investigate the impact of ADC/DAC on OFDM signals.

Although the impact of ADC/DAC has been numerically simulated,[2,29] we review the analytical approach developed in Mestdagh et al.[30] As reviewed in Chapter 2, OFDM occasionally exhibits very high amplitude; to generate these rare events of high peaks without modification will waste the precious resource of the limited resolution of DAC/ADC. Therefore, it is necessary to bound the range of the signal amplitude within an appropriate level. Denote $x[n]$ and $x'[n]$ as the unclipped and clipped real or imaginary components of an OFDM signal, respectively. Clipping through digital signal processing can be expressed as

$$x'[n] = \begin{cases} x[n], & -k\sigma \leq x[n] \leq k\sigma \\ -k\sigma, & x[n] < -k\sigma \\ k\sigma, & x[n] > k\sigma \end{cases} \tag{4.67}$$

where σ is the standard deviation of $x[n]$, and k is the clipping ratio factor to be determined. Figure 4.9 shows an arbitrary OFDM waveform that is bounded between $-k\sigma$ and $k\sigma$ via clipping. This applies to both DAC and ADC. Obviously, there is a penalty associated with the clipping. For a large number of subcarriers (>10), the ensemble average of the amplitude $x[n]$ can be accurately modeled as a Gaussian random process by the argument of the central limit theorem as a zero mean and a variance σ^2. Thus, the probability density function of $x[n]$ is given by

$$p(x) = \frac{1}{\sqrt{2\pi\sigma^2}} e^{-\frac{x^2}{2\sigma^2}} \tag{4.68}$$

From Figure 4.9, the clipping noise n_{clip} could be estimated as the power lost due to the clipping given by

$$n_{\text{clip}}^2 = 2 \int_{k\delta}^{\infty} (x - k\delta)^2 \frac{1}{\sqrt{2\pi\sigma^2}} e^{-\frac{x^2}{2\sigma^2}} dx \approx 2\sqrt{\frac{2}{\pi}} \sigma^2 k^{-3} e^{-k^2/2} \tag{4.69}$$

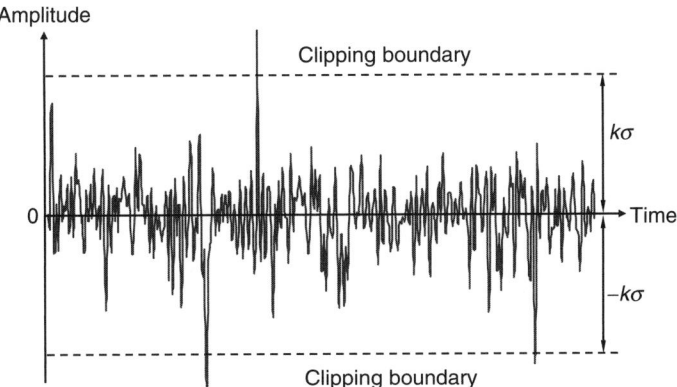

Figure 4.9: An arbitrary waveform of an OFDM signal that is clipped at the boundary of $-k\sigma$ and $k\sigma$.

It has been shown that for q-bit resolution ADC or DAC, the quantization error n_q is given by[24]

$$n_q^2 = \frac{1}{12} \times 2^{-2q} = \frac{(k\delta)^2 2^{-2q}}{3} \tag{4.70}$$

The signal to the combined signal processing noise, n_c^2, from both the clipping and quantization is thus given by

$$\text{SNR} = \frac{\sigma^2}{n_c^2} = \frac{\sigma^2}{n_{\text{clip}}^2 + n_q^2} = \left(2\sqrt{\frac{2}{\pi}} k^{-3} e^{-k^2/2} + \frac{k^2 2^{-2q}}{3} \right)^{-1} \tag{4.71}$$

Figure 4.10 shows the SNR as a function of the clipping ratio when only the clipping noise and quantization noise are considered, which are expressed in Eq. (4.71). It can be seen that for a given number of bit resolution, there is an optimal clipping ratio k, which is respectively 2.4, 2.7, 3, and 3.4 for the bit resolution of 3, 4, 5, and 6. In the same way as we derived the SNR penalty from the phase noise in Chapter 2 (Eqs. 2.43–2.45), we can characterize the signal processing noise penalty as an effective SNR penalty $\Delta\gamma$(dB) given by

$$\Delta\gamma(\text{dB}) = \frac{10}{\ln(10)} \left(2\sqrt{\frac{2}{\pi}} k^{-3} e^{-k^2/2} + \frac{k^2 2^{-2q}}{3} \right) \gamma \tag{4.72}$$

where γ is the target SNR. Figure 4.11 shows the SNR penalty for 4-QAM and 16-QAM OFDM signals with the target SNR γ, respectively, at the values of 9.8 dB and 15.8 dB for a BER of 10^{-3}. To maintain the SNR penalty below 0.5 dB at a BER of 10^{-3}, the bit resolution for 4-QAM and 16-QAM should be higher than 4 and 6 bits, respectively.

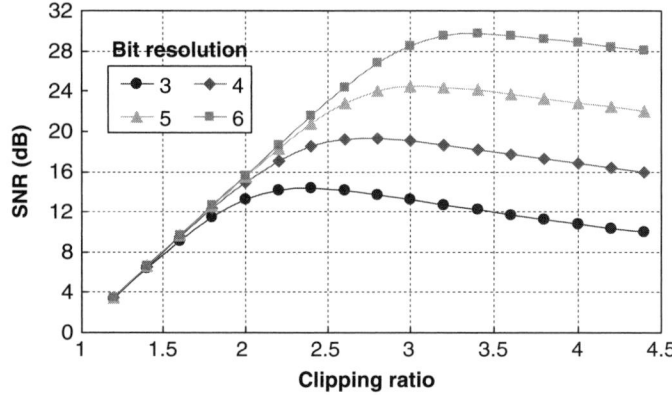

Figure 4.10: SNR as a function of clipping ratio with varying ADC/DAC resolution.

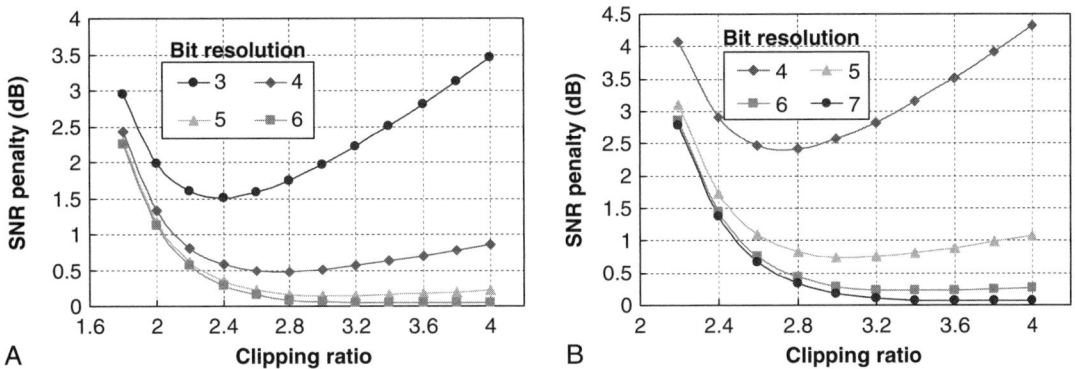

Figure 4.11: SNR penalty against the clipping ratio with varying ADC/DAC bit resolution for (a) 4-QAM and (b) 16-QAM OFDM systems.

4.8 MIMO-OFDM Perspective

MIMO systems emerged in wireless communications approximately a decade ago, based on the seminal works of Telatar[31] and Foschini.[32] The basic idea is simple but powerful. By using multiple transmitters (transmit diversity) and (or) multiple receivers (receive diversity), the channel capacity is dramatically enhanced or outage probability is greatly reduced, even under the hostile deep fading wireless environment. These early works were further strengthened by other fundamental research in MIMO coding schemes, such as space–time trellis coding by Tarokh et al.[33] and space–time coding by Alamouti.[34] These theoretical ideas were supported by the groundbreaking experimental demonstration from Bell Laboratories of its Bell Laboratories Layered Space–Time (BLAST) coding techniques, which showed approximately a 10-fold increase in capacity. Since then, MIMO systems have become a fast progressing and vibrant leading research topic in both industry and academia. There may be many competing proposals for the future mobile communication systems (4G), but common to all are the MIMO techniques to increase capacity or reduce the outage probability.

On the optical front, the earliest optical MIMO was introduced by Stuart[35] in 2000 for multimode fiber short-range communications, in which he argued that due to the large number of modes in multimode fiber, communications using MIMO techniques could provide large capacity in multimode fibers. However, further work was thwarted by the complexity of the optical MIMO systems due to lack of optoelectronics integration. A similar approach has been extended to coherent optical MIMO in multimode fibers in which high-level QAM modulation can be supported.[36] The MIMO concept has also been applied to the free-space optical communications.[37–40] However, the rapid advancement in photonic integrated circuits and high-speed digital signal processing reignited interest in optical

MIMO. There has been renewed interest in optical MIMO systems to use multimode fibers for optical interconnect.[41] Another major advancement is the optical MIMO in single-carrier systems proposed by Li and Han[42,43] and in multicarrier systems proposed by Shieh et al.[44,45] The 100 Gb/s polarization domain MIMO-OFDM has been demonstrated over 1000-km standard single-mode fiber (SSMF) transmission by various groups.[46,47] From a theoretical standpoint, multimode fiber (MMF) possesses more capacity than SMF and has less nonlinearity due to its larger core area. It is proposed that MMF replace the SMF in long reach (>100 km) transmission systems, in which MIMO-OFDM signal processing will be required for treating polarization multiplexing and mode multiplexing.[48] MMF based transmission systems with MIMO-OFDM signal processing can potentially break the 100 Tb/s per fiber barrier that exists in SMF based systems.

4.8.1 MIMO Fundamentals

The benefit of MIMO depends very much on the channel model. In this section, we introduce some basic MIMO concepts. Some of the concepts may be applicable to wireless channels or MMF channels but not to SMF systems. The application of the MIMO concept is discussed in detail later in this book.

4.8.1.1 Multiple-Transmit Multiple-Receive Systems

Figure 4.12 shows the generic multiple-transmit multiple-receive systems that consist of M transmitters and N receivers. The M transmit signals are presented in a column vector as $s = (s_1, s_2, s_3, \ldots, s_M)^T$, where s_i is the signal for the ith element. Similarly, the N receive signals are represented in a column vector as $r = (r_1, r_2, r_3, \ldots, r_M)^T$. The index in the time domain is temporarily ignored for the sake of simplicity. It can be shown that the transmit

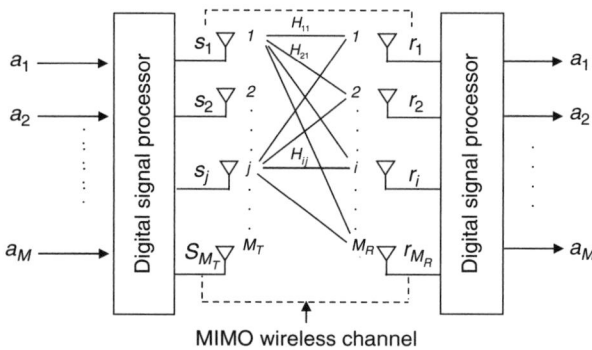

Figure 4.12: Conceptual diagram of a MIMO wireless system with M_T transmitters and M_R receivers.

and receive relationship is given by

$$r = Hs + n \tag{4.73}$$

where H is the $M \times N$ MIMO channel matrix, and n is the $N \times 1$ receiver noise matrix. H_{ij} is interpreted as the coupling coefficient of the jth transmit element to the ith receive element as shown in Figure 4.12.

Telatar[31] and Foschini[32] presented the original proposal of the MIMO technique, and the mathematical derivation of the channel capacity can be found in those references. We only show the well-known capacity formula for a MIMO channel, expressed as

$$C = \log_2 \left(\det \left(I_M + \frac{s}{Mn_0} HH^+ \right) \right) \tag{4.74}$$

where C is the capacity of the MIMO channel, "det" is the determinant of a matrix, s is the average transmitter power, n_0 is the noise power, and M is the number of transmitters. Note that HH^+ is an $N \times N$ positive defined Hermitian matrix. Using its eigen-decomposition, HH^+ can be expressed as

$$HH^+ = U\Lambda U^+ \tag{4.75}$$

where U is a unitary matrix, and $\Lambda = \text{diag}\{\lambda_1, \lambda_2, \ldots, \lambda_N\}$ is a diagonal matrix where we have assumed that the eigenvalues are already placed in order such that $\lambda_1 > \lambda_2 > \ldots > \lambda_N$. Using the following matrix manipulation,

$$\det \left(I_M + \frac{s}{Mn_0} HH^+ \right) = \det \left(I_M + \frac{s}{Mn_0} U\Lambda U^+ \right) = \det \left(U \left(I_M + \frac{s}{Mn_0} \Lambda \right) U^+ \right)$$
$$= \det \left(\left(I_M + \frac{s}{Mn_0} \Lambda \right) U^+ U \right) = \det \left(I_M + \frac{s}{Mn_0} \Lambda \right) = \prod_{I=1}^{r} \left(1 + \frac{s}{Mn_0} \lambda_i \right) \tag{4.76}$$

where r is the rank of H and is bounded by $\min(M, N)$. Using the result of Eq. (4.76), the channel capacity can be rewritten as

$$C = \log_2 \left(\det \left(I_M + \frac{s}{Mn_0} HH^+ \right) \right) = \sum_{i=1}^{r} \log_2 \left(1 + \frac{s}{N \cdot n_0} \lambda_i \right) \tag{4.77}$$

Equation (4.77) shows that the capacity of a MIMO channel can be considered as a summation of r parallel single-input single-output (SISO) channels. The general conclusion has been shown by Foschini[32] and Telatar[31] that it grows linearly with r, or $\min(M, N)$. Obviously, this enhancement of the capacity is dependent on the properties of the eigenvalues of the specific channel matrix, H. If they are dominated by a few large eigenvalues, then a linear increase with r does not occur. Nevertheless, for some common channels, the eigenvalues have a known limiting distribution[49] and tend to be spaced out along a range of this distribution. Therefore, it is unlikely that most of the eigenvalues are very small and thus the linear capacity enhancement through MIMO can be obtained.

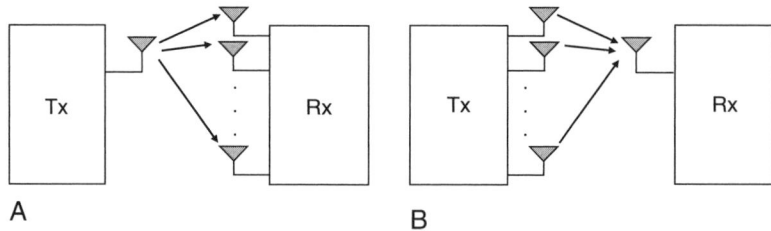

Figure 4.13: Architecture of (a) SIMO and (b) MISO systems for the purpose of illustration of array gain.

4.8.1.2 Array Gain

Array gain is defined as the increase in SNR in MIMO systems compared with that of SISO systems. It is a result of coherent combining of multiple signals, either from multiple transmitters or from multiple receivers. Although "array" as discussed here applies to generic MIMO systems, it is best illustrated by using two scenarios—single-input multiple-output (SIMO) and multiple-input single-output (MISO) systems, as shown in Figure 4.13. From Eq. (4.77), the capacity of the SIMO systems can be shown as

$$C = \log_2\left(1 + \frac{s}{n_0}\sum_{i=1}^{N}|H_i|^2\right) = \log_2\left(1 + N\frac{sh_1}{n_0}\right) \tag{4.78}$$

The channel H matrix is reduced to an $N \times 1$ column matrix in SIMO systems. We have assumed that each receive antenna has identical distribution for H_i. From Eq. (4.78), it follows that the array gain is N times that of the SISO systems; that is, the array gain is equal to the number of the receivers. For the MISO systems, there are two scenarios that give two different answers. In the first scenario, the channel is unknown to the transmitter, and thus equal power is used for each transmitter. The channel capacity has become

$$C = \log_2\left(1 + \frac{s}{Mn_0}\sum_{i=1}^{M}|H_i|^2\right) = \log_2\left(1 + \frac{s|H_1|^2}{n_0}\right) \tag{4.79}$$

Therefore, there is no increase in the array gain compared with SISO systems. The second scenario is that the channel is known to the transmitter, the transmitter will weigh the transmitting power in proportion to the channel coefficient, and the channel capacity becomes[50]

$$C = \log_2\left(1 + \sum_{i=1}^{M}|H_{i1}|^2\frac{s}{n_0}\right) = \log_2\left(1 + M\frac{s}{n_0}\right) \tag{4.80}$$

Therefore, the array gain is a factor of M, which is the number of the transmitters.

4.8.1.3 Diversity Gain

Wireless channel suffers from multipath fading, whereas optical channel is also susceptible to polarization fading. The principle of diversity is to transmit multiple copies of signals over the dimension of time, frequency, space, or polarization such that the fading could be mitigated, or even completely recovered in the case of fiber-optic systems. Diversity is categorized according to the diversity dimension used as follows:

Time diversity: Time is a natural dimension for signal transmission. This can be obtained through channel coding or time interleaving. However, to achieve diversity gain through time diversity, the communication channel has to vary between the signal coherence time and interleaving time; that is, the channel coherence time should be much smaller than the interleaving time.

Frequency diversity: Multiple copies of the signals are spread across different parts of the OFDM spectrum. As in the case of time diversity, it is effective only when frequency selectivity is significant where the multiple copies of the signal are spread; that is, the coherence bandwidth of the channel is narrower than the bandwidth in which the signal is spread.

Spatial diversity: This can be considered as antenna diversity in wireless systems or modal diversity in MMF fiber systems. This is effective when the antenna spacing is larger than the coherent distance of the signal.

Polarization diversity: Polarization provides another dimension for the signal to achieve diversity gain. Interestingly, the TE and TM polarizations in fiber do not satisfy the independent fading that is required by the conventional diversity gain. As a matter of fact, the correlation between the two polarization components enables the polarization diversity to completely eliminate the polarization fading as discussed in Chapter 5.

Base on where diversity is applied, it can also be classified into two categories:

Receive diversity: Multiple receivers are used to coherently combine the received signal. Maximum ratio combining is often used to achieve diversity gain. In the broadcasting network such as mobile networks, it may be cost-prohibitive to build complicated multiple antennas for the mobile user. Thus, receive diversity can be better explored in the base station.

Transmit diversity: Replicas of the signal are generated at transmit through the dimension of time, frequency, space, or polarization. The receiver can be as simple as one element, which is extremely attractive for mobile applications in which cost can be more easily absorbed in the base station. Alamouti's[34] scheme is one of the most famous examples of transmit diversity, for which no channel condition is required in the transmitter.

4.8.1.4 Multiplexing Gain

As discussed previously, MIMO systems can be used to increase the capacity of a communication channel. The capacity of MIMO systems is measured against that of SISO systems with the same power and the same bandwidth. To achieve a full multiplexing gain requires fully independent usage of the antennas or each data pipe; however, this sacrifices the diversity gain, and BER performance tends to suffer. Additional coding may incur some overhead, but the resulting diversity gain in fact can help improve BER performance, even though the data rate is kept at the same level. Therefore, it is of much interest to trade off the multiplexing gain with the diversity gain, which has been extensively investigated by Zheng and Tse[51] and Heath and Paulraj.[52]

References

1. Tang Y, Shieh W, Yi X, Evans R. Optimum design for RF-to-optical up-converter in coherent optical OFDM systems. *IEEE Photon Technol Lett* 2007;**19**:483–5.
2. Hara S, Prasad R. *Multicarrier Techniques for 4G Mobile Communications*. Boston: Artech House; 2003.
3. Hanzo L, Munster M, Choi BJ, Keller T. *OFDM and MC-CDMA for Broadband Multi-User Communications, WLANs and Broadcasting*. New York: Wiley; 2003.
4. Bao H, Shieh W. Transmission simulation of coherent optical OFDM signals in WDM systems. *Opt Express* 2007;**15**:4410–8.
5. Lowery J, Wang S, Premaratne M. Calculation of power limit due to fiber nonlinearity in optical OFDM systems. *Opt Express* 2007;**15**:13282–7.
6. Nazarathy M, Khurgin J, Weidenfeld R, et al. Phased-array cancellation of nonlinear FWM in coherent OFDM dispersive multi-span links. *Opt Express* 2008;**16**:15777–810.
7. Djordjevic IB. PMD compensation in fiber-optic communication systems with direct detection using LDPC-coded OFDM. *Opt Express* 2007;**15**:3692–701.
8. Schmidl TM, Cox DC. Robust frequency and timing synchronization for OFDM. *IEEE Trans Commun* 1997;**45**:1613–21.
9. Ip E, Kahn JP, Anthon D, Hutchins J. Linewidth measurements of MEMS-based tunable lasers for phase-locking applications. *IEEE Photon Technol Lett* 2005;**17**:2029–31.
10. Minn H, Bhargava VK, Letaief KB. A robust timing and frequency synchronization for OFDM systems. *IEEE Trans Wireless Commun* 2003;**2**:822–39.
11. Pollet T, Van Bladel M, Moeneclaey M. BER sensitivity of OFDM systems to carrier frequency offset and Wiener phase noise. *IEEE Trans Commun* 1995;**43**:191–3.
12. Armstrong J. Analysis of new and existing methods of reducing intercarrier interference due to carrier frequency offset in OFDM. *IEEE Trans Commun* 1999;**47**:365–9.
13. Sari H, Karam G, Jeanclaude I. Transmission techniques for digital terrestrial TV broadcasting. *IEEE Commun Magazine* 1995;**33**:100–9.
14. Moose P. A technique for orthogonal frequency division multiplexing frequency offset correction. *IEEE Trans Commun* 1994;**42**:2908–14.
15. Hara S, Mouri M, Okada M, Morinaga N. Transmission performance analysis of multi-carrier modulation in frequency selective fast Rayleigh fading channel. *Wireless Personal Commun* 1996;**2**:335–56.

16. Yi X, Shieh W, Tang Y. Phase estimation for coherent optical OFDM. *IEEE Photon Technol Lett* 2007;**19**:919–21.
17. Shieh W. Maximum-likelihood phase and channel estimation for coherent optical OFDM. *IEEE Photon Technol Lett* 2008;**20**:605–7.
18. Wu S, Bar-Ness Y. A phase noise suppression algorithm for OFDM-based WLANs. *IEEE Commun Lett* 2002;**6**:535–7.
19. Li Y, Cimini LJ, Sollenberger NR. Robust channel estimation for OFDM systems with rapid dispersive fading channels. *IEEE Trans Commun* 1998;**46**:902–15.
20. Edfors O, Sandell M, van de Beek JJ, Wilson SK. OFDM channel estimation by singular value decomposition. *IEEE Trans Commun* 1998;**46**:931–9.
21. Li Y. Simplified channel estimation for OFDM systems with multiple transmit antennas. *IEEE Trans Wireless Commun* 2002;**1**:67–75.
22. Morelli M, Mengali U. A comparison of pilot-aided channel estimation methods for OFDM systems. *IEEE Trans Signal Processing* 2001;**49**:3065–73.
23. Negi R, Cioffi J. Pilot tone selection for channel estimation in a mobile OFDM system. *IEEE Trans Consumer Elect* 1998;**44**:1122–8.
24. Proakis J. *Digital Communications*. Boston: WCB/McGraw-Hill; 1995.
25. Proakis J. *Digital Signal Processing: Principles, Algorithms, and Applications*. Upper Saddle River, NJ: Prentice Hall; 1996.
26. Hsieh M, Wei C. Channel estimation for OFDM systems based on comb-type pilot arrangement in frequency selective fading channels. *IEEE Trans Consumer Elect* 1998;**44**:217–25.
27. Scharf LL. *Statistical Signal Processing: Detection, Estimation, and Time Series Analysis*. Reading, MA: Addison-Wesley; 1991.
28. IEEE Std 802.11a-1999(R2003). Supplement to IEEE standard for information technology telecommunications and information exchange between systems local and metropolitan area networks specific requirements.
29. Tang Y, Ho KP, Shieh W. Coherent optical OFDM transmitter design employing predistortion. *IEEE Photon Technol Lett* 2008;**20**:954–6.
30. Mestdagh DJG, Spruyt PMP, Biran B. Effect of amplitude clipping in DMT-ADSL transceivers. *IET Elect Lett* 1993;**29**:1354–5.
31. Telatar E. Capacity of multi-antenna Gaussian channels. *Eur Trans Telecom* 1999;**10**:585–95.
32. Foschini G. On limits of wireless communications in a fading environment when using multiple antennas. *Wireless Personal Commun* 1998;**6**:311–35.
33. Tarokh V. Space–time codes for high data rate wireless communications: Performance criterion and code construction. *IEEE Trans Inform Theory* 1998;**44**:744–65.
34. Alamouti SM. A simple transmit diversity technique for wireless communications. *IEEE J Select Areas Commun* 1998;**16**:1451–8.
35. Stuart H. Dispersive multiplexing in multimode optical fiber. *Science* 2000;**289**(5477):281–3.
36. Tarighat A, Hsu R, Shah A, Sayed A, Jalali B. Fundamentals and challenges of optical multiple-input multiple-output multimode fiber links. *IEEE Commun Magazine* 2007;**45**:57–63.
37. Alqudah Y, Kavehrad M. MIMO characterization of indoor wireless optical link using a diffuse-transmission configuration. *IEEE Trans Commun* 2003;**51**:1554–60.
38. Jivkova S, Hristov BA, Kavehrad M. Power-efficient multispot-diffuse multiple-input-multiple-output approach to broad-band optical wireless communications. *IEEE Trans Vehicular Technol* 2004;**53**:882–9.
39. Wilson SG, Brandt-Pearce M, Cao Q, Baedke M. Optical repetition MIMO transmission with multipulse PPM. *IEEE J Selected Areas Commun* 2005;**23**:1901–10.

40. Djordjevic I. LDPC-coded MIMO optical communication over the atmospheric turbulence channel using Q-ary pulse-position modulation. *Opt Express* 2007;**15**:10026–32.

41. Agmon A, Nazarathy M. Broadcast MIMO over multimode optical interconnects by modal beamforming. *Opt Express* 2007;**15**:13123–8.

42. Han Y, Li G. Coherent optical communication using polarization multiple-input–multiple-output. *Opt Express* 2005;**13**:7527–34.

43. Han Y, Li G. Polarization diversity transmitter and optical nonlinearity mitigation using polarization-time coding. In: *Proc. COTA,* paper no. CThC7. Whistler, Canada; 2006.

44. Shieh W, Yi X, Ma Y, Tang Y. Theoretical and experimental study on PMD-supported transmission using polarization diversity in coherent optical OFDM systems. *Opt Express* 2007;**15**:9936–47.

45. Shieh W. Coherent optical MIMO-OFDM for optical fibre communication systems, In: *European Conf. Opt. Commun., workshop 5.* Berlin, Germany; 2007.

46. Yang Q, Ma Y, Shieh W. 107 Gb/s coherent optical OFDM reception using orthogonal band multiplexing, In: *Opt. Fiber Commun. Conf.,* paper no. PDP7. 2008.

47. Jansen SL, Morita I, Tanaka H. 10 × 121.9-Gb/s PDM-OFDM transmission with 2-b/s/Hz spectral efficiency over 1000 km of SSMF. In: *Opt. Fiber Commun. Conf.,* paper no. PDP2. 2008.

48. Tong Z, Yang Q, Ma Y, Shieh W. 21.4 Gb/s coherent optical OFDM transmission over 200 km multi-mode fiber, In: *Optoelectronics Commun. Conference (OECC),* paper no. PDP5. 2008.

49. Silverstein JW. Strong convergence of the empirical distribution of eigenvalues of large dimensional random matrices. *J Multivariate Anal* 1995;**55**:331–9.

50. Paulraj A, Nabar R, Gore D. *Introduction to Space–Time Wireless Communications.* Cambridge, UK: Cambridge University Press; 2003.

51. Zheng LZ, Tse D. Diversity and multiplexing: A fundamental tradeoff in multiple-antenna channels. *IEEE Trans Info Theory* 2003;**49**:1073–96.

52. Heath RW, Paulraj AJ. Switching between diversity and multiplexing in MIMO systems. *IEEE Trans Commun* 2005;**53**:962–8.

CHAPTER 5

Polarization Effects in Optical Fiber

5.1 Introduction

It is well known that single-mode fiber (SMF) supports two polarization modes. The asymmetry of optical fiber leads to polarization mode coupling or random polarization rotation along a sufficiently long fiber, which is called polarization mode dispersion (PMD).[1,2] This polarization effect has long been considered a nuisance rather than a benefit to take advantage of in the mainstream of optical communications. Random polarization rotation can be overcome by using a polarization maintain fiber (PMF). However, it is cost-prohibitive for long-haul applications. The rapid stride toward high-speed transmission at 40 Gb/s and beyond has spurred extensive research efforts to understand the PMD impact. The general conclusion is that PMD is a fundamental barrier to the continued increase in transmission speed. Consequently, strict PMD specifications have been written for the new-generation fiber in anticipation of future high-speed transmission systems. This indeed has led to a large deployment of new low-PMD fiber for several carriers throughout the world that decided that new fiber was the best solution to the PMD problem. However a breakthrough in the area of polarization effects came about from two advances in the field of optical communications: (1) the renaissance of coherent optical communications[3–6] and (2) the arrival of optical orthogonal frequency-division multiplexing (OFDM) in optical communications.[7–13] With these two techniques, PMD is no longer a fundamental barrier to high-speed transmission and can be effectively estimated and compensated. Nevertheless, it is instructive to note that the application of coherent and multicarrier technologies does not imply the end of PMD research in optical communications. In fact, PMD has re-emerged as a critically important channel characteristic that needs to be re-examined.

In this chapter, we perform a study on polarization effects from the standpoint of optical OFDM systems. The chapter is organized as follows: Section 4.2 deals with the fundamentals of the polarization dispersion effect, from its basic mathematical representation to its impact on conventional optical communication systems. Section 4.3 introduces the basics of polarization-dependent loss (PDL). The coherent optical (CO)-OFDM transmission system is in essence a 2 × 2 multiple-input multiple-output (MIMO)-OFDM channel in the presence of polarization dispersion effects, explained in Section 4.4. Section 4.5 reviews the experimental and theoretical work on various polarization domain MIMO-OFDM systems. Finally, Section 4.6 discusses nonlinear PMD effects in optical fiber.

5.2 Polarization Dispersion Effect in Optical Fiber

The polarization dispersion effect has been studied since the very early stages of optical fiber communications. However, it has only been considered synonymous with PMD since Poole and Wagner's seminal work.[1] In the context of optical OFDM, it seems that the channel transfer function of Jones matrix is more convenient for describing the dispersion effect. In this section, we discuss the fundamentals of the polarization dispersion effect, including its origin, the Jones matrix representation, the concept of PMD, the stochastic properties of PMD, the autocorrelation function of the Jones matrix, and polarization dispersion impairment in conventional transmission systems.

5.2.1 The Origin of Polarization Dispersion Effect

SMF supports only the fundamental mode, particularly the HE11 mode, and cuts off all the other modes. However, SMF is not truly "single-mode" but, rather, contains two degenerate modes of orthogonal polarization. The "degeneracy" is broken when there is asymmetry in the fiber geometry, either from the manufacture process or mechanical stress after installation. This is manifested as fiber birefringence, which is given by

$$B = |n_x - n_y| \tag{5.1}$$

where n_x and n_y are, respectively, the index of refraction for x and y polarizations. As a consequence of the birefringence, an optical signal will be coupled from one polarization mode to the other when passing through the fiber if the launch angle is not aligned with the eigen axis as shown in Figure 5.1. The period of this coupling, or beat length L_B, is given by[14]

$$L_B = \lambda/B \tag{5.2}$$

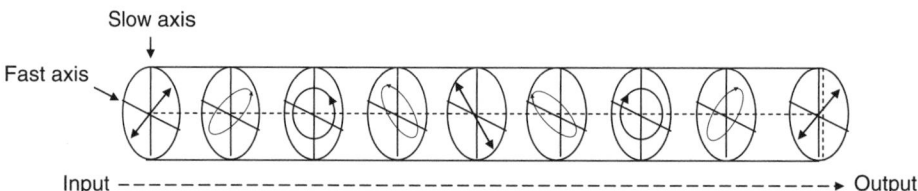

Figure 5.1: Illustration of the state of polarization evolution when an optical signal is launched at 45 degree with respect to the polarization eigen axes. Evolution of one beat length of the birefringence fiber is shown.

In a typical SMF, the birefringence coefficient B is approximately 10^{-7}, or the beat length is approximately 10 m at 1.5 μm. This implies that after passing through 10 m SMF, the signal will tend to lose its original polarization as shown in Figure 5.1. This is especially problematic for some applications in which the polarization needs to be maintained. For example, the connection between a laser and an external modulator is very polarization sensitive. Fortunately, such polarization rotation can be resolved by using PMF. The principle underlying PMF is that by artificially introducing a birefringence that is several orders of magnitude stronger than that of SMF, the eigen axis of the PMF is fixed irrespective of the mechanical stress introduced in handling. A relatively large birefringence is introduced via enduring stress in the core through a noncircular cladding cross section or through rods of another material included within the cladding. Various shapes of rod can be used, among which "panda" and "bow tie" are most well known. By launching the light along either of the eigen axes of the polarization, the signal will stay on the eigen axis, immune to the mechanical environment of the fiber.

5.2.2 Jones Vector and Jones Matrix Representation

Because there are two polarization components in an optical fiber, the electrical field of a signal can be readily represented with a Jones vector, given by[2,15]

$$e(t) = \begin{bmatrix} e_x(t) \\ e_y(t) \end{bmatrix} \tag{5.3}$$

Throughout the chapter, to distinguish from a scalar, bold font of a quality indicates a complex-valued matrix or vector, whereas an arrow bar over a quality indicates a real-valued Euclidian three-dimensional vector. For an optical medium with polarization dispersion, the time domain channel model that relates the output electrical field to the input one can be generalized as

$$e_{\text{out}}(t) = m(t) \otimes e_{\text{in}}(t) \tag{5.4}$$

where

$$e_{\text{in}}(t) = \begin{bmatrix} e_x(t) \\ e_y(t) \end{bmatrix}, \quad e_{\text{out}}(t) = \begin{bmatrix} e'_x(t) \\ e'_y(t) \end{bmatrix} \tag{5.5}$$

$$m(t) = \begin{bmatrix} m_{xx} & m_{xy} \\ m_{yx} & m_{yy} \end{bmatrix} \tag{5.6}$$

where $e_{\text{in}}(t)$ and $e_{\text{out}}(t)$ are, respectively, the input and output fields, $m(t)$ is the impulse response function of the polarization medium, and $\otimes$ stands for convolution. Although it is generally convenient to represent the channel model in the time domain in RF wireless systems, it is more natural to express the channel response in the frequency domain for optical communication systems. Taking the Fourier transform of Eq. (5.4), we obtain

$$E_{out}(\omega) = M(\omega)E_{in}(\omega) \tag{5.7}$$

where

$$E_{in}(\omega) = \begin{bmatrix} E_x(\omega) \\ E_y(\omega) \end{bmatrix}, \quad E_{out}(\omega) = \begin{bmatrix} E'_x(\omega) \\ E'_y(\omega) \end{bmatrix} \tag{5.8}$$

$$M(\omega) = \begin{bmatrix} M_{xx} & M_{xy} \\ M_{yx} & M_{yy} \end{bmatrix} \tag{5.9}$$

$E_{in}(\omega)$, $E_{out}(\omega)$, and $M(\omega)$ are the Fourier transform of $e_{in}(t)$, $e_{out}(t)$, and $m(t)$; for instance,

$$E_{in}(\omega) = \mathfrak{F}\{e_{in}(t)\} = \frac{1}{2\pi} \int_{-\infty}^{\infty} e_{in}(t)e^{-j\omega t}dt \; . \; M(\omega)$$

is the commonly used Jones matrix that has been extensively studied in optical communications.[1,2]

5.2.3 Principal State of Polarization and Differential Group Delay

When discussing the fiber birefringence property in Section 5.2.1, we assumed that the polarization eigen axis is invariant throughout a fiber. This is not the case for a typical fiber, in which the eigen axis becomes gradually decorrelated after a sufficiently long distance.[16] Nevertheless, as shown in Poole and Wagner's[1] seminal work, there always exist two eigen modes, for which the polarization dispersion is zero to first order. Since their work, the study of polarization dispersion has been focused on the investigation of PMD.

Ignoring the fiber loss, the Jones matrix $M(\omega)$ in Eq. (5.7) can be assumed unitary, which is in a Cayley–Klein representation given by[2]

$$M(\omega) = \begin{bmatrix} \alpha(\omega) & \beta(\omega) \\ -\beta^*(\omega) & \alpha^*(\omega) \end{bmatrix} \tag{5.10}$$

$$MM^+ = I \tag{5.11}$$

where the superscript "+" stands for the Hermitian conjugate or simultaneous complex conjugate and transpose operation of a matrix. Equation (5.11) simply states the unitary condition for the Jones matrix $M(\omega)$. We also define a matrix $D \equiv jM_\omega M^+$, where the ω is the derivation over ω. By differentiating Eq. (5.11) over ω, it is easy to show that

$$D = D^+ \tag{5.12}$$

Namely, D is a Hermitian matrix. Now differentiating Eq. (5.7) with respect to ω, we have

$$\frac{\partial}{\partial \omega} E_{\text{out}}(\omega) = -jDE_{\text{out}}(\omega) \tag{5.13}$$

Because D is a Hermitian matrix and its trace equals zero, it has two eigenvalues with the same magnitude but opposite sign, and subsequently it can be expressed as a linear superposition of the Pauli matrix[2] given by

$$D = \frac{1}{2}\vec{\Omega} \cdot \vec{\sigma} = \frac{1}{2}\sum_{k=1}^{3}\Omega_k \sigma_k \tag{5.14}$$

where $\vec{\Omega} = [\Omega_1 \quad \Omega_2 \quad \Omega_3]^T$, $\vec{\sigma} = [\sigma_1 \quad \sigma_2 \quad \sigma_3]^T$, and T stands for matrix transpose. $\vec{\sigma}$ is the Pauli spin vector, each element of which is expressed as[2]

$$\sigma_1 = \begin{bmatrix} 1 & 0 \\ 0 & -1 \end{bmatrix}, \ \sigma_2 = \begin{bmatrix} 0 & 1 \\ 1 & 0 \end{bmatrix}, \ \sigma_3 = \begin{bmatrix} 0 & -j \\ j & 0 \end{bmatrix} \tag{5.15}$$

The two eigenvalues of D are thus $\pm\frac{1}{2}\Omega$, a concise denotation for $|\vec{\Omega}|$. Therefore, for any arbitrary frequency ω_0, if the out polarization is along the direction of the eigen polarization $E_{\text{out}}^0(\omega_0)$, using Eq. (5.13) and Eq. (5.14), the output field at $\omega_0 + \Delta\omega$ can be approximated as

$$E_{\text{out}}(\omega_0 + \Delta\omega) \cong E_{\text{out}}^0(\omega_0) + \frac{\partial E_{\text{out}}}{\partial \omega}\Delta\omega = E_{\text{out}}^0(\omega_0) - jDE_{\text{out}}^0(\omega_0)$$

$$= E_{\text{out}}^0(\omega_0) \pm \frac{1}{2}j\Omega\Delta\omega E_{\text{out}}^0(\omega_0) \cong e^{\pm\frac{1}{2}j\Omega\Delta\omega} E_{\text{out}}^0(\omega_0) \tag{5.16}$$

This implies that the output field $E_{\text{out}}(\omega_0 + \Delta\omega)$ is only changed by a common phase $\pm\frac{1}{2}\Omega\Delta\omega$ in comparison to $E_{\text{out}}^0(\omega_0)$, and therefore the polarization of the Jones vector $E_{\text{out}}(\omega_0 + \Delta\omega)$ is not changed. The phase shift $\pm\frac{1}{2}\Omega\Delta\omega$ has the physical interpretation of the pulse delay. The two eigen states are called principal states of polarization (PSP), and the delay difference between the two principal states, Ω, is called the differential group delay (DGD). Shieh and Kogelnik[17] extend the concept of PSP and show the existence of the dynamic eigen states (DES) of polarization for a broad range of physical parameters, such as temperature and voltage.

5.2.4 Stokes Representation of PMD and Its Statistical Properties

Often in optical communications, the polarization state is represented as a Stokes vector. The conversion from Jones vector $E = [E_x \quad E_y]^T$ to Stokes vector $\vec{s} = [s_1 \quad s_2 \quad s_3]^T$ is given by[2]

$$s_1 = |E_x|^2 - |E_y|^2, \quad s_2 = 2\,\text{Re}\left[E_x E_y^*\right], \quad s_3 = 2\,\text{Im}\left[E_x E_y^*\right] \tag{5.17}$$

or, more concisely using the Pauli matrix vector, as

$$\vec{s} = E^+ \vec{\sigma} E \tag{5.18}$$

The transformation of the input Stokes vector to the output one is represented as

$$\vec{s}_{out}(\omega) = R(\omega) \vec{s}_{in} \tag{5.19}$$

where $R(\omega)$ is an orthogonal matrix belonging to the SO(3) Lie group that is mapped from the matrix M, which belongs to the SU(2) Lie group.[2] By substituting E_{out} for E in Eq. (5.18), and differentiating Eq. (5.19) over ω, we have

$$\frac{\partial}{\partial \omega} \vec{s}_{out} = E_{in}^+ \vec{\sigma} \left(\frac{\partial}{\partial \omega} E_{out} \right) + \text{c.c.} \tag{5.20}$$

where c.c. is the complex conjugate. Substituting Eq. (5.13) and Eq. (5.14) into Eq. (5.20), we have

$$\frac{\partial}{\partial \omega} \vec{s}_{out}(\omega) = - j\frac{1}{2} E_{out}^+ \vec{\sigma}(\vec{\Omega} \cdot \vec{\sigma}) E_{out} + \text{c.c.} \tag{5.21}$$

Substituting into Eq. (5.21) a Pauli matrix identity[2]

$$\vec{\sigma}(\vec{\Omega} \cdot \vec{\sigma}) = \vec{\Omega} I + j\vec{\Omega} \times \vec{\sigma} \tag{5.22}$$

and after simple rearrangement, we obtain

$$\frac{\partial}{\partial \omega} \vec{S}_{out}(\omega) = \vec{\Omega} \times \vec{S}_{out} \tag{5.23}$$

From Eq. (5.23), it follows that to the first order the output Stokes vector is insensitive to the frequency variation when the output polarization $\vec{S}_{out}$ is aligned with the PSP state, $\vec{\Omega}$.

A long span of fiber can be approximated as cascading of many small segments of the birefringence element, and therefore the instantaneous DGD is a stochastic process. It has been shown that the magnitude of DGD, Ω, follows the Maxwellian distribution[18,19] given by

$$\rho(\Omega) = \frac{32\Omega^2}{\pi^2 \Omega_0^3} \exp\left(-\frac{4\Omega^2}{\Omega_0^2 \pi} \right) \tag{5.24}$$

Where $\rho(\Omega)$ is the PDF of Ω, and Ω_0 is the mean DGD. The density distribution of $\rho(\Omega)$ is shown in Figure 5.2. It can be seen that although most of the probability distribution is concentrated around Ω_0, there is still a tail event that extends to a relatively large value. For instance, if we use the probability of 10^{-6} as the criterion, the instant DGD Ω can be 3.7 times larger than the mean DGD.

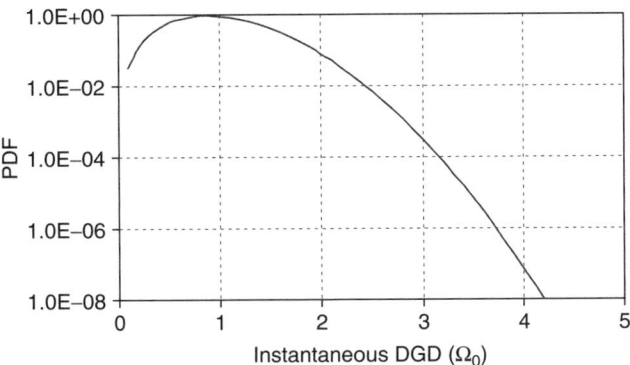

Figure 5.2: Probability density function (PDF) of instantaneous differential group delay (DGD). DGD is normalized to mean DGD, Ω_0.

Another important property of the PMD vector is its correlation bandwidth—that is, how far the frequency can deviate such that the PMD vector is still correlated to the original value. This can be determined by studying the autocorrelation function (ACF) defined as

$$R_{\vec{\Omega}\vec{\Omega}}(\Delta\omega) \equiv E\left\{\vec{\Omega}(\omega_1) \cdot \vec{\Omega}(\omega_1 + \Delta\omega)\right\} \tag{5.25}$$

The ACF has been investigated by Karlsson and Brentel,[20] and the result is given as

$$R_{\vec{\Omega}\vec{\Omega}}(\Delta\omega) = 3\frac{1 - \exp\left(-\frac{\Omega_0^2 \Delta\omega^2}{3}\right)}{\Delta\omega^2} \tag{5.26}$$

The correction bandwidth for the PMD vector can be defined as[20]

$$\Delta\omega_c = \int_{-\infty}^{\infty} R_{\vec{\Omega}\vec{\Omega}}(\Delta\omega)\mathrm{d}\Delta\omega / R_{\vec{\Omega}\vec{\Omega}}(0) = 4\sqrt{2}/\Omega_0 \tag{5.27}$$

The correlation bandwidth of the PMD vector is important for conventional non-return-to-zero (NRZ) or return-to-zero (RZ) direct detection modulation formats, where the PMD vector statistics within a signal spectrum determine the signal pulse broadening. However, in CO-OFDM systems, the correlation bandwidth for the transfer function, namely the Jones matrix, is of more direct relevance and is discussed in the following section.

5.2.5 Autocorrelation Function of the Channel Transfer Function and Coherence Bandwidth of an Optical Fiber

The ACF of the channel transfer function is important because it provides the correlation bandwidth information of the communication channel. It is critical for channel estimation in determining the number of pilot subcarriers that should be allocated.[21] For instance, for the first order of approximation, we could assign one pilot within each coherence bandwidth. The autocorrelation function of the transfer function has been studied by Bononi and Vannucci[22] and is given by

$$R_{MM} \equiv E\{\boldsymbol{M}^+(\omega)\,\boldsymbol{M}(\omega + \Delta\omega)\} = e^{-\Omega_0^2 \Delta\omega^2/8}\boldsymbol{I} \qquad (5.28)$$

where $\boldsymbol{I}$ is a 2×2 identity matrix. We use 3 dB bandwidth as the coherence bandwidth of the fiber channel. From Eq. (5.28), we have coherence bandwidth $(\Delta f)_c$ as

$$(\Delta f)_c = \frac{(\Delta\omega)_c}{2\pi} = \frac{4\sqrt{2\ln 2}}{2\pi}\Omega_0^{-1} = 0.75\Omega_0^{-1} \qquad (5.29)$$

For most of the deployed systems, the mean DGD Ω_0 is limited to 50 ps, resulting in a coherence bandwidth of 15 GHz. For the recently demonstrated 100 Gb/s transmission, the OFDM bandwidth used is only approximately 37 GHz,[23] which suggests that even three pilot subcarriers are sufficient for channel estimation and six pilot subcarriers are preferred for the purpose of oversampling. Figure 5.3 shows the coherence bandwidth as a function of the transmission reach. At a reach of 1000 km, the correlation bandwidths are, respectively, 50, 100, and 500 GHz for the fiber with PMD coefficients of 1, 0.2, and 0.05 ps/$\sqrt{\text{km}}$. Therefore, PMD not only contributes little to the total channel dispersion as opposed to the chromatic dispersion but also requires a small overhead in channel estimation because of its relatively broad correlation bandwidth.

5.2.6 Why PMD Has Been Considered the Fundamental Barrier

An optical signal traversing through PMD will split into two pulses with different delay, resulting in pulse distortion, namely intersymbol interference. The first evidence of the PMD impairment was reported by Poole et al.,[24] who showed that dispersion impairment is scaled

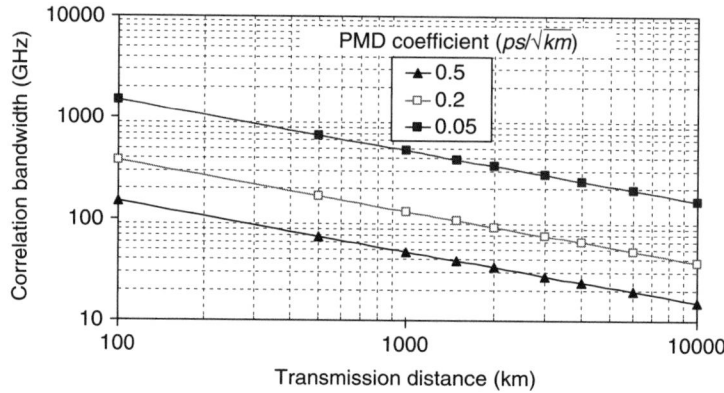

Figure 5.3: Correlation bandwidth of the fiber link transfer function against transmission distance with varying PMD coefficients.

with the square of the bit rate. As mean DGD increases relative to the bit period of the optical signal, higher order PMD contributes to the overall system penalty.[25–28] There have been extensive efforts to obtain higher order approximations for the PMD fiber transmission matrix.[29–32] Instead of using second-order Taylor expansion of the electric field, Karlsson[33] derived an analytical expression for the PMD pulse broadening valid to arbitrary order. A similar approach was also adapted by Shieh[34] where he extended the definition of PSP from the frequency domain to the time domain. The root mean square (RMS) width δ of any optical pulse can be defined as

$$\delta^2 = \langle t^2 \rangle - \langle t \rangle^2 \tag{5.30}$$

where, in the context of PMD, it has a form[26,33]

$$\langle t \rangle = -\tfrac{1}{2}\vec{\Omega}_p \cdot \vec{J} \tag{5.31}$$

$$\langle t^2 \rangle = -\int_{-\infty}^{\infty} \left[|\tilde{f}|^2 \left| \phi' \cdot \vec{J} + \tfrac{1}{2}\vec{\Omega} \right|^2 + \left(|\tilde{f}|' \right)^2 \right] \frac{d\omega}{2\pi} \tag{5.32}$$

$$|\tilde{f}|\exp(i\phi) = \tilde{f}\,\exp(i\beta), \quad t_0 = -\int_{-\infty}^{\infty} |\tilde{f}|^2 \phi' \frac{d\omega}{2\pi}, \quad \vec{\Omega}_p = \int_{-\infty}^{\infty} |\tilde{f}|^2 \vec{\Omega} \frac{d\omega}{2\pi} \tag{5.33}$$

where $\tilde{f}$ is the Fourier transform of the electric field of the input optical pulse; β is the net fiber chromatic dispersion; ϕ is the phase of $\tilde{f}\exp(i\beta)$; $\vec{J}$ is the input polarization as a Stokes vector; $\vec{\Omega}$ is the input PMD vector, with a magnitude of the DGD and a direction of the negative delay state; t_0 is the polarization-independent delay; and $\vec{\Omega}_p$ is the principal state of polarization for the pulse defined in Eq. (15) of Shieh.[34] The prime indicates the derivative over angular frequency ω.

A reasonable definition of the second-order PMD approximation is to retain all the terms up to $(\Delta\omega)^2$ in Eq. (5.30), where $\Delta\omega$ is the RMS width of the angular frequency of the input pulse. By expanding ϕ and Ω in a Taylor series around the center frequency ω_0, carrying out the integration, and keeping all the terms up to $(\Delta\omega)^2$ in Eq. (5.30), we obtain

$$\delta^2 = \delta_0^2 + \delta_1^2 + \delta_2^2 \tag{5.34}$$

$$\delta_0^2 = \int_{-\infty}^{\infty} \left(|\tilde{f}|' \right)^2 \cdot \frac{d\omega}{2\pi} \tag{5.35}$$

$$\delta_1^2 = \frac{1}{4} |\vec{\Omega}_0|^2 \left(1 - (\vec{J} \cdot \vec{e}_p)^2 \right) \tag{5.36}$$

$$\delta_2^2 = \left(\vec{J}\,\phi'' + \frac{1}{2}\vec{\Omega}' \right)^2 \cdot (\Delta\omega)^2 \tag{5.37}$$

where ϕ'' is the second derivative of the phase, $\vec{\Omega}_0$ is the PMD vector, and $\vec{\Omega}'/\vec{\Omega}''$ is the first/second derivative of PMD vector $\vec{\Omega}$ with respect to ω. δ_0^2 is the pulse dispersion without PMD. δ_1^2 in Eq. (5.34) is the first-order PMD penalty, the dominant impairment without PMD compensation. Assuming the power penalty is proportional to the pulse distortion normalized to the square of the bit period, ignoring the residual chromatic dispersion, from Eq. (5.36) and Eq. (5.37), the maximum penalties for the first- and second-order penalties ΔP_1 and ΔP_2 are

$$\Delta P_1 = A\Omega^2 T^{-2} \tag{5.38}$$

$$\Delta P_2 = B\Omega'^2 T^{-4} \tag{5.39}$$

where T is the signal bit period, or pulse width, and A and B are empirical constants dependent on the modulation format. We use some typical empirical data assuming a 1 dB penalty for a 10 Gb/s NRZ signal at a DGD of 40 ps, which gives an A of 6.25.[24] Figure 5.4 shows the first-order PMD-induced penalty as a function of the instantaneous DGD with varying data rates. It can be seen that as the data rate increases, PMD impairment increases dramatically given a fixed DGD. For future-generation transmission systems at 100 and 400 Gb/s, the DGD has to be limited to 10 and 4 ps, respectively, to bound the penalty below 1 dB. Even for newly deployed fiber, it will be difficult to meet this specification. As such, PMD impairment has been recognized as a fundamental barrier to ultra-high-speed transmission systems. One focus of research is to compensate the dispersion, and various approaches have been investigated.[35-40] One of the common approaches is to optically compensate the PMD, namely undo the fiber PMD effect at the receiver. There are two fundamental drawbacks of the optical PMD compensation: (1) The optical approach is usually per-channel based and is thus expensive and bulky, and (2) most of the compensators can deal with the first-order compensation well. However, as can be seen from Eq. (5.39), the second-order penalty still exists after the first-order compensation, and it becomes the

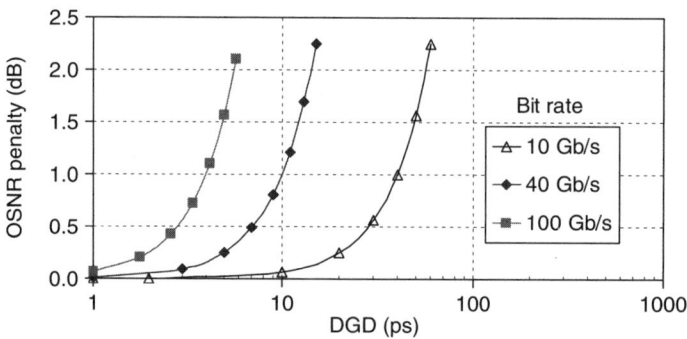

Figure 5.4: Optical signal-to-noise ratio (OSNR) penalty against DGD for NRZ systems with data rates of 10, 40, and 100 Gb/s. OSNR penalty is assumed to be 1 dB at a DGD of 40 ps for 10 Gb/s NRZ systems.

dominant effect. More troublesome is the fact that the second-order penalty acts like chromatic dispersion, scaling with the fourth power of the PMD value. Therefore, the complexity of the compensator increases dramatically for high-speed systems because a high-order or practically all-order optical compensator is needed.

5.3 Polarization-Dependent Loss

Optical components inevitably introduce insertion loss that is dependent on the input polarization. Analogous to fiber birefringence, there exist two eigen axes along which the minimum and maximum loss occur. The polarization-dependent loss/gain (PDL/G) has various physical origins. It may result from the asymmetry of the devices or the polarization-dependent gain of an erbium-doped fiber amplifier (EDFA) due to the polarization hole-burning effect. The amplitude of maximum and minimum loss can be described as[15,41]

$$\Gamma_{\max} = e^{\alpha/2}, \quad \Gamma_{\min} = e^{-\alpha/2} \tag{5.40}$$

Then the PDL can be expressed as

$$\text{PDL} = \frac{\Gamma_{\max}}{\Gamma_{\min}} = 20 * \log_{10}(e^{\alpha}) \tag{5.41}$$

The Jones matrix of a PDL element is described as[15]

$$M_{\text{PDL}} = \exp\left(-\frac{1}{2}\vec{\alpha} \cdot \vec{\sigma}\right), \vec{\alpha} = \alpha \vec{e}_{\text{PDL}} \tag{5.42}$$

where $\vec{e}_{\text{PDL}}$ is the Stokes vector along the direction of the eigen axis of PDL.

In general, any communication system can be considered as a cascade of multiple sections, each containing both birefringence and PDL elements as shown in Figure 5.5. From Eqs. (5.14) and (5.42), we derive the Jones matrix comprising N sections of PMD and PDL elements expressed as

$$T(\omega) = \prod_{l=1}^{N} \exp\left[\left(-\frac{1}{2}j\vec{\gamma}_l\omega - \frac{1}{2}\vec{\alpha}_l\right) \cdot \vec{\sigma}\right] \tag{5.43}$$

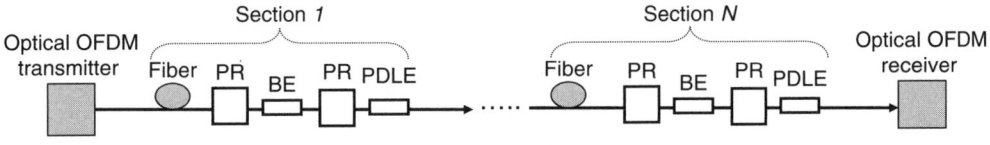

BE: Birefringence element   PDLE: PDL element   PR: Polarization rotator

Figure 5.5: A generic fiber-optic system consisting of N segments of PMD and PDL elements. Polarization rotators are used to arbitrarily control the eigen axes of birefringence and PDL elements.

where $\vec{\gamma}_l$ and $\vec{\alpha}_l$ are, respectively, the birefringence vector (equivalent to $\vec{\Omega}$ in Eq. 5.14) and the PDL vector (Eq. 5.42) for the lth section of an optical communication link.[15] Gisin and Huttner[15] showed that the eigen equation for the generalized PSP, Eqs. (5.13) and (5.14), is still valid, but $D = jT_\omega T^+$ is no longer unitary. Furthermore, the vector $\vec{\Omega}$ defined in Eq. (5.14) is no longer a real value. There are two fundamental complications associated with the addition to PDL elements:

1. Because of the introduction of PDL, the Jones matrix $T(\omega)$ is no longer unitary, and therefore the conventional concept of the polarization mode does not exist. Although the generalized polarization PSP still can be defined, its physical interpretation loses most of the significance. For a PDL-less system, the two eigen polarizations represent the slowest and fastest axes, and they do not suffer the polarization dispersion to the first order. These unique properties cannot extend to the generalized concept of PSP.[15] To the best of our knowledge, the ACF of the Jones matrix in the presence of PDL is still not known analytically. However, we expect that the coherence bandwidth derived in Eq. (5.29) should stand practically even in the presence of PDL.

2. For coherent systems that take benefit of dual polarization modes in the fiber, the existence of PDL will unavoidably weaken one of the modes and could significantly degrade the system performance. In addition, the interaction between PMD and PDL becomes a very interesting and important problem for coherent systems.[9]

5.4 Theoretical Model for Coherent Optical MIMO-OFDM Signals in the Presence of Polarization Effects

As discussed previously, propagation of an optical signal is influenced by both the PMD and PDL effects. Similar to the scalar OFDM signal model introduced in Eqs. (2.1)–(2.3), the transmitted OFDM time domain signal, $s(t)$, is described using the Jones vector given by

$$s(t) = \sum_{i=-\infty}^{+\infty} \sum_{k=-\frac{1}{2}N_{sc}+1}^{\frac{1}{2}N_{sc}} c_{ik} \Pi(t - iT_s) \exp(j2\pi f_k(t - iT_s)) \tag{5.44}$$

$$s(t) = \begin{bmatrix} s_x \\ s_y \end{bmatrix}, \quad c_{ik} = \begin{bmatrix} c_x^{ik} \\ c_y^{ik} \end{bmatrix} \tag{5.45}$$

$$f_k = \frac{k - 1}{t_s} \tag{5.46}$$

$$s_k(t) = \Pi(t) \exp(j2\pi f_k t) \tag{5.47}$$

$$\Pi(t) = \begin{cases} 1, & (0 < t \leq T_s) \\ 0, & (t \leq 0, t > T_s) \end{cases} \tag{5.48}$$

where s_x and s_y are the two polarization components for $s(t)$ in the time domain; c_{ik} is the transmitted OFDM information symbol in the form of Jones vector for the kth subcarrier in the ith OFDM symbol; c_x^{ik} and c_y^{ik} are the two polarization components for c_{ik}; f_k is the frequency for the kth subcarrier; N_{sc} is the number of OFDM subcarriers; and T_s, ΔG, and t_s are the OFDM symbol period, guard interval length, and observation period, respectively. The Jones vector c_{ik} is employed to describe the generic OFDM information symbol regardless of any polarization configuration for the OFDM transmitter. In particular, c_{ik} encompasses various modes of the polarization generation, including single polarization, polarization multiplexing, and polarization modulation, because they all can be represented by a two-element Jones vector c_{ik}. The different scheme of polarization modulation for the transmitted information symbol is automatically dealt with in the initialization phase of OFDM signal processing by sending known training symbols.

We select a guard interval long enough to handle the fiber dispersion, including PMD and chromatic dispersion (CD). This timing margin condition is given by[10]

$$\frac{c}{f_O^2}|D_t| \cdot N_{sc} \cdot \Delta f + \text{DGD}_{max} \leq \Delta G \tag{5.49}$$

where f_O is the frequency of the optical carrier, c is the speed of light, D_t is the total accumulated chromatic dispersion in units of ps/pm, N_{sc} is the number of subcarriers, Δf is the subcarrier channel spacing, and DGD_{max} is the maximum budgeted DGD, which is approximately 3.5 times the mean DGD to have sufficient margin.

Following the same procedure as that of Shieh and Athaudage,[42] using the frequency domain channel transfer function for the polarization effects in Gisin and Huttner,[15] and assuming the use of a sufficiently long symbol period, we arrive at the received symbol given by

$$c'_{ki} = e^{j\phi_i} e^{j\Phi_D(f_k)} T_k c_{ki} + n_{ki} \tag{5.50}$$

$$T_k = \prod_{l=1}^{N} \exp\left\{\left(-\frac{1}{2}j\vec{\gamma}_l 2\pi f_k - \frac{1}{2}\vec{\alpha}_l\right) \cdot \vec{\sigma}\right\} \tag{5.51}$$

$$\Phi_D(f_k) = \pi \cdot c \cdot D_t \cdot f_k^2 / f_O^2 \tag{5.52}$$

where $c'_{ki} = [c_x^{'ki} \quad c_y^{'ki}]^T$ is the received information symbol in the form of the Jones vector for the kth subcarrier in the ith OFDM symbol, $n_{ki} = [n_x^{ki} \quad n_y^{ki}]^T$ is the noise for the two polarization components, $\Phi_D(f_k)$ is the phase dispersion due to the fiber CD,[41] and ϕ_i is the OFDM common phase error (CPE)[43] due to the phase noises from the lasers and RF local oscillators (LOs) at both the transmitter and receiver. ϕ_i is usually dominated by the laser phase noise.

The channel model and timing margin condition expressed in Eqs. (5.49)–(5.52) for a polarization dispersive optical channel was first derived by Shieh et al.,[10] who identified the optical fiber channel as a 2×2 MIMO-OFDM model. In general, in the context of MIMO systems, the architecture of the CO-OFDM system in the fiber optic channel is divided into four categories according to the number of the transmitters and receivers used in the polarization dimension:

Single-input single-output (SISO): As shown in Figure 5.6a, one optical OFDM transmitter and one optical OFDM receiver are used for CO-OFDM transmission. The optical OFDM transmitter includes an RF OFDM transmitter and an OFDM RF-to-optical up-converter, whereas the optical OFDM receiver includes an OFDM optical-to-RF down-converter and an RF OFDM receiver.[44] The architectures of the OFDM up/down-converters are thoroughly discussed in Tang.[44] For instance, in the direct up/down-conversion architecture, an optical I/Q modulator can be used as the up-converter and a coherent optical receiver including an optical 90 degree hybrid and a local laser can be used as the down-converter. The SISO configuration is susceptible to the polarization mode coupling in fiber, analogous to the multipath fading impairment in SISO wireless systems.[45] A polarization controller is needed before the receiver to align the input signal polarization with the local oscillator polarization.[46,47] More important, in the presence of large PMD, due to polarization rotation between subcarriers, even the polarization controller will not function well. This is because

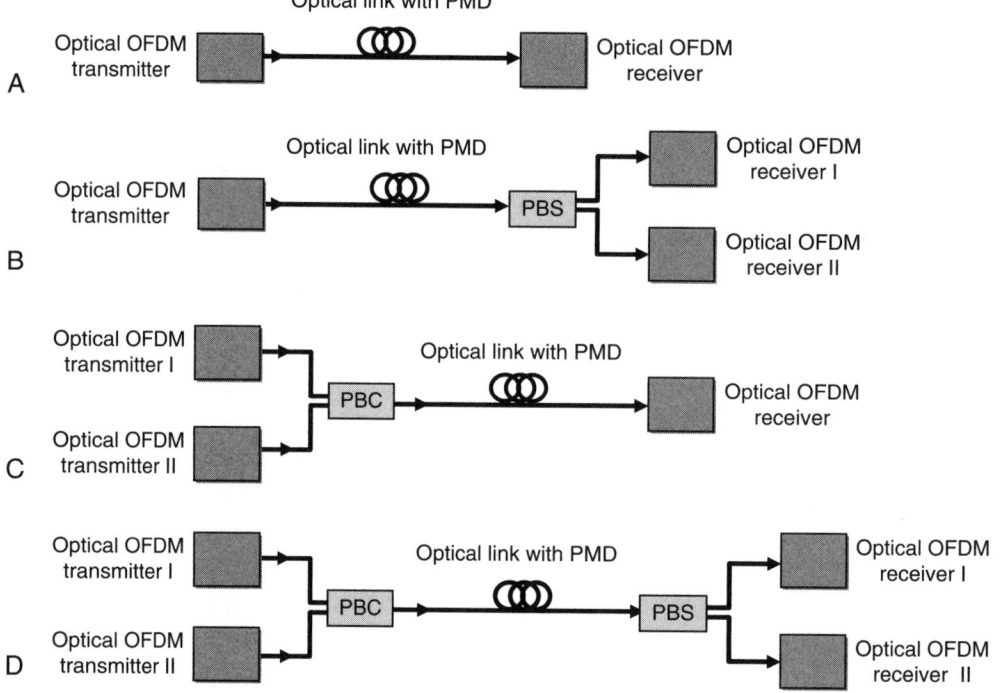

Figure 5.6: Variations of coherent optical MIMO OFDM (CO-MIMO-OFDM) models: (a) single-input single-output (SISO), (b) single-input two-output (SITO), (c) two-input single-output (TISO), and (d) two-input two-output (TITO). The optical OFDM transmitter includes an RF OFDM transmitter and an OFDM RF-to-optical up-converter. The optical OFDM receiver includes an OFDM optical-to-RF down-converter and an RF OFDM receiver. PBC/S, polarization beam combiner/splitter.

there is no uniform subcarrier polarization with which the local receiver laser can align its polarization. Subsequently, coherent optical SISO-OFDM is susceptible to polarization-induced fading and should not be implemented in the field application.

Single-input two-output (SITO): As shown in Figure 5.6b, at the transmit end, only one optical OFDM transmitter is used. However, compared with the SISO system, there are two optical OFDM receivers employed, one for each polarization. The receiver shown in Figure 5.6b is the so-called polarization diversity receiver as discussed in both single-carrier[6] and multicarrier systems.[9] As such, there is no need for optical polarization control. Furthermore, the impact of PMD on CO-OFDM transmission is simply a subcarrier polarization rotation, which can be easily treated through channel estimation and constellation reconstruction.[9] Therefore, coherent optical SITO-OFDM is resilient to PMD when the polarization diversity receiver is used. Most important, the introduction of PMD in the fiber link in fact will improve the system margin against PDL-induced fading,[9] analogous to the scenario for which the delay spread channel improves the wireless MIMO system performance.[44]

Two-input single-output (TISO): As shown in Figure 5.6c, at the transmit end, two optical OFDM transmitters are used, one for each polarization. However, at the receive end, only one optical OFDM receiver is used. This transmitter configuration is called the polarization diversity transmitter. By configuring the transmitted OFDM information symbols properly, the CO-OFDM transmission can be performed without a need for a polarization controller at the receiver. The following is one possible transmission scheme: At the transmitter, the same OFDM symbol is repeated in two consecutive OFDM symbols with orthogonal polarizations, which is essential polarization time coding,[48,49] similar to Alamouti space–time coding.[50] It can be easily shown that the polarization diversity transmitter can also achieve PMD resilience and there is no need for polarization tracking.[47] This seems to indicate that TISO has the same performance as SITO. However, in the TISO scheme, the same information symbol is repeated in two consecutive OFDM symbols, and subsequently the electrical and optical efficiency is reduced by half, and the OSNR requirement is doubled, compared with the SITO scheme.

Two-input two-output (TITO): As shown in Figure 5.6d, both a polarization diversity transmitter and a polarization diversity receiver are employed in the TITO scheme. First, in such a scheme, because the transmitted OFDM information symbol c_{ik} can be considered as polarization modulation or polarization multiplexing, the capacity is thus doubled compared with that of the SITO scheme. Because the impact of PMD is to simply rotate the subcarrier polarization and can be treated with channel estimation and constellation reconstruction, the doubling of the channel capacity will not be affected by PMD. Second, due to the polarization diversity receiver employed at the receive end, the TITO scheme does not need polarization tracking at the receiver.

Table 5.1: Performance Comparison of Four CO-MIMO-OFDM Configurations

Type of MIMO	Polarization Rotation Resilience	PMD Resilience	Relative Capacity	Relative Sensitivity (dB)
1 × 1 SISO-OFDM	No	No	Not applicable	Not applicable
1 × 2 SIMO-OFDM or SITO-OFDM	Yes	Yes	1	0
2 × 1 MISO-OFDM or TISO-OFDM	Yes	Yes	1	3
2 × 2 MIMO-OFDM or TITO-OFDM	Yes	Yes	2	0

MISO, multiple-input single-output; SIMO, single-input multiple-output; SITO, single-input two-output; TISO, two-input single-output; TITO, two-input two-output.

The performance of various CO-MIMO-OFDM models is summarized in Table 5.1. It is observed that from the framework of CO-MIMO-OFDM models, all schemes except the SISO scheme are capable of PMD-resilient transmission. However, as previously discussed, the TISO scheme has intrinsic penalties in spectral efficiency (electrical and optical) and OSNR sensitivity. Consequently, SITO- and TITO-OFDM transmission are the preferred configurations.

5.5 Simulation and Experimental Study of MIMO-OFDM Systems

Shieh et al.[7] made an early attempt to take advantage of OFDM to combat PMD by exploiting polarization receiver diversity, which in essence is a 1 × 2 SIMO-OFDM system. It was shown that the PMD in the installed fiber network can be gracefully mitigated without an optical compensator. Another report[9] suggested that it may be advantageous to introduce PMD in transmission systems to achieve polarization diversity against PDL impairment. This may present an intriguing paradigm shift in the study of PMD effects because before this work, PMD was considered an "impairment" that should be removed from the fiber as much as possible. Shieh et al.[10] performed the first experiment of 1 × 2 SIMO-OFDM, in which a CO-OFDM signal at 10.7 Gb/s was successfully recovered after 900 ps DGD and 1000 km transmission through SSMF fiber without optical dispersion compensation. A 2 × 2 MIMO-OFDM using direct up/down-conversion has been reported for 1000 km SSMF transmission at 21.4 Gb/s[51] and for 4160 km transmission of 16 channels at 52.5 Gb/s.[8] PMD mitigation using direct detection optical OFDM has also been proposed and investigated.[52,53] The most exciting development is 2×2 CO-MIMO-OFDM transmission at 100 Gb/s by three groups from the University of Melbourne, KDDI, and NTT.[23,54,55] In the remainder of this section, we review the progress of CO-MIMO-OFDM transmission in both simulation analysis and experimental demonstration.

5.5.1 *Polarization Mode Dispersion: Detriment or Benefit?*

Here, we review the interaction between PMD and PDL in a 1×2 SIMO-OFDM system. The system parameters for the simulation can be found in Shieh.[9] In a polarization impaired system, we use outage probability to characterize the system performance, which is defined as the probability that the Q penalty exceeds a certain value. Figure 5.7 shows the outage probability as a function of the Q penalty. A mean PDL of 5 dB is assumed for all simulations. It can be seen that without PMD in the system, there is a system Q penalty of 4.4 dB at the outage probability of 10^{-3}. However, if we introduce a mean PMD of 150 ps into otherwise the same system, to maintain the same outage probability of 10^{-3}, the system penalty is reduced to 3.0 dB for a 10 Gb/s system, which is a 1.4 dB improvement over the case of no PMD in the fiber link. A higher bit rate system at 40 Gb/s further reduces the Q penalty to 2.4 dB with a PMD of 150 ps, which is a 2 dB improvement over the case of no PMD. This further improvement is attributed to a higher degree of diversity with a larger number of subcarriers for a higher bit rate system. We can see that the PMD in the fiber improves the system margin for PDL-induced penalty, and the improvement is enhanced for a higher bit rate system. The similar PMD and PDL interaction in the CO-OFDM system has also been confirmed by Liu and Buchali.[21] We also note that a relatively large PMD is needed on the order of 150 ps to gain the benefit, which usually is not available in the fiber link. One possibility is to artificially introduce high PMD into the fiber or optical components. Mitigation of a large PDL in CO-OFDM systems may significantly loosen the PDL specification for various components and subsequently bring appreciable cost savings. Finally, the fiber nonlinearity including self-phase modulation and cross-phase modulation (XPM) will be reduced in PMD-supported CO-OFDM systems,

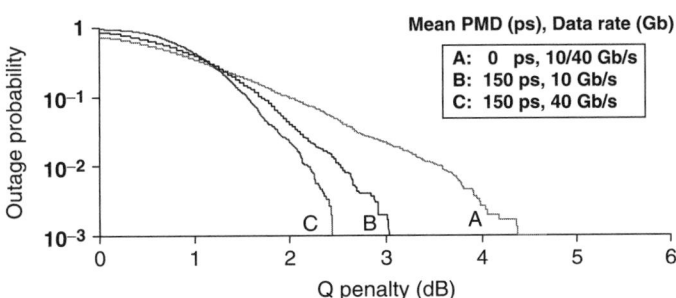

Figure 5.7: Outage probability as a function of Q penalty with varying PMD and data rate.

as previously reported for single-carrier systems.[56] Equivalently, the interchannel and intrachannel OFDM subcarrier four-wave-mixing (FWM) can be greatly reduced by a large phase mismatch between subcarriers in the presence of a large PMD.

The results can be better understood by drawing a comparison with RF MIMO-OFDM systems.[44] It is well known that the performance of MIMO-OFDM in the space domain can be improved through introduction of a delay spread into the system.[44] This is because it will bring frequency selectivity into the channel. Similarly, introducing PMD into a MIMO-OFDM channel in the polarization domain will bring frequency selectivity into the optical channel, and subsequently the performance is expected to improve. Although whether any PMD and how much PMD should be reintroduced into the fiber will be influenced by other factors, the work presented here adopts a fresh angle to explore this "old" PMD problem. Moreover, the extreme resilience to PMD for CO-OFDM transmission is extremely significant and timely to those carriers that hold a significant amount of old fiber with large PMD and still wish to upgrade to 40 Gb/s systems and beyond. This finding shows that these fibers may not be necessarily "bad" for future high-speed transmission systems.

5.5.2 1 × 2 SIMO-OFDM Experiment: Polarization Diversity Detection

This and the subsequent two sections encompass three variations of optical MIMO-OFDM systems, each organized to comprise complete experimental setup, OFDM parameter description, experimental results, and associated discussion, which can be read independently without referring to other sections.

Figure 5.8 shows the experimental setup for verifying the PMD-resilient CO-OFDM systems using 1 × 2 SIMO-OFDM architecture. The OFDM signal is generated by using a Tektronix arbitrary waveform generator (AWG) as an RF OFDM transmitter. The time domain waveform is first generated with a Matlab program including mapping $2^{15}-1$ PRBS into corresponding 77 subcarriers with QPSK encoding within multiple OFDM symbols, which are subsequently converted into a time domain using IFFT and inserted with guard interval (GI). The number of OFDM subcarriers is 128 and GI is one-eighth of the observation period. Only the middle 87 of 128 subcarriers are filled, from which 10 pilot subcarriers are used for phase estimation. This filling is done to achieve tighter spectral control by oversampling and should not be confused with the selective carrier filling due to channel fading. The BER performance is measured using all 77 data bearing channels. The baseband OFDM digital waveform is of complex value. Its real and imaginary parts are uploaded into the AWG operated at 10 GS/s, and two-channel analog signals each representing the real and imaginary components of the complex OFDM signal are generated synchronously. The generated OFDM waveform carries 10.7 Gb/s net data. These two signals are fed into I and Q ports of an optical I/Q modulator, respectively, to perform direct up-conversion of OFDM baseband signals from the RF domain to the optical domain.[43] The optical

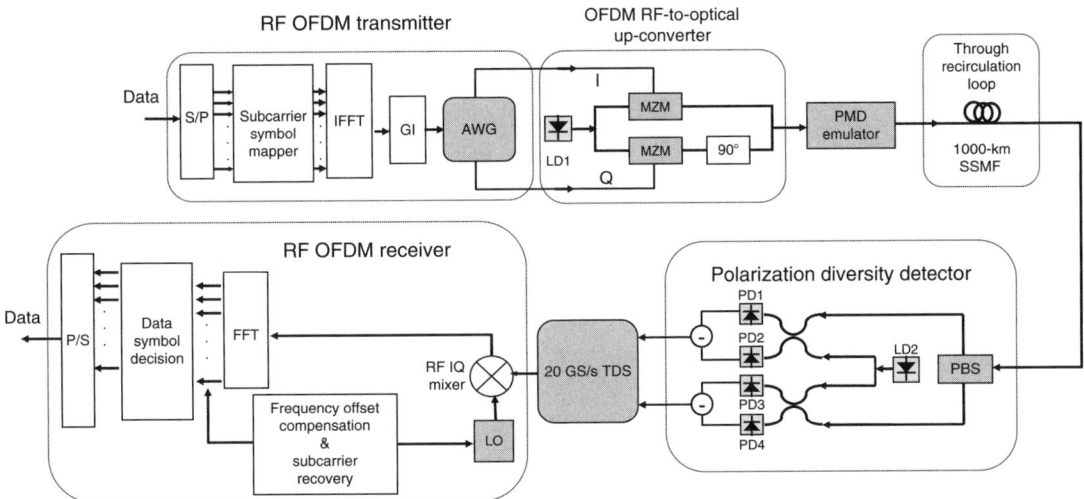

AWG: Arbitrary waveform generator MZM: Mach-Zehnder modulator PBS: Polarization beam splitter
PD: Photodiode LD: Laser diode TDS: Time-domain sampling scope

Figure 5.8: Experimental setup for 1 × 2 SIMO-OFDM transmission system.

OFDM signal from the I/Q modulator is first inserted into a home-built PMD emulator and then fed into a recirculation loop, which includes one span of 100 km SSMF and an EDFA to compensate for the loss. The advantages of such a direct up-conversion scheme are that (1) the required electrical bandwidth is less than half of that of the intermediate frequency (IF) counterpart and (2) there is no need for an image-rejection optical filter.[43] The launch power into each fiber span is set at −8 dBm to avoid nonlinearity, and the received OSNR is 14 dB after 1000 km transmission. At the receive end, the polarization diversity detection is employed. IF down-conversion is used where LO laser frequency is offset with that of the signal laser. The output optical signal from the loop is first split into two polarizations, each fed into an OFDM optical-to-RF down-converter that includes a balanced receiver and a local laser. The two RF signals for the two polarizations are then input into a Tektronix time domain sampling scope (TDS) and acquired synchronously. The RF signal traces corresponding to the 1000 km transmission are acquired at 20 GS/s and processed with a Matlab program as an RF OFDM receiver. The RF OFDM receiver signal processing involves (1) FFT window synchronization using the Schmidl format to identify the start of the OFDM symbol, (2) software down-conversion of the OFDM RF signal to baseband using a complex pilot subcarrier tone, (3) phase estimation for each OFDM symbol, (4) channel estimation in terms of a Jones vector and a Jones matrix, and (5) constellation construction for each carrier and BER computation.

The RF OFDM signal is down-converted to baseband by simply multiplying a complex residual carrier tone in software, eliminating the need for a hardware RF LO. This complex carrier tone can be supplied with the pilot symbols or pilot subcarriers as discussed in

Section 5.4. The down-converted baseband signal is segmented into blocks of 400 OFDM symbols with the cyclic prefix removed, and the individual subcarrier symbol in each OFDM symbol is recovered by using FFT.

The associated channel model after removing the phase noise ϕ_i in Eq. (5.50) is given by

$$r_{ki} = H_k c_{ki} + n_{ki}^p \tag{5.53}$$

where $r_{ki} = c'_{ki} \cdot e^{-j\phi_i}$ is the received OFDM information symbol in a Jones vector for the kth subcarrier in the ith OFDM symbol, with the phase noise removed; $H_k = e^{j\Phi_D(f_k)} T_k$ is the channel transfer function; and $n_{ki}^p = n_{ki} e^{-j\phi_i}$ is the random noise with the superscript "p" representing the common phase noise compensation. The expectation values for the received phase-corrected information symbols r_{ki} are obtained by averaging over a running window of 400 OFDM symbols. The expectation values for 4 QPSK symbols are computed separately by using received symbols r_{ki}, respectively. An error occurs when a transmitted QPSK symbol in a particular subcarrier is closer to the incorrect expectation values at the receiver. The process described here is the same as finding the channel transfer function H_k first and using Eq. (5.9) for transmitted symbol estimation and decision.

Figure 5.9 shows the BER performance of the CO-OFDM signal after 900 ps DGD and 1000 km SSMF transmission. The optical power is evenly launched into the two principal states of the PMD emulator. The measurements using other launch angles show an insignificant difference. Compared with the back-to-back case, it has less than 0.5 dB penalty at the BER of 10^{-3}. The magnitude of the PMD tolerance is shown to be independent of the data rate.[7] Therefore, we expect the same PMD resilience in absolute magnitude will hold for 40 Gb/s if fast analog-to-digital and digital-to-analog converters at 20 GS/s are available.

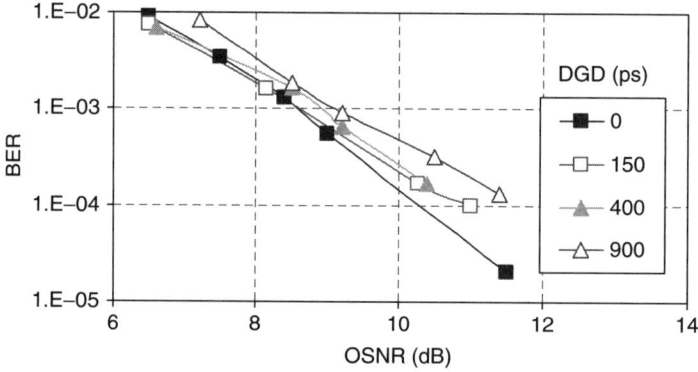

Figure 5.9: BER performance of a CO-OFDM signal in the presence of PMD.

5.5.3 2 × 1 MISO-OFDM Experiment: Polarization Time Coding for Optical Broadcast Networks

As discussed in Section 5.4, the CO-OFDM transmission system is in essence a 2 × 2 MIMO-OFDM channel. Consequently, it seems to be prohibitively expensive for use in a broadcast network. However, the complexity of the receiver structure can be significantly reduced by employing transmit diversity. This is analogous to a wireless network in which multiple antennas in the base station are used to enhance the system performance without significantly increasing the cost of the mobile user.[44,49] In the transmit diversity scheme for CO-OFDM, namely for a 2 × 1 MISO-OFDM system, two OFDM transmitters are used, one for each polarization. In the remainder of this section, we discuss the theoretical basis for the polarization transmit diversity coding and its associated proof-of-principle experiment.

For notational simplicity, we drop the subcarrier index k in Eq. (5.53), implying that the polarization time coding is processed on the per-subcarrier basis, and we also drop the superscript "p" for the noise component in Eq. (5.53). Thus, we have

$$r_i = Hc_i + n_i \tag{5.54}$$

where $r_i = [r_x^i \quad r_y^i]^T$ and $c_i = [c_x^i \quad c_y^i]^T$ are, respectively, the received and transmitted OFDM symbol, and H is the transfer function describing the polarization effects and chromatic dispersion and assumes a form of

$$H = \begin{bmatrix} H_{xx} & H_{xy} \\ H_{yx} & H_{yy} \end{bmatrix} \tag{5.55}$$

In the transmit diversity scheme, two consecutive OFDM symbols are sent with orthogonal polarization. By processing the OFDM received signals properly, both polarization signals can be recovered without resorting to a polarization diversity receiver. Mathematically, the two consecutive OFDM symbols—numbers $2j - 1$ and $2j$—with orthogonal polarization in the form of a Jones vector are given by

$$c_{2j-1} = [c_x^j \quad c_y^j]^T, \; c_{2j} = \left[-c_y^{j*} \quad c_x^{j*}\right]^T \tag{5.56}$$

The polarizations of the subcarriers in two consecutive OFDM symbols are orthogonal by examining the inner product of these two vectors; that is,

$$c_{2j-1} + c_{2j} = 0 \tag{5.57}$$

To simplify the receiver architecture, only one polarization of the received signal, along the polarization of the local laser, is detected in the receiver. Without loss of generality, we assume that the polarization of the local laser is x-polarization. Substituting Eq. (5.56) into Eq. (5.54), the two received OFDM symbols r_x^{2j-1} and r_x^{2j} are, respectively, given by

$$r_x^{2j-1} = T_{xx}c_x^j + T_{xy}c_y^j + n_x^{2j-1} \tag{5.58}$$

$$r_x^{2j} = -T_{xx}c_y^{*j} + T_{xy}c_x^{*j} + n_x^{2j} \tag{5.59}$$

Solving Eqs. (5.58) and (5.59), we can recover the $c_{2j-1} = [c_x^j \quad c_y^j]^T$ as

$$[c_x^j \quad c_y^j]^T = H'\left([r_x^{2j-1} \quad r_x^{2j*}]^T + [n_x^{2j-1} \quad n_x^{2j*}]^T\right) \tag{5.60}$$

$$H' = \begin{bmatrix} T_{xx} & T_{xy} \\ T_{xy}^* & -T_{xx}^* \end{bmatrix}^{-1} \tag{5.61}$$

The superscript "−1" stands for matrix inversion. From Eq. (5.60), it follows that the estimated OFDM symbol of $c_{2j-1}, \hat{c}_{2j-1}$, is given by

$$\hat{c}_{2j-1} = H'[r_x^{2j-1} \quad r_x^{2j*}]^T \tag{5.62}$$

The two elements of $\hat{c}_{2j-1}$ in Eq. (5.62) will be demapped to the nearest constellation points to obtain the estimated/detected symbols. This polarization time (PT) coding is equivalent to Alamouti coding for the space–time coding in wireless systems.[49] The uniqueness of the PT coding in optical fiber systems is that in the absence of PDL, the fading for each individual transmitter is completely recovered due to the complementarity of T_{xx} and T_{xy}, irrespective of polarization rotation and PMD in the fiber link. PT coding has been studied in CO-OFDM systems[47,57] and in single-carrier systems.[48]

OFDM is inherently spectral efficient due to spectral overlapping between subcarriers, allowing for approximately 6 GHz optical spectrum for 10 Gb/s data using QPSK modulation.[10] By using direct up/down-conversion architecture, where the transmitter and receiver lasers are placed at the center frequency of the optical OFDM spectrum, the electrical bandwidth required for both transmitter and receiver is only half of the optical spectrum bandwidth,[43] which is 3 GHz for a 10 Gb/s system. This is fundamentally different from other modulation approaches, in which the electrical bandwidth can be reduced through excessive electrical filtering or through higher order modulation, both of which will inevitably result in system penalty. The high electrical spectral efficiency of CO-OFDM signals signifies that the 10 Gb/s system can potentially use cheap 2.5 GHz components. For some applications for which optical preamplification at the receiver may be cost-prohibitive, the receiver sensitivity can be greatly enhanced through coherent detection, which has been a known fact for more than a decade. Considering all the advantages, we conclude that CO-OFDM may have a significant role for downstream systems in a high-speed and long-reach PON beyond 10 Gb/s to 100 Gb/s, where the dispersion and receiver sensitivity are of critical importance.[58]

Figure 5.10 shows the experimental setup for verifying the transmitter diversity scheme for a broadcast network. The OFDM signal is generated by using a Tektronix AWG as an RF OFDM transmitter. The time domain waveform is first generated with a Matlab program including mapping $2^{15}-1$ PRBS into corresponding 77 subcarriers with QPSK encoding within multiple OFDM symbols, which are subsequently converted into time domain using IFFT and inserted with GI. The number of OFDM subcarriers is 128 and GI is one-eighth of

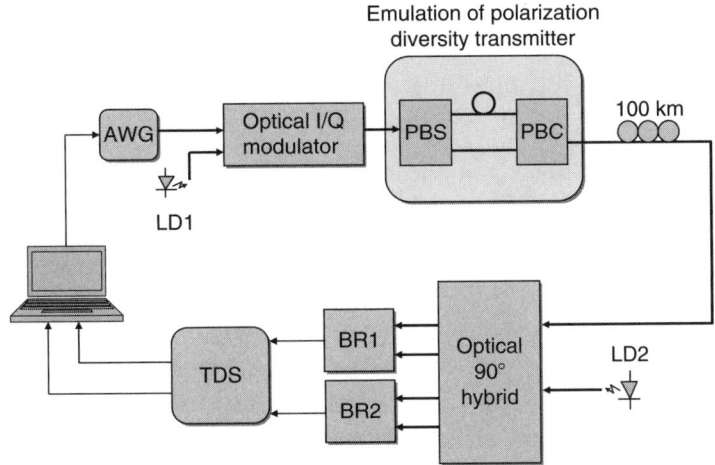

AWG: Arbitrary waveform generator PBS/C: Polarization beam splitter/combiner
BR: Balanced receiver   LD: Laser diode TDS: Time-domain sampling scope

Figure 5.10: Experimental setup of the proof-of-concept PT-coded transmitter for broadcast networks.

the observation period. Only the middle 87 subcarriers out of 128 are filled, among which 10 pilot subcarriers are used for phase estimation. The real and imaginary parts of the OFDM waveforms are uploaded into the AWG operated at 10 GS/s, and two-channel analog signals, each representing the real and imaginary components of the complex OFMD signal, are generated synchronously. The so-generated OFDM waveform carries 10.7 Gb/s data. These two signals are fed into I and Q ports of an optical I/Q modulator, respectively, to perform direct up-conversion of OFDM baseband signals from RF domain to optical domain. The polarization transmit diversity is emulated by using a polarization beam splitter to generate two OFDM streams. The two polarization streams are delayed with respect to each other by an OFDM symbol period of 14.4 ns, and then they are combined into one optical signal. We use one span of 100 km to emulate the fiber plant. At the receive end, optical-to-RF direct down-conversion is performed. In particular, the LO laser center frequency is tuned approximately to that of the incoming signal, and both the signal and LO are fed into an optical 90 degree hybrid. The I/Q optical signals are then input into two balanced detectors. A single-ended detector has been shown to have almost the same performance.[4] The two RF signals from the photodetectors are then input into a Tektronix TDS and are acquired at 20 GS/s and processed with a Matlab program. No phase locking and polarization control between LO and signal are needed. The RF OFDM receiver signal processing again involves (1) FFT window synchronization using the Schmidl format to identify the start of the OFDM symbol, (2) software down-conversion of the OFDM RF signal to baseband to remove residual frequency offset, (3) phase estimation and channel estimation, and (4) constellation construction for each carrier and BER computation. The channel matrix H is estimated by sending 50 OFDM symbols. The OFDM symbols are partitioned into four groups with proper coding similar to Eq. (5.56), resulting in two of the four OFDM symbols with orthogonal polarizations, which are subsequently used for system performance characterization.

Figure 5.11 shows the electrical spectrum for the received 10 Gb/s signal. The OFDM spectrum is tightly bounded within 3 GHz. More important, although our system shows 3 dB roll-off at 3 GHz, this does not seem to affect the performance because the OFDM spectrum is limited to only 3 GHz. Such a high spectral efficiency will prove critical for the future high-speed access network. Figure 5.12 shows receiver sensitivity without an optical amplifier after 100 km transmission at 10 Gb/s. It is expected that OFDM may have an error floor of 10^{-6} because of intentional clipping. Subsequently, error correction is required for OFDM systems. We perform measurement at a BER of 10^{-1} to 10^{-4}. The lower end of BER is limited by the number of OFDM symbols processed (approximately 500 each time). We do not see much penalty after 100 km transmission. This is expected because much longer transmission has been demonstrated without optical dispersion compensation but with a more sophisticated polarization diversity receiver.[8–10] This implies that

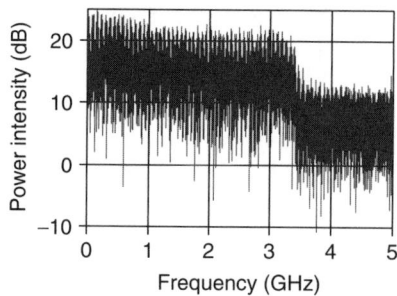

Figure 5.11: Electrical spectrum of 10 Gb/s PT-coded CO-OFDM signal.

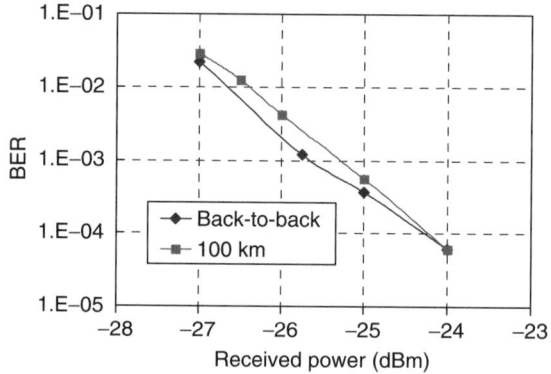

Figure 5.12: BER performance for PT-coded CO-OFDM signal.

CO-OFDM could enable much longer reach. It is noted that in the experiment, the polarization in the link is free-running and the system is immune to the polarization rotation in the link due to the polarization diversity transmitter.

5.5.4 2 × 2 MIMO-OFDM in Polarization Domain

The two polarization modes supported in the fiber present an attractive avenue to double the fiber capacity. Another major benefit is that by using polarization multiplexing, the effective symbol rate is reduced to half that of single-polarization transmission, which enables high-speed transmission systems using lower speed electronics. Here, we review 2 × 2 MIMO-OFDM transmission that achieves doubling of system capacity without sacrificing the receiver sensitivity. The signal processing involved is similar to that of 2 × 1 MISO-OFDM described previously. Equation (5.54) can be rearranged as

$$c_k = H_k^{-1} r_k - H_k^{-1} n_k \qquad (5.63)$$

where $r_i = [r_x^i \quad r_y^i]^T$ and $c_i = \left[c_x^i \quad c_y^i \right]^T$ are, respectively, the received and transmitted OFDM symbol, and H is the transfer function describing polarization effects and chromatic dispersion. From Eq. (5.63), it follows that the estimated transmitted symbol $c_k, \hat{c}_k$, is given by

$$\hat{c}_k = H_k^{-1} r_k \qquad (5.64)$$

From Eq. (5.64), it can be seen that once the received symbol r_k and the channel transfer function H_k are known, the estimated transmitted symbols $\hat{c}_k$ will be readily computed. The two elements of $\hat{c}_k$ will be then demapped to the nearest constellation points to obtain estimated/detected symbols.

Early experiments of 2 × 2 MIMO-OFDM were reported in direct up/down-conversion configuration for 1000 km SSMF transmission at 21.4 Gb/s[50,59] and for 4160 km transmission of 16 channels at 52.5 Gb/s.[8,60] Figure 5.13 shows the transmitter configuration for the experimental demonstration of 2 × 2 MIMO-OFDM systems at 52.5 Gb/s reported by Jansen et al.[60] As shown in Figure 5.13, the transmitter adopted an IF architecture, where the OFDM baseband signal is first up-converted to IF frequencies at 11.5 and 18.7 GHz. The two-band OFDM signals from two independent optical modulators are combined with a polarization combiner. The two groups (odd and even channel) of polarization division multiplexed (PDM) signals are coupled with an optical interleaver. Long-haul transmission is emulated with a recirculation loop shown in Figure 5.14a, which consists of four spans of 80 km SSMF. A PMF fiber with a DGD of 86 ps is also included in the loop, which is used to emulate a 300 ps mean DGD after 13 loops. The optical OFDM signal is split in two random polarizations and detected with a polarization diversity heterodyne receiver shown in Figure 5.14b. A real-time digital storage oscilloscope with a bandwidth of 16 GHz is used

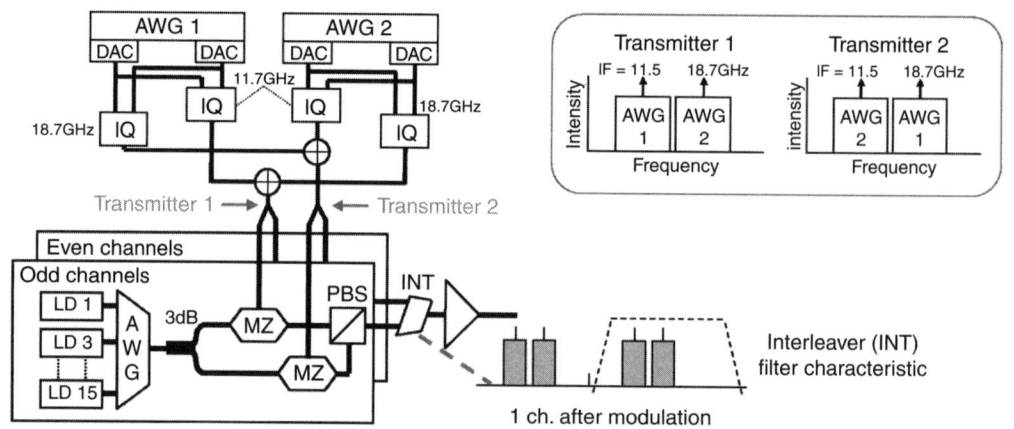

Figure 5.13: Experimental configuration of the 16 × 52.5 Gb/s PDM-OFDM transmitter. After Jansen et al.[60]

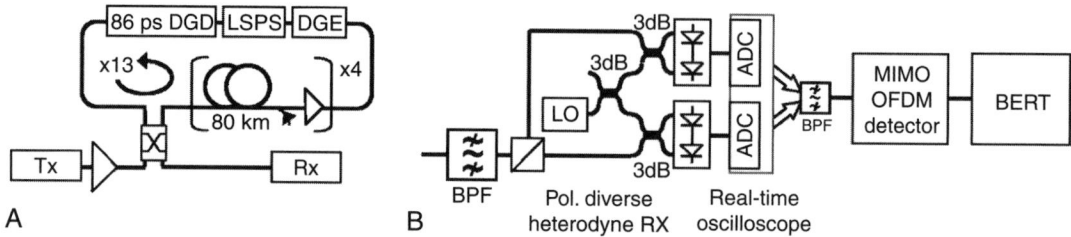

Figure 5.14: Experimental setup of (a) the recirculating loop and (b) the polarization diverse heterodyne receiver. After Jansen et al.[60]

to sample the two outputs of the heterodyne receiver. After detection, the data are postprocessed using the 2 × 2 MIMO-OFDM procedure.[60]

Figure 5.15a shows the measured back-to-back sensitivities for 26.3 Gb/s (single polarization) and 52.5 Gb/s (PDM). The simulated sensitivity for 26.3 Gb/s (single polarization) is also shown. It is observed that the 52.5 Gb/s PDM-OFDM signal has an OSNR sensitivity 3.0 dB higher than that of the 26.3 Gb/s single-polarization signal at a BER of 10^{-3}. This implies that the polarization multiplexing introduces insignificant penalty. For lower BER values, an increase in performance penalty between the simulated and experimental curves is observed. This penalty increase at the lower BER is likely attributed to the residual carrier frequency offset and phase noise. The influence of the interchannel XPM is evaluated by comparing the performance between 50 and 200 GHz spaced systems. Figure 5.15b shows the BER as a function of the fiber launch power after 3200 km transmission. For this experiment, a simple common phase rotation (CPR) compensator (also called a CPE, ϕ_i in Eq. 5.50) was implemented by multiplying the received constellation with the inverse of the CPR. The optimal BER performance and the optimal

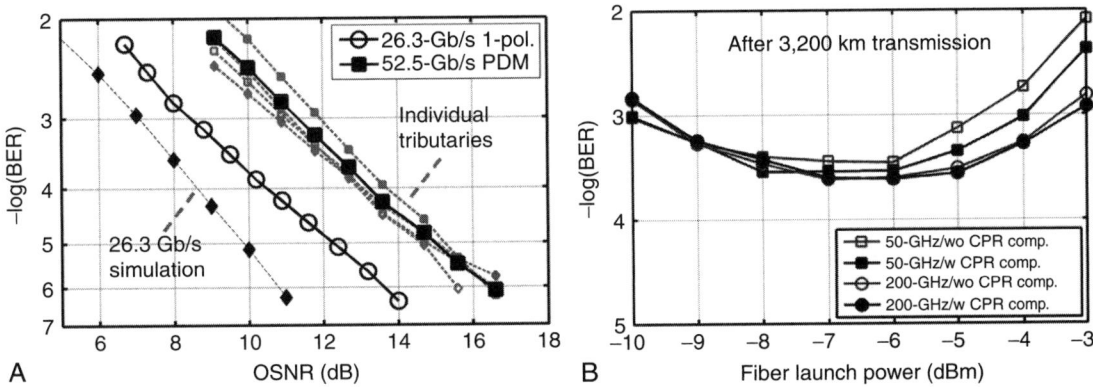

Figure 5.15: (a) Back-to-back BER performance for single-polarization and PDM-OFDM signals and (b) BER performance of PDM-OFDM signals against the launch power for 50 and 200 GHz channel spacings. After Jansen et al.[60]

launch power exhibit only a small difference between 50 and 200 GHz channel spacing when CPR is compensated for. Therefore, in the demonstrated PDM-OFDM system, the intrachannel FWM is dominant over interchannel XPM for both 200 and 50 GHz WDM channel spacing.

5.6 Nonlinear Polarization Effects

The primary difference between a fiber-optic channel and a wireless one is nonlinearity in the fiber. The previously discussed polarization effects, both PMD and PDL, are linear system properties. Under nonlinear conditions, in a strict sense, the concept of end-to-end link PMD and PDL is not valid anymore. Nevertheless, we most often operate optical transmission systems at an optimum level so that nonlinearity is about to onset and the linear concepts of PMD and PDL are still important descriptions of the system behavior. However, in the strong nonlinear regime, we have to rely on the detailed nonlinear Schrödinger transmission model to simulate the system performance, and often an analytically close form is not achievable.

In contrast with single-carrier systems for which each data bit is encoded across the entire signal spectrum, OFDM transmits the data with many orthogonal subcarriers. Therefore, we study OFDM nonlinearity in terms of individual subcarriers instead of a wavelength channel in single-carrier systems.

5.6.1 Nonlinear Polarization Effects in a Birefringence Fiber

Fiber nonlinearity effect for single polarization such as FWM and XPM was discussed in Section 3.4.5. Here, we turn our attention to polarization-dependent nonlinearity effects.[61] It is instructive to consider a short piece of optical fiber where the birefringence is constant. The nonlinearity can be comprehended by studying the third-order nonlinearity. We first investigate the nonlinearity interaction between two subcarriers with the electrical field written as

$$E = \frac{1}{2}(E_x \hat{x} + E_y \hat{y}) + \text{c.c.} \tag{5.65}$$

$$E_x = c_1^x e^{j\omega_1 t} + c_2^x e^{j\omega_2 t}, \quad E_y = c_1^y e^{j\omega_1 t} + c_2^y e^{j\omega_2 t} \tag{5.66}$$

where $\hat{x}$ and $\hat{y}$ are two eigen axes of the polarization, $c_1^x(c_1^y)$ and $c_2^x(c_2^y)$ are, respectively, the amplitude for the $x(y)$ component of the first and the second OFDM subcarrier, and c.c. is the complex conjugate.

The third-order polarization can be generally written as[62]

$$P_{\text{NL}} = \frac{1}{2}(\hat{x}P_x + \hat{y}P_y) + \text{c.c.} \tag{5.67}$$

where P_x and P_y are given by

$$P_i = \frac{3\varepsilon_0}{2} \sum_j \left(\chi_{xxyy}^{(3)} E_i E_j E_j^* + \chi_{xyxy}^{(3)} E_j E_i E_j^* + \chi_{xyyx}^{(3)} E_j E_j E_i^* \right) \tag{5.68}$$

where ε_0 is the vacuum permeability, $\chi^{(3)}$ is the third-order nonlinearity coefficient,[60] and i and j stand for x or y. Substituting Eq. (5.66) into Eq. (5.68), we group the result in terms of the frequency components instead of showing the complete expansion of Eq. (5.68). At the frequency of ω_1, we obtain

$$P_1 = \frac{1}{2}(\hat{x}P_x + \hat{y}P_y)e^{j\omega_1 t} \tag{5.69}$$

$$P_{1x} = \frac{3\varepsilon_0}{4} \chi_{xxxx}^{(3)} \left(|c_{1x}|^2 + 2|c_{2x}|^2 + \frac{2}{3}|c_{1y}|^2 + \frac{2}{3}|c_{2y}|^2 \right) c_{1x} \tag{5.70}$$

$$P_{1y} = \frac{3\varepsilon_0}{4} \chi_{xxxx}^{(3)} \left(|c_{1y}|^2 + 2|c_{2y}|^2 + \frac{2}{3}|c_{1x}|^2 + \frac{2}{3}|c_{2x}|^2 \right) c_{1y} \tag{5.71}$$

Therefore, the nonlinearity-induced refraction coefficients for the x and y components are, respectively,

$$\Delta n_x = n_2 \left(|c_{1x}|^2 + 2|c_{2x}|^2 + \frac{2}{3}|c_{1y}|^2 + \frac{2}{3}|c_{2y}|^2 \right) \tag{5.72}$$

$$\Delta n_y = n_2 \left(|c_{1y}|^2 + 2|c_{2y}|^2 + \frac{2}{3}|c_{1x}|^2 + \frac{2}{3}|c_{2x}|^2 \right) \tag{5.73}$$

where the nonlinear index coefficient n_2 is defined as[60]

$$n_2 = \frac{3}{8n} \text{Re}\left(\chi_{xxxx}^{(3)} \right) \tag{5.74}$$

Then we arrive at the nonlinearity birefringence given by

$$B_{\text{NL}} = |\Delta n_x - \Delta n_y| = n_2 \left(\frac{1}{3} \left(|c_{1x}|^2 - |c_{1y}|^2 \right) + \frac{4}{3} \left(|c_{2x}|^2 - |c_{2y}|^2 \right) \right) \tag{5.75}$$

Because the OFDM signal consists of a large number of subcarriers, from the standpoint of the subcarrier "1," there are many other subcarriers. If we extend the birefringence due to all the other subcarriers, we have

$$B_{\mathrm{NL}} = n_2 \left| \frac{1}{3} \left(|c_1^x|^2 - |c_1^y|^2 \right) + \frac{4}{3} \sum_{k \neq 1} \left(|c_k^x|^2 - |c_k^y|^2 \right) \right|$$

$$\approx \frac{4}{3} n_2 \left| \sum_{k=1}^{N_{\mathrm{sc}}} \left(|c_k^x|^2 - |c_k^y|^2 \right) \right| \tag{5.76}$$

where k is the subcarrier number. We have assumed a large number of subcarriers such that the power of one subcarrier is negligible compared to the overall power. The conclusion of the exercise is that for an OFDM symbol consisting of many subcarriers, there is a nonlinear birefringence equal to the amount expressed in Eq. (5.76) that is common to all subcarriers. This is an important conclusion because such a common nonlinear polarization rotation can be estimated just as the linear or nonlinear CPE by using a few pilot tones on a per-symbol basis.

The other third-order nonlinearity terms in Eq. (5.68) represent polarization-dependent FMW. However, spelling them out does not give a picture as clear as the nonlinear birefringence expression of Eq. (5.76), and therefore we rely on the nonlinear Schrödinger equation to identify the nonlinearity impact, which is discussed in Section 5.6.2.

5.6.2 Nonlinear Polarization Effects in Randomly Varying Birefringence Fiber

Because of random fluctuation of the birefringence eigen axis in the fiber, an optical OFDM signal will go through many small segments of birefringence fiber as discussed in the previous section, and resultant nonlinear birefringence will vary on the symbol basis. As we alluded to in the previous section, in addition to the nonlinear birefringence, FWM is also an important impairment, not to mention the need to include other linear effects such as PMD and PDL. Subsequently, it is difficult to develop analytical formulas for signal nonlinear transmission over optical fiber channel. Wai et al.[63,64] developed a nonlinear Schrödinger equation that includes both linear and nonlinear polarization effects. The nonlinear Schrödinger equation is given by

$$i\frac{\partial A}{\partial z} + i\frac{1}{2}\vec{\alpha}'(z) \cdot \vec{\sigma}' A - i\frac{1}{2}\vec{\gamma}'(z) \cdot \vec{\sigma}' \frac{\partial A}{\partial t} - \frac{1}{2}\beta'' \frac{\partial^2 A}{\partial t^2} + \gamma \frac{8}{9}|A|^2 A = 0 \tag{5.77}$$

where $A(z, t)$ is the time domain signal given by

$$A(z, t) = \begin{pmatrix} A_x(z,t) \\ A_y(z,t) \end{pmatrix} \tag{5.78}$$

$\vec{\gamma}'(z)$ and $\vec{\alpha}'(z)$ are, respectively, the PMD and PDL vectors in the direction of the eigen axis of the large segment of the fiber. $A(z, t)$ is the time domain signal using the rotating local polarization coordinate and is subsequently different from the signal using the fixed coordinate. Nevertheless, these two representations are only different by a linear transform in the frequency domain; therefore, with regard to system performance, it does not matter which coordinate is used.[62] Furthermore, it is not practical to compute Eq. (5.77) at the microscale of a few meters of local birefringence; instead, it should be applied in a macroscale of a few kilometers valid for the common optical communication systems. For instance, we may partition the fiber into many 10 km long sections and rescale $\vec{\gamma}'(z)$ and $\vec{\alpha}'(z)$ so that the composite PMD and PDL effects are in agreement with targeted mean values. To clarify this point, we first ignore the nonlinearity and perform a Fourier transform. Thus, we have

$$i\frac{\partial A}{\partial z} + \frac{1}{2}i\vec{\alpha}'(z)\cdot\vec{\sigma}A - \omega\frac{1}{2}\vec{\gamma}'(z)\cdot\vec{\sigma}A - \frac{1}{2}\beta''\omega^2 A = 0 \tag{5.79}$$

We assume that the fiber is partitioned into N sections, and that inside each section $\vec{\alpha}'(z)$ and $\vec{\gamma}'(y)$ are constant. Equation (5.79) can be solved as

$$A(z, \omega) = \left[\prod_{l=1}^{N}\exp\left\{\left(-\frac{1}{2}j\cdot\vec{\beta}_l'\cdot f_k\Delta z - \frac{1}{2}\vec{\alpha}_l'\Delta z\right)\cdot\vec{\sigma}\right\}\right]A_0 \tag{5.80}$$

Compared with the linear transmission Jones matrix $T(\omega)$ in Eq. (5.43), and enforcing the consistency between Eqs. (5.43) and (5.80), we obtain

$$\vec{\gamma}_l'\cdot\Delta z = \vec{\gamma}_l', \quad \vec{\alpha}_l'\cdot\Delta z = \vec{\alpha}_l \tag{5.81}$$

It is worth noting that $\vec{\alpha}_l$ and $\vec{\gamma}_l$ are assumed to be randomly distributed on the Poincare sphere. They are generated in a way that the composite link PMD and PDL conform to target values.

The nonlinear Schrödinger equation is commonly solved using the so-called split-step algorithm[60] as an interleaved operation of linear (polarization effects and chromatic dispersion) and nonlinear operation (fiber nonlinearity). Figure 5.16 shows the conceptual diagram of split-step algorithm, with its corresponding mathematical formulation given by

$$A(z, t) = \left(\prod_{l=1}^{N}\exp(\hat{N}_l)\exp(\hat{D}_l)\right)A(0, t) \tag{5.82}$$

$$\hat{N}_i A(t) = \gamma\frac{8}{9}|A|^2 \tag{5.83}$$

$$\hat{D}_l A(t) = \left\{\mathfrak{F}^{-1}\left(e^{-\frac{1}{2}\beta''\omega^2}M_l\right)\mathfrak{F}\right\}A(t) \tag{5.84}$$

$$M_l = \exp\left\{\left(-\frac{1}{2}j\cdot\vec{\beta}_l'\cdot\omega - \frac{1}{2}\vec{\alpha}_l\right)\cdot\vec{\sigma}\right\} \tag{5.85}$$

where $\hat{N}_i$ and $\hat{D}_i$, respectively, are the nonlinearity operator and linear dispersion operator for the lth segment,[60] $\mathfrak{F}$ is the Fourier transfer function, M_l is the Jones matrix for the lth

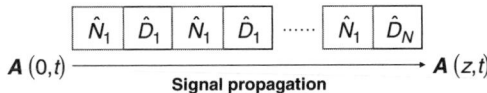

Figure 5.16: Conceptual diagram of a split-step algorithm in which the fiber link is partitioned into *N* sections, each comprising nonlinear and linear operations.

segment, and $A(0, t)$ is the time-domain input signal. In summary, the system performance, taking consideration of all the linear and nonlinear polarization effects, can be evaluated by obtaining the time domain output signal $A(z, t)$ via the split-step algorithm laid out in Eqs. (5.82)–(5.85). Once the received signal $A(z, t)$ is known, the standard OFDM signal processing procedure discussed in Chapter 4 can be initiated to recover the OFDM information symbol, and subsequently the BER performance can be computed.

References

1. Poole CD, Wagner RE. Phenomenological approach to polarization dispersion in long single-mode fibers. *Electron Lett* 1986;**22**:1029–30.
2. Gordon JP, Kogelnik H. PMD fundamentals: Polarization mode dispersion in optical fibers. *Proc Natl Acad Sci U S A* 2000;**97**:4541–50.
3. Ly-Gagnon DS, Tsukarnoto S, Katoh K, Kikuchi K. Coherent detection of optical quadrature phase-shift keying signals with carrier phase estimation. *J Lightwave Technol* 2006;**24**:12–21.
4. Savory SJ, Gavioli G, Killey RI, Bayvel P. Electronic compensation of chromatic dispersion using a digital coherent receiver. *Opt Express* 2007;**15**:2120–6.
5. Laperle C, Villeneuve B, Zhang Z, et al. Wavelength division multiplexing (WDM) and polarization mode dispersion (PMD) performance of a coherent 40Gbit/s dual-polarization quadrature phase shift keying (DP-QPSK) transceiver. In: *Opt. Fiber Commun. Conf.*, paper no. PDP16. Anaheim, CA; 2007.
6. Charlet G, Renaudier J, Salsi M, et al. Efficient mitigation of fiber impairments in an ultra-long haul transmission of 40Gbit/s polarization-multiplexed data, by digital processing in a coherent receiver. In: *Opt. Fiber Commun. Conf.*, paper no. PDP17. Anaheim, CA; 2007.
7. Shieh W, Chen W, Tucker RS. Polarization mode dispersion mitigation in coherent optical orthogonal frequency division multiplexed systems. *Electron Lett* 2006;**42**:996–7.
8. Jansen SL, Morita I, Tanaka H. 16 × 52.5-Gb/s, 50-GHz spaced, POLMUX-CO-OFDM transmission over 4160 km of SSMF enabled by MIMO processing KDDI R&D Laboratories. In: *Eur. Conf. Opt. Commun.*, paper no. PD1.3. Berlin, Germany; 2007.
9. Shieh W. PMD-supported coherent optical OFDM systems. *IEEE Photon Technol Lett* 2007;**19**:134–6.
10. Shieh W, Yi X, Ma Y, Tang Y. Theoretical and experimental study on PMD-supported transmission using polarization diversity in coherent optical OFDM systems. *Opt Express* 2007;**15**:9936–47.
11. Cvijetic N, Xu L, Wang T. Adaptive PMD compensation using OFDM in long-haul 10 Gb/s DWDM systems. In: *Opt. Fiber Commun. Conf.*, paper no. OTuA5. Anaheim, CA; 2007.
12. Djordjevic IB. PMD compensation in fiber-optic communication systems with direct detection using LDPC-coded OFDM. *Opt Express* 2007;**15**:3692–701.
13. Mayrock M, Haunstein H. PMD tolerant direct-detection optical OFDM system. In: *Eur. Conf. Opt. Commun.*, paper no. 5.2.5. Berlin, Germany; 2007.

14. Agrawal GP. *Fiber-Optic Communication Systems*. 2nd ed. New York: Wiley; 1997.

15. Gisin N, Huttner B. *Opt Commun* 1997;**142**:119–25.

16. Wai PKA, Menyuk CR. Polarization decorrelation in optical fibers with randomly varying birefringence. *Opt Lett* 1994;**19**:1517–9.

17. Shieh W, Kogelnik H. Dynamic eigenstates of polarization. *IEEE Photon Technol Lett* 2001;**13**:40–2.

18. Karlsson M. Probability density functions of the differential group delay in optical fiber communication systems. *J Lightwave Technol* 2001;**19**:324–31.

19. Foschini GJ, Poole CD. Statistical theory of polarization mode dispersion in single mode fibers. *J Lightwave Technol* 1991;**9**:1439–56.

20. Karlsson M, Brentel J. Autocorrelation function of the polarization-mode dispersion vector. *Opt Lett* 1999;**24**:939–41.

21. Liu X, Buchali F. Intra-symbol frequency-domain averaging based channel estimation for coherent optical OFDM. *Opt Express* 2008;**16**:21944–57.

22. Bononi A, Vannucci A. Statistics of the Jones matrix of fibers affected by polarization mode dispersion. *Opt Lett* 2001;**26**:675–7.

23. Shieh W, Yang Q, Ma Y. 107 Gb/s coherent optical OFDM transmission over 1000-km SSMF fiber using orthogonal band multiplexing. *Opt Express* 2008;**16**:6378–86.

24. Poole C, Tkach R, Chraplyvy A, Fishman D. Fading in lightwave systems due to polarization-mode dispersion. *IEEE Photon Technol Lett* 1991;**3**:68–70.

25. Bulow H. Limitation of optical first-order PMD compensation. In: *Opt. Fiber Commun. Conf.*, paper no. WE1-1. San Jose, CA; 1999.

26. Shieh W. On the second-order approximation of PMD. *IEEE Photon Technol Lett* 2000;**12**:290–2.

27. Pennickx D, Bruyere F. Impact of the statistics of second-order polarization mode dispersion on system performance. In: *Opt. Fiber Commun. Conf.*, paper no. ThR2. San Jose, CA; 1998.

28. Kogelnik H, Nelson LE, Winzer PJ. Second-order PMD outage of first-order compensated fiber systems. *IEEE Photon Technol Lett* 2004;**16**:1053–5.

29. Orlandini A, Vincetti L. A simple and useful model for Jones matrix to evaluate higher order polarization-mode-dispersion effects. *IEEE Photon Tech Lett* 2001;**13**:1176–8.

30. Forestieri E, Vincetti L. Exact evaluation of the Jones matrix of a fiber in the presence of polarization mode dispersion of any order. *J Lightwave Technol* 2001;**19**:1898–909.

31. Kogelnik H, Nelson LE, Gordon JP. Emulation and inversion of polarization-mode dispersion. *J Lightwave Technol* 2003;**21**:482–95.

32. Shtaif M, Boroditsky M. The effect of the frequency dependence of PMD on the performance of optical communications systems. *IEEE Photon Technol Lett* 2003;**15**:1369–71.

33. Karlsson M. Polarization mode dispersion-induced pulse broadening in optical fibers. *Opt Lett* 1998;**23**:688–90.

34. Shieh W. Principal states of polarization for an optical pulse. *IEEE Photon Technol Lett* 1999;**11**:677–9.

35. Ono T, Yamazaki S, Shimizu H, Emura H. Polarization control method for suppressing polarization mode dispersion in optical transmission systems. *J Lightwave Technol* 1994;**12**:891–8.

36. Noé R, Sandel D, Yoshida-Dierolf M, et al. Polarization mode dispersion compensation at 10, 20 and 40 Gbit/s with various optical equalizers. *J Lightwave Technol* 1999;**17**:1602–16.

37. Sunnerud H, Xie C, Karlsson M, Samuelsson R, Andrekson PA. A comparison between different PMD compensation techniques. *J Lightwave Technol* 2002;**20**:368–78.

38. Shieh W, Haunstein H, Mckay B, et al. Dynamic polarization-mode dispersion compensation in WDM systems. In: *Eur. Conf. Opt. Commun.*, vol. 2. Munich, Germany; 2000. pp. 41–3.

39. Heismann F, Fishman DA, Wilson DL. Automatic compensation of first-order polarization mode dispersion in a 10 Gb/s transmission system. In: *Eur. Conf. Opt. Commun.*, vol. 1. Madrid, Spain; 1998. pp. 529–30.

40. Yu Q, Yan L, Lee S, et al. Enhanced higher-order PMD compensation using a variable time delay between polarizations. In: *Eur. Conf. Opt. Commun.*, vol. 2. Munich, Germany; 2000. pp. 47–8.

41. Xie C, Mollenauer LF. Performance degradation induced by polarization dependent loss in optical fiber transmission systems with and without polarization mode dispersion. *J Lightwave Technol* 2003;**21**:1953–7.

42. Shieh W, Athaudage C. Coherent optical orthogonal frequency division multiplexing. *Electron Lett* 2006;**42**:587–9.

43. Shieh W. Maximum-likelihood phase and channel estimation for coherent optical OFDM. *IEEE Photon Technol Lett* 2008;**20**:605–7.

44. Tang Y, Shieh W, Yi X, Evans R. Optimum design for RF-to-optical up-converter in coherent optical OFDM systems. *IEEE Photon Technol Lett* 2007;**19**:83–485.

45. Bolcskei H, Gesbert D, Paulraj AJ. On the capacity of OFDM-based spatial multiplexing systems. *IEEE Trans Commun* 2002;**50**:225–34.

46. Shieh W, Yi X, Tang Y. Transmission experiment of multi-gigabit coherent optical OFDM systems over 1000 km SSMF fiber. *Electron Lett* 2007;**43**:183–5.

47. Jansen SL, Morita I, Takeda N, Tanaka H. 20-Gb/s OFDM transmission over 4160-km SSMF enabled by RF-pilot tone phase noise compensation. In: *Opt. Fiber Commun. Conf.*, paper no. PDP15. Anaheim, CA; 2007.

48. Shieh W, Yi X, Ma Y, Yang Q. Coherent optical OFDM: Has its time come? [Invited]. *J Opt Networking* 2008;**7**:234–55.

49. Han Y, Li G. Polarization diversity transmitter and optical nonlinearity mitigation using polarization-time coding. In: *Proc. COTA*, paper CThC7. Whistler, Canada; 2006.

50. Alamouti SM. A simple transmit diversity technique for wireless communications. *IEEE J Select Areas Commun* 1998;**16**:1451–8.

51. Shieh W. Coherent optical MIMO-OFDM for optical fibre communication systems. In: *Eur. Conf. Opt. Commun.*, workshop 5. Berlin, Germany; 2007.

52. Cvijetic N, Xu L, Wang T. Adaptive PMD compensation using OFDM in long-haul 10-Gb/s DWDM systems. In: *Opt. Fiber Commun. Conf.*, paper no. OTuA5. Anaheim, CA; 2007.

53. Xie C. PMD insensitive direct-detection optical OFDM systems using self-polarization diversity. In: *Opt. Fiber Commun. Conf.*, paper no. OMM2. San Diego; 2008.

54. Kobayash T, Sano A, Yamada E. Electro-optically subcarrier multiplexed 110 Gb/s OFDM signal transmission over 80 km SMF without dispersion compensation. *Elect Lett* 2008;**44**:225–6.

55. Jansen SL, Morita I, Tanaka H. 10 × 121.9-Gb/s PDM-OFDM transmission with 2-b/s/Hz spectral efficiency over 1000 km of SSMF. In: *Opt. Fiber Commun. Conf.*, paper no. PDP2. San Diego; 2008.

56. Moller L, Su Y, Raybon G. Polarization-mode-dispersion-supported transmission in 40-Gb/s long haul systems. *IEEE Photon Technol Lett* 2003;**15**:335–7.

57. Djordjevic IB, Xu L, Wang T. Alamouti-type polarization-time coding in coded-modulation schemes with coherent detection. *Opt Express* 2008;**16**:14163–72.

58. Lee SM, Mun SG, Kim MH, Lee CH. Demonstration of a long-reach DWDM-PON for consolidation of metro and access networks. *J Lightwave Technol* 2007;**25**:271–6.

59. Ma Y, Shieh W, Yang Q. Bandwidth-efficient 21.4 Gb/s coherent optical 2 × 2 MIMO OFDM transmission. In: *Opt. Fiber Commun. Conf.*, paper no. JWA59. San Diego; 2008.

60. Jansen SL, Morita I, Schenk TC, Tanaka H. Long-haul transmission of 16 × 52.5 Gbits/s polarization-division-multiplexed OFDM enabled by MIMO processing [Invited]. *J Opt Networking* 2008;**7**:173–82.

61. Xie C, Möller L, Kilper DC, Mollenauer LF. Impact of cross-phase modulation induced polarization scattering on optical PMD compensation performance in WDM systems. *Opt Lett* 2003;**28**:2303–5.

62. Agrawal GP. *Nonlinear Fiber Optics*. 2nd ed. New York: Academic Press; 1995.

63. Wai P, Menyuk CR. Polarization mode dispersion, decorrelation, and diffusion in optical fibers with randomly varying birefringence. *J Lightwave Technol* 1996;**14**:148–57.

64. Marcuse D, Menyuk CR, Wai PKA. Application of the Manakov-PMD equation to studies of signal propagation in optical fibers with randomly varying birefringence. *J Lightwave Technol* 1997;**15**:1735–46.

CHAPTER 6

Coding for Optical OFDM Systems

Orthogonal frequency-division multiplexing (OFDM) is an efficient approach to deal with intersymbol interference (ISI) due to chromatic dispersion and polarization mode dispersion (PMD). By providing that the guard interval is longer than the combined delay spread due to chromatic dispersion (CD) and maximum differential group delay (DGD), the ISI can be successfully eliminated. However, the four-wave mixing (FWM) between different subcarriers and its interplay with CD and PMD will result in different subcarriers being affected differently. Although most of the subcarriers are without errors, the overall bit error ratio (BER) will be dominated by the BER of the worst subcarriers. In order to avoid this problem, the use of forward error correction (FEC)[1–29] is essential. Advanced FEC schemes are needed to provide BER performance of an OFDM system that is determined by the average received power rather than by the power of the weakest subcarrier.

In this chapter, we describe different FEC schemes suitable for use in coded OFDM. We start with a description of standard block codes. Then we describe the use of interleavers to deal with burst errors caused by errors in deteriorated neighboring subcarriers. We also describe the use of concatenated and product codes as efficient codes to deal with burst errors. After the description of conventional block and convolutional codes, we discuss more advanced FEC schemes—iteratively decodable codes such as turbo codes, turbo product codes, and low-density parity-check (LDPC) codes. We further describe the M-ary phase-shift keying (PSK) and M-ary quadrature amplitude modulation (QAM) as the most commonly used modulation techniques in OFDM. In the last section, we describe how to combine coding and modulation with OFDM.

The state-of-the-art optical communication systems standardized by the International Telecommunication Union (ITU) employ concatenated Bose–Chaudhuri–Hocquenghem (BCH)/Reed–Solomon (RS) codes.[1,2] The RS(255, 239) in particular has been used in a broad range of long-haul communication systems,[1,2] and it is commonly considered as the first-generation of FEC.[5] The elementary FEC schemes (BCH, RS, or convolutional codes) may be combined to design more powerful FEC schemes, such as RS(255, 239) + RS(255, 233). Several classes of concatenation codes are listed in ITU-T G975.1. Different concatenation schemes, such as the concatenation of two RS codes or the concatenation of

RS and convolutional codes, are commonly considered as the second generation of FEC.[5] In recent years, iteratively decodable codes, such as turbo codes[3–6] and LDPC codes,[7–12] have generated significant attention. Sab and Lemarie[4] proposed an FEC scheme based on block turbo code for long-haul dense wavelength division multiplexing optical transmission systems. It has been shown that iteratively decodable LDPC codes outperform turbo product codes in BER performance.[7–11] The decoder complexity of these codes is comparable to (or lower than) that of turbo product codes and significantly lower than that of serial/parallel concatenated turbo codes. For reasons mentioned previously, LDPC code is a viable and attractive choice for the FEC scheme of a 40 Gb/s optical transmission system. The soft iteratively decodable codes, turbo and LDPC codes, are commonly referred to as the third generation of FEC.[5]

LDPC decoding is based on a suboptimal message-passing algorithm, and therefore one may expect to be able to further improve this coding scheme in terms of its distance from Shannon's limit. Generalized low-density parity-check (GLDPC) coding[7,23–25] provides a way to achieve this goal and improves the overall characteristics of LDPC codes. Two classes of GLDPC codes are considered: random-like and structured GLDPC codes.

The capacity approaching coding gains can only be achieved by employing extremely long binary LDPC codes, which would introduce an unacceptable processing delay in the decoder. On the other hand, based on theoretical results for the additive white Gaussian noise (AWGN) channel,[13,14] it follows that codes over higher order Galois fields may be able to provide required coding gains even for moderate lengths.

This chapter is organized as follows. In Section 6.1, we describe different standard FEC schemes, such as cyclic codes and BCH codes. The special subclass of BCH codes, the RS codes, is also described in this section. To deal with simultaneous burst and random errors, we describe the concatenated and interleaved codes. In Section 6.2, we describe powerful iteratively decodable codes, such as turbo codes, turbo product codes, and LDPC codes. Given the fact that LDPC codes outperform turbo and turbo product codes in terms of BER performance and do not exhibit error floor phenomena when properly designed, we describe them with more details. In Section 6.3, we describe the multilevel modulation formats suitable for use in optical OFDM systems, such as *M*-ary QAM and *M*-ary PSK. In Section 6.4, we describe how to jointly combine multilevel modulation and channel coding. Finally, in Sections 6.5 and 6.6 we describe how to use coded modulation schemes in optical OFDM systems.

Given the fact that most state-of-the-art fiber-optic communication systems essentially use intensity modulation/direct detection (IM/DD), we first describe the coded optical OFDM communication with direct detection in Section 6.5. In Section 6.6, we describe the coded OFDM with coherent detection, specifically several PMD compensation architectures: (1) polarization diversity coded OFDM, (2) BLAST-type coded OFDM, and (3) iterative polarization cancellation (IPC) coded OFDM schemes.

6.1 Standard FEC Schemes

Two key system parameters are transmitted power and channel bandwidth, which together with additive noise sources determine the signal-to-noise ratio (SNR) and corresponding BER. In practice, we often encounter situations in which the target BER cannot be achieved with a given modulation format. For the fixed SNR, the only practical option to change the data quality transmission from unacceptable to acceptable is by use of channel coding. Another practical use of the channel coding is to reduce the required SNR for a given target BER. The amount of energy that can be saved by coding is commonly described by coding gain. *Coding gain* refers to the savings attainable in the energy per information bit-to-noise spectral density ratio (E_b/N_0) required to achieve a given bit error probability when coding is used compared to that when no coding is used. A typical digital optical communication system employing channel coding is shown in Figure 6.1. The discrete source generates the information in the form of a sequence of symbols. The channel encoder accepts the message symbols and adds redundant symbols according to a corresponding prescribed rule. The channel coding is the act of transforming a length-k sequence into a length-n codeword. The set of rules specifying this transformation are called the channel code, which can be represented as the following mapping:

$$C : M \rightarrow X$$

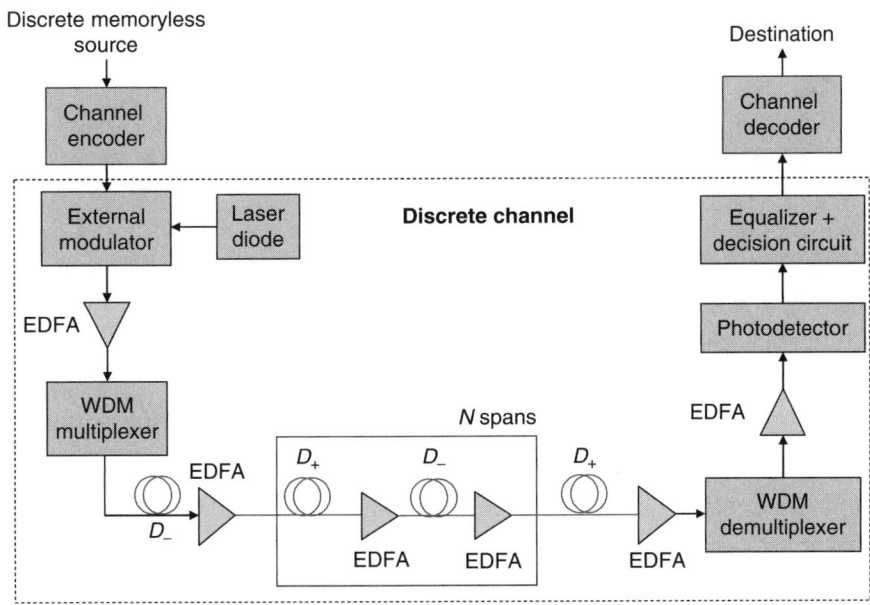

Figure 6.1: Block diagram of a point-to-point digital optical communication system.

where C is the channel code, M is the set of information sequences of length k, and X is the set of codewords of length n. The decoder exploits these redundant symbols to determine which message symbol was actually transmitted. Encoder and decoder consider the entire digital transmission system as a discrete channel. Other blocks shown in Figure 6.1 were explained in Chapter 3; here, we are concerned with channel encoders and decoders. Channel codes can be classified into three broad categories: (1) error detection, for which we are concerned only with detecting the errors that occurred during transmission (e.g., automatic request for transmission); (2) FEC, for which we are interested in correcting the errors that occurred during transmission; and (3) hybrid channel codes that combine the previous two approaches. In this chapter, we are concerned only with FEC.

The key idea behind the forward error correcting codes is to add extra redundant symbols to the message to be transmitted and use these redundant symbols in the decoding procedure to correct the errors introduced by the channel. The redundancy can be introduced in the time, frequency, or space domain. For example, the redundancy in the time domain is introduced if the same message is transmitted at least twice; this technique is known as the *repetition code*. The space redundancy is used as a means to achieve high spectrally efficient transmission, in which the modulation is combined with error control.

The codes commonly considered in fiber-optics communications belong either to the class of block codes or to the class of convolutional codes. In (n, k) block code, the channel encoder accepts information in successive k symbol blocks and adds $n - k$ redundant symbols that are algebraically related to the k message symbols, thereby producing an overall encoded block of n symbols $(n > k)$ known as a *codeword*. If the block code is systematic, the information symbols remain unchanged during the encoding operation, and the encoding operation may be considered as adding the $n - k$ generalized parity checks to k information symbols. Because the information symbols are statistically independent (a consequence of source coding or scrambling), the next codeword is independent of the content of the current codeword. The *code rate* of an (n, k) block code is defined as $R = k/n$, and *overhead* is defined by $\mathrm{OH} = (1/R - 1) \times 100\%$. In *convolutional code*, however, the encoding operation may be considered as the discrete-time convolution of the input sequence with the impulse response of the encoder. Therefore, the $n - k$ generalized parity checks are not only functions of k information symbols but also functions of m previous k-tuples, with $m + 1$ being the encoder impulse response length. The statistical dependence is introduced to the window of length $n(m + 1)$, the parameter known as *constraint length* of convolutional codes.

As mentioned previously, the channel code considers the entire transmission system as a discrete channel, in which the sizes of input and output alphabets are finite. Two examples of such channel are shown in Figure 6.2. Figure 6.2a shows an example of a discrete memoryless channel, which is characterized by channel (transition) probabilities. Let

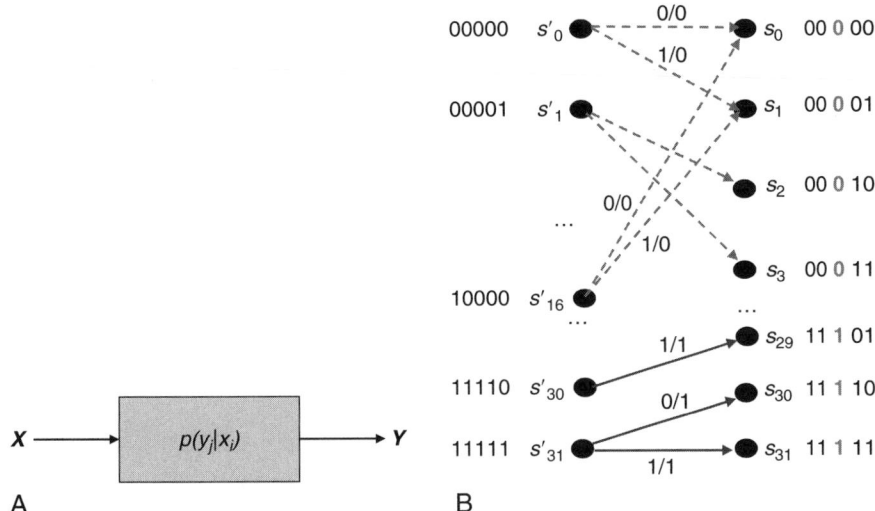

Figure 6.2: Two examples of discrete channels: (a) discrete memoryless channel (DMC) and (b) discrete channel with memory described as dynamic trellis.[30]

$X = \{x_0, x_1, \ldots, x_{I-1}\}$ denote the channel input alphabet, and $Y = \{y_0, y_1, \ldots, y_{J-1}\}$ denote the channel output alphabet. This channel is completely characterized by the following set of transition probabilities:

$$p(y_j|x_i) = P(Y = y_j|X = x_i),\ 0 \le p(y_j|x_i) \le 1,\ i \in \{0,1,\ldots,I-1\},\ j \in \{0,1,\ldots,J-1\} \quad (6.1)$$

where I and J denote the sizes of input and output alphabets, respectively. The transition probability $p(y_j|x_i)$ represents the conditional probability that channel output $Y = y_j$ given the channel input $X = x_i$. The channel introduces the errors, and if $j \ne i$ the corresponding $p(y_j|x_i)$ represents the conditional probability of error, whereas for $j = i$ it represents the conditional probability of correct reception. For $I = J$, the *average symbol error probability* is defined as the probability that output random variable Y_j is different from input random variable X_i, with averaging being performed for all $j \ne i$:

$$P_e = \sum_{i=0}^{I-1} p(x_i) \sum_{j=0,\ j\ne i}^{J-1} p(y_j|x_i) \quad (6.2)$$

where the inputs are selected from the distribution $\{p(x_i) = P(X = x_i);\ i = 0, 1, \ldots, I-1\}$, with $p(x_i)$ being the *a priori* probability of input symbol x_i. The corresponding probabilities of output symbols can be calculated by

$$p(y_j) = \sum_{i=0}^{I-1} P(Y = y_j|X = x_i)P(X = x_i) = \sum_{i=0}^{I-1} p(y_j|x_i)p(x_i);\ j = 0, 1, \ldots, J-1 \quad (6.3)$$

The decision rule that minimizes average symbol error probability (Eq. 6.2), denoted as $D(y_j) = x^*$, is known as the *maximum a posteriori* (MAP) rule, and it can be formulated as follows:

$$D(y_j) = x^*: \quad P(x^*|y_j) \geq P(x_i|y_j), \quad i = 0, 1, \ldots, I - 1 \tag{6.4}$$

Therefore, the symbol error probability P_e will be minimal when for every output symbol y_j the input symbol x^* is assigned the largest *a posteriori* probability $P(x^*|y_j)$. By using the Bayes' rule, Eq. (6.4) can be rewritten as

$$D(y_j) = x^*: \quad \frac{P(y_j|x^*)P(x^*)}{P(y_j)} \geq \frac{P(y_j|x_i)P(x_i)}{P(y_j)}, \quad i = 0, 1, \ldots, I - 1 \tag{6.5}$$

If all input symbols are equally likely, $P(x_i) = 1/I$ ($i = 0, \ldots, I - 1$), the corresponding rule is known as the maximum likelihood decision rule:

$$D(y_j) = x^*: \quad P(y_j|x^*) \geq P(y_j|x_i), \quad i = 0, 1, \ldots, I - 1 \tag{6.6}$$

Figure 6.2b shows a discrete channel model with memory,[30] which is more suitable for optical communications because the optical channel is essentially the channel with memory. We assume that the optical channel has memory equal to $2m + 1$, where $2m$ is the number of bits that influence the observed bit from both sides. This dynamical trellis is uniquely defined by the set of the previous state, the next state, in addition to the channel output. The state (the bit pattern configuration) in the trellis is defined as $s_j = (x_{j-m}, x_{j-m+1}, \ldots, x_j, x_{j+1}, \ldots, x_{j+m}) = x[j - m, j + m]$, where $x_k \in X = \{0, 1\}$. An example trellis of memory $2m + 1 = 5$ is shown in Figure 6.2b. The trellis has $2^5 = 32$ states ($s_0, s_1, \ldots, s_{31}$), each of which corresponds to a different 5-bit pattern. For the complete description of the trellis, the transition probability density functions (PDFs) $p(y_j|x_j) = p(y_j|s)$, $s \in S$ can be determined from collected histograms, where y_j represents the sample that corresponds to the transmitted bit x_j, and S is the set of states in the trellis.

One important figure of merit for DMCs is the amount of information conveyed by the channel, which is known as the *mutual information* and defined as

$$I(X; Y) = H(X) - H(X|Y) = \sum_{i=0}^{I-1} p(x_i) \log_2 \left[\frac{1}{p(x_i)}\right] - \sum_{j=0}^{J-1} p(y_j) \sum_{i=0}^{I-1} p(x_i|y_j) \log_2 \left[\frac{1}{p(x_i|y_j)}\right] \tag{6.7}$$

where $H(X)$ denotes the uncertainty about the channel input before observing the channel output, also known as entropy; whereas $H(X|Y)$ denotes the conditional entropy or the amount of uncertainty remaining about the channel input after the channel output has been received. Therefore, the mutual information represents the amount of information (per symbol) that is conveyed by the channel—that is, the uncertainty about the channel input that is resolved by observing the channel output. The mutual information can be interpreted by means of the Venn diagram shown in Figure 6.3a. The left circle represents the entropy of channel input, the right circle represents the entropy of channel output, and the mutual information is obtained in intersection of these two circles. Another interpretation from Ingels[31] is shown in Figure 6.3b. The mutual information (i.e., the information conveyed by the channel) is obtained as output information minus information lost in the channel.

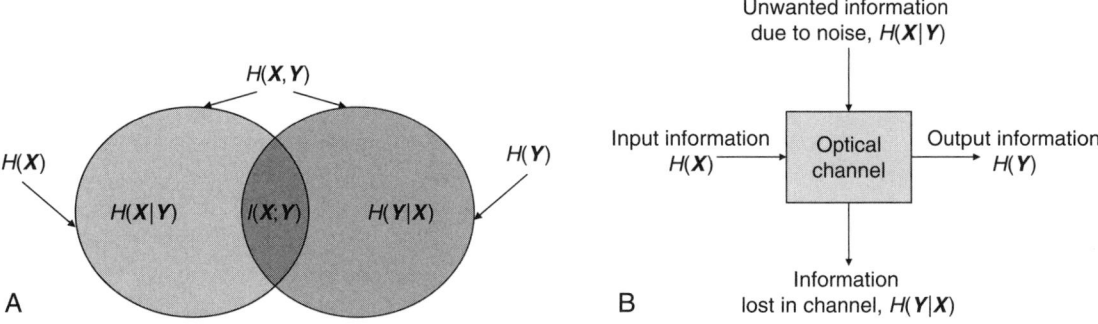

Figure 6.3: Interpretation of the mutual information: (a) using Venn diagrams and (b) using Ingels'[31] approach.

It is clear from Eq. (6.7) that mutual information is independent of the channel, and one may try to maximize the information conveyed by the channel to obtain the so-called *channel capacity*:

$$C = \max_{\{p(x_i)\}} I(X; Y); \text{ subject to: } p(x_i) \geq 0 \quad \sum_{i=0}^{I-1} p(x_i) = 1 \tag{6.8}$$

Now we have enough knowledge to formulate the *channel coding theorem*[32,33]:

Let a discrete memoryless source with an alphabet S have entropy $H(S)$ and emit the symbols every T_s seconds. Let a discrete memoryless channel have capacity C and be used once in T_c seconds. Then, if

$$H(S)/T_s \leq C/T_c \tag{6.9}$$

there exists a coding scheme for which the source output can be transmitted over the channel and reconstructed with an arbitrary small probability of error. The parameter $H(S)/T_s$ is related to the average information rate, whereas the parameter C/T_c is related to the channel capacity per unit time. For a binary symmetric channel ($I = J = 2$), the inequality (6.9) simply becomes

$$R \leq C \tag{6.10}$$

where R is the code rate introduced previously.

Another very important theorem is Shannon's third theorem, also known as the *information capacity theorem*, which can be formulated as discussed next.[32,33]

The information capacity of a continuous channel of bandwidth B Hz, perturbed by AWGN of PSD $N_0/2$ and limited in bandwidth B, is given by

$$C = B \log_2 \left(1 + \frac{P}{N_0 B}\right) \text{ [bits/s]} \tag{6.11}$$

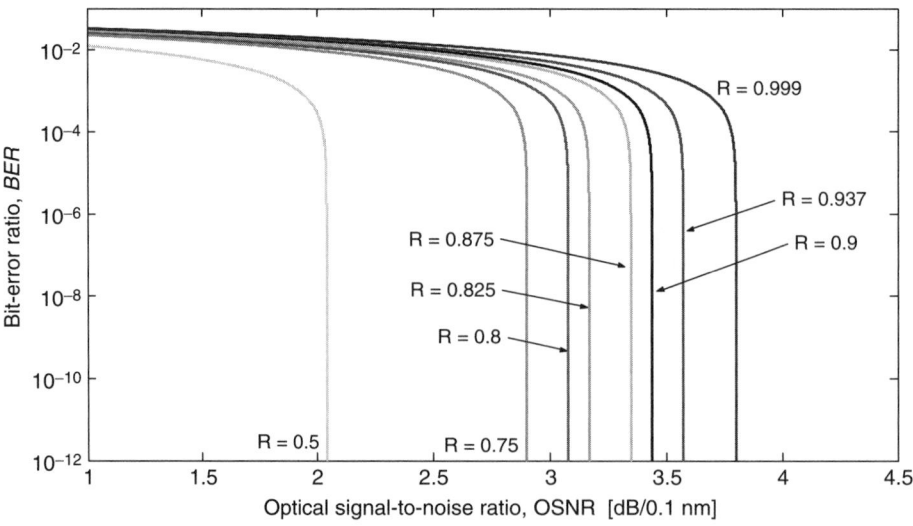

Figure 6.4: Minimum BER against optical SNR for different code rate values (for BPSK at 40 Gb/s).

where P is the average transmitted power. This theorem represents a remarkable result of information theory because it combines all important system parameters (transmitted power, channel bandwidth, and noise power spectral density) in one formula. Also interesting is the fact that LDPC codes can approach the Shannon's limit within 0.0045 dB.[12] By using Eq. (6.11) and Fano's inequality[33]

$$H(X|Y) \leq H(P_e) + P_e \log_2(I - 1), H(P_e) = -P_e \log_2 P_e - (1 - P_e) \log_2(1 - P_e) \qquad (6.12)$$

for amplified spontaneous emission (ASE) noise-dominated scenario and binary phase-shift keying (BPSK) at 40 Gb/s, Figure 6.4 shows the minimum BERs against optical SNR for different code rates.

In the following sections, an elementary introduction to linear block codes, BCH codes, RS codes, concatenated codes, and turbo product codes is given. These classes of codes are already employed in fiber-optics communication systems. For a detailed treatment of different error-control coding schemes, the interested reader is referred to references 22, 32, and 34–44.

6.1.1 Linear Block Codes

The *linear block code* (n, k), using the language of vector spaces, can be defined as a subspace of a vector space over finite field GF(q), where q is the prime power. Every space is described by its *basis*—a set of linearly independent vectors. The number of vectors in the basis determines the dimension of the space. Therefore, for an (n, k) linear block code the dimension of the space is n, and the dimension of the code subspace is k. The basis vectors may be put in a matrix of dimensions $k \times n$, known as the *generator matrix*, in which every row represents a vector from the coding subspace, and the columns represent corresponding vector components:

$$G = \begin{bmatrix} g_{11} & g_{12} & \cdots & g_{1n} \\ g_{21} & g_{22} & \cdots & g_{2n} \\ \cdots & \cdots & \cdots & \cdots \\ g_{k1} & g_{k2} & \cdots & g_{kn} \end{bmatrix} \tag{6.13}$$

In the rest of this section, the codes over GF(2) (binary field, F_2) are considered. The *linearity property* of a linear block code over GF(2) means that the linear combination of any two codewords is another codeword, and it represents a necessary and sufficient condition for any subset of F_2^n (the set of all n-tuples) to be a subspace. To encode, the message vector $m(m_1, m_2, \ldots, m_k)$ has to be multiplied with a generator matrix G to get $c = mG$, where $c(c_1, c_2, \ldots, c_n)$ is a codeword.

By elementary operations on rows in the generator matrix, the code may be transformed into systematic form

$$G_s = \begin{bmatrix} 1 & 0 & \cdots & 0 & p_{11} & \cdots & p_{1,n-k} \\ 0 & 1 & \cdots & 0 & p_{21} & \cdots & p_{2,n-k} \\ \cdots & \cdots & \cdots & \cdots & \cdots & \cdots & \cdots \\ 0 & 0 & \cdots & 1 & p_{k1} & \cdots & p_{k,n-k} \end{bmatrix} = \begin{bmatrix} I_k & P \end{bmatrix} \tag{6.14}$$

where I_k is the unity matrix of dimensions $k \times k$, and P is the matrix of dimensions $k \times (n - k)$ with columns denoting the positions of parity checks. The codeword of a systematic code is obtained by

$$c = m \begin{bmatrix} I_k & P \end{bmatrix} = \begin{bmatrix} mI_k & mP \end{bmatrix} = \begin{bmatrix} m & b \end{bmatrix} \tag{6.15}$$

Therefore, during encoding the message vector remains unchanged and the elements of vector of parity checks b are obtained by

$$b_j = m_1 p_{1j} + m_2 p_{2j} + \ldots + m_k p_{k,j}, (j = 1, 2, \ldots, n - k) \tag{6.16}$$

During transmission, an optical channel introduces the errors so that the received vector r can be written as $r = c + e$, where e is the error vector (pattern) with element components determined by

$$e_i = \begin{cases} 1 & \text{if an error occurred in the } i\text{th location} \\ & 0 \quad \text{otherwise} \end{cases}$$

To determine whether the received vector r is a codeword vector, we introduce the concept of a *parity-check matrix*. The parity-check matrix of an (n, k) linear block code H is a matrix of rank $n - k$ and dimensions $(n - k) \times n$ whose null space is k-dimensional vector with the basis being the generator matrix G. Therefore, the matrices G and H are orthogonal $GH^{\mathrm{T}} = 0$, and for systematic code H becomes

$$H = [-P^T \ I_{n-k}] \ \text{(in GF(2) the sign "} - \text{" is to be neglected.)}$$

The received vector r is a codeword if the following syndrome equation is satisfied $s = rH^{\mathrm{T}} = 0$. For example, for Hamming $(7, 4)$ code, the generator and parity-check matrices are given, respectively, as

$$G = \begin{bmatrix} 1000|110 \\ 0100|011 \\ 0010|111 \\ 0001|101 \end{bmatrix} \qquad H = \begin{bmatrix} 1011|100 \\ 1110|010 \\ 0111|001 \end{bmatrix}$$

To determine the error correction capability of the code, we have to introduce the concept of Hamming distance and Hamming weight. *Hamming distance* between two codewords c_1 and c_2, $d(c_1, c_2)$, is defined as the number of locations in which their respective elements differ. *Hamming weight*, $w(c)$, of a code vector c is defined as the number of nonzero elements in the vectors. The minimum distance, $d_{\min}$, of a linear block code is defined as the smallest Hamming distance between any pair of code vectors in the code. Because the zero vector is a codeword, the minimum distance of a linear block code can be determined simply as the smallest Hamming weight of the nonzero code vectors in the code. Let the parity-check matrix be written as $H = [h_1 \ h_2, \ldots, h_n]$, where h_i is the ith column in H. Because every codeword c must satisfy the syndrome equation, $cH^{\mathrm{T}} = 0$, the minimum distance of a linear block code is determined by the minimum number of columns of the H matrix whose sum is equal to the zero vector. For example, $(7, 4)$ Hamming code in the previous example has the minimum distance $d_{\min} = 3$ since the addition of the first, fifth, and sixth columns leads to zero vector.

It is straightforward to show that an (n, k) linear block code of minimum distance $d_{\min}$ can correct up to t errors if, and only if, $t \le \lfloor 1/2(d_{\min} - 1) \rfloor$ (where $\lfloor \ \rfloor$ denotes the largest integer less than or equal to the enclosed quantity) or, equivalently, $d_{\min} \ge 2t + 1$. If we are only interested in detecting e_d errors, then $d_{\min} \ge e_d + 1$. Finally, if we are interested in detecting e_d errors and correcting e_c errors, then $d_{\min} \ge e_d + e_c + 1$. The Hamming $(7, 4)$ code is therefore a single error-correcting and double error-detecting code. More generally, a family of (n, k) linear block codes with parameters

- Block length: $n = 2^m - 1$

- Number of message bits: $k = 2^m - m - 1$

- Number of parity bits: $n - k = m$

- $d_{\min} = 3$

where $m \geq 3$, are known as *Hamming codes*. Hamming codes belong to the class of perfect codes—the codes that satisfy the following Hamming inequality with equality sign[32,33]

$$2^{n-k} \geq \sum_{i=0}^{t} \binom{n}{i} \tag{6.17}$$

This bound gives the number of errors t that can be corrected with an (n, k) linear block code by using the syndrome decoding.

A very important characteristic of an (n, k) linear block code is the so-called coding gain, which was introduced previously in this chapter as being the savings attainable in the energy per information bit-to-noise spectral density ratio (E_b/N_0) required to achieve a given bit error probability when coding is used compared to when no coding is used. Let E_c denote the transmitted bit energy and E_b denote the information bit energy. Because the total information word energy kE_b must be the same as the total codeword energy nE_c, we obtain the following relationship between E_c and E_b:

$$E_c = (k/n)E_b = RE_b \tag{6.18}$$

The probability of error for BPSK on an AWGN channel, when a coherent hard decision (bit-by-bit) demodulator is used, can be obtained as follows:

$$p = \frac{1}{2}\text{erfc}\left(\sqrt{\frac{E_c}{N_0}}\right) = \frac{1}{2}\text{erfc}\left(\sqrt{\frac{RE_b}{N_0}}\right) \tag{6.19}$$

where $\text{erfc}(x)$ function is defined by

$$\text{erfc}(x) = \frac{2}{\sqrt{\pi}} \int_{x}^{+\infty} e^{-z^2} dz$$

For high SNRs, the word error probability (remained upon decoding) of a t-error correcting code is dominated by a $t + 1$ error event:

$$P_w(e) \approx \binom{n}{t+1} p^{t+1}(1-p)^{n-t+1} \approx \binom{n}{t+1} p^{t+1} \tag{6.20}$$

The bit error probability P_b is related to the word error probability by

$$P_b \approx \frac{2t+1}{n} P_w(e) \approx c(n,t) p^{t+1} \tag{6.21}$$

because $2t + 1$ and more errors per codeword cannot be corrected and they can be located anywhere on n codeword locations, and $c(n, t)$ is a parameter dependent on error correcting capability t and codeword length n. By using the upper bound on $\text{erfc}(x)$, we obtain

$$P_b \approx \frac{c(n,t)}{2}\left[\exp\left(\frac{-RE_b}{N_0}\right)\right]^{t+1} \tag{6.22}$$

The corresponding approximation for the uncoded case is

$$P_{b,\text{uncoded}} \approx \frac{1}{2}\exp\left(-\frac{E_b}{N_0}\right) \tag{6.23}$$

By equating Eqs. (6.22) and (6.23) and ignoring the parameter $c(n, t)$, we obtain the following expression for hard decision decoding coding gain:

$$\frac{(E_b/N_0)_{\text{uncoded}}}{(E_b/N_0)_{\text{coded}}} \approx R(t + 1) \tag{6.24a}$$

The corresponding soft decision asymptotic coding gain of convolutional codes is[22,34–36,45]

$$\frac{(E_b/N_0)_{\text{uncoded}}}{(E_b/N_0)_{\text{coded}}} \approx Rd_{\min} \tag{6.24b}$$

and it is approximately 3 dB better than hard decision decoding (because $d_{\min} \geq 2t + 1$).

In optical communications, it is very common to use the Q-factor as the figure of merit instead of SNR, which is related to the BER on an AWGN, as shown in Chapter 3, as follows:

$$\text{BER} = \frac{1}{2}\text{erfc}\left(\frac{Q}{\sqrt{2}}\right) \tag{6.25}$$

Let BER_{in} denote the BER at the input of FEC decoder, let BER_{out} denote the BER at the output of FEC decoder, and let BER_{ref} denote the target BER (such as either 10^{-12} or 10^{-15}). The corresponding coding gain (CG) and net coding gain (NCG) are, respectively, defined as[46]

$$\text{CG} = 20\log_{10}\left[\text{erfc}^{-1}(2\text{BER}_{\text{ref}})\right] - 20\log_{10}\left[\text{erfc}^{-1}(2\text{BER}_{\text{in}})\right] \quad [\text{dB}] \tag{6.26}$$

$$\text{NCG} = 20\log_{10}\left[\text{erfc}^{-1}(2\text{BER}_{\text{ref}})\right] - 20\log_{10}\left[\text{erfc}^{-1}(2\text{BER}_{\text{in}})\right] + 10\log_{10}R \quad [\text{dB}] \tag{6.27}$$

All coding gains reported in this chapter are in fact NCG, although they are sometimes called the coding gains only, because this is the common practice in the coding theory literature.[22,34–36,45]

6.1.2 Cyclic Codes

The most commonly used class of linear block codes is the cyclic codes. Examples of cyclic codes include BCH codes, Hamming codes, and Golay codes. RS codes are also cyclic but nonbinary codes. Even LDPC codes can be designed in cyclic or quasi-cyclic fashion.

Let us observe the vector space of dimension n. The subspace of this space is *cyclic code* if for any codeword $c(c_0, c_1, \ldots, c_{n-1})$ arbitrary cyclic shift $c_j(c_{n-j}, c_{n-j+1}, \ldots, c_{n-1}, c_0, c_1, \ldots, c_{n-j-1})$ is another codeword. With every codeword $c(c_0, c_1, \ldots, c_{n-1})$ from a cyclic code, we associate the *codeword polynomial*

$$c(x) = c_0 + c_1 x + c_2 x^2 + \ldots + c_{n-1}x^{n-1} \tag{6.28}$$

The jth cyclic shift, observed $\mod(x^n - 1)$, is also a codeword polynomial

$$c^{(j)}(x) = x^j c(x) \mod(x^n - 1) \tag{6.29}$$

It is straightforward to show that observed subspace is cyclic if composed of polynomials divisible by a polynomial $g(x) = g_0 + g_1 x + \ldots + g_{n-k} x^{n-k}$ that divides $x^n - 1$ at the same time. The polynomial $g(x)$, of degree $n - k$, is called the *generating polynomial* of the code. If $x^n - 1 = g(x)h(x)$, then the polynomial of degree k is called the *parity-check polynomial*. The generating polynomial $g(x)$ and the parity-check polynomial $h(x)$ serve the same role as the generating matrix G and parity-check matrix H of a linear block code. n-Tuples pertaining to the k polynomials $g(x)$, $xg(x)$, $\ldots$, $x^{k-1}g(x)$ may be used in rows of the $k \times n$ generator matrix G, whereas n-tuples pertaining to the $(n - k)$ polynomials $x^k h(x - 1)$, $x^{k+1} h(x)$, $\ldots$, $x^{n-1} h(x)$ may be used in rows of the $(n - k) \times n$ parity-check matrix H.

To encode, we simply have to multiply the message polynomial $m(x) = m_0 + m_1 x + \ldots + m_{k-1} x^{k-1}$ by the generating polynomial $g(x)$; that is, $c(x) = m(x)g(x)\mod(x^n - 1)$, where $c(x)$ is the codeword polynomial. To encode in systematic form, we have to find the remainder of $x^{n-k} m(x)/g(x)$ and add it to the shifted version of message polynomial $x^{n-k} m(x)$; that is, $c(x) = x^{n-k} m(x) + \mathrm{rem}[x^{n-k} m(x)/g(x)]$, where rem[ ] is denoted the remainder of a given entity. The general circuit for generating the codeword polynomial in systematic form is given in Figure 6.5. The encoder operates as follows: When the switch S is in position 1 and the gate is closed (on), the information bits are shifted into the shift register and at the same time transmitted onto the channel. After all the information bits are shifted into register in k shifts, with the gate open (off), the switch S is moved to position 2, and the content of $(n - k)$ shift register is transmitted onto the channel.

To check if the received word polynomial is the codeword polynomial $r(x) = r_0 + r_1 x + \ldots + r_{n-1} x^{n-1}$, we have simply to determine the syndrome polynomial $s(x) = \mathrm{rem}[r(x)/g(x)]$. If $s(x)$ is zero, then there is no error introduced during transmission. The corresponding circuit is given in Figure 6.6. For more details on decoding of cyclic code, see references.[22,34–36]

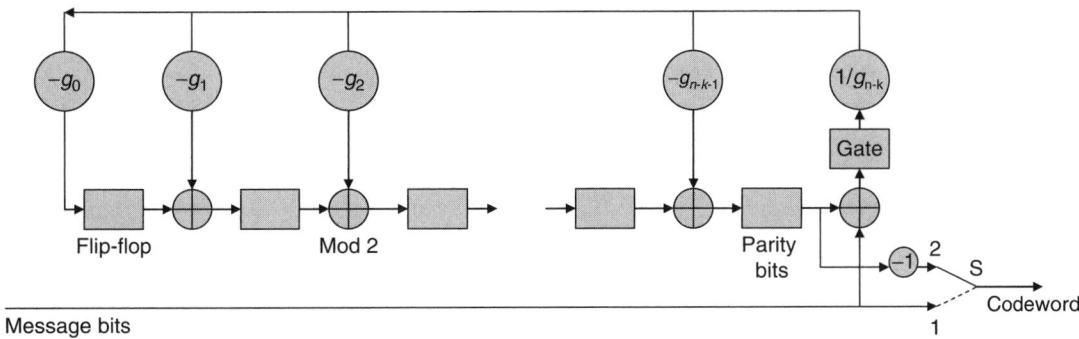

Figure 6.5: Systematic cyclic encoder.

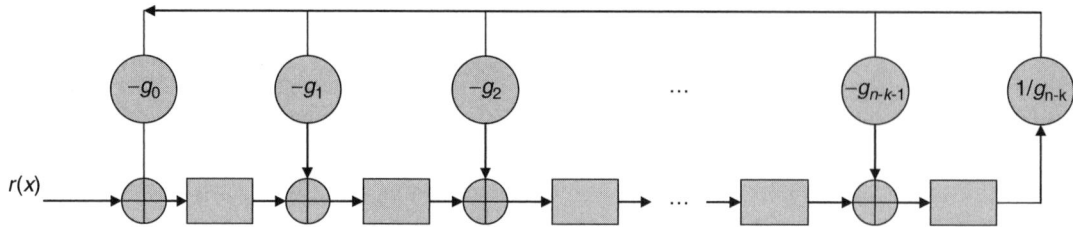

Figure 6.6: Syndrome calculator.

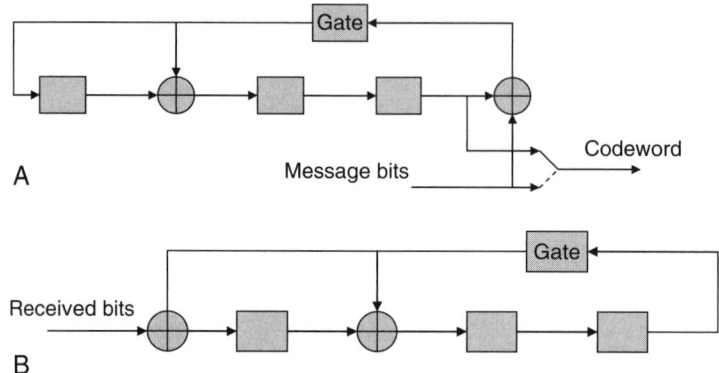

Figure 6.7: (a) Hamming (7, 4) encoder and (b) syndrome calculator.

For example, the encoder and syndrome calculator for (7, 4) Hamming code are given in Figures 6.7a and 6.7b, respectively. The generating polynomial is given by $g(x) = 1 + x + x^3$.

6.1.3 Bose–Chaudhuri–Hocquenghem Codes

The BCH codes, the most famous cyclic codes, were discovered by Hocquenghem in 1959 and by Bose and Chaudhuri in 1960. Among many different decoding algorithms, the most important are the Massey–Berlekamp algorithm and Chien's search algorithm. An important subclass of BCH is the RS codes proposed in 1960.

Let the finite field GF(q) (symbol field) and extension field GF(q^m) (locator field), $m \geq 1$, be given. For every m_0 ($m_0 \geq 1$) and Hamming distance d there exists a BCH code with the generating polynomial $g(x)$, if and only if it is of the smallest degree with coefficients from GF(q) and the roots from the extension field GF(q^m) as follows:

$$\alpha^{m_0}, \alpha^{m_0+1}, \ldots, \alpha^{m_0+d-2} \tag{6.30}$$

where α is from GF(q^m). The codeword length is determined as the least common multiple of orders of roots. (The order of an element β from the finite field is the smallest positive integer j such that $\beta^j = 1$.)

It can be shown that for any positive integer $m(m \geq 3)$ and $t(t < 2^{m-1})$ there exists a binary BCH code with the following properties:

- Codeword length: $n = 2^m - 1$

- Number of parity bits: $n - k \leq mt$

- Minimum Hamming distance: $d \geq 2t + 1$

This code is able to correct up to t errors. The generator polynomial can be found as the least common multiple of the minimal polynomials of α^i:

$$g(x) = \text{LCM}\left[\phi_{\alpha^1}(x), \phi_{\alpha^3}(x), \ldots, \phi_{\alpha^{2t-1}}(x)\right] \tag{6.31}$$

where α is a primitive element in GF(2^m), and $\phi_{\alpha^i}(x)$ is the minimal polynomial of α^i. The primitive element is defined as a field element that generates all of the nonzero elements as its successive powers, and the minimal polynomial is defined as the smallest degree polynomial over GF(2) having a field element β from GF(2^m) as its root. For more details on finite fields, see Anderson and Mohan.[34] The parity-check matrix of this code can be written as[22,34–36]

$$H = \begin{bmatrix} \alpha^{n-1} & \alpha^{n-2} & \ldots & \alpha & 1 \\ \alpha^{3(n-1)} & \alpha^{3(n-2)} & \ldots & \alpha^3 & 1 \\ \alpha^{5(n-1)} & \alpha^{5(n-2)} & \ldots & \alpha^5 & 1 \\ \ldots & \ldots & \ldots & \ldots & \ldots \\ \alpha^{(2t-1)(n-1)} & \alpha^{(2t-1)(n-2)} & \ldots & \alpha^{2t-1} & 1 \end{bmatrix} \tag{6.32}$$

For example, (15, 7) 2-error correcting BCH code has the generating matrix[34]

$$g(x) = \text{LCM}[\phi_\alpha(x), \phi_{\alpha^3}(x)]$$

$$= \text{LCM}\left[x^4 + x + 1, (x + \alpha^3)(x + \alpha^6)(x + \alpha^9)(x + \alpha^{12})\right] = x^8 + x^7 + x^6 + x^4 + 1$$

and the parity-check matrix[34]

$$H = \begin{bmatrix} \alpha^{14} & \alpha^{13} & \alpha^{12} & \alpha^{11} & \ldots & \alpha & 1 \\ \alpha^{42} & \alpha^{39} & \alpha^{36} & \alpha^{33} & \ldots & \alpha^3 & 1 \end{bmatrix}$$

$$= \begin{bmatrix} \alpha^{14} & \alpha^{13} & \alpha^{12} & \alpha^{11} & \alpha^{10} & \alpha^9 & \alpha^8 & \alpha^7 & \alpha^6 & \alpha^5 & \alpha^4 & \alpha^3 & \alpha^2 & \alpha & 1 \\ \alpha^{12} & \alpha^9 & \alpha^6 & \alpha^3 & 1 & \alpha^{12} & \alpha^9 & \alpha^6 & \alpha^3 & 1 & \alpha^{12} & \alpha^9 & \alpha^6 & \alpha^3 & 1 \end{bmatrix}$$

In the previous expression, we used the fact that in GF(2^4), $\alpha^{15} = 1$. The primitive polynomial used to design this code was $p(x) = x^4 + x + 1$. (The primitive polynomial is an

Table 6.1: GF(2^4) Generated by $x^4 + x + 1$

Power of α	Polynomial of α	4-Tuple
0	0	0000
α^0	1	0001
α^1	α	0010
α^2	α^2	0100
α^3	α^3	1000
α^4	$\alpha + 1$	0011
α^5	$\alpha^2 + \alpha$	0110
α^6	$\alpha^3 + \alpha^2$	1100
α^7	$\alpha^3 + \alpha + 1$	1011
α^8	$\alpha^2 + 1$	0101
α^9	$\alpha^3 + \alpha$	1010
α^{10}	$\alpha^2 + \alpha + 1$	0111
α^{11}	$\alpha^3 + \alpha^2 + \alpha$	1110
α^{12}	$\alpha^3 + \alpha^2 + \alpha + 1$	1111
α^{13}	$\alpha^3 + \alpha^2 + 1$	1101
α^{14}	$\alpha^3 + 1$	1001

irreducible polynomial that has a primitive element as its root.) Every element in GF(2^4) can be represented as 4-tuple, as shown in Table 6.1. To create the second column, we used the relation $\alpha^4 = \alpha + 1$, and the 4-tuples are obtained by reading off coefficients in the second column. By replacing the powers of α in the previous parity-check matrix by corresponding 4-tuples, the parity-check matrix can be written in the following binary form:

$$H = \begin{bmatrix} 1 & 1 & 1 & 1 & 0 & 1 & 0 & 1 & 1 & 0 & 0 & 1 & 0 & 0 & 0 \\ 0 & 1 & 1 & 1 & 1 & 0 & 1 & 0 & 1 & 1 & 0 & 0 & 1 & 0 & 0 \\ 0 & 0 & 1 & 1 & 1 & 1 & 0 & 1 & 0 & 1 & 1 & 0 & 0 & 1 & 0 \\ 1 & 1 & 1 & 0 & 1 & 0 & 1 & 1 & 0 & 0 & 1 & 0 & 0 & 0 & 1 \\ 1 & 1 & 1 & 1 & 0 & 1 & 1 & 1 & 1 & 0 & 1 & 1 & 1 & 1 & 0 \\ 1 & 0 & 1 & 0 & 0 & 1 & 0 & 1 & 0 & 0 & 1 & 0 & 1 & 0 & 0 \\ 1 & 1 & 0 & 0 & 0 & 1 & 1 & 0 & 0 & 0 & 1 & 1 & 0 & 0 & 0 \\ 1 & 0 & 0 & 0 & 1 & 1 & 0 & 0 & 0 & 1 & 1 & 0 & 0 & 0 & 1 \end{bmatrix}$$

In general, there is no need for q to be a prime; it could be a prime power. However, the symbols must be taken from GF(q) and the roots from GF(q^m). Of the nonbinary BCH codes, the RS codes are the most famous, and these are briefly explained in the next section. For RS codes, the symbol field and locator field are the same ($m = 1$).

6.1.4 Reed–Solomon Codes, Concatenated Codes, and Product Codes

The generator polynomial for a t-error correcting RS code is given by[22,34–36]

$$g(x) = (x - \alpha)(x - \alpha^2) \ldots (x - \alpha^{2t}) \tag{6.33}$$

The generator polynomial degree is $2t$, and it is the same as the number of parity symbols $n - k = 2t$, whereas the block length of the code is $n = q - 1$. Because the minimum distance of BCH codes is $2t + 1$, the minimum distance of RS codes is $n - k + 1$, satisfying the Singleton bound ($d_{\min} \leq n - k + 1$) with equality and belonging to the class of *maximum-distance separable* codes.

The RS codes may be considered as burst error-correcting codes, and as such they are suitable for high-speed transmission at 40 Gb/s or higher because the fiber-optics channel at 40 Gb/s is burst-error prone due to intrachannel nonlinearities, especially intrachannel FWM. Consider an RS code over GF(q), with $q = 2^m$. Each symbol in binary notation can be represented by m bits, and therefore the original code can be considered as binary with codeword length $n = m(2^m - 1)$ and number of parity bits $n - k = 2mt$. This binary code is able to correct up to t bursts of length m. Equivalently, this binary code is able to correct a single burst of length $(t - 1)m + 1$.

To improve the burst error correction capability of an RS code, the code can be combined with an inner binary block code in a concatenation scheme as shown in Figure 6.8. The key idea behind the concatenation scheme can be explained as follows[34]: Consider the codeword generated by inner (n, k, d) code (where d is the minimum distance of the code) and transmitted over the bursty channel. The decoder processes the erroneously received codeword and decodes it correctly. However, occasionally the received codeword is decoded incorrectly. Therefore, the inner encoder, the channel, and the inner decoder may be considered as a superchannel whose input and output alphabets belong to GF(2^k). The outer encoder (N, K, D) (where D is the minimum distance of outer code) encodes input K symbols and generates N output symbols transmitted over the superchannel. The length of each symbol is k information digits. The resulting scheme, known as concatenated code and initially proposed by Forney,[37] is an (Nn, Kk, $\geq Dd$) code with the minimum distance of at least Dd. For example, RS(255, 239, 8) code can be combined with the (12, 8, 3) single parity-check code in the concatenation scheme ($12 \cdot 255$, $239 \cdot 8$, ≥ 24). The outer RS decoder can be implemented using the Massey–Berlekamp algorithm,[34–36] whereas the inner decoder

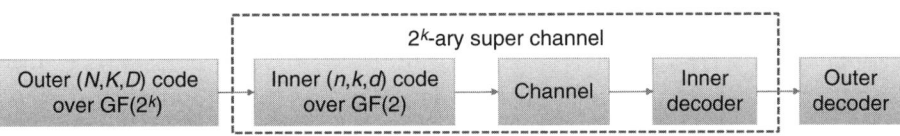

Figure 6.8: The concatenated (Nn, Kk, $\geq Dd$) code.

can be implemented using the maximum *a posteriori* decoding based on the Bahl–Cocke–Jelinek–Raviv (BCJR) algorithm.[38] The concatenated scheme from Figure 6.8 can be generalized to q-ary channels, with the inner code operating over $GF(q)$ and the outer code operating over $GF(q^k)$.

Two RS codes can be combined in a concatenated scheme by interleaving. An *interleaved code* is obtained by taking L codewords (of length N) of a given code $x_j = (x_{j1}, x_{j2}, \ldots, x_{jN})$ ($j = 1, 2, \ldots, L$) and forming the new codeword by interleaving the L codewords as follows: $y_i = (x_{11}, x_{21}, \ldots, x_{L1}, x_{12}, x_{22}, \ldots, x_{L2}, \ldots, x_{1N}, x_{2N}, \ldots, x_{LN})$. The process of interleaving can be visualized as the process of forming an $L \times N$ matrix of L codewords written row by row and transmitting the matrix column by column:

$$x_{11}\ x_{12} \ldots x_{1N}$$

$$x_{21}\ x_{22} \ldots x_{2N}$$

$$\cdots \cdots \cdots$$

$$x_{L1}\ x_{L2} \ldots x_{LN}$$

The parameter L is known as the interleaving *degree*. The transmission must be postponed until L codewords are collected. To be able to transmit a column whenever a new codeword becomes available, the codewords should be arranged down diagonals as given here, and the interleaving scheme is known as *delayed interleaving* (1-frame delayed interleaving):

$$x_{i-(N-1),1} \quad \cdots \quad x_{i-2,1} \qquad x_{i-1,1} \qquad x_{i,1}$$

$$x_{i-(N-1),2} \quad \cdots \qquad x_{i-2,2} \qquad x_{i-1,2} \qquad x_{i,2}$$

$$\cdots \cdots \cdots \cdots$$

$$x_{i-(N-1),N-1} \quad x_{i-(N-2),N-1}$$

$$x_{i-(N-1),N} \qquad x_{i-(N-2),N}$$

Each new codeword completes one column of this array. In the previous example, the codeword x_i completes the column (frame) $x_{i,1}, x_{i-1,2}, \ldots, x_{i-(N-1),N}$. A generalization of this scheme, in which the components of ith codeword x_i such as $x_{i,j}$ and $x_{i,j+1}$ are spaced λ frames apart, is known as λ-frame delayed interleaved.

Another way to deal with burst errors is to arrange two RS codes in a *product* manner as shown in Figure 6.9. A product code[3–6] is an ($n_1 n_2, k_1 k_2, d_1 d_2$) code in which codewords form an $n_1 \times n_2$ array such that each row is a codeword from an (n_1, k_1, d_1) code C_1 and each column is a codeword from an (n_2, k_2, d_2) code C_2, with n_i, k_i, and d_i ($i = 1, 2$) being the codeword length, dimension, and minimum distance, respectively, of the ith component code. Turbo product codes were proposed by Elias.[28] In turbo product codes, we perform several iterations; wherein one iteration is defined as decoding per rows followed by decoding per columns. Both binary (e.g., binary BCH codes) and nonbinary codes (e.g., RS codes) may be arranged in the product manner. It is possible to show that the minimum distance of a product

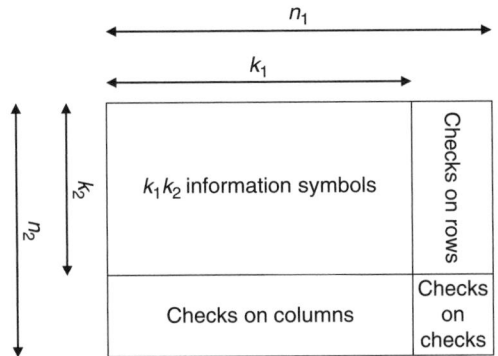

Figure 6.9: The structure of a codeword of a turbo product code.

code is the product of minimum distances of component codes.[34] It is straightforward to show that the product code is able to correct the burst error of length $b = \max(n_1 b_2, n_2 b_1)$, where b_i is the burst error capability of component code $i = 1, 2$.

The results of Monte Carlo simulations for different RS concatenation schemes and an AWGN channel are shown in Figure 6.10. Interestingly, the concatenation scheme RS(255, 239) + RS(255, 223) of code rate $R = 0.82$ outperforms the concatenation scheme RS(255, 223) + RS(255, 223) of lower code rate $R = 0.76$, as well as the concatenation scheme RS(255, 223) + RS(255, 239) of the same code rate.

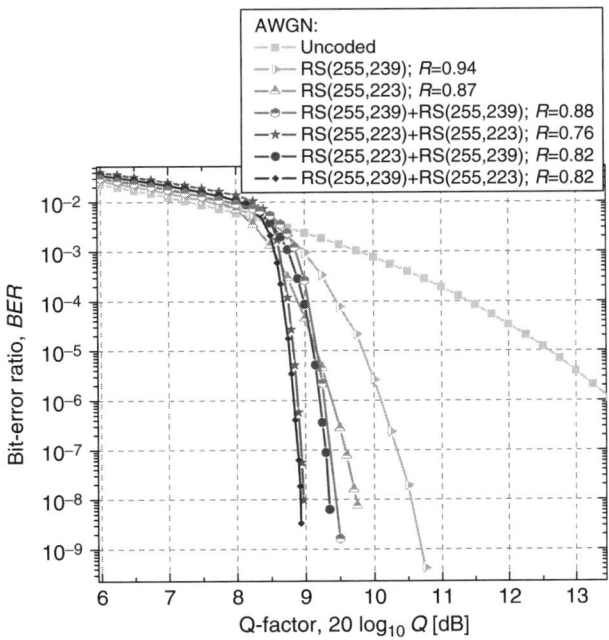

Figure 6.10: BER performance of concatenated RS codes.

6.2 Codes on Graphs

The codes on graphs of interest in optical communications, such as turbo codes, turbo product codes, and LDPC codes, are described in this section.

6.2.1 Turbo Codes

Turbo codes[39,40] can be considered as the generalization of the concatenation of codes in which, during iterative decoding, the decoders interchange the soft messages a certain number of times.[39-42] One possible implementation of turbo encoder and decoder based on systematic convolutional codes is given in Figure 6.11. Let both systematic convolutional encoders have the code rate ½. Since the encoders are systematic, the information bit can be sent directly to the channel. Encoder 1 generates the parity bits operating on the information sequence directly, whereas encoder 2 generates parity bits operating on the interleaved sequence. The resulting code rate, without puncturing, is 1/3. Notice that encoder 2 operates on information bits only, whereas the inner concatenation coder from the previous section operates on both information and parity bits from the outer encoder. The interleaver is introduced to prevent protecting the same information bits using the same parity checks and as a means to deal with burst errors. Since the code rate 1/3 is unacceptably low for

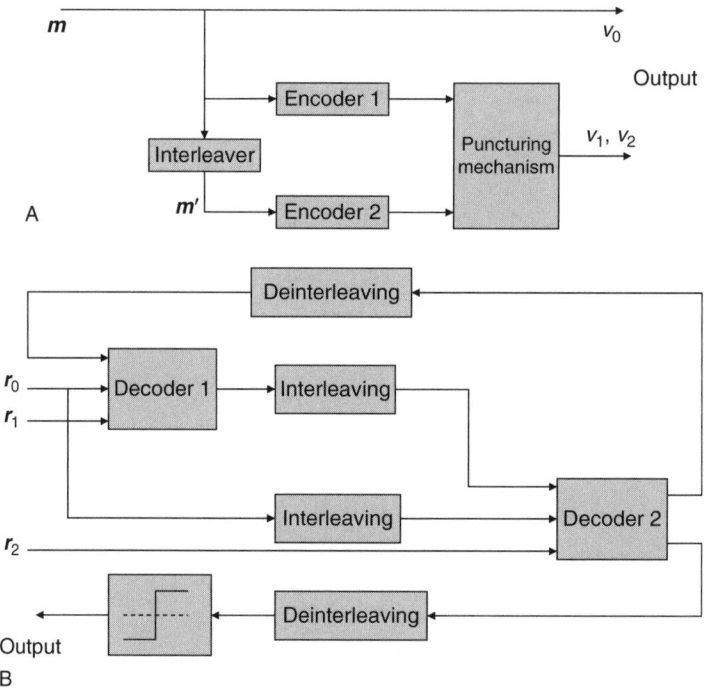

Figure 6.11: Parallel concatenated convolutional encoder (a) and decoder (b) configurations.

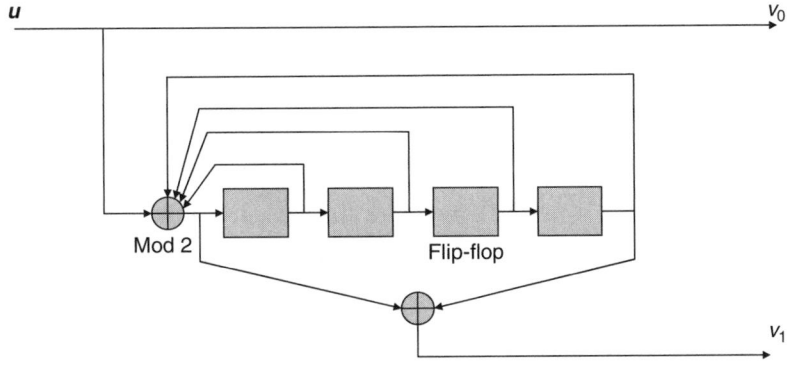

Figure 6.12: RSC encoder with $G_R(D) = [1\ (1 + D^4)/(1 + D + D^2 + D^3 + D^4)]$.

fiber-optics communication systems, the puncturing mechanism is introduced. The resulting code rate after puncturing is $R = L/(L + P)$, where L is the interleaver size, and P is the number of bits remaining after the puncturing. To attain the exceptional performance of turbo codes, the constituent encoders are commonly designed as recursive systematic convolutional (RSC) codes.[42] In principle, the encoders can be connected in a serial manner.[42] The generator matrix for a rate ½ RSC code can be written as $G_R(D) = [1 g_1(D)/g_2(D)]$, where $g_1(D)$ describes the transform of the parity sequence, and $g_2(D)$ denotes the transform of the information sequence. For example, the code described by the generator matrix $G_R(D) = [1(1 + D^4)/(1 + D + D^2 + D^3 + D^4)]$ is given in Figure 6.12.

Due to the presence of interleaver, the maximum likelihood sequence decoder (MLSD) would be too complicated to implement; however, a suboptimum iterative decoder (turbo decoder) involving two soft-in/soft-output decoders, proposed in references 39 and 40. provides near MLSD BER performance. The idea behind the iterative decoder is to estimate the *a posteriori* probabilities $\Pr(u_k|r)$ iteratively (where u_k is the kth information bit, and r is the received codeword in the presence of noise $r = v + n$) and apply the maximum *a posteriori* rule:

$$\frac{P(u_k = 0|r)}{P(u_k = 1|r)} \begin{array}{c} 0 \\ > \\ < \\ 1 \end{array} 1 \Leftrightarrow L(u_k) = \log\left(\frac{P(u_k = 0|r)}{P(u_k = 1|r)}\right) \begin{array}{c} 0 \\ > \\ < \\ 1 \end{array} 0 \qquad (6.34)$$

Each component decoder receives the extrinsic information for each codeword bit from its companion decoder, which is used as *a priori* information. The key idea behind the extrinsic information is that the observed decoder provides the soft information about reliability of each codeword bit using only the information not available to the other decoder. Decoder 1, using the extrinsic information from decoder 2 and its own estimate about the reliabilities of information bits, provides the extrinsic information for decoder 2 for the next iteration.

Decoder 2, using the extrinsic information provided from decoder 1 and its own estimate about the reliabilities of information bits, provides the extrinsic information for decoder 1 for the next iteration. The procedure is repeated until the predetermined number of iterations is reached. To determine the extrinsic reliabilities, the decoders employ the BCJR algorithm.[38] For more details on the turbo decoding principle, see Ryan.[42]

Turbo codes, although perfect for wireless communications, exhibit strong error floors in the region of interest for fiber-optics communications (see Figures 4 and 5 in Ryan[42]), and other alternative iterative soft decoding approaches are to be sought. As shown,[4–11] the turbo product codes and LDPC codes can provide excellent coding gains, and when properly designed, they do not exhibit error floor in the region of interest for fiber-optics communications.

6.2.2 Turbo Product Codes

Turbo product codes were introduced in Section 6.1, see Figure 6.9. For fiber-optics communications, turbo product codes based on BCH component codes have been intensively studied.[4–6] As indicated by Elias,[28] turbo product codes can be decoded by sequentially decoding the columns and rows using maximum *a posteriori* decoding of component codes. However, due to the high complexity of the BCJR algorithm,[29] the low-complexity Chase II algorithm[3,17] is commonly used in practical applications.[5,6] An efficient realization of the Chase II algorithm proposed by Hurst et al.[18] on an AWGN channel is adopted; however, it has been modified and improved to be applicable in fiber-optics channel. Unlike in references 18 and 3–5, our version of the Chase II algorithm does not require scaling and correction factors. It is briefly outlined here:

1. Determine ($p = 5$) least reliable positions starting from the initial likelihood ratios. Generate 2^p test patterns to be added to the hard-decision word.

2. Determine the ith ($i = 1, \ldots, 2^p$) perturbed sequence by adding (modulo-2) the test pattern to the hard-decision word (on least reliable positions).

3. Perform the algebraic or hard decoding to create the set of candidate codewords. Simple syndrome decoding is suitable for high-speed transmission.

4. Calculate the candidate codeword log-likelihood ratios (LLRs).

5. Calculate the extrinsic bit reliabilities for the next decoding stage.

Extrinsic reliabilities from the previous decoding stage are used as inputs processed by the next decoding stage, and so on. The iterative procedure is terminated when either a valid codeword has been generated or a predetermined number of iterations has been reached. In the calculation of candidate codewords, LLRs, and extrinsic reliabilities, no approximation is made so there is no need to introduce any correction factor.

6.2.3 LDPC Codes

LDPC codes, created by Gallager[47] in 1960, are linear block codes for which the parity-check matrix has a low density of ones.[7–16,20,21,23–25,43,47–56] LDPC codes have generated great interest in the coding community,[7–16] resulting in a great deal of understanding of the different aspects of the code and the decoding process. An iterative LDPC decoder based on the sum-product algorithm (SPA) has been shown to achieve a performance as close as 0.0045 dB to the Shannon limit.[12] The inherent low complexity[11–16] of this decoder opens up avenues for its use in different high-speed applications, such as optical communications. If the parity-check matrix has low density, and the number of 1's both per row and per column are constant, the code is said to be a *regular* LDPC code.

The graphical representation of LDPC codes, known as bipartite (Tanner) graph representation, is helpful in efficient description of LDPC decoding algorithms. A *bipartite (Tanner) graph* is a graph whose nodes may be separated into two classes (variable and check nodes) and in which undirected edges may only connect two nodes not residing in the same class. The Tanner graph of a code is drawn according to the following rule: Check (function) node c is connected to variable (bit) node v whenever element h_{cv} in parity-check matrix H is a 1. Therefore, there are $m = n - k$ check nodes and n variable nodes. As an illustrative example, consider the H matrix of the affine geometry, AG(2, 2)-based LDPC code,

$$H = \begin{bmatrix} 1 & 0 & 1 & 0 & 1 & 0 \\ 1 & 0 & 0 & 1 & 0 & 1 \\ 0 & 1 & 1 & 0 & 0 & 1 \\ 0 & 1 & 0 & 1 & 1 & 0 \end{bmatrix}$$

For any valid codeword $v = [v_0 \, v_1, \ldots, v_{n-1}]$ the checks used to decode the codeword are written as follows:

- Equation (c_0): $v_0 + v_2 + v_4 = 0 \pmod 2$

- Equation (c_1): $v_0 + v_3 + v_5 = 0 \pmod 2$

- Equation (c_2): $v_1 + v_2 + v_5 = 0 \pmod 2$

- Equation (c_4): $v_1 + v_3 + v_4 = 0 \pmod 2$

The bipartite graph (Tanner graph) representation of this code is given in Figure 6.13a. The circles represent the bit (variable) nodes, whereas squares represent the check (function) nodes. For example, the variable nodes v_0, v_2, and v_4 are involved in Eq. (c_0) and therefore connected to the check node c_0. A closed path in a bipartite graph comprising l edges that closes back on itself is called a *cycle* of length l. The shortest cycle in the bipartite graph is called the *girth*. The girth influences the minimum distance of LDPC codes and correlates the extrinsic LLRs, thus affecting the decoding process. The use of large-girth LDPC codes is

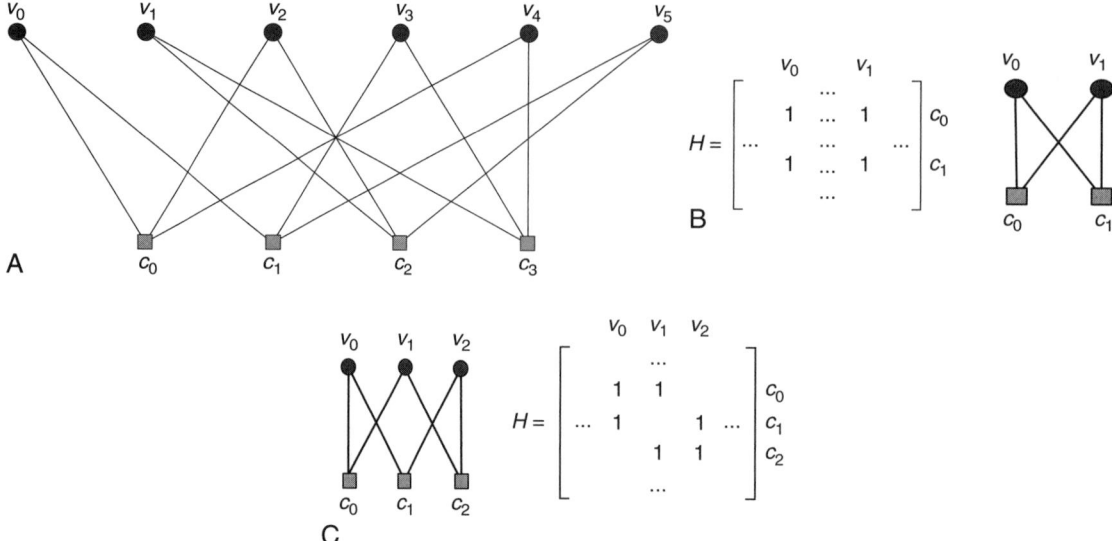

Figure 6.13: (a) Bipartite graph of LDPC(6, 2) code from AG(2, 2) described by the *H* matrix above. (b, c) Cycles in a Tanner graph: (b) cycle of length 4 and (c) cycle of length 6.

preferable because the large girth increases the minimum distance and decorrelates the extrinsic information in the decoding process. To increase the iterative decoding efficiency, we have to avoid LDPC codes of length 4, and preferably 6 as well. To check an LDPC code for those short cycles, one can examine the parity-check matrix *H* for the patterns shown in Figures 6.13b and 13c.

To facilitate the implementation at high speed, we prefer the use of regular rather than irregular LDPC codes. The code description can be done by the degree distribution polynomials $\lambda(x)$ and $\rho(x)$ for the v node and the c node, respectively[57]:

$$\lambda(x) = \sum_{d=1}^{d_v} \lambda_d x^{d-1}, \qquad \rho(x) = \sum_{d=1}^{d_c} \rho_d x^{d-1} \tag{6.35}$$

where λ_d and ρ_d denote the fraction of the edges that are connected to degree-d v nodes and c nodes, respectively, and d_v and d_c denote the maximum v-node and c-node degrees, respectively.

6.2.3.1 Design of Large-Girth LDPC Codes

Based on Tanner's bound for the minimum distance of an LDPC code [21]

$$d \geq \begin{cases} 1 + \dfrac{r}{r-2}\left((r-1)^{\lfloor(g-2)/4\rfloor} - 1\right), & g/2 = 2m+1 \\[3mm] 1 + \dfrac{r}{r-2}\left((r-1)^{\lfloor(g-2)/4\rfloor} - 1\right) + (r-1)^{\lfloor(g-2)/4\rfloor}, & g/2 = 2m \end{cases} \tag{6.36}$$

(where g and r denote the girth of the code graph and the column weight, respectively, and d is the minimum distance of the code) it follows that large girth leads to an exponential increase in the minimum distance, provided that the column weight is at least 3. ($\lfloor\ \rfloor$ denotes the largest integer less than or equal to the enclosed quantity.) For example, the minimum distance of girth-10 codes with column weight $r = 3$ is at least 10. The structured LDPC codes introduced in this section belong to the class of quasi-cyclic (QC) codes[10,58,59] also known as array codes or block-circulant codes. Their parity-check matrix can be represented by

$$H = \begin{bmatrix} I & I & I & \cdots & I \\ I & P^{S[1]} & P^{S[2]} & \cdots & P^{S[c-1]} \\ I & P^{2S[1]} & P^{2S[2]} & \cdots & P^{2S[c-1]} \\ \cdots & \cdots & \cdots & \cdots & \cdots \\ I & P^{(r-1)S[1]} & P^{(r-1)S[2]} & \cdots & P^{(r-1)S[c-1]} \end{bmatrix} \quad (6.37)$$

where I is $p \times p$ (p is a prime number) identity matrix, P is $p \times p$ permutation matrix

$$P = \begin{bmatrix} 0 & 1 & 0 & 0 & \cdots & 0 \\ 0 & 0 & 1 & 0 & \cdots & 0 \\ \cdots & \cdots & \cdots & \cdots & \cdots & \cdots \\ 0 & 0 & 0 & 0 & \cdots & 1 \\ 1 & 0 & 0 & 0 & \cdots & 0 \end{bmatrix}$$

and r and c represent the number of rows and columns in Eq. (6.37), respectively. The set of integers S is to be carefully chosen from the set $\{0, 1, \ldots, p-1\}$ so that the cycles of short length, in corresponding Tanner (bipartite) graph representation of Eq. (6.37), are avoided. According to Theorem 2.1 in Fossorier,[58] we have to avoid the cycles of length $2k$ ($k = 3$ or 4) defined by the following equation:

$$i_1(S[j_1] - S[j_2]) + i_2(S[j_2] - S[j_3]) + \ldots + i_k(S[j_k] - S[j_1]) = 0 \bmod p \quad (6.38)$$

where p is the dimension of the permutation matrix P, and the closed path is defined by $(i_1, j_1), (i_1, j_2), (i_2, j_2), (i_2, j_3), \ldots, (i_k, j_k), (i_k, j_1)$, with the pair of indices denoting row–column indices of permutation blocks in Eq. (6.37) such that $l_m \neq l_{m+1}, l_k \neq l_1$ ($m = 1, 2, \ldots, k; l \in \{i, j\}$). Therefore, we have to identify the sequence of integers $S[i] \in \{0, 1, \ldots, p-1\}$ ($I = 0, 1, \ldots, r-1; r < p$) not satisfying Eq. (6.38), which can be done either by computer search or in a combinatorial manner. For example, to design the QC LDPC codes in Milenkovic et al.,[10] we introduced the concept of the cyclic-invariant difference set (CIDS). The CIDS-based codes come naturally as girth-6 codes, and to increase the girth we had to selectively remove certain elements from a CIDS. The design of LDPC codes of rate above 0.8, column weight 3, and girth-10 using the CIDS approach is a very challenging and open problem. Instead, here we solve this problem by developing an efficient computer search algorithm, which begins with an initial set. We add an additional integer at the time from the set $\{0, 1, \ldots, p-1\}$ (not used before) to the initial set S and check if Eq. (6.38) is satisfied. If Eq. (6.38) is satisfied, we remove that integer from the set S and continue our search with another integer from set $\{0, 1, \ldots, p-1\}$ until we exploit all the elements from $\{0, 1, \ldots, p-1\}$. The code rate R is lower bounded by

$$R \geq \frac{np - rp}{np} = 1 - r/n \qquad (6.39)$$

and the code length is np, where n denotes the number of elements from S being used. The parameter n is determined by desired code rate R_0 by $n = r/(1 - R_0)$. If the desired code rate is set to $R_0 = 0.8$ and column weight to $r = 3$, the parameter $n = 5r$.

Example

By setting $p = 8311$, the set of integers to be used in Eq. (6.37) is obtained as $S = \{0, 3, 7, 26, 54, 146, 237, 496, 499, 1087, 1294, 5326\}$. The corresponding LDPC code has rate $R_0 = 1 - 3/12 = 0.75$, column weight 3, girth-12, and length $nq = 12 \times 8311 = 99{,}732$. In the previous example, the initial set of integers was $S = \{0, 3, 7, 26\}$. The use of a different initial set will result in a different set from that obtained previously. Notice that turbo product codes of similar rates were considered in Mizuochi et al.[5,6]

6.2.3.2 Decoding of LDPC Codes

Throughout our simulations, we utilized the sum-product with the correction term algorithm described by Xiao-Yu et al.[16] as a decoding algorithm for LDPC codes. It is a simplified version of the original algorithm proposed by Gallager.[47] Gallager proposed a near optimal iterative decoding algorithm for LDPC codes that computes the distributions of the variables in order to calculate the *a posteriori* probability (APP) of a bit v_i of a codeword $v = [v_0 \, v_1, \ldots, v_{n-1}]$ to be equal to 1 given a received vector $y = [y_0 \, y_1, \ldots, y_{n-1}]$. This iterative decoding scheme involves passing the message bits back and forth among the c nodes and the v nodes over the edges to update the distribution estimation.

The sum-product algorithm is an iterative decoding scheme, as mentioned previously. Each iteration in this scheme is processed in two half iterations. The first half includes a v node processing the received message and sending the output to the c nodes connected to it. The second half includes a c node processing the received message and sending the output back to the v nodes connected to it.

Figure 6.14 illustrates both the first and the second halves of an iteration of the algorithm. For example, Figure 14a shows the message sent from v node v_0 to c node c_2. v_0 collects the information from the y_0 sample, in addition to c_0 and c_1, to send it to c_2. This is called the extrinsic information because the information collected from c_2 is not taken into consideration. This message contains the probability $\Pr(v_0 = b|y_0)$, where $b \in \{0, 1\}$. On the other hand, Figure 14b shows the message sent from c node c_0 to v node v_1; c_0 sends

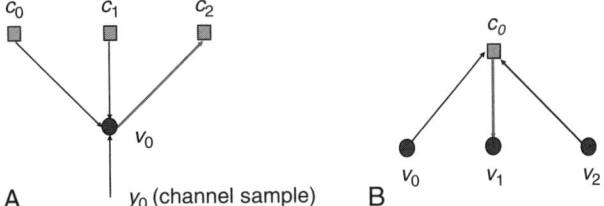

Figure 6.14: The two halves of the sum-product algorithm iteration: (a) first half: message sent from v nodes to c nodes; (b) second half: message sent from c nodes to v nodes.

the extrinsic information to v_1. This message contains the probability $\Pr(c_0$ equation is satisfied $\mid \mathbf{y})$.

Here we describe the sum-product with a correction term algorithm because of its simplicity and suitability for high-speed implementation. Generally, we can either compute the APP $\Pr(v_i|\mathbf{y})$ or the APP ratio $l(v_i) = \Pr(v_i = 0|\mathbf{y})/\Pr(v_i = 1|\mathbf{y})$, which is also referred to as the likelihood ratio. For the final decision, if $l(v_i) > 1$, we decide in favor of 0, whereas for $l(v_i) < 1$ we decide in favor of 1. In the log-domain version of the sum-product algorithm, we replace these likelihood ratios with LLRs due to the fact that the probability domain includes many multiplications, which leads to numerical instabilities; moreover, LLRs will involve addition only. Furthermore, the log-domain representation is more suitable for finite precision representation. Thus, we compute the LLRs by $L(v_i) = \log[\Pr(v_i = 0|\mathbf{y})/\Pr(v_i = 1|\mathbf{y})]$. Hence, for the final decision, if $L(v_i) > 0$, we decide in favor of 0, whereas if $L(v_i) < 0$, we decide in favor of 1.

To further explain the algorithm, we introduce the following notations from Ryan[43] and MacKay[60]:

$V_j = \{v$ nodes connected to c node $c_j\}$

$V_{\mathcal{N}} = \{v$ nodes connected to c node $c_j\} \backslash \{v$ node $v_i\}$

$C_i = \{c$ nodes connected to v node $v_i\}$

$C_{\mathcal{N}} = \{c$ nodes connected to v node $v_i\} \backslash \{c$ node $c_j\}$

$M_v(\sim i) = \{$messages from all v nodes except node $v_i\}$

$M_c(\sim j) = \{$messages from all c nodes except node $c_j\}$

$P_i = \Pr(v_i = 1|y_i)$

$S_i = $ event that the check equations involving c_i are satisfied

$q_{ij}(b) = \Pr(v_i = b \mid S_i, y_i, M_c(\sim j))$

$r_{ji}(b) = \Pr(\text{check equation } c_j \text{ is satisfied} \mid v_i = b, M_v(\sim i))$

As mentioned previously, we use the log-domain version of the sum-product algorithm, and thus all the calculations are performed in the log-domain as follows:

$$L(v_i) = \log\left(\frac{\Pr(v_i = 0|y_i)}{\Pr(v_i = 1|y_i)}\right) \tag{6.40}$$

$$L(r_{ji}) = \log\left(\frac{r_{ji}(0)}{r_{ji}(1)}\right) \tag{6.41a}$$

$$L(q_{ij}) = \log\left(\frac{q_{ij}(0)}{q_{ij}(1)}\right) \tag{6.41b}$$

The algorithm starts with the initialization step, where we set $L(q_{ij})$ as follows:

$$L(v_i) = (-1)^{y_i} \log\left(\frac{1-\varepsilon}{\varepsilon}\right), \quad \text{for BSC}$$

$$L(v_i) = 2^{y_i}/_{\sigma^2}, \quad \text{for binary input AWGN}$$

$$L(v_i) = \log\left(\frac{\sigma_1}{\sigma_0}\right) - \frac{(y_i - \mu_0)^2}{2\sigma_0^2} + \frac{(y_i - \mu_1)^2}{2\sigma_1^2}, \quad \text{for BA-AWGN} \tag{6.42}$$

$$L(v_i) = \log\left(\frac{\Pr(v_i = 0|y_i)}{\Pr(v_i = 1|y_i)}\right), \quad \text{for arbitrary channel}$$

where ε is the probability of error in the BSC, σ^2 is the variance of the Gaussian distribution of the AWGN, and μ_j and σ_j^2 ($j = 0, 1$) represent the mean and the variance of the Gaussian process corresponding to the bits $j = 0, 1$ of a binary asymmetric (BA)-AWGN channel.

After initialization of $L(q_{ij})$, we calculate $L(r_{ji})$ as follows:

$$L(r_{ji}) = L\left(\sum_{i' \in V_j \setminus i} b_i'\right) = L(\ldots \oplus b_k \oplus b_l \oplus b_m \oplus b_n \ldots) = \ldots L_k \boxplus L_l \boxplus L_m \boxplus L_n \boxplus \ldots \tag{6.43}$$

where $\oplus$ denotes the modulo-2 addition, and $\boxplus$ denotes a pairwise computation defined by[16,43]

$$\begin{aligned} L_a \boxplus L_b &= \text{sign}(L_a)\text{sign}(L_b) \min(|L_a|, |L_b|) + S(L_a, L_b) \\ s(L_a, L_b) &= \log(1 + e^{-|L_a + L_b|}) - \log(1 + e^{-|L_a - L_b|}) \end{aligned} \tag{6.44}$$

The term $s(L_a, L_b)$ is the correction term, and it is implemented as a lookup table. After we calculate $L(r_{ji})$, we update

$$L(q_{ij}) = L(v_i) + \sum_{j' \in C_i \setminus j} L(r_{j'i})$$

and

$$L(Q_i) = L(v_i) + \sum_{j \in C_i} L(r_{ji}) \tag{6.45}$$

to check whether we can make a final decision for v_i. In other words, we decide

$$\hat{v}_i = \begin{cases} 1, & L(Q_i) < 0 \\ 0, & \text{otherwise} \end{cases} \tag{6.46}$$

and check whether the syndrome equation is satisfied $\hat{v}H^T = 0$. If the syndrome equation is satisfied for all i or the maximum number of iterations is reached, we stop; otherwise, we recalculate $L(r_{ji})$ and update $L(q_{ij})$ and $L(Q_i)$ and check again. It is important to set the number of iterations high enough to ensure that most of the codewords are decoded correctly and low enough not to affect the processing time. It is important to mention that good LDPC code requires fewer iterations to guarantee successful decoding.

The results of simulations for an AWGN channel model are given in Figure 6.15, which compares the large-girth LDPC codes (Figure 6.15a) against RS codes, concatenated RS codes, turbo product codes (TPCs), and other classes of LDPC codes. The girth-10 LDPC (24015, 19212) code of rate 0.8 outperforms the concatenation RS(255, 239) + RS(255, 223) (of rate 0.82) by 3.35 dB and RS(255, 239) by 4.75 dB, both at a BER of 10^{-7}. The same LDPC code outperforms projective geometry $(2, 2^6)$-based LDPC(4161, 3431) (of rate 0.825) of girth-6 by 1.49 dB at BER of 10^{-7}, and it outperforms CIDS-based LDPC(4320, 3242) of rate 0.75 and girth-8 LDPC codes by 0.25 dB. At a BER of 10^{-10}, it outperforms lattice-based LDPC(8547, 6922) of rate 0.81 and girth-8 by 0.44 dB and BCH(128, 113) × BCH(256, 239) TPC of rate 0.82 by 0.95 dB. The net effective coding gain at a BER of 10^{-12} is 10.95 dB.

In Figure 6.15b, different LDPC codes are compared against RS(255, 223) code, concatenated RS code of rate 0.82, and convolutional code (CC) (of constraint length 5). It can be seen that LDPC codes, both regular and irregular, offer much better performance than the conventional codes. It should be noticed that pairwise balanced design (PBD)-[61,62] based irregular LDPC code of rate 0.75 is only 0.4 dB away from the concatenation of convolutional RS codes (denoted in Figure 6.12 as RS + CC) with significantly lower code rate $R = 0.44$ at a BER of 10^{-6}. As expected, irregular LDPC codes outperform regular LDPC codes.

6.2.4 Generalized LDPC Codes

GLDPC coding[7,23–25,119] can further improve the overall characteristics of LDPC codes by (1) decreasing the complexity of the decoder and (2) approaching closer to the Shannon's limit. To design a GLDPC code, one replaces each single parity-check equation of a *global* LDPC code by the parity-check matrix of a simple linear block code, known as the *constituent (local)* code,[24] such as a Hamming code, a BCH, or a Reed–Muller (RM) code. Decoding of local codes is based on a MAP probability decoding, known as BCJR,[29] which provides very accurate estimates for the variable nodes in the global LDPC graph after a very small number of iterations. Two classes of GLDPC codes are considered in this chapter:

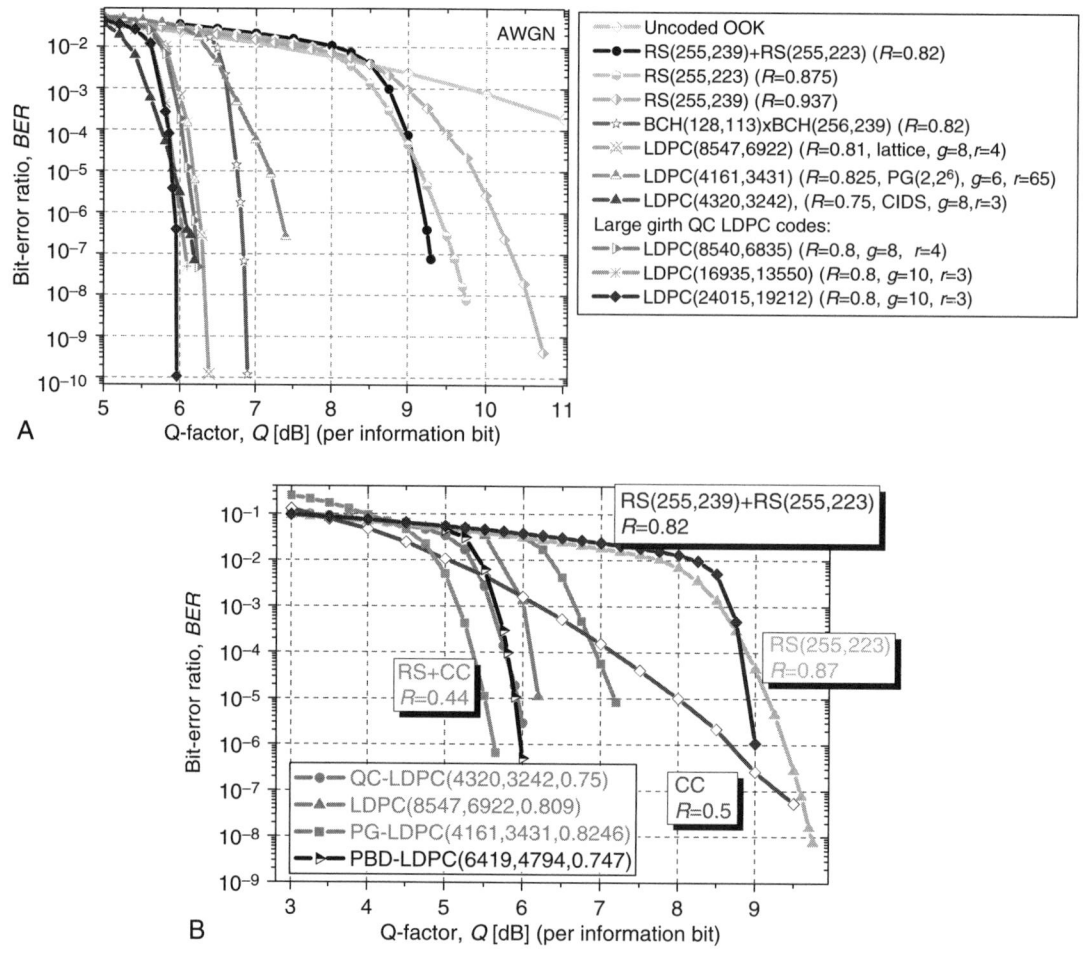

Figure 6.15: (a) Large-girth block-circulant LDPC codes compared to RS codes, concatenated RS codes, TPCs, and previously proposed LDPC codes on an AWGN channel model, and (b) LDPC codes compared to convolutional, concatenated RS, and concatenation of convolutional and RS codes on an AWGN channel.

random-like and structured GLDPC codes. For the random-like codes, decoding on the local level is accomplished through the use of an appropriate number of BCJR decoders operating in parallel; for the structured codes, low-complexity decoding based on a suboptimal message-passing algorithm is used instead.

In general, one seeks GLDPC codes for which the local codes have large minimum distance and a rate as high as possible and for which the global code has very large girth. This is a consequence of the following lower bound on the minimum distance D of a GLDPC code[21]:

$$D \geq \begin{cases} d\dfrac{[(d-1)(J-1)]^{(g-2)/4} - 1}{(d-1)(J-1) - 1} + \dfrac{d}{J}[(d-1)(J-1)]^{(g-2)/4}, \text{for } g/2 \text{ odd} \\[2em] d\dfrac{[(d-1)(J-1)]^{g/4} - 1}{(d-1)(J-1) - 1}, \text{for } g/2 \text{ even} \end{cases}$$

(6.47)

where g and J denote the girth (shortest cycle length) of the global code graph and the column weight of the global code, respectively, and d is the minimum distance of the local code. Obviously, a large girth leads to an exponential increase in the minimum distance, whereas large values of d lead to an increase in the bases of this exponent. GLDPC codes can be designed in such a way that the minimum distance D grows linearly with the code length.[24]

Depending on the structure of the local components, GLDPC codes may be classified as follows:

- GLDPC codes with algebraic local codes of short length, such as Hamming codes, BCH codes, RS codes, or RM codes[22]

- GLDPC codes for which the local codes are high-rate regular or irregular LDPC codes with a large minimum distance

- Fractal GLDPC codes in which the local code is in fact another GLDPC code

Each of these constructions offers some specific advantages. For example, in the first scenario, the MAP decoders of the local codes[29] allow for fast propagation of accurate estimates of the LDPC variables. The second construction is very well suited for longer length codes. Because the local components in this case can have a large length and therefore a large minimum distance as well, the minimum distance of the resulting GLDPC codes is very high. The third approach is attractive in terms of possible practical partly parallel very large scale integration (VLSI) implementations because the same code structure can be reused on different levels. For more details on GLDPC codes, see references 7 and 63.

To construct a GLDPC code, one can replace each single parity-check equation of a global LDPC code by the parity-check matrix of a simple linear block code, known as the constituent (local) code. This construction is proposed by Lentmaier and Zigangirov,[24] and we refer to this as LZ-GLDPC code construction. In a construction proposed by Boutros et al.,[23] referred to as B-GLDPC code construction, the parity-check matrix, H, is a sparse matrix partitioned into W submatrices $H_1, \ldots, H_W$. H_1 is a block-diagonal matrix generated from an identity matrix by replacing the ones by a parity-check matrix of a local code of codeword-length n and dimension k. Each submatrix H_j is derived from H_1 by random column permutations. For more details on LZ-GLDPC- and B-GLDPC-like codes, and their generalization-fractal GLDPC codes (in which a local code is another GLDPC code), see Djordjevic et al.[7] The code rate of a GLDPC code is lower bounded by $R = K/N \geq 1 - W$ $(1 - k/n)$, where K and N denote the dimension and the codeword length of a GLDPC code, W is the column weight of a global LDPC code, and k/n is the code rate of a local code (k and n denote the dimension and the codeword length of a local code). The GLDPC codes with component Hamming codes are considered in Djordjevic et al.[7]; here, we are concerned with GDLPC codes with component codes based on RM or BCH codes.[119] An RM code RM(r, m) of order r and codeword length $n = 2^m$ [45] is the set of all binary vectors associated with coefficients of Boolean polynomials of degree at most r in m variables. The minimum distance of RM(r, m) code is 2^{m-r}, and the dimension is determined as[45]

$$k = 1 + \binom{m}{1} + \ldots + \binom{m}{r} \tag{6.48}$$

An interesting property of RM codes is that they can be defined recursively[45]: RM(r, m) = $\{(a|a + b): a \in RM(r, m - 1), b \in RM(r - 1, m - 1)\}$, where $(x|y)$ denotes the concatenation operation. The generator matrix of RM(r, m) code, denoted as $G(r, m)$, can therefore be defined recursively by

$$G(r, m) = \begin{bmatrix} G(r, m - 1) & G(r, m - 1) \\ 0 & G(r - 1, m - 1) \end{bmatrix} \tag{6.49}$$

and can be observed as two-level generalized concatenated code.[45] RM($0, m$) is a repetition code, RM($m - 1, m$) is a parity-check code, and RM(m, m) corresponds to 2^m-tuples of a vector space. The generator matrix of RM(m, m) can be represented by

$$G(m, m) = \begin{bmatrix} G(m - 1, m) \\ 0 \ 0 \ \ldots \ 0 \ 1 \end{bmatrix} \tag{6.50}$$

Another interesting property of RM codes is that the dual of RM(r, m) code is another RM($m - r - 1, m$) code. Therefore, the generator matrix of RM($m - r - 1, m$) code can be used as the parity-check matrix of RM(r, m) code. If the recursion in Eq. (6.49) is applied successively several times, the RM(r, m) can be decomposed into several parity-check codes RM($m' - 1, m'$), repetition codes RM($0, m'$), and the first-order RM($1, m'$) codes. The MAP decoding of parity-check or repetition codes is trivial, whereas the first-order RM($1, m'$) codes can be decoded using an efficient MAP decoding algorithm proposed in Ashikhmin and Lytsin[64] based on the fast Hadamard–Walsh transform. The overall complexity of this algorithm is of order $n' \log_2 n'$ (where $n' = 2^{m'}$), which is significantly lower than the complexity of the BCJR algorithm that requires approximately n^{n-k+1} operations.[64] Therefore, the complexity of GLDPC codes with RM component codes is of order $N \log_2 n$. Because the complexity of the sum-product algorithm is of order $(N_{\text{LDPC}} - K_{\text{LDPC}})w_r$, where w_r is the row weight of the LDPC code parity-check matrix, by proper selection of global LDPC code length N and local RM code length n, the complexity of GLDPC codes is approximately $(N_{\text{LDPC}} - K_{\text{LDPC}})w_r/$ $[(N/n)\sum(n'\log_2 n')](n' < n)$ times lower. For example, RM(4, 6) code can be decomposed using Eq. (6.49) on RM(1, 2), RM(1, 3), RM(2, 2), RM(3, 3), and RM(4, 4) component codes. Decoding of RM(m', m') ($m' = 1, 2, 3, 4$) is trivial, whereas the complexity of RM(4, 6) is dominated by complexity of RM(1, 3) decoding block and three RM(1, 2) blocks, which is of order $\sum(n'\log_2 n') = 8\log_2 8 + 3 \times 4\log_2 4 = 48$. B-GLDPC code with $W = 2$ and length $N = 4096$ based on RM(4, 6) code therefore has complexity 11 times lower than girth-8 column weight-4 LDPC code of length 8547 (and row weight 21). Notice also that the GLDPC decoder for 4096 code contains $4096/64 = 64$ decoder blocks (composed of RM(1, 2), RM(2, 3), and RM(m', m') ($m' = 1, \ldots, 4$) decoders), operating in parallel, and this structure is suitable for field-programmable gate array (FPGA) or VLSI implementation. In order to keep the code rate high, we select the column weight of a global code to be $W = 2$. In this case, if the girth of global code is $g = 8$, the minimum distance of GLDPC code is simply $D \geq d^2$.

The BCH codes were introduced in Section 6.1. Notice that BCH codes can be decoded using an efficient MAP algorithm proposed in Ashikhmin and Lytsin[64] with complexity $n/[(n - k) \log_2 n]$ times lower than that of the BCJR algorithm (for more details about the BCJR algorithm, see references 29 and 50). The GLDPC codes described here can be put in systematic form so that the efficient encoding algorithm by Zhang and Parhi[65] can be employed. This algorithm can efficiently be performed in general-purpose digital signal processors, as shown in Zhang and Parhi.[65]

The results of simulation for the AWGN channel model are shown in Figure 6.16 and are obtained by maintaining the double precision. GLDPC codes based on BCH(63, 57) and RM(4, 6) component codes for $W = 2$ perform comparably. The RM(4, 6)-based GLDPC code outperforms the BCH(128, 113) $\times$ BCH(256, 239) TPC based with Chase II decoding algorithm on $p = 3$ least reliable bit positions by 0.93 dB at a BER of 10^{-9} (see Figure 6.16a). Note that similar TPC was implemented in LSI technology.[5] The TPC codeword is significantly longer, and the decoding complexity of GLDPC code based on RM(4, 6) is at least 10 times lower. During decoding, the TPC decoder employs 239 Chase II blocks operating in parallel, whereas GLDPC code on RM(4, 6) code requires only 64 low-complexity MAP decoders, as explained previously. In simulations presented here, we have employed an efficient realization of the Chase II algorithm described in Section 6.2.2. In Figure 6.16b, BER performance of several classes of iteratively decodable codes (TPCs, LDPC, and GLDPC codes) of high code rate are compared against conventional RS and concatenated RS codes. B-GLDPC code of rate 0.88 outperforms concatenated RS code of rate 0.82 by 2.47 dB (also at a BER = 10^{-9}). RM(6, 8) $\times$ RM(5, 7) TPC of rate 0.905 outperforms concatenated RS code ($R = 0.82$) by 0.53 dB at a BER of 10^{-9}. LDPC code of rate 0.93, designed using the concept of product of orthogonal arrays, outperforms the same RS concatenated code by 1.15 dB at a BER of 10^{-9}. $R = 0.93$ LDPC code outperforms RS code of rate 0.937 by 2.44 dB at a BER of 10^{-9}.

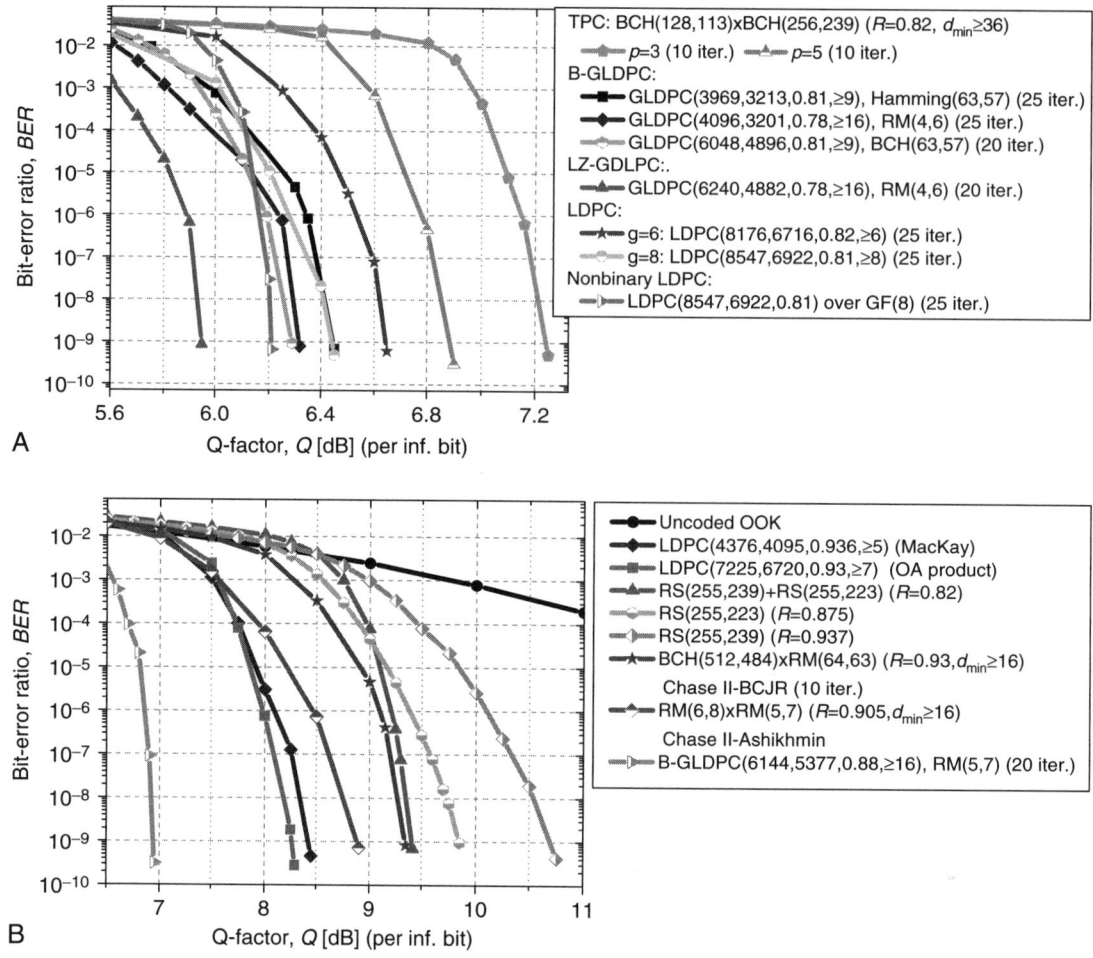

Figure 6.16: BER performance on an AWGN channel: (a) GLDPC codes against TPCs and LDPC codes, and (b) high-rate codes.

6.2.5 FPGA Implementation of Large-Girth LDPC Codes

We use the min-sum algorithm, which is a further simplified version of the min-sum-with-correction-term algorithm detailed in the previous subsection. The only difference is that the min-sum algorithm omits the correction term in Eq. (6.44). Among various alternatives, in Arabaci and Djordjevic[66] we adopted a partially parallel architecture in our implementation because it is a natural choice for quasi-cyclic codes. In this architecture, a processing element (PE) is assigned to a group of nodes of the same kind instead of a single node. A PE mapped to a group of bit nodes is called a bit-processing element (BPE), and a PE mapped to a group of check nodes is called a check-processing element (CPE). BPEs (CPEs)

	BPEs				
	0	**1**	**2**	**...**	***c*-1**
0	I	I	I	...	I
1	I	$PS[1]$	$PS[2]$	...	$PS[c-1]$
CPEs 2	I	$P^2S[1]$	$P^2S[2]$	...	$P^2S[c-1]$
...	...	...	...	...	...
***r*-1**	I	$P^{(r-1)}S[1]$	$P^{(r-1)}S[2]$	...	$P^{(r-1)}S[c-1]$

Figure 6.17: The assignment of bit nodes and check nodes to BPEs and CPEs, respectively.

process the nodes assigned to them in a serial manner. However, all BPEs (CPEs) carry out their tasks simultaneously. Thus, by changing the number of elements assigned to a single BPE and CPE, one can control the level of parallelism in the hardware. Figure 6.17 depicts a convenient method for assigning BPEs and CPEs to the nodes in a QC-LDPC code. This method is not only easy to implement but also advantageous because it simplifies the memory addressing.

The messages between BPEs and CPEs are exchanged via memory banks. Table 6.2 summarizes the memory allocation in our implementation, in which we used the following notation: MEM B and MEM C denote the memories used to store bit node and check node edge values, respectively; MEM E stores the codeword estimate; MEM I stores the initial log-likelihood ratios; and MEM R holds the state of the random number generator, which is based on the Mersenne twister algorithm.

In our initial design, we used the Mitrion-C hardware programming language, which is "an intrinsically parallel C-family language" developed by Mitrionics, Inc.[67] Using Mitrion-C syntax, we provided a pseudo-code in Figure 6.18 showing how the data are transferred from MEM B to MEM C after being processed by BPEs. The code features three loop expressions of two types. The *for* loop sequentially executes its loop body for every bit node, i, in a BPE. On the contrary, the *for each* loop is a parallel loop, and hence, the operations in the loop body are applied to all the elements in its declaration simultaneously. To expatiate, due to the first *for each* loop, all BPEs perform their operations on their ith bit nodes in parallel.

Table 6.2: Memory Allocation of the Implementation

MEM name	MEM B	MEM C	MEM E	MEM I	MEM R
Data word (bits)	8	11	1	8	32
Address word (bits)	16	16	15	15	10
Memory block size (words)	50,805	50,805	16,935	16,935	625

```
for i = 0 to p–1 do
    for each b = 0 to c–1 do
        - Read r data values from MEM B located in the range
        [b * p * r + i * r, b * p * r + i * r + r – 1].
        - Sum them up and store the sum.
        - Update MEM E at location (b * p + i).
        for each k = 0 to r–1 do
            - Subtract from the computed sum the value
            located at  (b * p * r + i * r + k – 1) in MEM B.
            - Use this value to update MEM C at location
            (k * p + ((p – k * b) % p)) * c + b).
        end
    end
end
```

Figure 6.18: The pseudo-code describing assignment of bit nodes and check nodes to BPEs and CPEs.

Because we are using a single memory in our implementation to store the edge values of all check nodes, the second *for each* loop causes a BPE to update its connections in MEM C in a pipelined manner. As also shown in Figure 6.18, we compute the memory addresses to read/write data from/to "on-the-fly" using the bit node ID (i), BPE ID (b), and CPE ID (k). This convenient calculation of addresses is possible because of the quasi-cyclic nature of the code and the way we assigned BPEs and CPEs.

Figure 6.19 presents a BER performance comparison of FPGA and software implementations for LDPC(16935, 13550) code. We observe a close agreement between the two BER curves. Furthermore, the performance of the min-sum algorithm is only 0.2 dB worse than

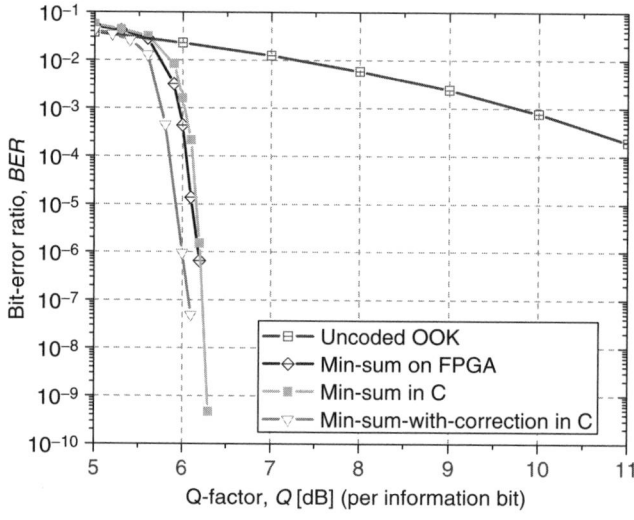

Figure 6.19: BER performance comparison of FPGA and software implementations of the min-sum algorithm.

that of the min-sum-with-correction-term algorithm at the BER of 10^{-6} and the gap narrows as the Q-factor increases. The net coding gain of the min-sum algorithm for the same LDPC code at a BER of 10^{-6} is found to be 10.3 dB.

The main problem in decoder implementation for large-girth *binary* LDPC codes is the excessive codeword length, and a fully parallel implementation on a single FPGA is quite a challenging problem. To solve this problem, in the next section we consider large-girth *nonbinary* LDPC codes over GF(2^m). By designing codes over higher order fields, we aim to achieve the coding gains comparable to those for binary LDPC codes but for shorter codeword lengths.

6.2.6 Nonbinary Quasi-Cyclic LDPC Codes

Here, we describe a two-stage design technique for constructing nonbinary regular high-rate LDPC codes proposed in Arabaci et al.[68] We show that the complexity of the nonbinary decoding algorithm over GF(4) used to decode this code is 91.28% of complexity of the Jacobian-based sum-product algorithm,[69] used for decoding a bit-length-matched binary LDPC code. Furthermore, we demonstrate that by enforcing the nonbinary LDPC codes to have the same nonzero field element in a given column in their parity-check matrices, we can reduce the hardware implementation complexity of their decoders without incurring any degradation in the error-correction performance.

A q-ary LDPC code is a linear block code defined as the null space of a sparse parity-check matrix H over a finite field of q elements that is denoted by GF(q), where q is a power of a prime. Davey and MacKay[13,70] devised a q-ary sum-product algorithm (QSPA) to decode q-ary LDPC codes, where $q = 2^p$ and p is an integer. They also proposed an efficient way of conducting QSPA via fast Fourier transform (FFT-QSPA). FFT-QSPA is further analyzed and improved in Declercq and Fossorier.[71] A mixed-domain version of the FFT-QSPA (MD-FFT-QSPA) that reduces the computational complexity by transforming the multiplications into additions with the help of logarithm and exponentiation operations is proposed in Spagnol et al.[72] Due to the availability of efficient decoding algorithms, we consider q-ary LDPC codes where q is a power of 2.

In the first step of our two-stage code design technique, we design binary QC LDPC codes of girth-6 using the algebraic construction method based on the multiplicative groups of finite fields.[73] Let α be a primitive element of GF(q) and let $W = [w_{i,j}]$ be a $(q-1)$-by-$(q-1)$ matrix given as follows:

$$W = \begin{bmatrix} \alpha^0 - 1 & \alpha - 1 & \dots & \alpha^{q-2} - 1 \\ \alpha - 1 & \alpha^2 - 1 & \dots & \alpha^{q-1} - 1 \\ \dots & \dots & \dots & \dots \\ \alpha^{q-2} - 1 & \alpha^{q-1} - 1 & \dots & \alpha^{2(q-2)} - 1 \end{bmatrix} \tag{6.51}$$

We can transform W into a quasi-cyclic parity-check matrix $H^{(1)}$ of the following form:

$$H^{(1)} = \begin{bmatrix} A_{0,0} & A_{0,1} & \cdots & A_{0,n-1} \\ A_{1,0} & A_{1,1} & \cdots & A_{1,n-1} \\ \cdots & \cdots & \cdots & \cdots \\ A_{m-1,0} & A_{m-1,1} & \cdots & A_{m-1,n-1} \end{bmatrix} \tag{6.52}$$

where every submatrix $A_{i,j}$ is related to the field element $w_{i,j}$ by

$$A_{i,j} = \left[z(w_{i,j}) \, z(\alpha w_{i,j}) \, z(\alpha^2 w_{i,j}), \ldots, z(\alpha^{q-2} w_{i,j}) \right]^{\mathrm{T}} \tag{6.53}$$

where $z(\alpha^i) = (z_0, z_1, \ldots, z_{q-2})$ is a $(q-1)$-tuple over GF(2) whose ith component $z_i = 1$ and all other $q-2$ components are zero, and the superscript "T" denotes transposition. Using Theorem 1 in Lan et al.,[73] we can show that the parity-check matrix, $H^{(1)}$, given in Eq. (6.52), which is a $(q-1)$-by-$(q-1)$ array of circulant permutation and zero matrices of size $(q-1)$-by-$(q-1)$, has a girth of at least 6. We use this quasi-cyclic, girth-6 parity-check matrix $H^{(1)}$ in the second stage.

If we simply choose γ rows and ρ columns from $H^{(1)}$ while avoiding the zero matrices, we obtain a (γ, ρ)-regular parity-check matrix whose null space yields a (γ, ρ)-regular LDPC code with a rate of at least $(\rho - \gamma)/\rho$. Instead of a simple, random selection, however, if we choose rows and columns of $H^{(1)}$ while avoiding performance-degrading short cycles, we can boost the performance of the resulting LDPC code. Hence, following the guidelines in Lan et al.,[73] the first step in the second stage is to select γ rows and ρ columns from $H^{(1)}$ in such a way that the resulting binary quasi-cyclic code has a girth of 8. In the second step, we replace the 1's in the binary parity-check matrix with nonzero elements from GF(q) either by completely random selection or by enforcing each column to have the same nonzero element from GF(q) while letting the nonzero element of each column be determined again by a random selection. We denote the final q-ary (γ, ρ)-regular, girth-8 matrix by $H^{(2)}$.

Following the two-stage design discussed previously, we generated (3, 15)-regular, girth-8 LDPC codes over the fields GF(2^p), where $0 < p \leq 7$. All the codes had a code rate (R) of at least 0.8 and hence an OH($= 1/R - 1$) of 25% or less. We compared the BER performances of these codes against each other and against some other well-known codes, namely the ITU-standard RS(255, 239), RS(255, 223), and RS(255, 239) + RS(255, 223) codes and the BCH(128, 113) × BCH(256, 239) TPC. We used the BI-AWGN channel model in our simulations and set the maximum number of iterations at 50.

Figure 6.20 presents the BER performances of the set of nonbinary LDPC codes discussed previously. From the figure, it can be concluded that when we fix the girth of a nonbinary regular, rate-0.8 LDPC code at 8, increasing the field order above 8 exacerbates the BER performance. In addition to having better BER performance than codes over higher order fields, codes over GF(4) have smaller decoding complexities when decoded using the MD-FFT-QSPA algorithm because the complexity of this algorithm is proportional to the field

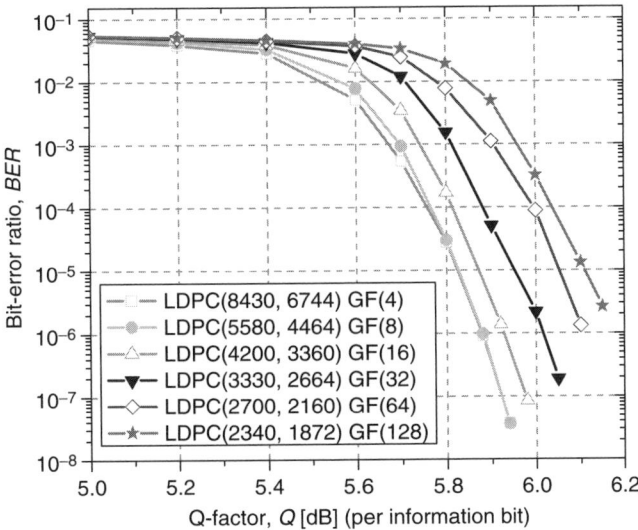

Figure 6.20: Comparison of nonbinary, (3, 15)-regular, girth-8 LDPC codes over the binary (BI)-AWGN channel.

order. Thus, we focus on nonbinary, regular, rate-0.8, girth-8 LDPC codes over GF(4) in the rest of this section.

Figure 6.21 compares the BER performance of the LDPC(8430, 6744) code over GF(4) shown in Figure 6.15 against that of the RS(255, 239) code, RS(255, 223) code, RS(255, 239) + RS(255, 223) concatenation code, and BCH(128, 113) × BCH(256, 239) TPC. We observe that the LDPC code over GF(4) outperforms all of these codes with a significant margin. In particular, it provides an additional coding gain of 3.363 and 4.401 dB at a BER of 10^{-7} compared to the concatenation code RS(255, 239) + RS(255, 223) and the RS(255, 239) code, respectively. Its coding gain improvement over BCH(128, 113) × BCH(256, 239) TPC is 0.886 dB at a BER of 4×10^{-8}. Finally, we computed the NCG of the 4-ary, regular, rate-0.8, girth-8 LDPC code over GF(4) to be 10.784 dB at a BER of 10^{-12}.

Figure 6.21 also shows a competitive, binary, (3, 15)-regular, girth-10 LDPC(16935, 13550) code that was proposed in Djordjevic et al.[74] We can see that the 4-ary, (3, 15)-regular, girth-8 LDPC(8430, 6744) code beats the bit-length-matched binary LDPC code with a margin of 0.089 dB at a BER of 10^{-7}. More important, the complexity of the MD-FFT-QSPA used for decoding the nonbinary LDPC code is lower than the Jacobian-based LLR-SPA algorithm[69] used for decoding the corresponding binary LDPC code. The complexity of MD-FFT-QSPA for a q-ary, bit-length matched (γ, ρ)-regular LDPC code with M check nodes is given by $(M/\log q)2\rho q(\log q + 1 - 1/(2\rho))$ additions. Thus, a (3, 15)-regular 4-ary nonbinary LDPC code requires 91.28% of the computational resources

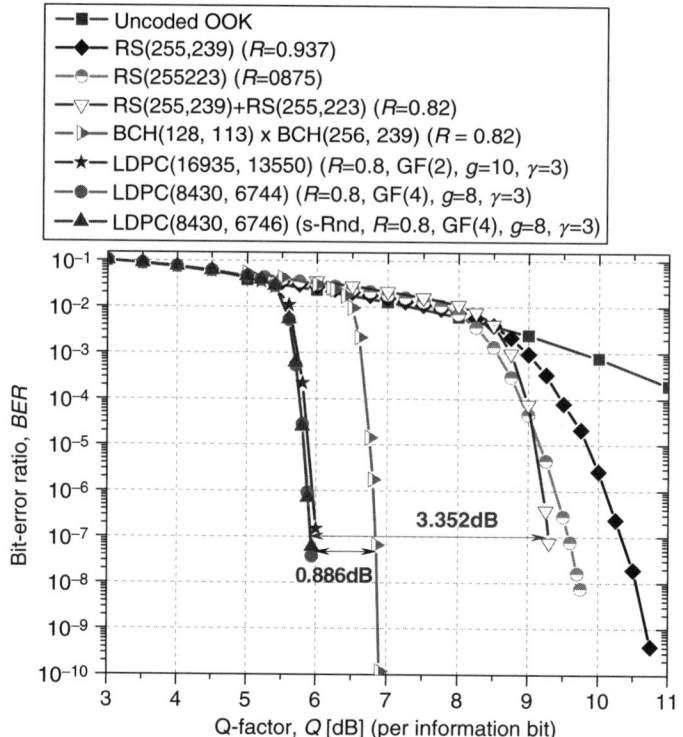

Figure 6.21: Comparison of 4-ary (3, 15)-regular, girth-8 LDPC codes; a binary, girth-10 LDPC code; three RS codes; and a TPC code.

required in decoding a bit-length matched (3, 15)-regular LDPC code of the same rate and bit length.

6.3 *M*-ary QAM and *M*-ary PSK

M-ary QAM is a two-dimensional generalization of *M*-ary pulse amplitude modulation, and its formulation involves two orthogonal passband basis functions:

$$
\begin{aligned}
\phi_I(t) &= \sqrt{\frac{2}{T}} \cos(2\pi f_c t), \quad 0 \le t \le T \\
\phi_Q(t) &= \sqrt{\frac{2}{T}} \sin(2\pi f_c t), \quad 0 \le t \le T
\end{aligned}
\tag{6.54}
$$

where f_c denotes the carrier frequency, and T denotes the symbol duration. The message point s_i can be described in terms of basis functions by $(I_k d_{\min}, Q_k d_{\min}/2)$, where $d_{\min}$ is related to the energy of the signal with the smallest amplitude E_0 by $d_{\min}/2 = \sqrt{E_0}$. The coordinates of the kth signal constellation point for square M-ary QAM signal constellation are given by

$$\{I_k, Q_k\} = \begin{bmatrix} (-\sqrt{M}+1, \sqrt{M}-1) & (-\sqrt{M}+3, \sqrt{M}-1) & \cdots & (\sqrt{M}-1, \sqrt{M}-1) \\ (-\sqrt{M}+1, \sqrt{M}-3) & (-\sqrt{M}+3, \sqrt{M}-3) & \cdots & (\sqrt{M}-1, \sqrt{M}-3) \\ \cdots & \cdots & \cdots & \cdots \\ (-\sqrt{M}+1, -\sqrt{M}+1) & (-\sqrt{M}+3, -L+1) & \cdots & (\sqrt{M}-1, -\sqrt{M}+1) \end{bmatrix}$$

(6.55)

For example, for $M = 16$ signal constellation points are given by

$$\{I_k, Q_k\} = \begin{bmatrix} (-3, 3) & (-1, 3) & (1, 3) & (3, 3) \\ (-3, 1) & (-1, 1) & (1, 1) & (3, 1) \\ (-3, -1) & (-1, -1) & (1, -1) & (3, -1) \\ (-3, -3) & (-1, -3) & (1, -3) & (3, -3) \end{bmatrix}$$

The transmitted M-ary QAM signal for the kth symbol can therefore be expressed as

$$s_k(t) = \sqrt{\frac{2E_0}{T}} I_k \cos(2\pi f_c t) - \sqrt{\frac{2E_0}{T}} Q_k \sin(2\pi f_c t), 0 < t \le T; k = 1, \ldots, M \quad (6.56)$$

with constellation points coordinates given by Eq. (6.55). On the other hand, using the same basis functions, the transmitted M-ary PSK signal for the kth symbol can be expressed as

$$s_k(t) = \begin{cases} \sqrt{\frac{2E}{T}} \cos\left(2\pi f_c t + (k-1)\frac{2\pi}{M}\right), & 0 \le t \le T \\ 0, & \text{otherwise} \end{cases} ; k = 1, 2, \ldots, M \quad (6.57)$$

As an illustration, the 8-PSK, 8-QAM, and 64-QAM constellation diagrams are given in Figure 6.22. The corresponding Gray mapping rule is given in Table 6.3.

6.4 Coded Modulation

M-ary PSK, M-ary QAM, and M-ary differential phase-shift keying (DPSK) achieve the transmission of $\log_2 M(= m)$ bits per symbol, providing bandwidth-efficient communication.[56,75] In coherent detection, the data phasor $\phi_l \in \{0, 2\pi/M, \ldots, 2\pi(M-1)/M\}$ is sent at each lth transmission interval. In direct detection, the modulation is differential, and the data phasor $\phi_l = \phi_{l-1} + \Delta\phi_l$ is sent instead, where $\Delta\phi_l \in \{0, 2\pi/M, \ldots, 2\pi(M-1)/M\}$ is determined by the sequence of $\log_2 M$ input bits using an appropriate mapping rule. Let us now introduce the transmitter architecture employing LDPC codes as channel codes. We have two options: (1) to use the same LDPC code for different data streams—this scheme is known as bit-interleaved coded modulation (BICM)[75]; and (2) to use different code rate LDPC codes (of the same length)—this scheme is known as multilevel coding (MLC).[56] The use of MLC allows optimally to allocate code rates. To keep the exposition simple, we describe only the bit-interleaved LDPC-coded modulation scheme in combination with the

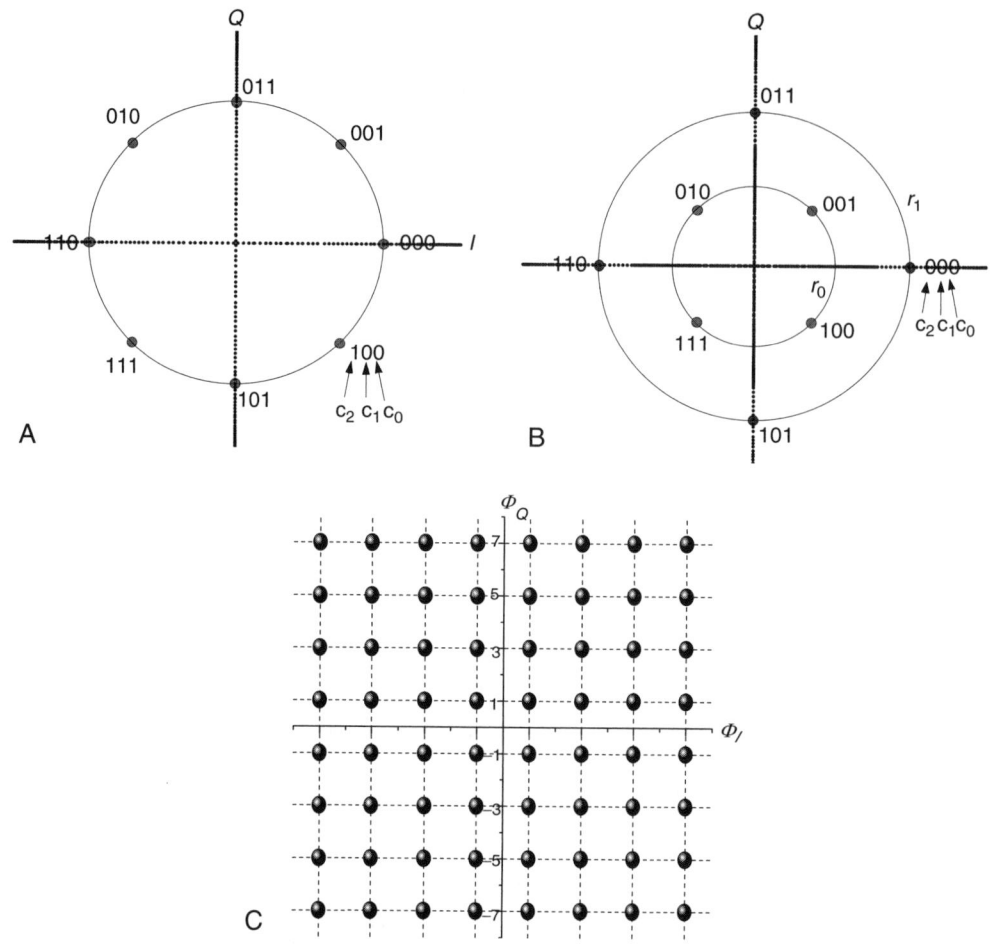

Figure 6.22: The constellation diagrams for (a) 8-PSK, (b) 8-QAM, and (c) 64-point square QAM.

Table 6.3: 8-PSK/8-QAM Gray Mapping Rule

Input bits $(c_2c_1c_0)$	ϕ_k	8-PSK		8-QAM	
		I_k	Q_k	I_k	Q_k
000	0	1	0	$1+\sqrt{3}$	0
001	$\pi/4$	$\sqrt{2}/2$	$\sqrt{2}/2$	1	1
011	$\pi/2$	0	1	0	$1+\sqrt{3}$
010	$3\pi/4$	$-\sqrt{2}/2$	$\sqrt{2}/2$	-1	1
110	π	-1	0	$-(1+\sqrt{3})$	0
111	$5\pi/4$	$-\sqrt{2}/2$	$-\sqrt{2}/2$	-1	-1
101	$3\pi/2$	0	-1	0	$-(1+\sqrt{3})$
100	$7\pi/4$	$\sqrt{2}/2$	$-\sqrt{2}/2$	1	-1

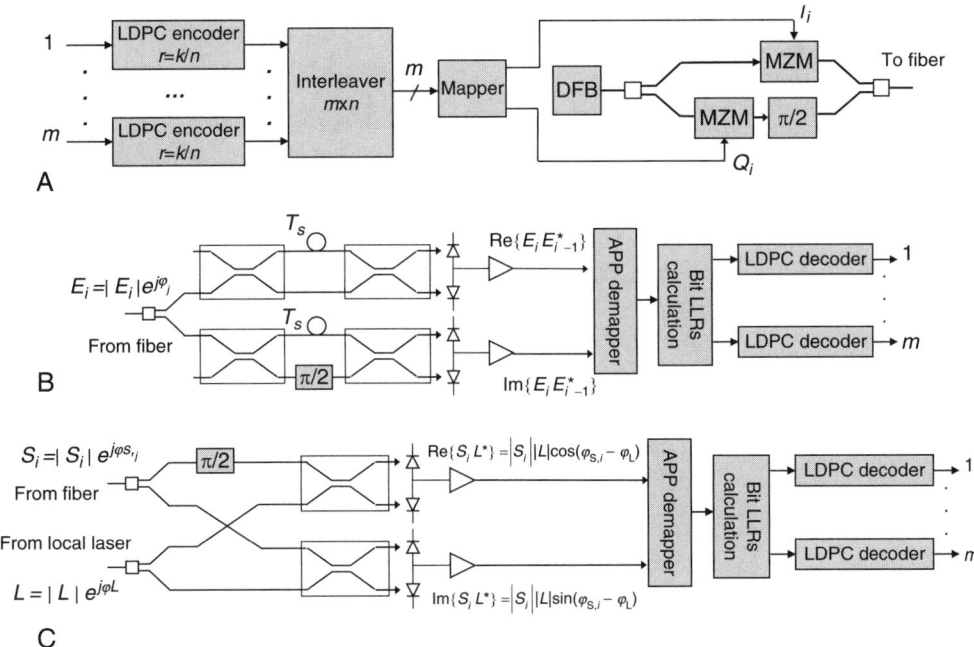

Figure 6.23: Bit-interleaved LDPC-coded modulation scheme: (a) transmitter architecture, (b) direct detection architecture, and (c) coherent detection receiver architecture. $T_s = 1/R_s$, where R_s is the symbol rate.

coherent detection scheme. Note that in this section we also analyze the direct detection and use it as a reference case (Figure 6.23). The source bitstreams coming from m information sources (e.g., carrying 40 Gb/s traffic) are encoded by using identical (n, k) LDPC codes of code rate $r = k/n$ (where k is the number of information bits, and n is the codeword length). The LDPC decoder outputs are written to the $m \times n$ block interleaver row-wise. The mapper accepts m bits, $c = (c_1, c_2, \ldots, c_m)$, at time instance i from the $(m \times n)$ interleaver columnwise and determines the corresponding M-ary ($M = 2^m$) constellation point $s_i = (I_i, Q_i) = |s_i|\exp(j\phi_i)$ (see Figure 6.23a). The mapper outputs for M-ary PSK/DPSK I_i and Q_i are proportional to $\cos\phi_i$ and $\sin\phi_i$, respectively.

The receiver input electrical field at time instance i for an optical M-ary DPSK receiver configuration from Figure 6.23b is denoted by $E_i = |E_i|\exp(j\varphi_i)$. The outputs of I and Q branches (upper and lower branches in Figure 6.23b) are proportional to $\text{Re}\{E_i E*_{i-1}\}$ and $\text{Im}\{E_i E*_{i-1}\}$, respectively. The corresponding coherent detector receiver architecture is shown in Figure 6.23c, where $S_i = |S|e^{j\varphi_{S,i}} \left(\varphi_{S,i} = \omega_S t + \varphi_i + \varphi_{S,\text{PN}}\right)$ is the coherent receiver input electrical field at time instance i, whereas $L = |L|e^{j\varphi_L} \left(\varphi_L = \omega_L t + \varphi_{L,\text{PN}}\right)$ is the local laser electrical field. For homodyne detection, the frequency of the local laser (ω_L) is the same as that of the incoming optical signal (ω_L), so the balanced outputs of I and Q channel branches (upper and lower branches of Figure 6.23c) can be written as

$$
\begin{aligned}
v_I(t) &= \Re|S_k||L|\cos\big(\varphi_i + \varphi_{S,\,PN} - \varphi_{L,\,PN}\big) \quad (i-1)T_s \leq t < iT_s \\
v_Q(t) &= \Re|S_k||L|\sin\big(\varphi_i + \varphi_{S,\,PN} - \varphi_{L,\,PN}\big) \quad (i-1)T_s \leq t < iT_s
\end{aligned}
\tag{6.58}
$$

where $\Re$ is photodiode responsivity, and $\varphi_{S,PN}$ and $\varphi_{L,PN}$ represent the laser phase noise of the transmitting and receiving (local) laser, respectively. These two noise sources are commonly modeled as the Wiener–Lévy process, which is a zero-mean Gaussian process with variance $2\pi(\Delta v_S + \Delta v_L)|t|$, where Δv_S and Δv_L are the laser linewidths of the transmitting and receiving laser, respectively. The transmitted signal constellation point $s_i = |s_i|\exp(j\phi_i)$ can be perfectly recovered in accordance with Eq. (6.58) only in the absence of laser phase noise. The outputs at I and Q branches (in either the coherent or the direct detection case) are sampled at the symbol rate, whereas the symbol LLRs are calculated in APP demapper block as follows:

$$
\lambda(s) = \log\frac{P(s = s_0|r)}{P(s \neq s_0|r)}
\tag{6.59}
$$

where $P(s|r)$ is determined by using Bayes' rule

$$
P(s|r) = \frac{P(r|s)P(s)}{\sum_{s'}P(r|s')P(s')}
\tag{6.60}
$$

where $s = (I_i, Q_i)$ denotes the transmitted signal constellation point at time instance i, whereas $r = (r_I, r_Q)$ denotes the received point. $r_I = v_I(t = iT_s)$, and $r_Q = v_Q(t = iT_s)$ are the samples of I and Q detection branches from Figures 6.23b and 6.23c. $P(r|s)$ from Eq. (6.60) is estimated by evaluation of histograms, employing sufficiently a long training sequence. With $P(s)$, we denoted the APP of symbol s, whereas s_0 is a referent symbol. Also note that normalization in Eq. (6.59) is introduced to equalize the denominator from Eq. (6.60). The bit LLRs c_j $(j = 1, 2, \ldots, m)$ are determined from symbol LLRs of Eq. (6.59) as

$$
L(\hat{c}_j) = \log\frac{\sum_{s:c_j=0}\exp[\lambda(s)]}{\sum_{s:c_j=1}\exp[\lambda(s)]}
\tag{6.61}
$$

The APP demapper extrinsic LLRs (the difference of demapper bit LLRs and LDPC decoder LLRs from the previous step) for LDPC decoders become

$$
L_{M,\,e}(\hat{c}_j) = L(\hat{c}_j) - L_{D,\,e}(c_j)
\tag{6.62}
$$

With $L_{D,e}(c)$, we denote the LDPC decoder extrinsic LLR, which is initially set to zero value. The LDPC decoder is implemented by employing the sum-product algorithm. The LDPC decoder extrinsic LLRs (the difference between LDPC decoder output and the input LLRs), $L_{D,e}$, are forwarded to the APP demapper as *a priori* bit LLRs ($L_{M,a}$) so that the symbol *a priori* LLRs are calculated as

$$
\lambda_a(s) = \log P(s) = \sum_{j=0}^{m-1}(1 - c_j)L_{D,\,e}(c_j)
\tag{6.63}
$$

By substituting Eq. (6.63) into Eq. (6.60) and then Eq. (6.59), we are able to calculate the symbol LLRs for the subsequent iteration. The iteration between the APP demapper and LDPC decoder is performed until the maximum number of iterations is reached or the valid codewords are obtained.

The results of simulations for 30 iterations in the sum-product algorithm and 10 APP demapper–LDPC decoder iterations for an AWGN channel model are shown in Figure 6.24a.

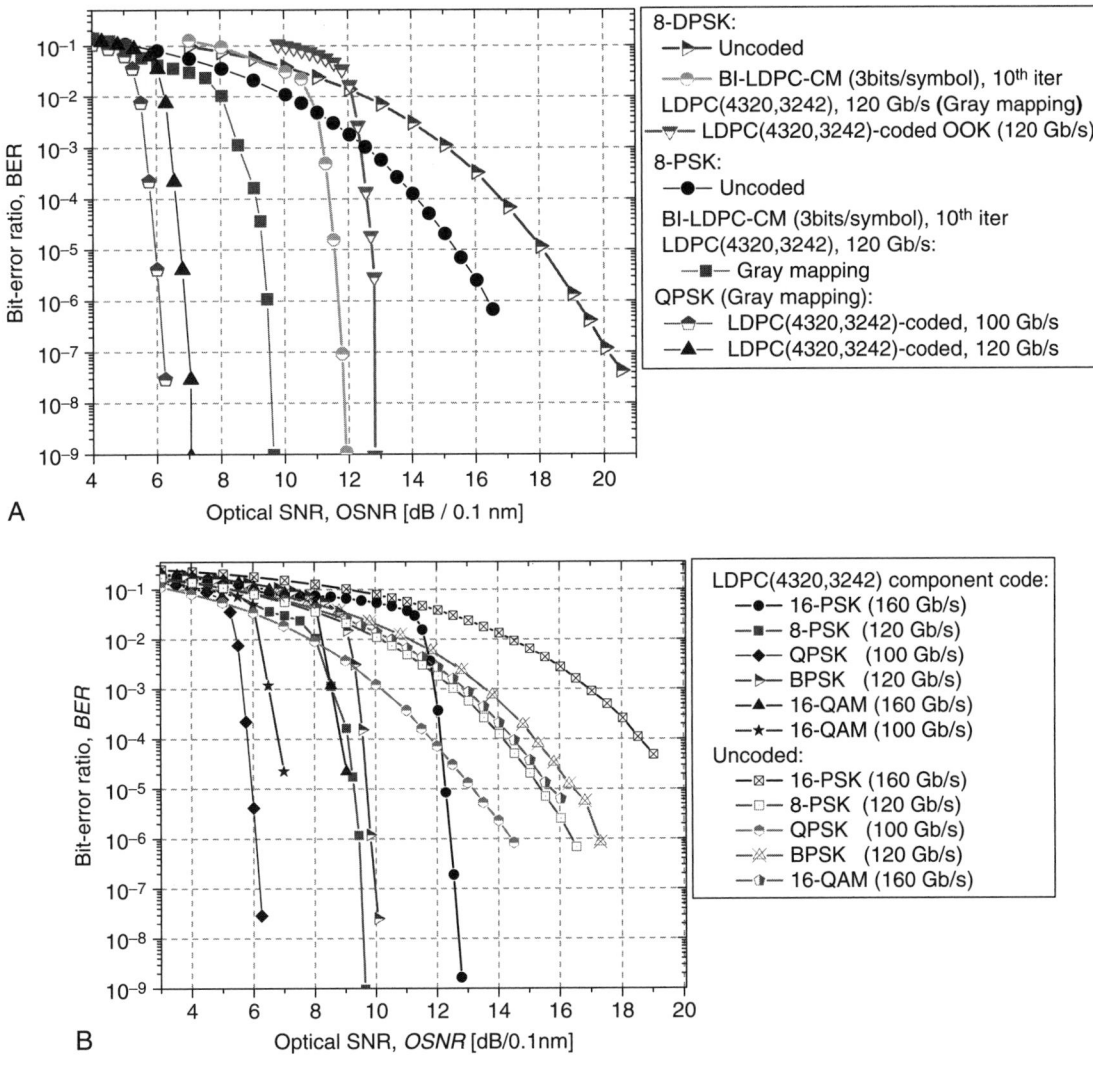

Figure 6.24: (a) BER performance of coherent detection bit-interleaved LDPC-coded modulation scheme versus the direct detection scheme on an AWGN channel model, and (b) performance comparison for different modulation schemes (the Gray mapping rule is applied).

The information symbol rate is set to 40 Giga symbols/s, whereas 8-PSK is employed so that the aggregate bit rate becomes 120 Gb/s. Two different mappers are considered: Gray and natural mapping. The coding gain for 8-PSK at a BER of 10^{-9} is approximately 9.5 dB, and much larger coding gain is expected at a BER below 10^{-12}. The coherent detection scheme offers an improvement of at least 2.3 dB compared to the corresponding direct detection scheme. The BER performance of coherent BICM with LDPC(4320, 3242) code, employed as component code for different modulations, is shown in Figure 6.24b. We can see that 16-QAM (with an aggregate rate of 160 Gb/s) outperforms 16-PSK by more than 3 dB. It is also interesting that 16-QAM slightly outperforms 8-PSK scheme of lower aggregate data rate (120 Gb/s). The 8-PSK scheme of aggregate rate of 120 Gb/s outperforms the BPSK scheme of data rate 120 Gb/s. Moreover, because the transmission symbol rate for 8-PSK is 53.4 Giga symbols/s, the impact of PMD and intrachannel nonlinearities is much less important than that at 120 G/s. Consequently, for 100 Gb/s Ethernet transmission, it is better to multiplex two 50 Gb/s channels than four 25 Gb/s channels.

The comparison of different LDPC component codes is given in Figure 6.25. The BICM scheme employing the balanced-incomplete block design-based girth-8 LDPC code[8] of rate 0.81 performs slightly worse or comparable to the quasi-cyclic-based scheme of lower code rate ($R = 0.75$). The BICM scheme of rate 0.75, based on PBD irregular LDPC code, outperforms the schemes based on regular LDPC codes.

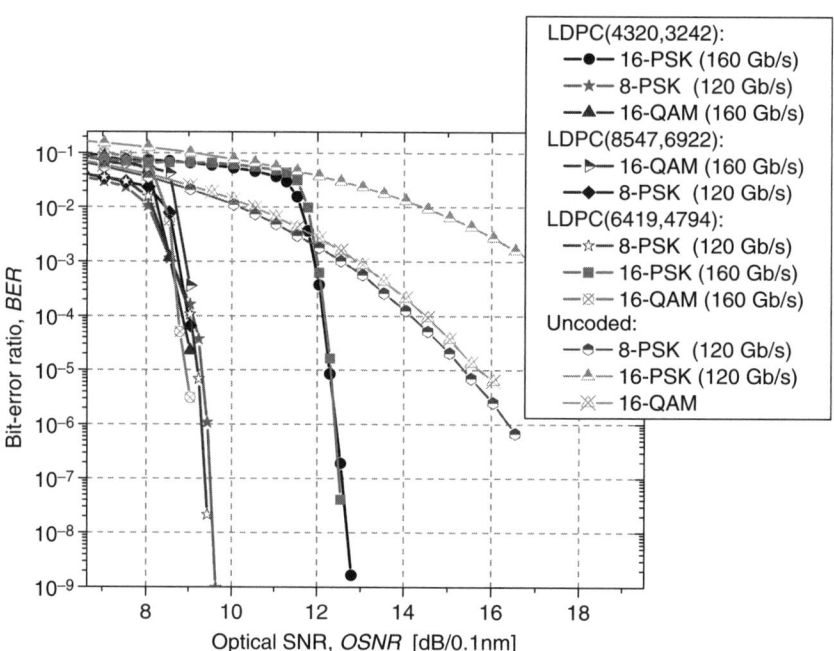

Figure 6.25: Comparison of different component LDPC codes (the Gray mapping rule is applied).

6.5 Coded OFDM in Fiber-Optics Communication Systems with Direct Detection

The transmitter and receiver configurations of an OFDM system with direct detection[76–78] and the format of the transmitted OFDM symbol are shown in Figures 6.26a–6.26c, respectively. On the transmitter side, the information-bearing streams at 10 Gb/s are encoded using identical LDPC codes. The outputs of these LDPC encoders are demultiplexed and parsed into groups of B bits corresponding to one OFDM frame. The B bits in each OFDM frame are subdivided into K subchannels with the ith subcarrier carrying b_i bits so that $B = \sum b_i$. The b_i bits from the ith subchannel are mapped into a complex-valued signal from a 2^{b_i}-point signal constellation such as M-ary QAM and M-ary PSK. For example, $b_i = 2$ for QPSK and $b_i = 4$ for 16-QAM. The complex-valued signal points from subchannels are considered to be the values of the FFT of a multicarrier OFDM signal. By selecting the number of subchannels K sufficiently large, the OFDM symbol interval can be made significantly larger than the dispersed pulse width of an equivalent single-carrier system, resulting in ISI reduction. The OFDM symbol, shown in Figure 6.26c, is generated as follows: $N_{QAM}(= K)$ input QAM symbols are zero-padded to obtain N_{FFT} input samples for inverse FFT (IFFT),

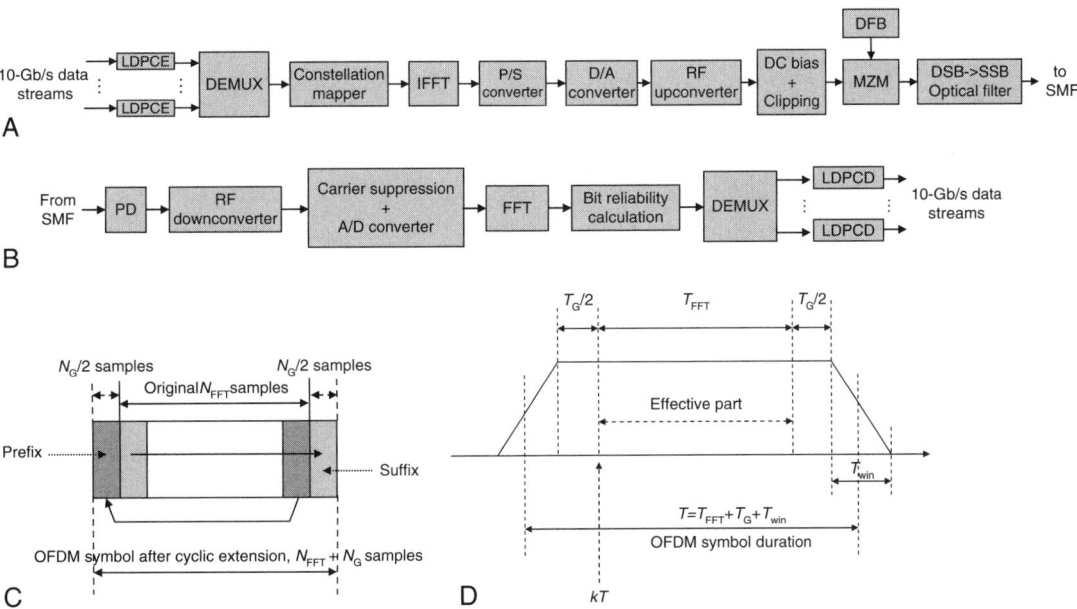

Figure 6.26: (a) Transmitter configuration, (b) receiver configuration, (c) OFDM symbol after cyclic extension, and (d) OFDM symbol after windowing. DSB, double sideband; LDPCD, LDPC decoder; LDPCE, LDPC encoder; MZM, Mach–Zehnder modulator; PD, photodetector; SMF, single-mode optical fiber; S/P, serial-to-parallel converter; SSB, single sideband.

N_G nonzero samples are inserted (as explained later) to create the guard interval, and the OFDM symbol is multiplied by the window function. The OFDM symbol windowing is illustrated in Figure 6.26d. The purpose of cyclic extension is to preserve the orthogonality among subcarriers even when the neighboring OFDM symbols partially overlap due to dispersion, and the role of windowing is to reduce the out-of-band spectrum. For efficient chromatic dispersion and PMD compensation, the length of the cyclically extended guard interval should be longer than the delay spread due to chromatic dispersion and PMD.

The cyclic extension, illustrated in Figure 6.26c, is accomplished by repeating the last $N_G/2$ samples of the effective OFDM symbol part (N_{FFT} samples) as a prefix and repeating the first $N_G/2$ samples as a suffix. After D/A conversion and RF up-conversion, the RF signal can be mapped to the optical domain using one of two options: (1) The OFDM signal can directly modulate a DFB laser, or (2) the OFDM signal can be used as the RF input of a Mach–Zehnder modulator (MZM). A *DC bias component* is added to the OFDM signal to enable recovery of the incoherent QAM symbols.

Here, three different OFDM schemes are described. The first scheme is based on direct modulation and is referred to as the "biased-OFDM" (B-OFDM) scheme. Because bipolar signals cannot be transmitted over an IM/DD link, it is assumed that the bias component is sufficiently large so that when added to the OFDM signal, the resulting sum is non-negative. The main disadvantage of the B-OFDM scheme is its poor power efficiency. To improve the OFDM power efficiency, two alternative schemes can be used. The first scheme, which we refer to as the "clipped-OFDM" (C-OFDM) scheme, is based on single-sideband (SSB) transmission and clipping of the OFDM signal after bias addition. The bias is varied in order to find the optimum one for fixed optical launched power. It has been found that the optimum bias is one in which approximately 50% of the total electrical signal energy before clipping is allocated for transmission of a carrier. The second power-efficient scheme, which we refer to as the "unclipped-OFDM" (U-OFDM) scheme, is based on SSB transmission using a LiNbO$_3$ MZM. To avoid distortion due to clipping at the transmitter, the information is mapped into the optical domain by modulating the electrical field of the optical carrier (instead of intensity modulation employed in the B-OFDM and C-OFDM schemes). In this way, both positive and negative portions of the electrical OFDM signal are transmitted to the photodetector. Distortion introduced by the photodetector, caused by squaring, is successfully eliminated by proper filtering, and recovered signal does not exhibit significant distortion. It is important to note, however, that the U-OFDM scheme is slightly less power efficient than the C-OFDM scheme. The SSB modulation can be achieved either by appropriate optical filtering of the double-sideband signal at MZM output (see Figure 6.26a) or by using the Hilbert transformation of the in-phase component of the OFDM RF signal. The first version requires the use of only the in-phase component of the RF OFDM signal, providing that zero padding is done in the middle of the OFDM symbol rather than at the edges.

The transmitted OFDM signal can be written as

$$s(t) = s_{OFDM}(t) + D \qquad (6.64)$$

where

$$s_{OFDM}(t) = \text{Re}\left\{ \sum_{k=-\infty}^{\infty} w(t - kT) \sum_{i=-N_{FFT}/2}^{N_{FFT}/2 - 1} X_{i,k} \cdot e^{j2\pi \frac{i}{T_{FFT}} \cdot (t - kT)} e^{j2\pi f_{RF} t} \right\} \qquad (6.65)$$

is defined for $t \in [kT - T_G/2 - T_{win}, kT + T_{FFT} + T_G/2 + T_{win}]$. In the previous expression, $X_{i,k}$ denotes the ith subcarrier of the kth OFDM symbol, $w(t)$ is the window function, and f_{RF} is the RF carrier frequency. T denotes the duration of the OFDM symbol, T_{FFT} denotes the FFT sequence duration, T_G is the guard interval duration (the duration of cyclic extension), and T_{win} denotes the windowing interval duration. D denotes the DC bias component, which is introduced to enable the OFDM demodulation using the direct detection.

The PIN photodiode output current can be written as

$$i(t) = R_{PIN}\left\{ \left[s_{OFDM}(t) + D \right] * h(t) + N(t) \right\}^2 \qquad (6.66)$$

where $s_{OFDM}(t)$ denotes the transmitted OFDM signal in the RF domain given by Eq. (6.65). D was previously introduced, and R_{PIN} denotes the photodiode responsivity. The impulse response of the optical channel is represented by $h(t)$. We modeled this channel by solving the nonlinear Schrödinger equation (see Section 6.5.1) numerically using the split-step Fourier method as described in Agrawal.[79] The signal after RF down-conversion and appropriate filtering can be written as

$$r(t) = \left[i(t) k_{RF} \cos\left(\omega_{RF} t \right) \right] * h_e(t) + n(t) \qquad (6.67)$$

where $h_e(t)$ is the impulse response of the low-pass filter, $n(t)$ is electronic noise in the receiver, and k_{RF} is the RF down-conversion coefficient. Finally, after the A/D conversion and cyclic extension removal, the signal is demodulated by using the FFT algorithm. The soft outputs of the FFT demodulator are used to estimate the bit reliabilities that are fed to identical LDPC iterative decoders implemented based on the sum-product with a correction term algorithm as explained previously.

For the sake of illustration, let us consider the signal waveforms and power spectral densities (PSDs) at various points in the OFDM system given in Figure 6.27. These examples were generated using SSB transmission in a back-to-back configuration. The bandwidth of the OFDM signal is set to B GHz and the RF carrier to $0.75\,B$, where B denotes the aggregate data rate. The number of OFDM subchannels is set to 64, the OFDM sequence is zero-padded, and the FFT is calculated using 128 points. The guard interval is obtained by a cyclic extension of 2×16 samples. The average transmitted launch power is set to 0 dBm. The OFDM transmitter

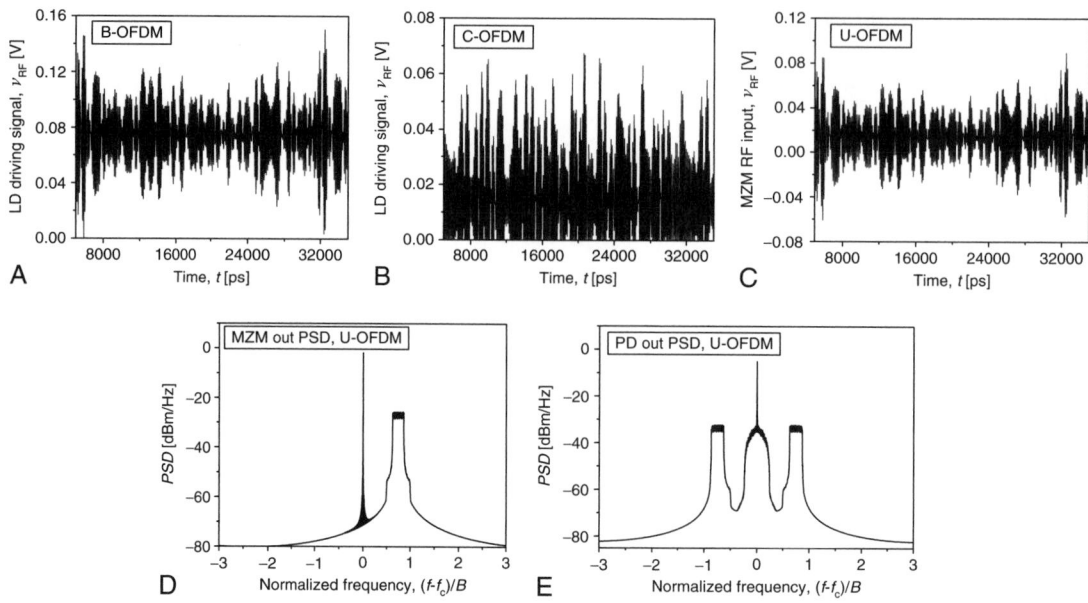

Figure 6.27: Waveforms and PSDs of SSB QPSK-OFDM signal at different points during transmission for electrical SNR (per bit) of 6 dB. f_c, **optical carrier frequency; LD, laser diode.**

parameters are carefully chosen such that RF driver amplifier and MZM operate in linear regime (see Figure 6.27a–6.27c). The PSDs of MZM output signal and the photodetector output signal are shown in Figures 6.27d and 6.27e, respectively. The OFDM term after beating in the PD, the low-pass term, and the squared OFDM term can easily be identified.

6.5.1 Performance Assessment of LDPC-Coded OFDM Fiber-Optics Communications

OFDM is suitable for long-haul transmission for three main reasons: (1) improvement of spectral efficiency, (2) simplification of the chromatic dispersion compensation engineering, and (3) PMD compensation.[78,80] We have shown[77] that U-OFDM provides the best power efficiency–BER performance compromise, and as such it is adopted here. The BER performance of this scheme (with an aggregate data rate of 40 Gb/s) against the conventional LDPC-coded RZ-OOK scheme (operating at 40 Gb/s) is given in Figure 6.28 for the thermal noise-dominated scenario. The LDPC-coded QPSK U-OFDM provides more than 2dB coding gain improvement over LDPC-coded RZ-OOK at a BER of 10^{-8}. We have shown that LDPC-coded OFDM provides much higher spectral efficiency than LDPC-coded RZ-OOK.[78]

Another possible application in fiber-optics communications is 100 Gb/s Ethernet.[81] For example, for QPSK OFDM transmission, two 1 Gb/s streams create a QPSK signal constellation point; with 50 subcarriers carrying 2 Gb/s traffic, the aggregate rate of 100 Gb/s can be achieved.

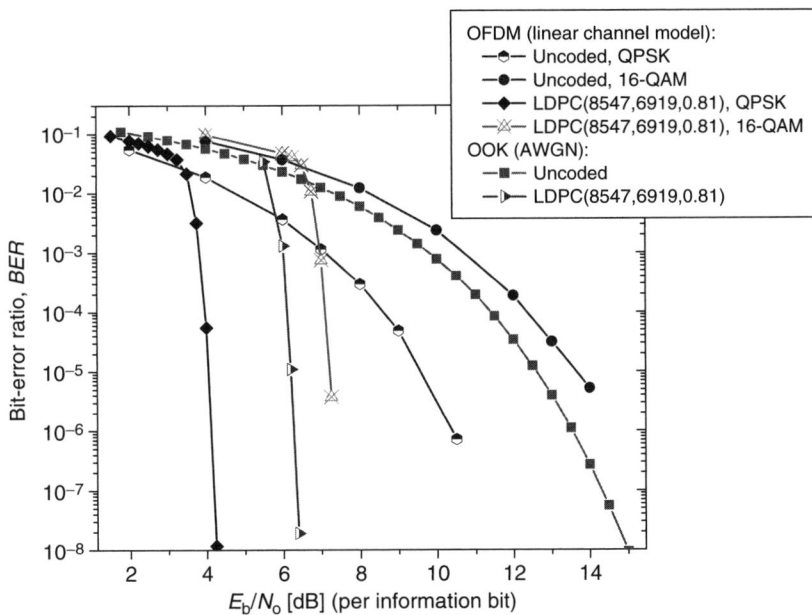

Figure 6.28: BER performance of OFDM against OOK for the thermal noise-dominated scenario.

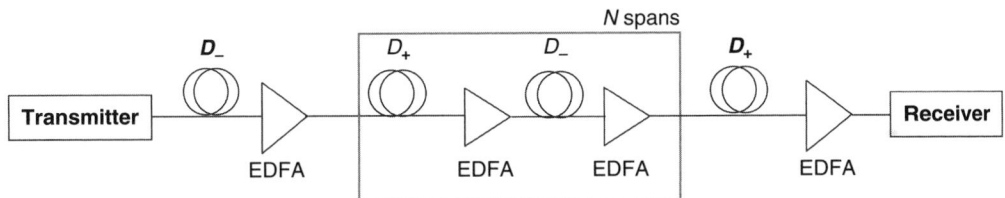

Figure 6.29: Dispersion map under study.

For the dispersion map described in Figure 6.29, the BER curves for the uncoded 100 Gb/s OFDM SSB transmission using QPSK are shown in Figure 6.30, and fiber parameters are given in Table 6.4. The dispersion map is composed of N spans of length $L = 120$ km, consisting of $2L/3$ km of D_+ fiber followed by $L/3$ km of D_- fiber, with precompensation of -1600 ps/nm and corresponding postcompensation. The propagation of a signal through the transmission media is modeled by a nonlinear Schrödinger equation (NLSE)[79]

$$\frac{\partial A}{\partial z} = -\frac{\alpha}{2} A - \frac{i}{2} \beta_2 \frac{\partial^2 A}{\partial T^2} + \frac{\beta_3}{6} \frac{\partial^3 A}{\partial T^3} + i\gamma \left(|A|^2 - T_R \frac{\partial |A|^2}{\partial T} \right) A \qquad (6.68)$$

where z is the propagation distance along the fiber; relative time, $T = t - z/v_g$, gives a frame of reference moving at the group velocity v_g; $A(z, T)$ is the complex field amplitude of the

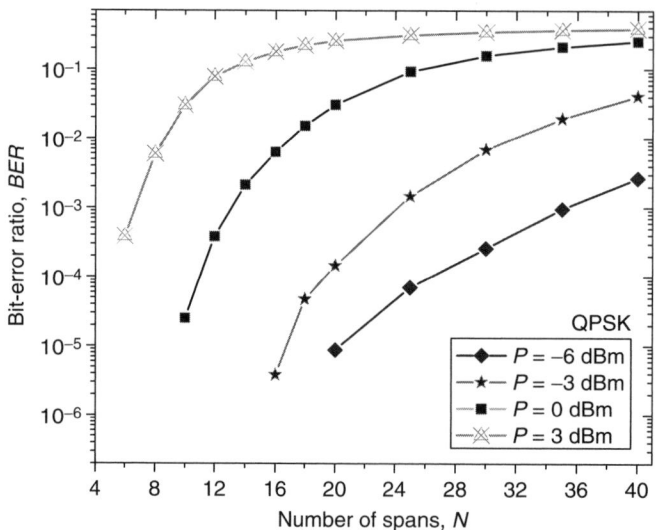

Figure 6.30: Uncoded BER versus the number of spans for SSB QPSK OFDM transmission for the dispersion map in Figure 6.29.

Table 6.4: Fiber Parameters

	D_+ Fiber	D_- Fiber
Dispersion (ps/(nm km))	20	−40
Dispersion slope (ps/(nm² km))	0.06	−0.12
Effective cross-sectional area [μm²]	110	50
Nonlinear refractive index (m²/W)	$2.6 \cdot 10^{-20}$	$2.6 \cdot 10^{-20}$
Attenuation coefficient (dB/km)	0.19	0.25

pulse; α is the attenuation coefficient of the fiber; β_2 is the group velocity dispersion (GVD) coefficient; β_3 is the second-order GVD; and γ is the nonlinearity coefficient giving rise to Kerr effect nonlinearities. Such nonlinearities include self-phase modulation, intrachannel FWM, intrachannel cross-phase modulation, cross-phase modulation, and FWM. T_R is the Raman coefficient describing the stimulated Raman scattering. The NLSE was solved using the split-step Fourier method.[79]

From Figure 6.30, it is evident that 100 Gb/s transmission over 3840 km is possible using OFDM and LDPC codes with a threshold BER of 10^{-2} (for launched power of −3 dBm). For more details, see Djordjevic and Vasic.[81]

In the presence of PMD, the PIN photodiode output current can be written as follows:

$$i(t) = R\left\{\left|\sqrt{k}(s_{\text{OFDM}}(t) + D) * h_V(t)\right|^2 + \left|\sqrt{1-k}(s_{\text{OFDM}}(t) + D) * h_H(t)\right|^2\right\} \qquad (6.69)$$

where k denotes the power-splitting ratio between two principal states of polarizations (PSPs). For the first-order PMD, the optical channel responses $h_H(t)$ and $h_V(t)$ of horizontal and vertical PSPs are given as $h_H(t) = \delta(t + \Delta\tau/2)$ and $h_V(t) = \delta(t - \Delta\tau/2)$, respectively, where $\Delta\tau$ is the DGD of two PSPs.[80] From Djordjevic,[80] it follows that in the presence of the first-order PMD, the photodiode output signal, after appropriate filtering to remove the squared and DC terms, is proportional to

$$i(t) \sim 2\text{Rb}[k \cdot s_{\text{OFDM}}(t - \Delta\tau/2) + (1-k) \cdot s_{\text{OFDM}}(t + \Delta\tau/2)] \qquad (6.70)$$

Equation (6.70) is similar to a two-ray multipath wireless model[82] and, therefore, the channel estimation techniques similar to those employed in wireless communications are straightforwardly applicable here. It can be shown that the received QAM symbol of the ith subcarrier of the kth OFDM symbol is related to transmitted QAM symbol $X_{i,k}$ by

$$Y_{i,k} = h_i e^{j\theta_k} X_{i,k} + n_{i,k} \qquad (6.71)$$

where h_i is channel distortion introduced by PMD (and chromatic dispersion), and θ_k is the phase shift of the kth OFDM symbol due to chromatic dispersion and self-phase modulation (SPM). The channel coefficients can be determined by training sequence-based channel estimation. In a decision-directed mode, the transmitted QAM symbols are estimated by

$$\hat{X}_{i,k} = \left(h_i^* / |h_i|^2\right) e^{-j\theta_k} Y_{i,k} \qquad (6.72)$$

The signal constellation diagrams, after the demapper from Figure 6.26 and before and after applying the channel estimation, are shown in Figure 6.31. They correspond to the worst-case scenario ($k = 1/2$) and 10 Gb/s aggregate data rate. Therefore, the channel estimation-based OFDM is able to compensate for DGD of 1600 ps.

The results of PMD simulations for U-OFDM and thermal noise-dominated scenario are shown in Figure 6.32 for different DGDs and the worst-case scenario ($k = 1/2$), assuming that the channel state information is known on a receiver side. The OFDM signal bandwidth is set to BW $= 0.25 B$ (where B is the aggregate bit rate set to 10 Gb/s), the number of subchannels is set to $N_{\text{QAM}} = 64$, FFT/IFFT is calculated in $N_{\text{FFT}} = 128$ points, the RF carrier frequency is set to $0.75 B$, the bandwidth of optical filter for SSB transmission is set to $2 B$, and the total averaged launched power is set to 0 dBm. The guard interval is obtained by cyclic extension of $N_G = 2 \times 16$ samples. The Blackman–Harris windowing function is applied. The 16-QAM OFDM with and without channel estimation is observed in simulations. The effect of PMD is reduced by (1) using a sufficient number of subcarriers so that the OFDM symbol rate is significantly lower than the aggregate bit rate and (2) using the training sequence

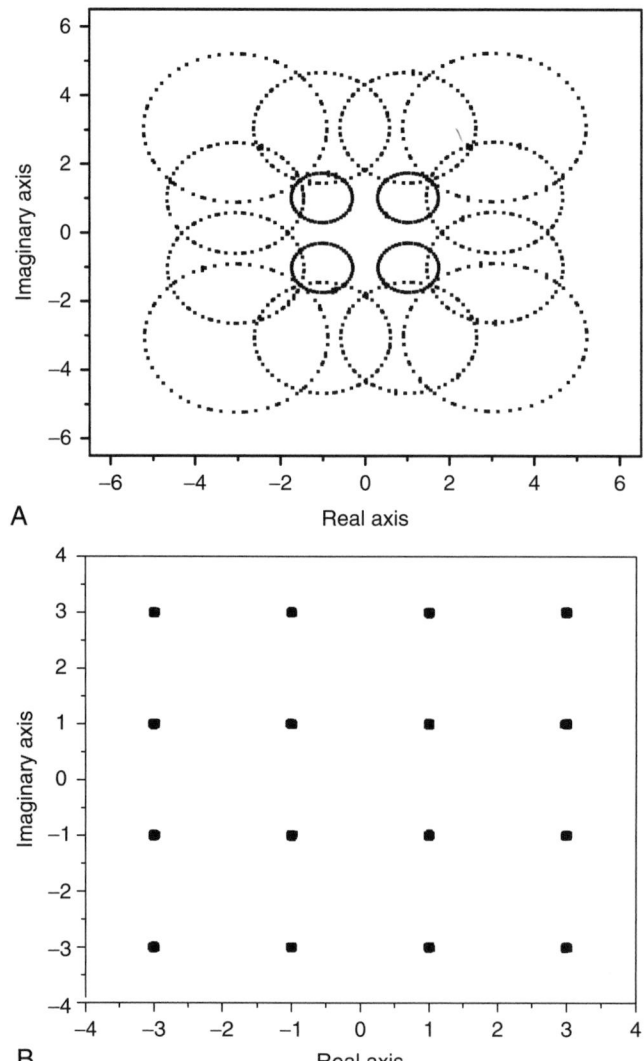

Figure 6.31: (a) Signal constellation diagram before PMD compensation for DGD of 1600 ps and (b) signal constellation diagram after PMD compensation (other effects except PMD have been ignored).

to estimate the PMD distortion. For DGD of 1/BW, the RZ-OOK threshold receiver is not able to operate properly because it enters the BER error floor. Note that 16-QAM OFDM without channel estimation enters the BER floor, and even advanced FEC cannot help much in reducing the BER.

The similar channel estimation technique can be used to compensate for the phase rotation due to SPM.[78,81] For example, the signal constellation diagrams before and after SPM phase

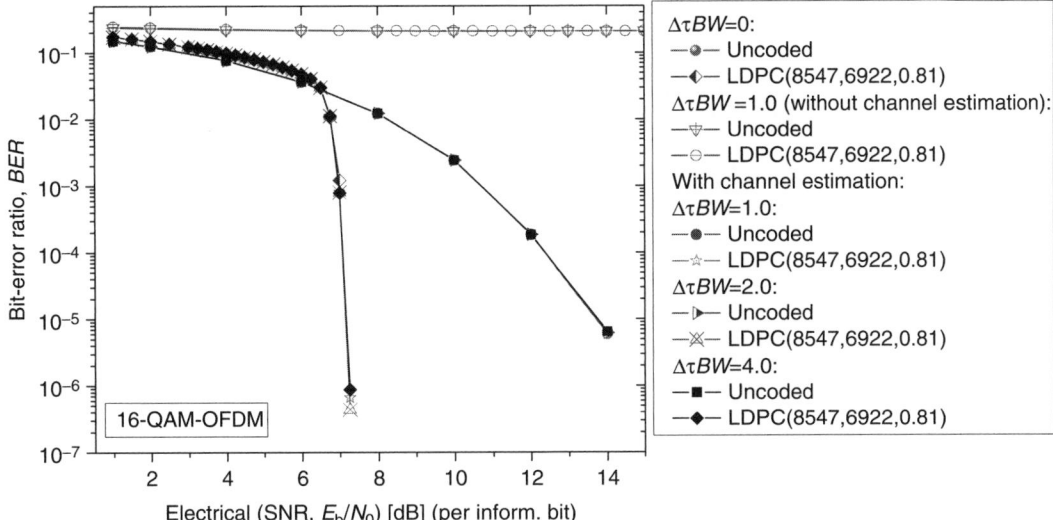

Figure 6.32: BER performance of LDPC(8547, 6922)-coded 16-QAM OFDM for different DGD values, and thermal noise-dominated scenario assuming that channel state information is known on the receiver side. The power-splitting ratio between two PSPs was set to $k = 1/2$.

correction are shown in Figure 6.33 for a transmission distance of 1200 km for the dispersion map given in Figure 6.29 and an aggregate data rate of 100 Gb/s. The four pilot tones were sufficient to cancel the phase rotation due to SPM.

6.5.2 Simultaneous Chromatic Dispersion and PMD Compensation via LDPC-Coded OFDM

The received electrical field, at the input of the transimpedance amplifier (TA), in the presence of chromatic dispersion and first-order PMD can be represented by

$$E(t) = \mathrm{FT}^{-1}\left\{\mathrm{FT}\left\{E_0 \begin{bmatrix} \sqrt{1-\kappa} \\ \sqrt{\kappa}e^{j\delta} \end{bmatrix} [s_{\mathrm{OFDM}}(t) + D]\right\} \exp\left[j\left(\frac{\beta_2\omega^2}{2} - \frac{\beta_3\omega^3}{6}\right)L_{tot}\right]\right\} + \begin{bmatrix} N_x(t) \\ N_y(t) \end{bmatrix}$$

(6.73)

where β_2 and β_3 represent the GVD and second-order GVD parameters, L_{tot} is the total SMF length, k is the splitting ratio between two PSPs, δ is the phase difference between PSPs, E_0 is transmitted laser electrical field amplitude, and N_x and N_y represent x- and y-polarization ASE noise components. FT and FT^{-1} denote the Fourier transform and inverse Fourier transform, respectively. The TA output signal can be represented by $v(t) = R_F R_{\mathrm{PIN}}|E(t)|^2 + n(t)$, where R_{PIN} is the photodiode responsivity, R_F is the TA feedback resistor, and $n(t)$ is TA

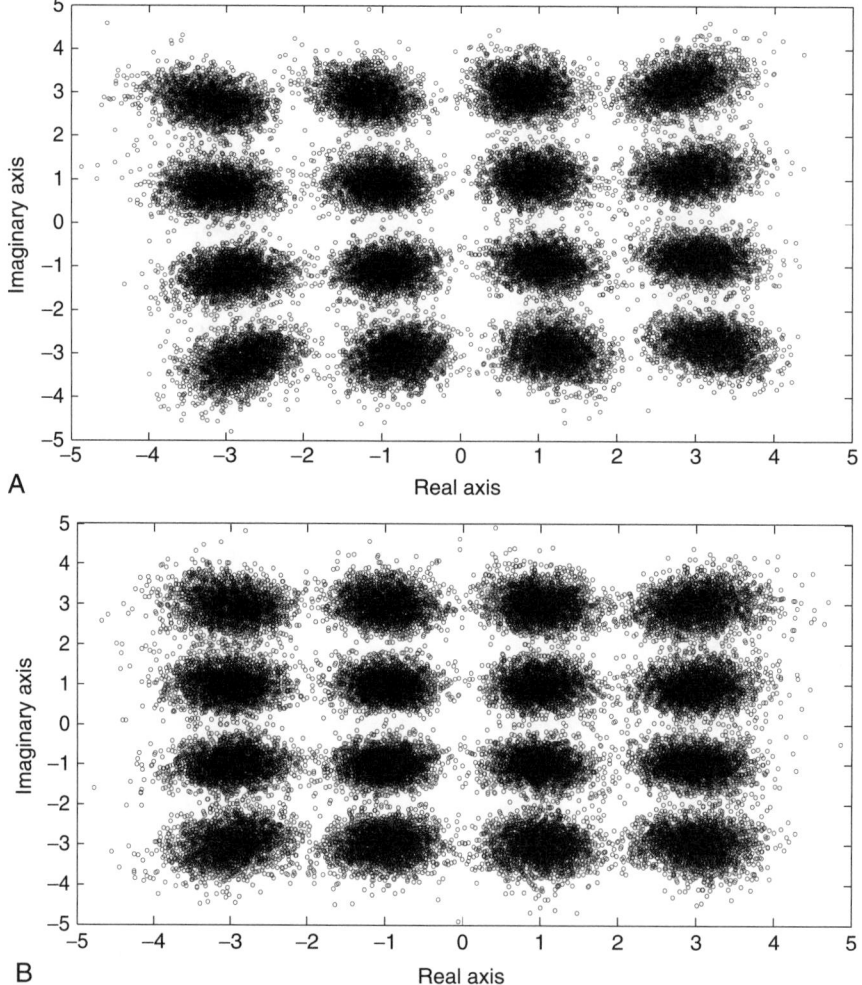

Figure 6.33: The received signal constellation for 16-QAM SSB transmission after 1200 km for the dispersion map in Figure 6.29 and aggregate data rate of 100 Gb/s: (a) before the SPM phase correction and (b) after the SPM phase correction.

thermal noise. For complete elimination of ISI, the total delay spread due to chromatic dispersion and DGD should be shorter than the guard time:

$$|\beta_2|L_{tot}\Delta\omega + \text{DGD}_{max} = \frac{c}{f^2}|D_t|N_{FFT}\Delta f + \text{DGD}_{max} \leq T_G \tag{6.74}$$

where D_t is the accumulated dispersion, Δf is the subcarrier spacing, c is the speed of the light, and f is the central frequency set to 193.1 THz. The number of subcarriers N_{FFT}, the guard interval T_G, GVD, and second-order GVD parameters were introduced previously. The received QAM symbol of the ith subcarrier in the kth OFDM symbol is related to transmitted QAM symbol $X_{i,k}$ by

$$Y_{i,k} = h_i e^{j\theta_i} e^{j\phi_k} X_{i,k} + n_{i,k} \tag{6.75}$$

where h_i is channel distortion introduced by PMD and chromatic dispersion, and θ_i is the phase shift of the ith subcarrier due to chromatic dispersion. ϕ_k represents the OFDM symbol phase noise due to SPM and RF down-converter, and it can be eliminated by pilot-aided channel estimation. Note that in the direct detection case, the laser phase noise is completely cancelled by photodetection. To estimate the channel distortion due to PMD, h_i, and phase shift due to chromatic dispersion, θ_i, we need to pretransmit the training sequence. Because in the ASE noise-dominated scenario (considered here) the channel estimates are sensitive to ASE noise, the training sequence should be sufficiently long to average the noise. For DGDs up to 100 ps, the training sequence composed of several OFDM symbols is sufficient. For larger DGDs, a longer OFDM training sequence is required; alternatively, the channel coefficients can be chosen to maximize the LLRs or the polarization beam splitter can be used to separate the x and y polarization components and consequently process them. The phase shift of the ith subcarrier due to chromatic dispersion can be determined from the training sequence as the difference of transmitted and received phase averaged over different OFDM symbols. Once the channel coefficients and phase shifts due to PMD and chromatic dispersion are determined, in a decision-directed mode, the transmitted QAM symbols are estimated by

$$\hat{X}_{i,k} = \left(h_i^* / |h_i|^2 \right) e^{-j\theta_i} e^{-j\phi_k} Y_{i,k} \tag{6.76}$$

The symbol LLRs $\lambda(q)$ $(q = 0, 1, \ldots, 2^b - 1)$ can be determined by

$$\lambda(q) = -\frac{\left(\mathrm{Re}\left[\hat{X}_{i,k}\right] - \mathrm{Re}\left[\mathrm{QAM}(\mathrm{map}(q))\right] \right)^2}{N_0} - \frac{\left(\mathrm{Im}\left[\hat{X}_{i,k}\right] - \mathrm{Im}\left[\mathrm{QAM}\left(\mathrm{map}(q)\right)\right] \right)^2}{N_0}; \tag{6.77}$$

$$q = 0, 1, \ldots, 2^b - 1$$

where Re[ ] and Im[ ] denote the real and imaginary part of a complex number, QAM denotes the QAM constellation diagram, N_0 denotes the power spectral density of an equivalent Gaussian noise process, and map(q) denotes a corresponding mapping rule (Gray mapping is applied here). (b denotes the number of bits per constellation point.) Let us denote by v_j the jth bit in an observed symbol q binary representation $v = (v_1, v_2, \ldots, v_b)$. The bit LLRs needed for LDPC decoding are calculated from symbol LLRs by

$$L(\hat{v}_j) = \log \frac{\sum_{q:v_j = 0} \exp[\lambda(q)]}{\sum_{q:v_j = 1} \exp[\lambda(q)]} \tag{6.78}$$

Therefore, the jth bit reliability is calculated as the logarithm of the ratio of a probability that $v_j = 0$ and probability that $v_j = 1$. In the nominator, the summation is done over all symbols q having 0 at the position j, whereas in the denominator it is done over all symbols q having 1 at the position j.

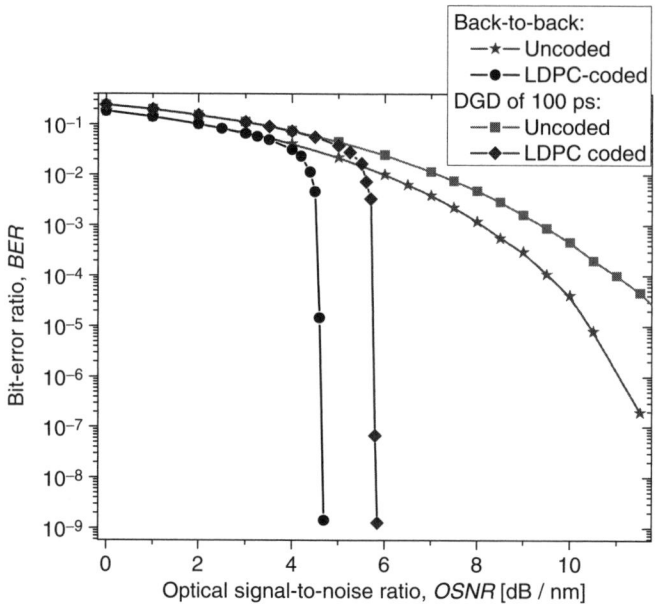

Figure 6.34: BER performance of an LDPC-coded OFDM system with an aggregate rate of 10 Gb/s for DGD of 100 ps.

The results of simulation for the ASE noise-dominated scenario and single-wavelength channel transmission are shown in Figures 6.34–6.36 for the LDPC-coded SSB OFDM system with an aggregate rate of 10 Gb/s, 512 subcarriers, RF carrier frequency of 10 GHz, oversampling factor of 2, and cyclic extension with 512 samples. The modulation format being applied is QPSK. The LDPC(16935, 13550) code of girth-10, code rate 0.8, and column weight 3, designed as explained in Section 6.2.3.1, is used. Figure 6.34 shows the BER performance for DGD of 100 ps, without residual chromatic dispersion. It can be seen that the uncoded case faces significant performance degradation at low BERs. On the other hand, the LDPC-coded case has degradation of 1.1 dB at a BER of 10^{-9} (compared to the back-to-back configuration). Figure 6.35 shows the BER performance after 6500 km of SMF (without optical dispersion compensation) for a dispersion map composed of 65 sections of SMF 100 km in length. The noise figure of erbium-doped fiber amplifiers (EDFAs), deployed periodically after every SMF section, was set to 5 dB. To achieve the desired optical signal-to-noise ratio (OSNR), the ASE noise loading was applied on the receiver side, whereas the launch power was kept below 0 dBm. We see that LDPC-coded OFDM is much less sensitive to chromatic dispersion compensation than PMD. Therefore, even 6500 km can be reached without optical dispersion compensation with a penalty within 0.4 dB at a BER of 10^{-9} when LDPC-coded OFDM is used.

Figure 6.36 shows the efficiency of LDPC-coded OFDM in simultaneous chromatic dispersion and PMD compensation. After 6500 km of SMF (without optical dispersion

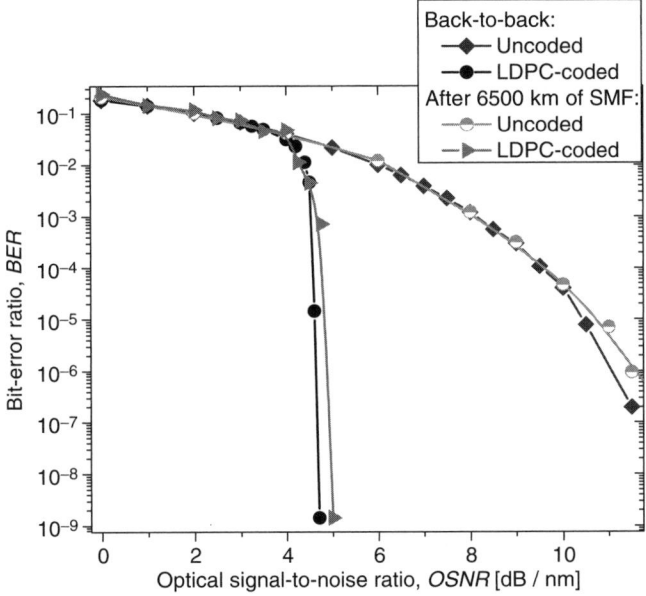

Figure 6.35: BER performance of an LDPC-coded OFDM system with an aggregate rate of 10 Gb/s after 6500 km of SMF.

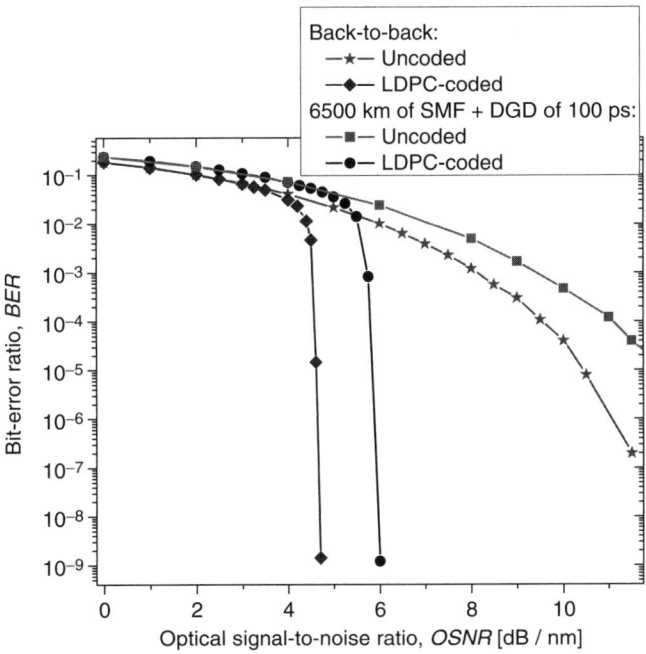

Figure 6.36: BER performance of an LDPC-coded OFDM system with an aggregate rate of 10 Gb/s after 6500 km of SMF and for DGD of 100 ps.

compensation) and for DGD of 100 ps, the LDPC-coded OFDM has the penalty within 1.5 dB. Notice that coded turbo equalization cannot be used for this level of residual chromatic dispersion and DGD. Also, based on the numerical results presented here, the major factor of performance degradation in LDPC-coded OFDM with direct detection in the ASE noise-dominated scenario is PMD. To improve the tolerance to PMD, one may use longer training sequences and redistribute the transmitted information among the subcarriers less affected by DGD or use the polarization beam splitter and separately process x and y PSPs in a manner similar to that proposed for OFDM with coherent detection as described in the next section. However, the complexity of such a scheme would be at least two times higher. Notice that for this level of DGD, the redistribution of power among subcarriers not being faded away is not needed. For larger values of DGDs, the penalty due to DGD grows as DGD increases, if the redistribution of subcarriers is not performed.

6.6 Coded OFDM in Fiber-Optics Communication Systems with Coherent Detection

The transmitter and receiver configurations for coherent detection and format of the transmitted OFDM symbol are shown in Figures 6.37a–6.37c, respectively. The bitstreams originating from m different information sources are encoded using different (n, k_l) LDPC codes of code rate $r_l = k_l/n$. k_l denotes the number of information bits of lth ($l = 1, 2, \ldots, m$) component LDPC code, and n denotes the codeword length, which is the same for all LDPC codes.

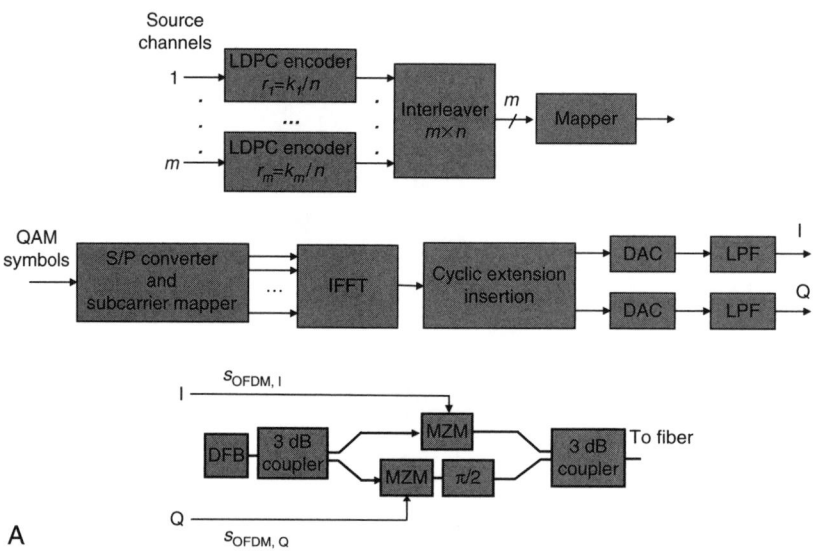

Figure 6.37: (a) Transmitter configuration,

(continued)

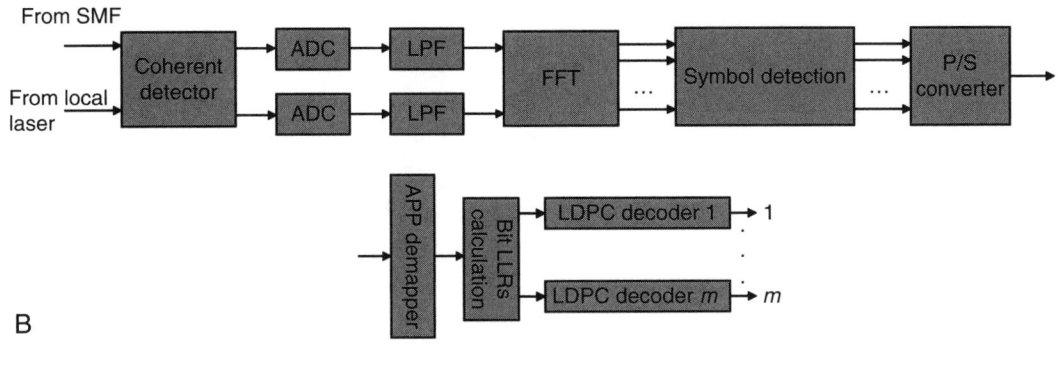

B

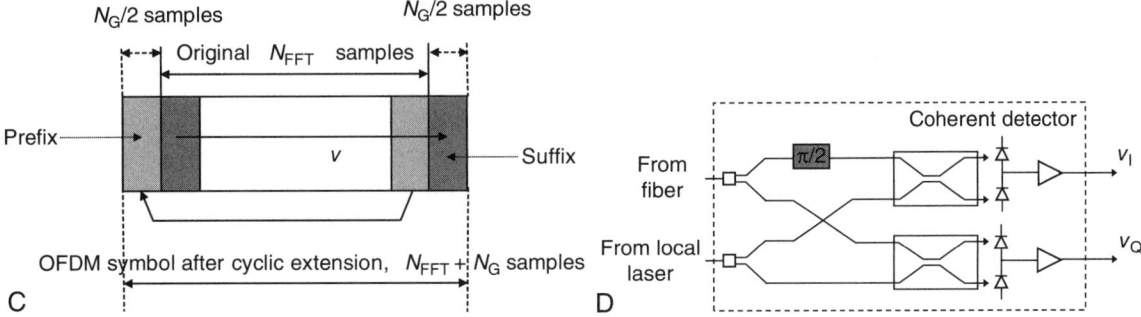

C

D

Figure 6.37—cont'd (b) receiver configuration, (c) OFDM symbol after cyclic extension, and (d) coherent detector architecture. DAC, D/A converter; MZM, Mach–Zehnder modulator; SMF, single-mode optical fiber; S/P, serial-to-parallel converter.

The use of different LDPC codes allows us to optimally allocate the code rates. The bit-interleaved coded modulation scheme can be considered as a special multilevel coding scheme in which all of the component codes are of the same rate.[75] The outputs of m LDPC encoders are written row-wise into a block-interleaver block. The mapper accepts m bits at time instance i from the $(m \times n)$ interleaver columnwise and determines the corresponding M-ary $(M = 2^m)$ signal constellation point $(\phi_{I,i}, \phi_{Q,i})$ in a two-dimensional (2D) constellation diagram such as M-ary PSK or M-ary QAM. The coordinates correspond to in-phase (I) and quadrature (Q) components of M-ary 2D constellation. The 2D constellation points, after serial-to-parallel (S/P) conversion, are used as the data values for the FFT of a multicarrier OFDM signal. By selecting the number of subcarriers, K, sufficiently large, the OFDM symbol interval can be made significantly larger than the dispersed pulse width of an equivalent single-carrier system, resulting in ISI reduction. The OFDM symbol shown in Figure 6.37c is generated as follows: $N_{QAM}(\leq K)$ input 2D symbols are zero-padded to obtain N_{FFT} input samples for IFFT, N_G nonzero samples are inserted (as explained later) to create the guard interval, and the OFDM symbol is multiplied by the window function. The purpose of cyclic extension is to preserve the orthogonality among subcarriers even when the neighboring OFDM symbols partially overlap due to dispersion, and the role of windowing is to reduce the out-of-band spectrum. For efficient chromatic dispersion and PMD compensation, the length of the cyclically extended guard interval should be larger than the spread due to chromatic dispersion and PMD. The cyclic extension, illustrated in Figure 6.37c, is accomplished by repeating the last $N_G/2$ samples of the effective OFDM symbol part (N_{FFT} samples) as a prefix and repeating the first $N_G/2$ samples as a suffix. After D/A conversion and low-pass filtering, the real part of OFDM signal (I) and the imaginary part of OFDM signal (Q) are mapped to the optical domain using two single-drive MZMs.

The complex envelope of a transmitted OFDM signal can be written as

$$s_{OFDM}(t) = \sum_{k=-\infty}^{\infty} w(t - kT) \sum_{i=-N_{FFT}/2}^{N_{FFT}/2-1} X_{i,k} \cdot e^{j2\pi\frac{i}{T_{FFT}}\cdot(t-kT)} \tag{6.79}$$

and it is defined for $t \in [kT - T_G/2 - T_{win}, kT + T_{FFT} + T_G/2 + T_{win}]$. In the previous expression, $X_{i,k}$ denotes the ith subcarrier of the kth OFDM symbol, $w(t)$ is the window function (the windowing function was explained in a previous section), T denotes the duration of the OFDM symbol, T_{FFT} denotes the FFT sequence duration, T_G is the guard interval duration (the duration of cyclic extension), and T_{win} is the windowing interval duration. One DFB laser is used as the CW source, and a 3 dB coupler is used to split the CW signal for both MZMs. After the appropriate sampling, the sum in Eq. (6.79) (obtained by ignoring the windowing function) can be written as follows:

$$x_{m,k} = \sum_{i=-N_{FFT}/2}^{N_{FFT}/2-1} X_{i,k} \exp\left(j2\pi \frac{im}{N_{FFT}}\right), \quad m = 0, 1, \ldots, N_{FFT} - 1 \tag{6.80}$$

and corresponds to the discrete Fourier transform (except for the normalization factor $1/N$).

The received electrical field, in the presence of chromatic dispersion, can be written as

$$E_s = \exp\left[j(2\pi f_{LD}t + \phi_{PN,S})\right] \sum_{k=1}^{N_{FFT}} X_k e^{j2\pi f_k t} e^{j\phi_{CD}(k)} \tag{6.81}$$

where $\phi_{CD}(k)$ denotes the phase factor of the kth subcarrier

$$\phi_{CD}(k) = \frac{\omega_k^2 |\beta_2| L}{2} = \frac{4\pi^2 f_k^2}{2} \frac{\lambda_{LD}^2}{2\pi c} DL = \frac{\pi c}{f_{LD}^2} D_t f_k^2 \tag{6.82}$$

where f_k is the kth subcarrier frequency (corresponding angular frequency is $\omega_k = 2\pi f_k$), f_{LD} is the transmitting laser emitting frequency (the corresponding wavelength is λ_k), D_t is the total dispersion coefficients ($D_t = DL$; D is the dispersion coefficient, and L is the fiber length), and c is the speed of light. (β_2 denotes the GVD parameter.) The coherent detector output (see Figure 6.37d) in complex notation can be written as

$$v(t) \simeq R_{PIN} r_F \exp\left\{j\left[2\pi(f_{LD} - f_{LO})t + \phi_{PN,S} - \phi_{PN,LO}\right]\right\} \sum_{k=1}^{N_{FFT}} X_k e^{j2\pi f_k t} e^{j\phi_{CD}(k)} + N \qquad (6.83)$$

where f_{LO} denotes the local laser emitting frequency, R_{PIN} is the photodetector responsivity, r_F is the feedback resistor from transimpedance amplifier configuration, and $\varphi_{S,PN}$ and $\varphi_{L,PN}$ represent the laser phase noise of the transmitting and receiving (local) laser, respectively. These two noise sources are commonly modeled as the Wiener–Lévy process,[83] which is a zero-mean Gaussian process with variance $2\pi(\Delta v_S + \Delta v_L)|t|$, where Δv_S and Δv_L are the laser linewidths of the transmitting and receiving laser, respectively. $N = N_I - jN_Q$ represents the noise process, mostly dominated by ASE noise.

The kth subcarrier received symbol Y_k can therefore be represented by

$$Y_k = X_k \exp\left[j(\phi_{PN,S} - \phi_{PN,LO})\right] \exp[j\phi_{CD}(k)] + N_k, \quad N_k = N_{k,I} + N_{k,Q} \qquad (6.84)$$

where N_k represents the circular complex Gaussian process due to ASE noise.
The transmitted symbol on the kth subcarrier can be estimated by

$$\tilde{X}_k = Y_k \exp\left[-j(\phi_{PN,S} - \phi_{PN,LO})\right] \exp[-j\phi_{CD}(k)] \qquad (6.85)$$

where the chromatic dispersion phase factor of the kth subcarrier $\phi_{CD}(k)$ is estimated by training-based channel estimation, whereas the phase factor $\phi_{PN,S} - \phi_{PN,LO}$ is estimated by pilot-aided channel estimation. The soft outputs of the FFT demodulator are used to estimate the bit reliabilities that are fed to identical LDPC iterative decoders implemented based on the sum-product-with-correction-term algorithm as explained previously. The chromatic dispersion is therefore quite straightforward to compensate for, whereas for PMD compensation there are different options, which are described in Section 6.6.2. Before we turn our attention to different PMD compensation schemes of high spectral efficiency, we briefly describe the PMD channel model.

6.6.1 Description of PMD Channel Model

For the first-order PMD study, the Jones matrix, neglecting the polarization-dependent loss and depolarization effects, can be represented by[84]

$$\mathbf{H} = \begin{bmatrix} h_{xx}(\omega) & h_{xy}(\omega) \\ h_{yx}(\omega) & h_{yy}(\omega) \end{bmatrix} = \mathbf{R}\mathbf{P}(\omega)\mathbf{R}^{-1}, \quad \mathbf{P}(\omega) = \begin{bmatrix} e^{-j\omega\tau/2} & 0 \\ 0 & e^{j\omega\tau/2} \end{bmatrix} \qquad (6.86)$$

where τ denotes DGD, ω is the angular frequency, and $R = R(\theta, \varepsilon)$ is the rotational matrix[84]

$$\mathbf{R} = \begin{bmatrix} \cos\left(\dfrac{\theta}{2}\right) e^{j\varepsilon/2} & \sin\left(\dfrac{\theta}{2}\right) e^{-j\varepsilon/2} \\ -\sin\left(\dfrac{\theta}{2}\right) e^{j\varepsilon/2} & \cos\left(\dfrac{\theta}{2}\right) e^{-j\varepsilon/2} \end{bmatrix}$$

with θ being the polar angle, and ε being the azimuth angle. For OFDM with coherent detection, the received symbol vector of the kth subcarrier in the ith OFDM symbol $r_{i,k} = [r_{x,i,k} \ r_{y,i,k}]^T$ can be represented by

$$r_{i,k} = H_k s_{i,k} e^{j[\phi_{CD}(k)+\phi_T-\phi_{LO}]} + n_{i,k} \tag{6.87}$$

where $s_{i,k} = [s_{x,i,k} \ s_{y,i,k}]^T$ denotes the transmitted symbol vector of the kth subcarrier in the ith OFDM symbol for both polarizations; $n_{i,k} = [n_{x,i,k} \ n_{y,i,k}]^T$ denotes the noise vector dominantly determined by the ASE noise; ϕ_T and ϕ_{LO} denote the laser phase noise processes of transmitting and local lasers, respectively; $\phi_{CD}(k)$ denotes the phase distortion of the kth subcarrier due to chromatic dispersion; and the Jones matrix of kth subcarrier H_k was introduced in Eq. (6.86). The transmitted/received symbols are complex-valued, with the real part corresponding to the in-phase coordinate and the imaginary part corresponding to the quadrature coordinate. The equivalent mathematical model is similar to that for wireless communications, and it is shown in Figure 6.38.

Figure 6.39 shows the magnitude responses of h_{xx} and h_{xy} coefficients of the Jones channel matrix against normalized frequency $f\tau$ (the frequency is normalized with DGD τ so that the conclusions are independent of the data rate) for two different cases: (1) $\theta = \pi/2$ and $\varepsilon = 0$ and (2) $\theta = \pi/3$ and $\varepsilon = 0$. In the first case, channel coefficient h_{xx} completely fades away for certain frequencies, whereas in the second case it never completely fades away, suggesting that the first case represents the worst-case scenario. To avoid this problem, in OFDM systems one can redistribute the transmitted power among subcarriers not fading or use the polarization diversity. Next, we describe several alternative approaches that can be used for a number of modulation formats, including M-ary PSK, M-ary QAM, and OFDM.

6.6.2 PMD Compensation by Coded OFDM in Fiber-Optics Communication Systems with Coherent Detection

In this section, we describe several PMD compensation schemes based on coded OFDM: (1) polarization diversity coded-OFDM, (2) Bell Laboratories Layered Space–Time Architecture

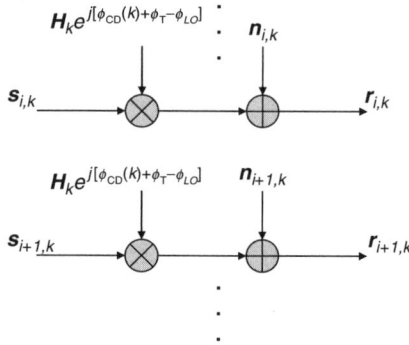

Figure 6.38: Equivalent OFDM channel model.

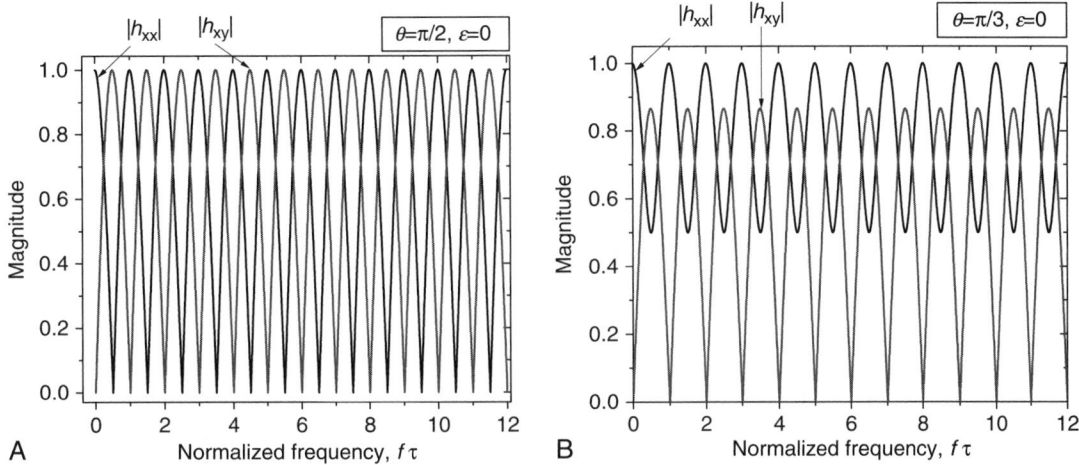

Figure 6.39: Magnitude response of h_{xx} and h_{xy} Jones matrix coefficients against the normalized frequency for (a) $\theta = \pi/2$ and $\epsilon = 0$ and (b) $\theta = \pi/3$ and $\epsilon = 0$.

(BLAST)-type polarization interference cancellation, (3) iterative polarization cancellation, and (4) Alamouti-type polarization-time (PT) coding. The performance assessment of different PMD compensation schemes is also given. This section is based on several publications.[85–87]

6.6.2.1 PMD Compensation by Polarization Diversity Coded OFDM

The polarization diversity receivers are similar to diversity receivers used in wireless communication systems.[82] The transmitter configuration is essentially the same as that in Figure 6.37a, except for the insertion of a polarization beam splitter (PBS) after the DFB laser. The receiver configuration of polarization diversity-coded OFDM is shown in Figure 6.40, whereas the coherent detector was shown in Figure 6.37d.

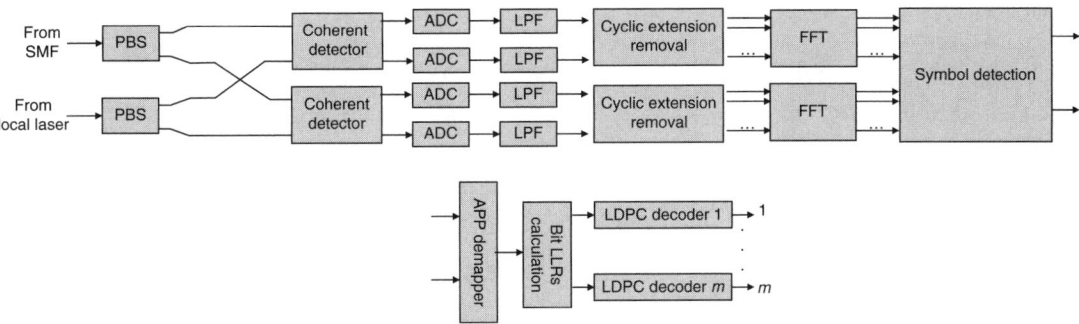

Figure 6.40: The receiver configuration for polarization diversity-coded OFDM with coherent detection. PBS, polarization beam splitter.

The operational principle of OFDM receivers, for x- and y-polarizations, is similar to that of the single OFDM receiver; the only difference is in the symbol detector. Assuming that x-polarization is used on a transmitter side, the transmitted symbol $s_{k,i}$ can be estimated by

$$\tilde{s}_{k,i} = \frac{r_{i,k,x}h_{xx}^*(k) + r_{i,k,y}h_{xy}^*(k)}{|h_{xx}(k)|^2 + |h_{xy}(k)|^2} \tag{6.88}$$

where h_{xx} and h_{xy} are channel coefficients introduced in Eq. (6.86), and $r_{i,k,x}$ and $r_{i,k,y}$ represent the received QAM symbol in the kth subcarrier of the ith OFDM symbol corresponding to x- and y-polarizations, respectively.

The symbol-detector soft estimates of symbols carried by the kth subcarrier in the ith OFDM symbol are forwarded to the APP demapper, which determines the symbol LLRs $\lambda(q)$ ($q = 0$, $1, \ldots, 2^b - 1$) by

$$\lambda(q) = -\frac{\left(\text{Re}\left[\tilde{s}_{i,k}\right] - \text{Re}\left[\text{QAM}(\text{map}(q))\right]\right)^2}{2\sigma^2}$$

$$-\frac{\left(\text{Im}\left[\tilde{s}_{i,k}\right] - \text{Im}\left[\text{QAM}(\text{map}(q))\right]\right)^2}{2\sigma^2}; \quad q = 0, 1, \ldots, 2^b - 1 \tag{6.89}$$

where Re[ ] and Im[ ] denote the real and imaginary parts of a complex number, QAM denotes the QAM constellation diagram, σ^2 denotes the variance of an equivalent Gaussian noise process originating from ASE noise, and map(q) denotes a corresponding mapping rule (Gray mapping is applied here). (b denotes the number of bits per constellation point.) Let us denote by v_j the jth bit in an observed symbol q binary representation $v = (v_1, v_2, \ldots, v_b)$. The bit LLRs needed for LDPC decoding are calculated from symbol LLRs by

$$L\left(\hat{v}_j\right) = \log\frac{\sum_{q:v_j=0}\exp[\lambda(q)]}{\sum_{q:v_j=1}\exp[\lambda(q)]} \tag{6.90}$$

Therefore, the jth bit reliability is calculated as the logarithm of the ratio of a probability that $v_j = 0$ and probability that $v_j = 1$. In the nominator, the summation is done over all symbols q having 0 at the position j, whereas in the denominator the summation is done over all symbols q having 1 at the position j. The extrinsic LLRs are iterated backward and forward until convergence or a predetermined number of iterations is reached. The LDPC code used here belongs to the class of quasi-cyclic codes of large girth ($g \geq 10$) described previously so that the corresponding decoder complexity is low compared to random LDPC codes and does not exhibit the error floor phenomena in the region of interest in fiber-optics communications ($\leq 10^{-15}$).

6.6.2.2 BLAST-Type Polarization Interference Cancellation and Iterative Polarization Cancellation in LDPC-Coded OFDM Fiber-Optics Communication Systems

Here, we discuss two schemes suitable for PMD compensation, which do not require the increase in complexity as DGD increases. The first scheme is based on BLAST,[88] originally proposed to deal with spatial interference in wireless communications. We consider two versions of this scheme[86]: (1) the zero-forcing vertical-BLAST scheme (ZF V-BLAST) and (2) the minimum-mean-square-error vertical-BLAST scheme (MMSE V-BLAST). Because the ZF V-BLAST scheme is derived by ignoring the influence of ASE noise, we proposed the second scheme that uses the output of the ZF V-BLAST scheme as a starting point and removes the remaining polarization interference in an iterative manner. This approach also leads to reducing the influence of ASE noise. We evaluate the performance of these schemes when used in combination with coherent detection-based OFDM. We describe how to use these schemes together with multilevel modulation and FEC. The arbitrary FEC scheme can be used with proposed PMD compensation schemes; however, the use of LDPC codes leads to channel capacity achieving performance.[89]

The polarization interference cancellation scheme based on the V-BLAST algorithm, which uses an LDPC code as a channel code, is shown in Figure 6.41. The 2D signal constellation points, generated in a manner similar to that reported in Section 6.3, are split into two streams for OFDM transmitters (see Figure 6.41a) corresponding to the x- and y-polarizations. The QAM constellation points are considered to be the values of the FFT of a multicarrier OFDM signal. The OFDM symbol is generated similarly as reported previously: N_{QAM} input QAM symbols are zero-padded to obtain N_{FFT} input samples for IFFT, N_G nonzero samples are inserted to create the guard interval, and the OFDM symbol is multiplied by the window function. For efficient chromatic dispersion and PMD compensation, the length of the cyclically extended guard interval should be smaller than the total spread due to chromatic dispersion and DGD. The cyclic extension is accomplished by repeating the last $N_G/2$ samples of the effective OFDM symbol part (N_{FFT} samples) as a prefix and repeating the first $N_G/2$ samples as a suffix. After DAC, the RF OFDM signal is converted into the optical domain using the dual-drive MZM. Two MZMs are needed, one for each polarization. The outputs of MZMs are combined using the polarization beam combiner (PBC). One DFB laser is used as the CW source, with x- and y-polarization separated by a PBS.

The receiver architecture is shown in Figure 6.41b. The configuration of the polarization interference cancellation scheme by the BLAST algorithm is shown in Figure 6.42a. The received symbol vector in the kth subcarrier of the ith OFDM symbol in both polarizations is linearly processed, the processing is described by matrix C_k related to channel matrix H_k as shown later, and the estimate of polarization interference obtained from preliminary decisions $\tilde{s}_{i,k}$, denoted as $D_k\tilde{s}_{i,k}$, is removed from received symbol $r_{i,k}$. The Euclidean

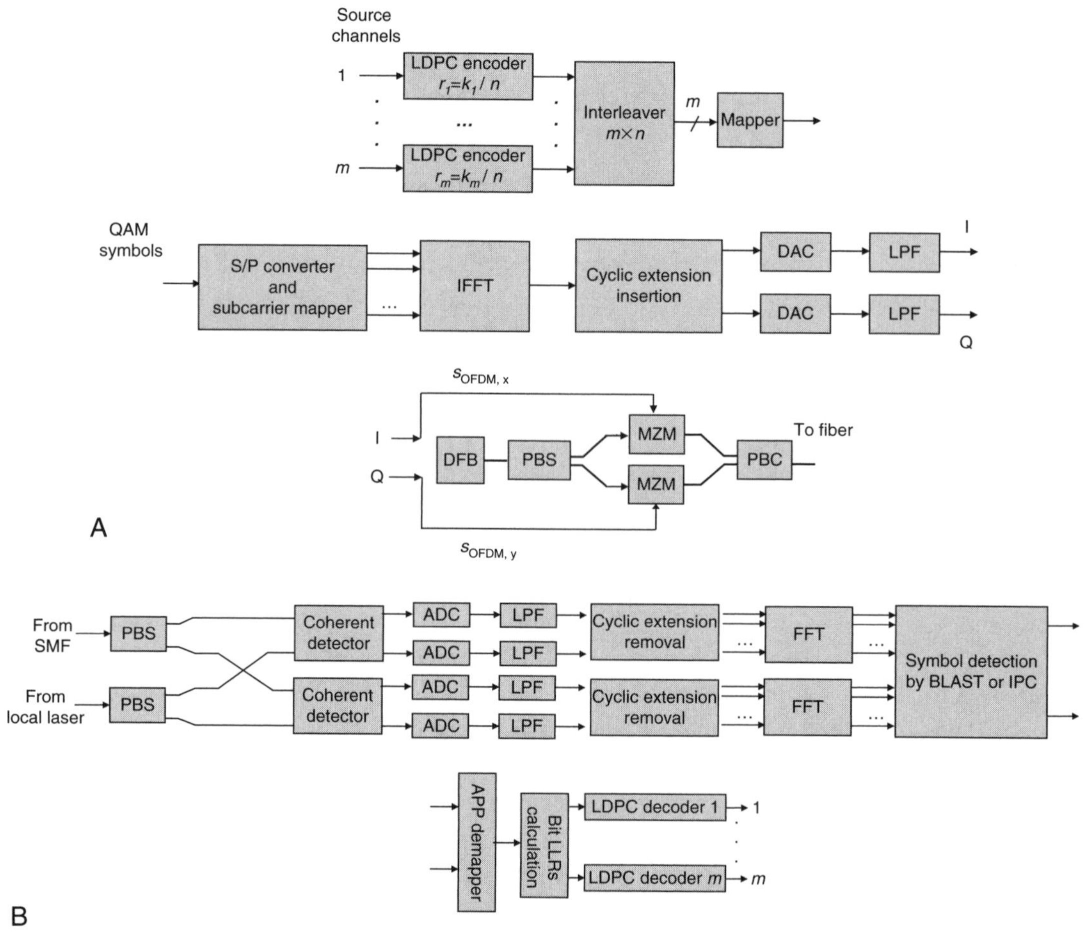

Figure 6.41a (transmitter architecture) labels:

Source channels
1 → LDPC encoder $r_1 = k_1/n$ → Interleaver $m \times n$ → m → Mapper →
m → LDPC encoder $r_m = k_m/n$

QAM symbols → S/P converter and subcarrier mapper → IFFT → Cyclic extension insertion → DAC → LPF → I
→ DAC → LPF → Q

$s_{OFDM, x}$
I → DFB → PBS → MZM → PBC → To fiber
Q → MZM
$s_{OFDM, y}$

A

Figure 6.41b (receiver architecture) labels:

From SMF → PBS → Coherent detector → ADC → LPF → Cyclic extension removal → FFT → Symbol detection by BLAST or IPC
From local laser → PBS → Coherent detector → ADC → LPF → Cyclic extension removal → FFT

APP demapper → Bit LLRs calculation → LDPC decoder 1 → 1
→ LDPC decoder m → m

B

Figure 6.41: The architecture of the polarization interference cancellation scheme in combination with LDPC-coded OFDM: (a) transmitter architecture and (b) receiver architecture. APP, *a posterior* probability; DFB, distributed feedback laser; LLRs, log-likelihood ratios; MZM, dual-drive Mach–Zehnder modulator; PBS(C), polarization beam splitter (combiner).

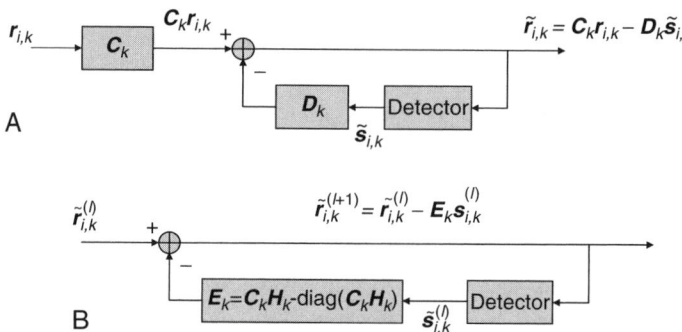

Figure 6.42: The configurations of polarization interference cancellation schemes: (a) BLAST-type polarization interference cancellation scheme and (b) iterative polarization cancellation scheme.

detector can be used to create the preliminary decisions. When the presence of ASE noise is ignored, the zero-forcing V-BLAST polarization interference cancellation scheme results. The matrices $\mathbf{C}_k$ and $\mathbf{D}_k$ can be determined from QR factorization of channel matrix $\mathbf{H}_k = \mathbf{Q}_k\mathbf{R}_k$ as follows:

$$C_k = \text{diag}^{-1}(R_k)Q_k^\dagger, \quad D_k = \text{diag}^{-1}(R_k)R_k - I \tag{6.91}$$

where $\mathbf{I}$ is the identity matrix, and diag( ) denotes the diagonal elements of $\mathbf{R}_k$. Notice that elements at the main diagonal in $\mathbf{D}_k$ are set to zero in order to have only polarization interference be removed. (We use $\dagger$ to denote the simultaneous transposition and complex conjugation.) In the presence of ASE noise, the matrices $\mathbf{C}_k$ and $\mathbf{D}_k$ can be determined by minimizing the MSE, which leads to

$$C_k = \text{diag}^{-1}(S_k)\left(S_k^\dagger\right)^{-1} H_k^\dagger, \quad D_k = \text{diag}^{-1}(S_k)S_k - I \tag{6.92}$$

where $\mathbf{S}_k$ is the upper triangular matrix obtained by Cholesky factorization of

$$H_k^\dagger H_k + I/\text{SNR} = S_k^\dagger S_k \tag{6.93}$$

where SNR denotes the corresponding electrical SNR. The derivation of Eq. (6.91), Eq. (6.92) is equivalent to that for wireless communications[90] and, as such, is omitted here. Because the ZF V-BLAST is derived by ignoring the influence of ASE noise, we propose to use ZF V-BLAST as a starting point and perform the polarization interference cancellation in an iterative manner as shown in Figure 6.42b. If $\tilde{r}_{i,k}^{(1)}$ denotes the processed received symbol of the kth subcarrier in the ith OFDM symbol (for both polarizations) in the lth iteration, then the corresponding received symbol in the $(l + 1)$th iteration can be found by

$$\tilde{r}_{i,k}^{(l+1)} = \tilde{r}_{i,k}^{(1)} - [C_kH_k - \text{diag}(C_kH_k)]\tilde{s}_{i,k}^{(1)} \tag{6.94}$$

where $\tilde{s}_{i,k}^{(1)}$ denotes the transmitted symbol (of the kth subcarrier in the ith OFDM symbol (for both polarizations)) estimate in the lth iteration. The matrices $\mathbf{C}_k$ and $\mathbf{D}_k$ were previously introduced. Notice that different matrix operations applied previously are trivial because the dimensionality of matrices is small, 2×2.

The BLAST-detector soft estimates of symbols carried by the kth subcarrier in the ith OFDM symbol, $S_{i,k,x(y)}$ are forwarded to the APP demapper, which determines the symbol LLRs $\lambda_{x(y)}(q)$ ($q = 0, 1, \ldots, 2^b - 1$) of x− (y−) polarization by

$$
\begin{aligned}
\lambda_{x(y)}(q) = \; & -\frac{\left(\text{Re}\left[\tilde{s}_{i,k,x(y)}\right] - \text{Re}\left[\text{QAM}(\text{map}(q))\right]\right)^2}{2\sigma^2} \\
& -\frac{\left(\text{Im}\left[\tilde{s}_{i,k,x(y)}\right] - \text{Im}\left[\text{QAM}(\text{map}(q))\right]\right)^2}{2\sigma^2}; \quad q = 0,1,\ldots,2^b - 1
\end{aligned}
\tag{6.95}
$$

where Re[ ] and Im[ ] denote the real and imaginary parts of a complex number, QAM denotes the QAM constellation diagram, σ^2 denotes the variance of an equivalent Gaussian noise process originating from ASE noise, and map(q) denotes a corresponding mapping rule (Gray mapping is applied here). (b denotes the number of bits per constellation point.) Let us denote by $v_{j,x(y)}$ the jth bit in an observed symbol q binary representation $\boldsymbol{v} = (v_1, v_2, \ldots, v_b)$ for x$-$ (y$-$) polarization. The bit LLRs needed for LDPC decoding are calculated from symbol LLRs by Eq. (6.90).

6.6.2.3 PMD Compensation by Polarization-Time Coding

In this section, we describe the PT encoder and PT decoder based on the Alamouti-type scheme and describe how to interplay them with coded OFDM. The PT encoder operates as follows[85]: In the first half of ith time instance ("the first channel use"), it sends symbol s_x to be transmitted using x-polarization channel and symbol s_y to be transmitted using y-polarization channel. In the second half of ith time instance ("the second channel use"), it sends symbol $-s_y^*$ to be transmitted using x-polarization channel and symbol s_x^* to be transmitted using y-polarization. Therefore, the PT-coding procedure is similar to the Alamouti scheme.[91] Note that the Alamouti-type PT-coding scheme has the spectral efficiency comparable to coherent OFDM with polarization diversity scheme. When the channel is used twice during the same symbol period, the spectral efficiency of this scheme is twice as high as that of polarization diversity OFDM. Note that the hardware complexity of the PT encoder/decoder is trivial compared to that of OFDM. The transmitter complexity is slightly higher than that reported in Section 6.6.2.1; it requires an additional PT encoder, a PBS, and a PBC. On the receiver side, we have the option to use only one polarization or to use both polarizations. The receiver architecture employing both polarizations is shown in Figure 6.43b.

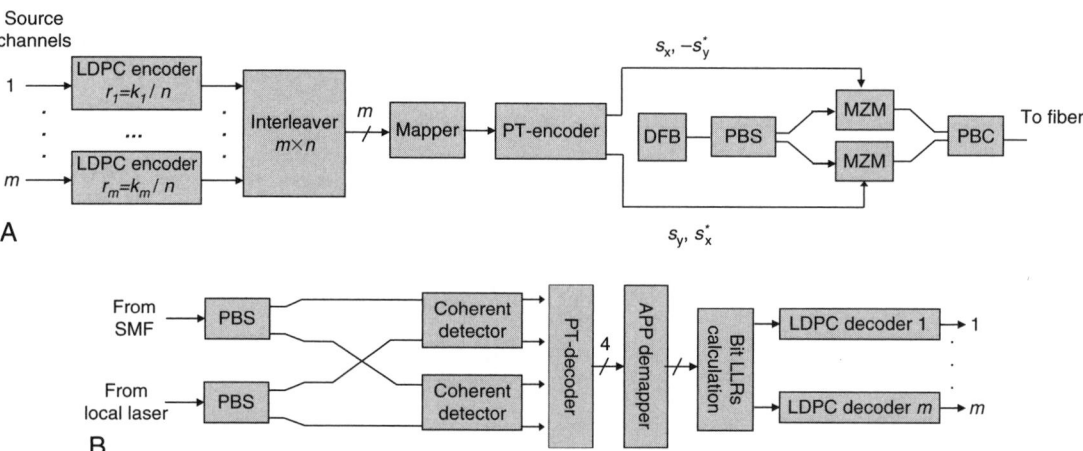

Figure 6.43: The architecture of a PT-coding scheme concatenated with LDPC coding for multilevel modulations: (a) transmitter architecture and (b) receiver architecture.

The operations of all blocks, except the PT decoder, are similar to those reported previously. Hence, we concentrate only on the description of the PT decoder. The received symbol vectors in ith time instance for the first $(r_{i,1})$ and second channel use $(r_{i,2})$ can be written, respectively, as follows:

$$r_{i,1} = Hs_{i,1}e^{j(\phi_T - \phi_{LO}) + j\phi_{CD}(k)} + n_{i,1} \tag{6.96}$$

$$r_{i,2} = Hs_{i,2}e^{j(\phi_T - \phi_{LO}) + j\phi_{CD}(k)} + n_{i,2} \tag{6.97}$$

where the Jones (channel) matrix H, $r_{i,1(2)} = [r_{x,i,1(2)}\ r_{y,i,1(2)}]^T$ denotes the received symbol vector in the first (second) channel use of ith time instance, whereas $n_{i,1(2)} = [n_{x,i,1(2)}\ n_{y,i,1(2)}]^T$ is the ASE noise vector corresponding to the first (second)channel use in ith time instance. We use $s_{i,1} = [s_{x,i}\ s_{y,i}]^T$ to denote the symbol transmitted in the first channel use of ith time instance, and $s_{i,2} = [-s_{y,i}^*\ s_{x,i}^*]^T$ to denote the symbol transmitted in the second channel use (of the same time instance). Because the symbol vectors transmitted in the first and the second channel use of ith time instance are orthogonal, $s_{i,1}^\dagger s_{i,2} = 0$, Eqs. (6.96) and (6.97) can be rewritten by grouping separately x- and y-polarizations as follows:

$$\begin{bmatrix} r_{x,i,1} \\ r_{x,i,2}^* \end{bmatrix} = \begin{bmatrix} h_{xx} & h_{xy} \\ h_{xy}^* & -h_{xx}^* \end{bmatrix} \begin{bmatrix} s_{x,i} \\ s_{y,i} \end{bmatrix} e^{j(\phi_T - \phi_{LO}) + j\phi_{CD}(k)} + \begin{bmatrix} n_{x,i,1} \\ n_{x,i,2}^* \end{bmatrix} \tag{6.98}$$

$$\begin{bmatrix} r_{y,i,1} \\ r_{y,i,2}^* \end{bmatrix} = \begin{bmatrix} h_{yx} & h_{yy} \\ h_{yy}^* & -h_{yx}^* \end{bmatrix} \begin{bmatrix} s_{x,i} \\ s_{y,i} \end{bmatrix} e^{j(\phi_T - \phi_{LO}) + j\phi_{CD}(k)} + \begin{bmatrix} n_{y,i,1} \\ n_{y,i,2}^* \end{bmatrix} \tag{6.99}$$

If only one polarization is to be used, we can solve either Eq. (6.98) or Eq. (6.99). However, the use of only one polarization results in a 3dB penalty with respect to the case when both polarizations are used. Following the derivation similar to that performed by Alamouti,[91] it can be shown that the optimum estimates of transmitted symbols at the output of the PT decoder (for the ASE noise-dominated scenario) can be obtain as follows:

$$\tilde{s}_{x,i} = h_{xx}^* r_{x,1} + h_{xy}r_{x,2}^* + h_{yx}^* r_{y,1} + h_{yy}r_{y,2}^* \tag{6.100}$$

$$\tilde{s}_{y,i} = h_{xy}^* r_{x,1} - h_{xx}r_{x,2}^* + h_{yy}^* r_{y,1} - h_{yx}r_{y,2}^* \tag{6.101}$$

where $\tilde{s}_{x,i}$ and $\tilde{s}_{y,i}$ denote the PT decoder estimates of symbols $s_{x,i}$ and $s_{y,i}$ transmitted in ith time instance. In the case in which only one polarization is to be used, such as x-polarization, then the last two terms in Eqs. (6.100) and (6.101) are to be omitted. The PT decoder estimates are forwarded to the APP demapper, which determines the symbol LLRs in a manner similar to Eq. (6.95). The bit LLRs are calculated from symbol LLRs using Eq. (6.90) and forwarded to the LDPC decoders. The LDPC decoders employ the sum-product-with-correction-term algorithm and provide the extrinsic LLRs to be used in the APP demapper. The extrinsic LLRs are iterated backward and forward until convergence or a predetermined number of iterations is reached.

6.6.2.4 *Evaluation of Different PMD Compensation Schemes*

We turn our attention to the BER performance evaluation of the schemes described in previous sections. The results of simulation for uncoded OFDM for different PMD

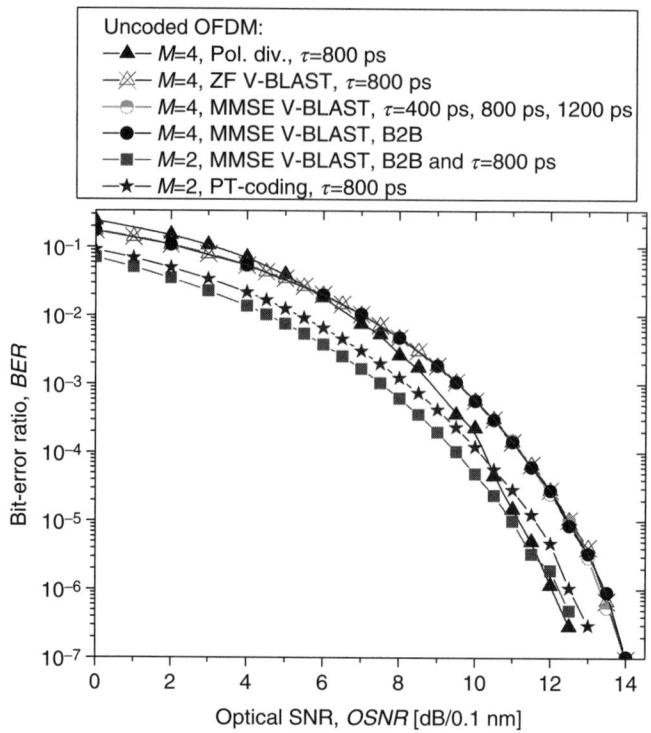

Figure 6.44: BER performance of BLAST algorithm-based PMD compensation schemes against polarization diversity OFDM and PT-coding-based OFDM. B2B, back-to-back.

compensation schemes are shown in Figure 6.44. The OFDM system parameters are chosen as follows: $N_{QAM} = 512$, oversampling is two times, OFDM signal bandwidth is set to 10 GHz, and $N_G = 256$ samples. The MMSE V-BLAST and iterative polarization cancellation schemes (with ZF V-BLAST as the starting point) perform identically (only the MMSE curve is shown because the curves overlap each other), whereas ZF V-BLAST is slightly worse. Polarization diversity OFDM outperforms MMSE V-BLAST at low BERs, but it performs comparable at BERs above 10^{-2}, which is the threshold region of girth-10 LDPC codes employed here. Moreover, the spectral efficiency of MMSE V-BLAST is twice as high because in polarization diversity OFDM only one polarization is used for transmission. The MMSE V-BLAST OFDM outperforms the PT-coding-based OFDM at both low and high BERs, and it has two times higher spectral efficiency. The MMSE V-BLAST OFDM scheme is able to compensate for even 1200 ps of DGD with negligible penalty. Note that for corresponding turbo equalization[92] or maximum-likelihood sequence estimation schemes, the detector complexity grows exponentially as DGD increases, and normalized DGD of 800 ps would require the trellis description[92] with 2^{17} states, which is too high for practical implementation. The schemes described here, although of lower complexity, are able to compensate up to 1200 ps of DGD with negligible penalty. The schemes described above

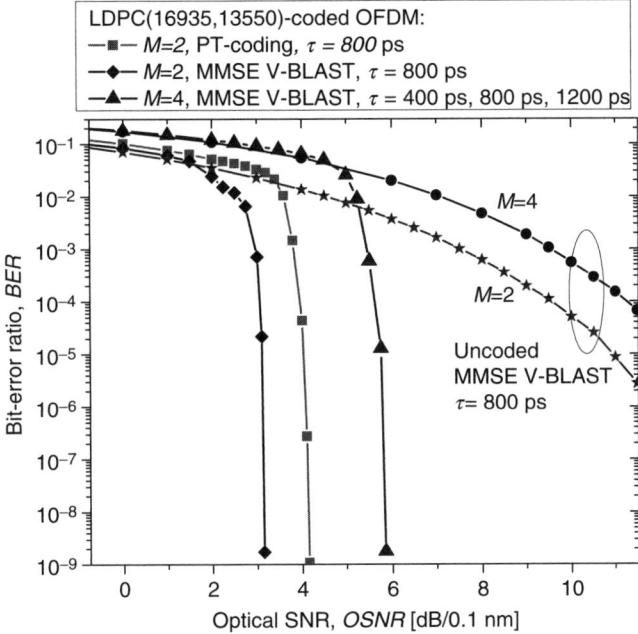

Figure 6.45: BER performance of the polarization interference cancellation schemes for LDPC-coded OFDM.

also outperform the scheme implemented by Nortel Networks researchers,[93] capable of compensating the rapidly varying first-order PMD with peak DGD of 150 ps.

The results of simulations for LDPC-coded OFDM when the MMSE V-BLAST polarization cancellation scheme is used are shown in Figure 6.45. The girth-10 LDPC(16935, 13550) code of rate 0.8 and column weight 3 is used in simulations. This code does not exhibit error floor phenomena for the region of interest in optical communications. At a BER of 10^{-9} the LDPC-coded OFDM with the MMSE V-BLAST polarization interference cancellation scheme outperforms PT-coding-based OFDM by approximately 1 dB and has twice as high spectral efficiency.

6.7 Summary

In this chapter, we described different FEC schemes suitable for use in optical OFDM systems. The schemes were classified into two categories—standard block codes and iteratively decodable codes. Given the fact that convolutional codes are of low rate, we prefer the use of block codes. One may use the puncturing to increase the code at the expense of BER performance degradation. Nevertheless, the convolutional codes are weak unless they are used in concatenation with RS codes. Two classes of standard block codes are cyclic codes and BCH codes. The cyclic codes are suitable for use in error detection, whereas the BCH codes can be used for error correction. RS codes, which are an important subclass of BCH codes, are standardized for use in optical communications.[1] To deal with simultaneous random errors and burst errors, we described the concatenated codes and interleaved codes.

The standard codes described in Section 6.1 belong to the class of hard-decision codes. In Section 6.2, we described more powerful soft iteratively decodable codes. We described several classes of iteratively decodable codes, such as turbo codes, turbo product codes, LDPC codes, GLDPC codes, and nonbinary LDPC codes. Because the BER performance of turbo product codes can be matched and outperformed in terms of both BER performance and decoder complexity, we described LDPC codes in more details, including LDPC code design and LDPC decoding by min-sum-with-correction-term algorithm. GLDPC codes are described as a possible alternative to both LDPC codes and turbo product codes to further decrease the decoding complexity. The nonbinary LDPC codes were described because they can achieve the coding gains comparable to or better than binary LDPC codes but with moderate codeword lengths.

In Section 6.3, we described different *M*-ary modulation formats, such as *M*-ary PSK and *M*-ary QAM, suitable for use in optical OFDM systems to achieve high spectral efficiency. For the best BER performance, coding and modulation should be jointly performed, which was described in Section 6.4.

The coded OFDM was considered in the context of long-haul optical communication systems and 100 Gb/s Ethernet. The LDPC codes are used as channel code because of their channel capacity-achieving capabilities. The OFDM is an excellent chromatic dispersion and PMD compensation candidate. Based on pilot-aided channel estimation, it is capable of compensating for DGD of even 1600 ps for thermal noise limited channel, when the channel state information is known on the receiver side. The similar channel estimation can be used to compensate for the common phase error due to self-phase modulation. However, the ICI due to FWM among subcarriers cannot be eliminated by pilot-aided channel estimation, and it remains an important performance degradation factor to be addressed in the future. Some initial results in this regard have been presented by Due and Lowery.[94–96]

We also discussed the LDPC-coded SSB OFDM as an efficient coded modulation scheme suitable for simultaneous chromatic dispersion and PMD compensation. We showed that residual dispersion over 6500 km of SMF and DGD of 100 ps, in a system with aggregate rate 10 Gb/s, can be simultaneously compensated with a penalty within 1.5 dB (with respect to the back-to-back configuration) when training sequence-based channel estimation and girth-10 LDPC codes of rate 0.8 are employed. The coded turbo equalization is not able to deal with this amount of residual chromatic dispersion and PMD if the complexity of the equalizer is to be kept reasonably low. We have found that in the ASE noise-dominated scenario, for coded OFDM with direct detection, the degradation mostly comes from PMD.

The PMD compensation schemes that can be used instead include: the turbo equalization scheme from Minkov et al.[92] and the FIR filter-based channel equalization scheme from Du and Lowery.[94] These schemes are suitable for use in multilevel ($M \geq 2$) block-coded modulation schemes with coherent detection. However, in contrast to the PMD turbo equalization scheme, whose complexity grows exponentially as DGD increases, the complexity of the schemes described in this chapter remains unchanged as DGD grows, except for the increase in cyclic extension interval. The spectral efficiency of the BLAST-type polarization cancellation scheme or IPC scheme is two times higher than that of polarization diversity OFDM and PT-coding-based OFDM. These two schemes (BLAST and IPC) perform comparable and are able to compensate up to 1200 ps of DGD with negligible penalty. When used in combination with girth-10 LDPC codes, these schemes outperform PT-coding-based OFDM.

References

1. International Telecommunication Union, Telecommunication Standardization Sector, Forward error correction for submarine systems. Tech. Recommendation G.975/G709.
2. Sab OA. FEC techniques in submarine transmission systems. *Proc Opt Fiber Commun Conf* 2001;**2**: TuF1-1–TuF1-3.
3. Pyndiah RM. Near optimum decoding of product codes. *IEEE Trans Commun* 1998;**46**:1003–1010.
4. Sab OA, Lemarie V. Block turbo code performances for long-haul DWDM optical transmission systems. *Proc Opt Fiber Commun Conf* 2001;**3**:280–2.
5. Mizuochi T, et al. Forward error correction based on block turbo code with 3-bit soft decision for 10 Gb/s optical communication systems. *IEEE J Selected Topics Quantum Electronics* 2004;**10**(2):376–86.
6. Mizuochi T, et al. Next generation FEC for optical transmission systems. *Proc Opt Fiber Commun Conf* 2003;**2**:527–8.
7. Djordjevic IB, Milenkovic O, Vasic B. Generalized low-density parity-check codes for optical communication systems. *J Lightwave Technol* 2005;**23**:1939–46.
8. Vasic B, Djordjevic IB, Kostuk R. Low-density parity check codes and iterative decoding for long haul optical communication systems. *J Lightwave Technol* 2003;**21**:438–46.
9. Djordjevic IB, et al. Projective plane iteratively decodable block codes for WDM high-speed long-haul transmission systems. *J Lightwave Technol* 2004;**22**:695–702.
10. Milenkovic O, Djordjevic IB, Vasic B. Block-circulant low-density parity-check codes for optical communication systems. *J Selected Topics Quantum Electronics* 2004;**10**:294–9.
11. Vasic B, Djordjevic IB. Low-density parity check codes for long haul optical communications systems. *IEEE Photon Technol Lett* 2002;**14**:1208–10.
12. Chung S, et al. On the design of low-density parity-check codes within 0.0045 db of the Shannon limit. *IEEE Commun Lett* 2001;**5**:58–60.
13. Davey MC, MacKay DJC. Low-density parity check codes over GF(q). *IEEE Commun Lett* 1998;**2**:165–7.
14. Wymeersch H, et al. Log-domain decoding of LDPC codes over GF(q). *Proc IEEE Int Con Commun* 2004;**2**:772–6.
15. Djordjevic IB, Vasic B. Nonbinary LDPC codes for optical communication systems. *IEEE Photon Technol Lett* 2005;**17**:2224–6.

16. Xiao-Yu H, et al. Efficient implementations of the sum-product algorithm for decoding of LDPC codes. *Proc IEEE Globecom* 2001;**2**:1036–1036E.

17. Chase D. A class of algorithms for decoding of block codes with channel measurements information. *IEEE Trans Inform Theory* 1972;**IT-18**:170–9.

18. Hirst SA, Honary B, Markarian G. Fast Chase algorithm with an application in turbo decoding. *IEEE Trans Commun* 2001;**49**:1693–9.

19. Ryan WE. Concatenated convolutional codes and iterative decoding. In: Proakis JG, editor. *Wiley Encyclopedia in Telecommunications*. New York: Wiley; 2003.

20. Montorsi G, Benedetto S. Design of fixed-point iterative decoders for concatenated codes with interleavers. *IEEE J Selected Areas Commun* 2001;**19**:871–82.

21. Tanner RM. A recursive approach to low complexity codes. *IEEE Trans Information Theory* 1981; **IT-27**:533–47.

22. Lin S, Costello DJ. *Error Control Coding: Fundamentals and Applications*. Englewood Cliffs, NJ: Prentice Hall; 1983.

23. Boutros J, Pothier O, Zemor G. Generalized low density (Tanner) codes. *Proc 1999 IEEE Int Conf Commun* 1999;**1**:441–5.

24. Lentmaier M, Zigangirov KSh. On generalized low-density parity-check codes based on Hamming component codes. *IEEE Commun Lett* 1999;**3**:248–50.

25. Zhang T, Parhi KK. A class of efficient-encoding generalized low-density parity-check codes. *Proc IEEE Int Conf Acoustics Speech Signal Processing* 2001;**4**:2477–80.

26. Vasic B, Rao V, Djordjevic IB, Kostuk R, Gabitov I. Ghost pulse reduction in 40 Gb/s systems using line coding. *IEEE Photon Technol Lett* 2004;**16**:1784–6.

27. Djordjevic IB, Chilappagari SK, Vasic B. Suppression of intrachannel nonlinear effects using pseudoternary constrained codes. *J Lightwave Technol* 2006;**24**(2):769–74.

28. Elias P. Error-free coding. *IRE Trans Inform Theory* 1954;**IT-4**:29–37.

29. Bahl LR, Cocke J, Jelinek F, Raviv J. Optimal decoding of linear codes for minimizing symbol error rate. *IEEE Trans Inform Theory* 1974;**IT-20**(2):284–7.

30. Djordjevic IB, Minkov LL, Batshon HG. Mitigation of linear and nonlinear impairments in high-speed optical networks by using LDPC-coded turbo equalization. *IEEE J Sel Areas Comm Optical Comm Netw* 2008;**26**(6):73–83.

31. Ingels FM. *Information and Coding Theory*. Scranton, PA: Intext Educational Publishers; 1971.

32. Haykin S. *Communication Systems*. New York: Wiley; 2004.

33. Cover TM, Thomas JA. *Elements of Information Theory*. New York: Wiley; 1991.

34. Anderson JB, Mohan S. *Source and Channel Coding: An Algorithmic Approach*. Boston: Kluwer; 1991.

35. MacWilliams FJ, Sloane NJA. *The Theory of Error-Correcting Codes*. Amsterdam, North Holland; 1977.

36. Wicker SB. *Error Control Systems for Digital Communication and Storage*. Englewood Cliffs, NJ: Prentice Hall; 1995.

37. Forney Jr GD. *Concatenated Codes*. Cambridge, MA: MIT Press; 1966.

38. Bahl LR, Cocke J, Jelinek F, Raviv J. Optimal decoding of linear codes for minimizing symbol error rate. *IEEE Trans Inform Theory* 1974;**IT-20**(2):284–7.

39. Berrou C, Glavieux A, Thitimajshima P. Near Shannon limit error-correcting coding and decoding: Turbo codes. *Proc 1993 Int Conf Commun*; 1993. pp. 1064–70.

40. Berrou C, Glavieux A. Near optimum error correcting coding and decoding: turbo codes. *IEEE Trans Commun* 1996;**44**:1261–71.

41. Drajic DB. *An Introduction to Information Theory and Coding*. 2nd ed. Belgrade, Serbia: Akademska misao; 2004 [In Serbian].

42. Ryan WE. Concatenated convolutional codes and iterative decoding. In: Proakis JG, editor. *Wiley Encyclopedia in Telecommunications*. New York: Wiley; 2003.

43. Ryan WE. An introduction to LDPC codes. In: Vasic B, editor. *CRC Handbook for Coding and Signal Processing for Recording Systems*. Boca Raton, FL: CRC Press; 2004.

44. Proakis JG. *Digital Communications*. New York: McGraw-Hill; 2001.

45. Morelos-Zaragoza RH. *The Art of Error Correcting Coding*. Chichester, UK: Wiley; 2002.

46. Mizuochi T. Recent progress in forward error correction and its interplay with transmission impairments. *IEEE J Sel Topics Quantum Electron* 2006;**12**(4):544–54.

47. Gallager RG. *Low Density Parity Check Codes*. Cambridge, MA: MIT Press; 1963.

48. Mansour M. Implementation of LDPC decoders. In: *IEEE Comm. Theory Workshop*. Park City, UT; 2005.

49. Richardson T. Error floors of LDPC codes. *Proc 41st Allerton Conf Commun Control Computing*; 2003.

50. Chernyak V. *Local theory of BER for LDPC codes: Instantons on a tree*. Paper presented at the LANL Workshop on Applications of Statistical Physics to Coding Theory, Santa Fe, New Mexico, Jan. 10–12; 2005. Available online at http://cnls.lanl.gov/~chertkov/FEC_agenda.htm.

51. Chilappagari SK, Sankaranarayanan S, Vasic B. Error floors of LDPC codes on the binary symmetric channel. *Proc IEEE Int Conf Commun*; 2006:1089–94.

52. Vasic B, Chilappagari SK, Sankaranarayanan S. *Error floors of LDPC codes on binary symmetric channel*. Paper presented at the IEEE Communication Theory Workshop, Park City, UT; 2005.

53. Wymeersch H, Steendam H, Moeneclay M. Computational complexity and quantization effects of decoding algorithms for non-binary LDPC Codes. *Proc ICASSP* 2004;**4**:iv-669–iv-672.

54. Agrawal GP. *Nonlinear Fiber Optics*. San Diego: Academic Press; 2001.

55. Sankaranarayanan S, Djordjevic IB, Vasic B. Iteratively decodable codes on m-flats for WDM high-speed long-haul transmission. *J Lightwave Technol* 2005;**23**(11):3696–701.

56. Djordjevic IB, Vasic B. Multilevel coding in M-ary DPSK/differential QAM high-speed optical transmission with direct detection. *J Lightwave Technol* 2006;**24**:420–8.

57. Richardson TJ, Shokrollahi MA, Urbanke RL. Design of the capacity-approaching irregular low-density parity-check codes. *IEEE Trans Inform Theory* 2001;**47**(2):619–37.

58. Fossorier MPC. Quasi-cyclic low-density parity-check codes from circulant permutation matrices. *IEEE Trans Inform Theory* 2004;**50**:1788–94.

59. Fan JL. Array codes as low-density parity-check codes. *Proc 2nd Int Symp. Turbo Codes Related Topics*. Brest, France; 2000. pp. 543–6.

60. MacKay DJC. Good error correcting codes based on very sparse matrices. *IEEE Trans Inform Theory* 1999;**45**:399–431.

61. Anderson I. *Combinatorial Designs and Tournaments*. Oxford, UK: Oxford University Press; 1997.

62. Lowery A. *Nonlinearity and its compensation in optical OFDM systems*. Paper presented at ECOC 2007 Workshop 5 (Electronic signal processing for transmission impairment mitigation: future challenges).

63. Djordjevic IB. Generalized LDPC codes and turbo-product codes with Reed–Muller component codes. *Proc 8th Int Conf Telecommun Modern Satellite Cable Broadcasting Services*; 2007. pp. 127–34 (Invited paper).

64. Ashikhmin A, Lytsin S. Simple MAP decoding of first-order Reed–Muller and Hamming codes. *IEEE Trans Inform Theory* 2004;**50**:1812–8.

65. Zhang T, Parhi KK. A class of efficient-encoding generalized low-density parity-check codes. *Proc ICASSP '01* 2001;**4**:2477–80.

66. Arabaci M, Djordjevic IB. An alternative FPGA implementation of decoders for quasi-cyclic LDPC codes. *Proc TELFOR* 2008;351–4.

67. Mitrionics, Inc. *Mitrion Users Guide*, v1.5.0-001. 2008.

68. Arabaci M, Djordjevic IB, Saunders R, Marcoccia R. A class of non-binary regular girth-8 LDPC codes for optical communication channels. In: *Proc. OFC/NFOEC*, paper no. JThA38. San Diego; 2009.

69. Chen J, Dholakia A, Eleftheriou E, Fossorier M, Hu X-Y. Reduced-complexity decoding of LDPC codes. *IEEE Trans Commun* 2005;**53**:1288–99.

70. Davey MC. *Error-correction using low-density parity-check codes*. PhD thesis, University of Cambridge; 1999.

71. Declercq D, Fossorier M. Decoding algorithms for nonbinary LDPC codes over GF(q). *IEEE Trans Commun* 2007;**55**:633–43.

72. Spagnol C, Marnane W, Popovici E. FPGA implementations of LDPC over GF(2^m) decoders. In: *IEEE Workshop Signal Proc. Sys.* Shanghai, China; 2007. pp. 273–8.

73. Lan L, Zeng L, Tai YY, et al. Construction of quasi-cyclic LDPC codes for AWGN and binary erasure channels: A finite field approach. *IEEE Trans Inform Theory* 2007;**53**:2429–58.

74. Djordjevic IB, Xu L, Wang T, Cvijetic M. Large girth low-density parity-check codes for long-haul high-speed optical communications. *Proc OFC/NFOEC*, paper no. JWA53. San Diego; 2008.

75. Djordjevic IB, Cvijetic M, Xu L, Wang T. Using LDPC-coded modulation and coherent detection for ultra high-speed optical transmission. *J Lightwave Technol* 2007;**25**:3619–25.

76. Djordjevic IB, Vasic B, Neifeld MA. LDPC coded orthogonal frequency division multiplexing over the atmospheric turbulence channel. In: *Proc. CLEO/QELS 2006*, paper no. CMDD5, 2006.

77. Djordjevic IB, Vasic B, Neifeld MA. LDPC coded OFDM over the atmospheric turbulence channel. *Opt Express* 2007;**15**(10):6332–46.

78. Djordjevic IB, Vasic B. Orthogonal frequency-division multiplexing for high-speed optical transmission. *Opt Express* 2006;**14**:3767–75.

79. Agrawal GP. *Nonlinear Fiber Optics*. San Diego: Academic Press; 2001.

80. Djordjevic IB. PMD compensation in fiber-optic communication systems with direct detection using LDPC-coded OFDM. *Opt Express* 2007;**15**:3692–701.

81. Djordjevic IB, Vasic B. 100 Gb/s transmission using orthogonal frequency-division multiplexing. *IEEE Photon Technol Lett* 2006;**18**(15):1576–8.

82. Goldsmith A. *Wireless Communications*. Cambridge, UK: Cambridge University Press; 2005.

83. Cvijetic M. *Coherent and Nonlinear Lightwave Communications*. Boston: Artech House; 1996.

84. Penninckx D, Morenás V. Jones matrix of polarization mode dispersion. *Opt Lett* 1999;**24**:875–7.

85. Djordjevic IB, Xu L, Wang T. PMD compensation in coded-modulation schemes with coherent detection using Alamouti-type polarization-time coding. *Opt Express* 2008;**16**(18):14163–72.

86. Djordjevic IB, Xu L, Wang T. PMD compensation in multilevel coded-modulation schemes with coherent detection using BLAST algorithm and iterative polarization cancellation. *Opt Express* 2008;**16**(19):14845–52.

87. Djordjevic IB, Xu L, Wang T. Beyond 100 Gb/s optical transmission based on polarization multiplexed coded-OFDM with coherent detection. *IEEE/OSA J. Opt. Commun. Netw.*, vol. 1, no. 1, pp. 50–56, June 2009.

88. Foschini GJ. Layered space–time architecture for wireless communication in a fading environment when using multi-element antennas. *Bell Labs Tech J* 1996;**1**:41–59.

89. Djordjevic IB, Denic S, Anguita J, Vasic B, Neifeld MA. LDPC-coded MIMO optical communication over the atmospheric turbulence channel. *J Lightwave Technol* 2008;**26**:478–87.

90. Biglieri E, Calderbank R, Constantinides A, et al. *MIMO Wireless Communications*. Cambridge, UK: Cambridge University Press; 2007.

91. Alamouti S. A simple transmit diversity technique for wireless communications. *IEEE J Sel Areas Commun* 1998;**16**:1451–8.

92. Minkov LL, Djordjevic IB, Batshon HG, et al. Demonstration of PMD compensation by LDPC-coded turbo equalization and channel capacity loss characterization due to PMD and quantization. *IEEE Photon Technol Lett* 2007;**19**:1852–4.

93. Sun H, Wu K-T, Roberts K. Real-time measurements of a 40 Gb/s coherent system. *Opt Express* 2008;**16**:873–9.

94. Du LB, Lowery AJ. Fiber nonlinearity compensation for CO-OFDM systems with periodic dispersion maps. In: *Proc. Optical Fiber Telecommunications*, paper no. OTuO1, San Diego; 2009.

95. Du LB, Lowery AJ. Improved nonlinearity precompensation for long-haul high-data-rate transmission using coherent optical OFDM. *Opt Express* 2008;**16**(24):19920–5.

96. Du LB, Lowery AJ. Improving nonlinear precompensation in direct-detection optical OFDM communications systems. In: *Proc. European Conference on Optical Communications*, paper no. P. 4.08. Brussels; 2008.

Uncited References

97. Gnauck AH, Wintzer PJ. Optical phase-shift keyed transmission. *J Lightwave Technol* 2005;**23**:115–30.

98. International Telecommunication Union. Telecommunication Standardization Sector, "Forward error correction for high bit rate DWDM submarine systems. *Tech Recommendation G* 2004;**975**:1.

99. Xiao-Yu H, Eleftheriou E, Arnold D-M, Dholakia A. Efficient implementations of the sum-product algorithm for decoding of LDPC codes. *Proc IEEE Globecom* 2001;**2**:1036–1036E.

100. Ghavami M, Michael LB, Kohno R. *Ultra Wideband Signals and Systems in Communication Engineering*. West Sussex, UK: Wiley; 2007.

101. Prasad R. *OFDM for Wireless Communications Systems*. Boston: Artech House; 2004.

102. Wu Y, Caron B. Digital television terrestrial broadcasting. *IEEE Commun Mag* 1994;**32**(5):46–52.

103. Kim A, Hun Joo Y, Kim Y. 60 ghz wireless communication systems with radio-over fiber links for indoor wireless lans. *IEEE Trans Consum Electron* 2004;**50**(2):517–20.

104. Shieh W, Athaudage C. Coherent optical frequency division multiplexing. *Electron Lett* 2006;**42**:587–9.

105. Lowery AJ, Du L, Armstrong J. Orthogonal frequency division multiplexing for adaptive dispersion compensation in long haul WDM systems. In: *Proc. OFC Postdeadline Papers*, paper no. PDP39. 2006.

106. Lowery AJ, Armstrong J. 10 Gb/s multimode fiber link using power-efficient orthogonal-frequency-division multiplexing. *Opt Express* 2005;**13**(25):10003–9.

107. Djordjevic IB. LDPC-coded OFDM transmission over graded-index plastic optical fiber links. *IEEE Photon Technol Lett* 2007;**19**(12):871–3.

108. Shieh W. PMD-supported coherent optical OFDM systems. *IEEE Photon Technol Lett* 2006;**19**:134–6.

109. Jansen SL, Morita I, Takeda N, Tanaka H. 20-Gb/s OFDM transmission over 4160-km SSMF enabled by RF-pilot tone phase compensation. In: *Proc. OFC/NFOEC 2007 Postdeadline Papers*, paper no. PDP15, Anaheim, CA; 2007.

110. Schmidt BJ, Lawery AJ, Amstrong J. Experimental demonstration of 20 Gbit/s direct-detection optical OFDM and 12 Gbit/s with a colorless transmitter. In: *Proc. OFC/NFOEC 2007 Postdeadline Papers*, paper no. PDP18. Anaheim, CA; 2007.

111. Djordjevic IB, Batshon HG, Cvijetic M, Xu L, Wang T. PMD compensation by LDPC-coded turbo equalization. *IEEE Photon Technol Lett* 2007;**19**(15):1163–5.

112. Djordjevic IB, Vasic B. Constrained coding techniques for the suppression of intrachannel nonlinear effects in high-speed optical transmission. *IEEE J Lightwave Technol* 2006;**24**:411–9.

113. Zapata A, Düser M, Spencer J, et al. Next-generation 100-gigabit metro Ethernet (100 gbme) using multiwavelength optical rings. *J Lightwave Technol* 2004;**22**:2420–34.

114. Duelk M. Next generation 100 G Ethernet. In: *Proc. ECOC 2005*, paper no. Tu3.1.2 Glasgow, Scotland; 2005.

115. Raybon G, Winzer PJ, Doerr CR. 10 × 107-Gb/s electronically multiplexed and optically equalized NRZ transmission over 400 km. In: *Proc. OFC Postdeadline Papers*, paper no. PDP32, 2006.

116. Lin C-C, Lin K-L, Chang H-Ch, Lee C-Y. A 3.33 Gb/s (1200,720) low-density parity check code decoder. In: *Proc. ESSCIRC.* Grenoble, France; 2005. p. 211–4.

117. Djordjevic IB, Sankaranarayanan S, Chilappagari SK, Vasic B. Low-density parity-check codes for 40 Gb/s optical transmission systems. *IEEE J Select Topics Quantum Electron* 2006;**12**(4):555–62.

118. Djordjevic IB, Vasic B. Macneish–Mann theorem based iteratively decodable codes for optical communication systems. *IEEE Commun Lett* 2004;**8**:538–40.

119. Djordjevic IB, Xu L, Wang T, Cvijetic M. GLDPC codes with Reed–Muller component codes suitable for optical communications. *IEEE Commun Lett* 2008;**12**:684–6.

120. Shieh W, Yi X, Ma Y, Tang Y. Theoretical and experimental study on PMD-supported transmission using polarization diversity in coherent optical OFDM systems. *Opt Express* 2007;**15**:9936–47.

CHAPTER 7

Various Types of Optical OFDM

7.1 Introduction

One of the major strengths of the orthogonal frequency-division multiplexing (OFDM) modulation format is its rich variation and ease of adaption to a wide range of applications. In wireless systems, OFDM has been incorporated in wireless local area network (LAN; IEEE 802.11a/g, better known as Wi-Fi), wireless WAN (IEEE 802.16e, better known as WiMAX), and digital radio/video systems (DAB/DVB) adopted in most areas of the world. In RF cable systems, OFDM has been incorporated in ADSL and VDSL broadband access via telephone copper wiring or power line. This rich variation results from the intrinsic advantages of OFDM modulation, including dispersion robustness, ease of dynamic channel estimation and mitigation, high spectral efficiency, and the capability of dynamic bit and power loading. Progress in optical OFDM has also benefited from this wide variation. We have witnessed many novel proposals and demonstrations of optical OFDM systems for different applications that aim to benefit from the aforementioned OFDM advantages. Despite the fact that OFDM has been extensively studied in the RF domain, it is rather surprising that the first report on optical OFDM in the open literature appeared only in 1996; Pan and Green[1] presented an in-depth performance analysis of hybrid AM/OFDM subcarrier multiplexed (SCM) fiber-optic systems. The lack of interest in optical OFDM in the past is largely due to the fact that the silicon signal processing power had not reached the point at which sophisticated OFDM signal processing could be performed in a CMOS integrated circuit (IC).

In this chapter, we classify the optical OFDM variations according to their underlying techniques and applications. Sections 7.2 and 7.3 provide an analysis of fundamentals of optical OFDM using coherent detection and direct detection, respectively. Although direct detection has been the mainstay of optical communications during the past two decades, progress in forward-looking research has unmistakably indicated that the future of optical communications is coherent detection. Section 7.4 briefly illustrates various applications for these two detection systems.

7.2 Coherent Optical OFDM

Coherent optical OFDM (CO-OFDM) represents the ultimate performance in receiver sensitivity, spectral efficiency, and robustness against polarization dispersion, but it requires the highest complexity in transceiver design. In the open literature, CO-OFDM was first proposed by Shieh and Athaudage,[2] and the concept of the coherent optical multiple-input multiple-output OFDM was formalized by Shieh et al.[3] The early CO-OFDM experiments were carried out by Shieh et al.[4] for a 1000 km standard single-mode fiber (SSMF) transmission at 8 Gb/s and by Jansen et al.[5] for 4160 km SSMF transmission at 20 Gb/s. Another interesting and important development is the proposal and demonstration of the no-guard interval CO-OFDM by Yamada et al.[6] in which optical OFDM is constructed using optical subcarriers without a need for the cyclic prefix. Nevertheless, the fundamental principle of CO-OFDM remains the same: to achieve high spectral efficiency by overlapping subcarrier spectrum while avoiding interference by using coherent detection and signal set orthogonality. In this section, we describe the detailed principle of CO-OFDM, including analysis of transmitter and receiver design.

7.2.1 Principle of CO-OFDM

The synergies between coherent optical communications and OFDM are twofold. OFDM brings coherent systems computation efficiency and ease of channel and phase estimation. The coherent systems bring OFDM a much needed linearity in RF-to-optical (RTO) up-conversion and optical-to-RF (OTR) down-conversion. Consequently, a linear transformation is the key goal for the OFDM implementation. A generic optical OFDM system as depicted in Figure 7.1 can be divided into five functional blocks: (1) RF OFDM transmitter, (2) RTO up-converter,

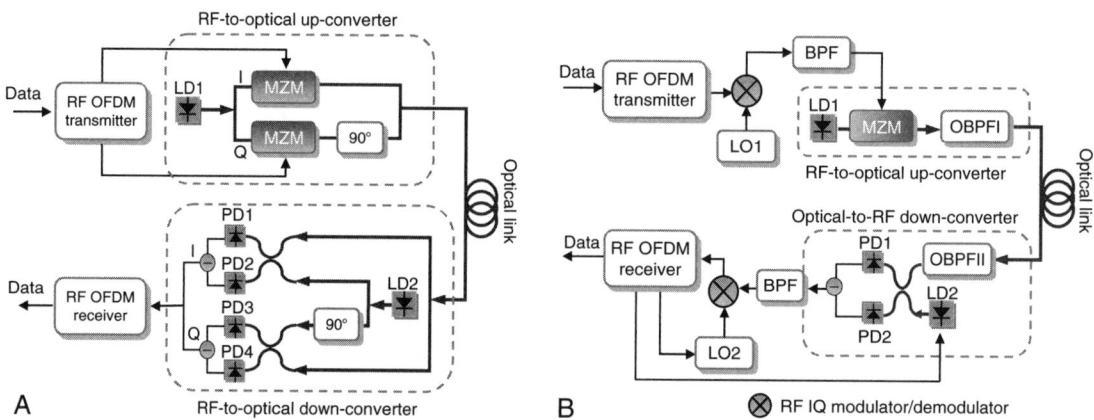

Figure 7.1: A CO-OFDM system in (a) direct up-/down-conversion architecture and (b) intermediate frequency architecture.

(3) optical channel, (4) OTR down-converter, and (5) RF OFDM receiver. The detailed architecture for the RF OFDM transmitter/receiver was shown in Figure 2.2, which generates/ recovers OFDM signals in either baseband or RF band. Let us assume a linear optical fiber channel where the fiber nonlinearity is not considered. It is apparent that the challenges for CO-OFDM implementation are to obtain a linear RTO up-converter and linear OTR down-converter. It has been proposed and analyzed that by biasing the Mach–Zehnder modulators (MZMs) at null point, a linear conversion between the RF signal and the optical field signal can be achieved.[2,7] We will further treat this important issue of (non)linearity of the RTO up-converter in Section 7.2.4. It has also been shown that by using coherent detection, a linear transformation from optical field signal to RF (or electrical) signal can be achieved,[2,7–9] which is discussed in Section 7.2.6. By creating such a composite system across the RF and optical domains,[2,4,5] a linear channel can be constructed in which OFDM can perform its best (i.e., mitigating channel dispersion impairment) in both the RF domain and the optical domain. In this chapter, we use the terms *RF domain* and *electrical domain* interchangeably.

7.2.2 Optical Transmitter Design for CO-OFDM

As discussed in Section 7.2.1, the primary design goal for CO-OFDM is to construct a linear transformation system. Consequently, investigation of the nonlinearity of a CO-OFDM transmitter is of vital importance.[7,10,11] The MZM characteristic has been extensively investigated by Auracher and Keil.[12] The nonlinearity of MZM on the system performance in the direct detected systems has been thoroughly studied.[13,14] It is shown that for conventional direct detected systems the optimal bias point is at quadrature and a large signal will add nonlinearity impairment to the systems. Here, we focus on the nonlinearity from the OFDM RTO up-converter. We first present the architecture choices for a generic CO-OFDM system. We then formulate the two-tone nonlinearity representation for the RTO up-converter that performs the linear transformation from electrical drive voltage to the optical field.

7.2.3 Up-/Down-Conversion Design Options for CO-OFDM Systems

Figures 7.1a and 7.1b show, respectively, a CO-OFDM system that uses direct up-/down-conversion architecture and intermediate frequency (IF) architecture. In the direct up-conversion architecture in Figure 7.1a, the optical transmitter uses an optical I/Q modulator that comprises two MZMs to up-convert the real/imaginary parts of the complex OFDM signal $s(t)$ in Eq. (2.1) from the RF domain to the optical domain; that is, each MZM is respectively driven by the real or the imaginary part of the $s(t)$. In the direct down-conversion architecture, the OFDM optical receiver uses two pairs of balanced receivers and an optical 90-degree hybrid to perform optical I/Q detection. The RF OFDM receiver performs OFDM baseband processing to recover the data. The advantages of such a direct-conversion architecture are (1) elimination of a need for an image rejection filter in both transmitter and receiver and (2) significant reduction of the required electrical bandwidth for both transmitter and receiver. In the IF up-conversion architecture in Figure 7.1b, the OFDM baseband signal is first up-converted to an intermediate frequency f_{LO1} in the electrical domain and then the OFDM IF signal is further up-converted to the optical domain with one MZM. In the IF down-conversion system, the optical OFDM signal is first down-converted to an intermediate frequency f_{LO2} and then the electrical I/Q detection is performed.

Because the transmitter can be of either direct or IF up-conversion architecture, and the receiver can be of either direct or IF down-conversion architecture, there are four design choices for a CO-OFDM system (only two are shown in Figure 7.1). Furthermore, at the receive end, direct down-conversion is synonymous with homodyne detection, and IF down-conversion is synonymous with heterodyne detection. However, the terms *direct conversion* and *IF conversion* are used in this chapter to encompass the architectures at both transmit and receive ends.

7.2.4 Optical I/Q Modulator for Linear RF-to-optical up Conversion

Two-tone analysis is widely used to characterize RF component and system nonlinearity. In this section, we use similar two-tone intermodulation analysis to study the optical I/Q modulator nonlinearity on the RF-to-optical conversion for a CO-OFDM transmitter. The result obtained here should apply to coherent analog multicarrier systems. For the direct up-/down-conversion architecture (Figure 7.1a), two complex subcarrier tones at $v_1 = v \cdot e^{j\omega_1 t}$ and $v_2 = v \cdot e^{j\omega_2 t}$ are applied to the input of the optical I/Q modulator. For the sake of simplicity, we only show the nonlinearity performance analysis for direct up-/down-conversion architecture. The optical signal at the output of the optical IQ modulator is

$$E(t) = A \cdot \cos\left(\frac{\pi}{2} \cdot \frac{V_I + V_{DC}}{V_\pi}\right) \cdot \exp(j\omega_{LD1}t + j\phi_{LD1})$$

$$+ A \cdot \cos\left(\frac{\pi}{2} \cdot \frac{V_Q + V_{DC}}{V_\pi}\right) \exp(j\omega_{LD1}t + \pi/2 + j\phi_{LD1})$$

(7.1)

where A is a proportionality constant. All the proportionality constants will be omitted in the remainder of the chapter. The optical field $E(t)$ is written in a complex form—a common practice in electromagnetism[18]—but the true electrical field should be considered as the real part of $E(t)$. V_I and V_Q are real and imaginary parts of the complex RF drive signal to each MZM, expressed as $V_I = v \cdot (\cos\omega_1 t + \cos\omega_2 t)$, $V_Q = v \cdot (\sin\omega_1 t + \sin\omega_2 t)$; V_{DC} is the DC bias voltage of the modulator; V_π is the half-wave switching voltage; and ω_{LD1}/ϕ_{LD1} is the frequency/phase for the transmitter laser. After simple rearrangement, the signal at the output of the optical IQ modulator becomes

$$E(t) = \exp(j\omega_{\text{LD1}}t + j\phi_{\text{LD1}}) \cdot E^{\text{B}}(t) \tag{7.2}$$

$$E^{\text{B}}(t) = \cos\left[\frac{M}{2}(\cos\omega_1 t + \cos\omega_2 t) + \frac{\phi}{2}\right] + j\cos\left[\frac{M}{2}(\sin\omega_1 t + \sin\omega_2 t) + \frac{\phi}{2}\right] \tag{7.3}$$

where $M = v\pi/V_\pi$ is defined as the modulation index, $\phi = V_{\text{DC}}\pi/V_\pi$ is a static phase shift (bias point), and $E^{\text{B}}(t)$ is the equivalent baseband (or frequency down-converted) version of the optical field $E(t)$. The so-defined modulation index M equals the optical modulation index of the optical signal if the optical modulator is biased at quadrature in a direct detected system. The modulation index M defined here is used to characterize the amplitude of the drive voltage, and subsequently it does not have the upper bound of 1 as the conventional optical modulation index. We perform the analysis on $E^{\text{B}}(t)$ for the sake of simplicity because it is a simple frequency translation of the optical field $E(t)$. The derived nonlinearity expression for $E^{\text{B}}(t)$ will apply to the optical field $E(t)$.

Expanding the cosine term in Eq. (7.3) using Bessel functions, the fundamental output component with frequency $\omega_{1,2}$ can be expressed as

$$E^{\text{B}}_{\omega_{1,2}}(t) = 2 \cdot \sin(\phi/2) \cdot J_0(M/2) \cdot J_1(M/2) \cdot e^{j\omega_{1,2}t} \tag{7.4}$$

The second-order intermodulation output with frequency $\omega_{1,2} - \omega_{2,1}$ can be expressed as

$$E^{\text{B}}_{\omega_{1,2} - \omega_{2,1}}(t) = 2 \cdot \cos(\phi/2) \cdot J_1^2(M/2) \cdot e^{j(\omega_{1,2} - \omega_{2,1})t} \tag{7.5}$$

The third-order intermodulation output with frequency $2\omega_{1,2} - \omega_{2,1}$ will be

$$E^{\text{B}}_{2\omega_{1,2} - \omega_{2,1}}(t) = 2 \cdot \sin(\phi/2) \cdot J_1(M/2) \cdot J_2(M/2) \cdot e^{j(2\omega_{1,2} - \omega_{2,1})t} \tag{7.6}$$

We employ the standard nth-order intercept point (IP$_n$) to characterize the modulator nonlinearity.[13] In particular, IP2 (IP3) is the point where the linear extension of the second-order (third-order) intermodulation output power intersects the linear extension of the fundamental output power. The intercept points are given in terms of fundamental output power. IP2 and IP3 are calculated from Eqs. (7.4)–(7.6) and can be expressed as

$$\text{IP2} = 2 \cdot \sin^4(\phi/2)/\cos^2(\phi/2) \tag{7.7}$$

$$\text{IP3} = 4 \cdot \sin^2(\phi/2) \tag{7.8}$$

The corresponding intercept points referenced at the input of the optical I/Q modulator are the modulation index given by

$$M_{\text{IP2}} = 2 \cdot \tan(\phi/2) \tag{7.9}$$

$$M_{\text{IP3}} = 4\sqrt{2} \tag{7.10}$$

It can be shown that Eqs. (7.8) and (7.9) are also valid for the IF architecture, but the second-order nonlinearity is avoided for the IF architecture.

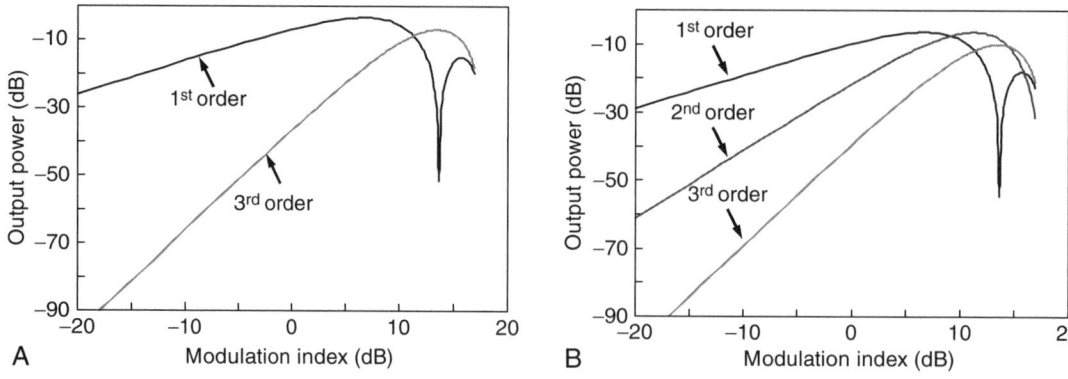

Figure 7.2: The first-, second-, and third-order output powers as a function of modulation index $M(\text{dB}) = 20\log M$ at the bias points of (a) π and (b) $\pi/2$.

Equation (7.8) shows that when the modulator is biased at the zero output ($\phi = \pi$), IP3 has the maximum value, or least nonlinearity, which means the optimum bias point should be π, the null point of the optical modulator. Figures 2.2a and 2.2b show the characteristics of the first-, second-, and third-order intermodulation as a function of the modulation index $M(\text{dB})$ for a bias point at null and quadrature, respectively. It can be seen that at the null bias point, the fundamental output is maximized while the second-order intermodulation product is eliminated in comparison with the quadrature bias point. In the CO-OFDM systems, any pair of the subcarriers will generate second- and third-order intermodulation products. The system penalty due to the optical I/Q modulator nonlinearity under various bias conditions is thoroughly discussed in Tang et al.[7]

In the optimal null bias condition $V_{DC} = V_\pi$, assume that V_I and V_Q are small and the nonlinearity is insignificant. Equation (7.1) becomes

$$E(t) = -\frac{AM}{2}(V_I + jV_Q) \cdot \exp(j\omega_{LD1}t + j\phi_{LD1}) = -\frac{AM}{2}S(t)\exp(j\omega_{LD1}t + j\phi_{LD1}) \quad (7.11)$$

where $S(t) = V_I + jV_Q$ is the complex baseband OFDM signal. It follows that the optical field at the output of the IQ modulator $E(t)$ is essentially a linear replica of the baseband signal $S(t)$ up-converted to the center frequency of $\omega_{LD1}/(2\pi)$.

7.2.5 Discussion of the Null Bias Point for CO-OFDM Systems

The quadrature bias point has been widely adopted for both analog and digital direct detection systems. The null bias point for CO-OFDM up-conversion signifies a fundamental difference between the optical intensity modulation and the optical field modulation. For instance, in coherent detection systems, the transformation between the electrical drive voltage and optical field is of concern, whereas in the conventional direct detection systems the transformation between the electrical drive voltage and the optical power is of concern.

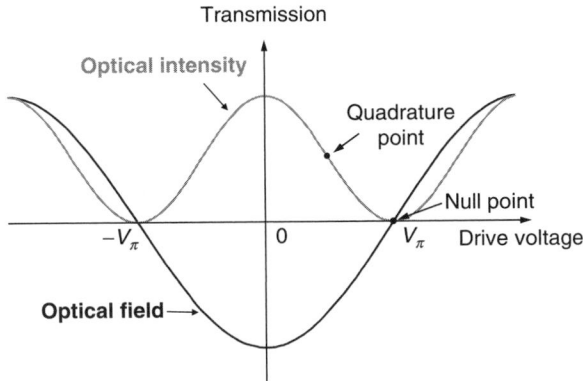

Figure 7.3: Transfer functions for the optical intensity and the optical field against the drive voltage.

Figure 7.3 shows the transfer functions for both the optical intensity and the optical field (I/Q component) against the drive voltage. In Figure 7.3, we assume zero initial phase offset between the two arms of MZM at zero bias voltage. The transfer function for the direct detection is the optical intensity against the drive voltage, whereas the transfer function for the coherent detection is the I or Q component of the optical field against the drive voltage. It is apparent from Figure 7.3 that the optimal MZM bias for optical intensity modulation is the quadrature point, whereas the optimal bias for optical field modulation is the null point.

Note that the conclusion of the optimal modulator null bias point for the RTO up-conversion is fundamentally independent of the detailed waveform of $V_{I/Q}$; that is, $V_{I/Q}$ can be any arbitrary waveform, not necessarily limited to that of CO-OFDM. For instance, for direct detection optical OFDM, it is still optimal to bias the modulator at the null point to minimize RTO up-conversion nonlinearity. However, the challenge is how to reinsert the main optical carrier, which is lost due to the bias at the null point. We anticipate a great deal of original research effort taking advantage of this linear transformation between the optical field and electrical drive voltage in coherent or incoherent, digital or analog systems.

7.2.6 Coherent Detection for Linear Down-Conversion and Noise Suppression

As shown in Figure 7.4, coherent detection uses a six-port 90-degree optical hybrid and a pair of balanced photo-detectors. The main purposes of coherent detection are (i) to linearly recover the I and Q components of the incoming signal, and (ii) to suppress or cancel the common mode noise. Using a six-port 90-degree hybrid for signal detection and analysis has been practiced in the RF domain for decades,[15,16] and its application to single-carrier coherent optical systems can be also found in Ly-Gagnon et al. and Savory et al.[8,9] In what

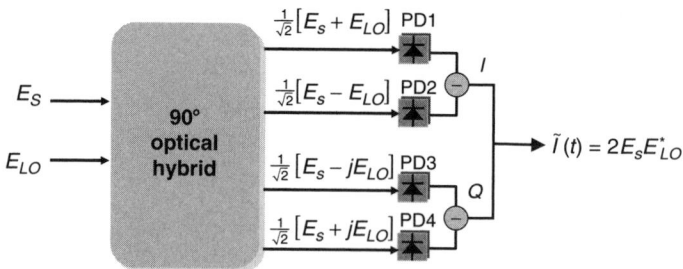

E_s : Incoming signal E_{LO} : Local oscillator signal

PD : Photo-detector $\tilde{I}(t)$: Complex photocurrent

Figure 7.4: Coherent detection using an optical hybrid and balanced photo-detection.

follows, in order to illustrate its working principle, we will perform an analysis of down conversion via coherent detection assuming ideal condition for each component in Figure 7.4.

The purpose of the four output ports of the 90-degree optical hybrid is to generate a 90-degree phase shift for I and Q components, and 180-degree phase shift for balanced detection. Ignoring imbalance and loss of the optical hybrid, the output signals E_{1-4} can be expressed as

$$E_1 = \frac{1}{\sqrt{2}}[E_s + E_{LO}], \; E_2 = \frac{1}{\sqrt{2}}[E_s - E_{LO}]$$

$$E_3 = \frac{1}{\sqrt{2}}[E_s - jE_{LO}], \; E_4 = \frac{1}{\sqrt{2}}[E_s + jE_{LO}]$$

(7.12)

where E_s and E_{LO} are, respectively, the incoming signal and local oscillator (LO) signal. We further decompose the incoming signal into two components: (i) the received signal when there is no amplified spontaneous noise (ASE), $E_r(t)$ and (ii) the ASE noise, $n_o(t)$, namely

$$E_s = E_r + n_o$$

(7.13)

We first study how the I component of the photo-detected current is generated, and the Q component can be derived accordingly. The I component is obtained by using a pair of the photo-detectors, PD1 and PD2 in Figure 7.2, whose photocurrent I_{1-2} can be described as

$$I_1 = |E_1|^2 = \frac{1}{2}\left\{|E_s|^2 + |E_{LO}|^2 + 2\,\mathrm{Re}\{E_s E_{LO}^*\}\right\}$$

(7.14)

$$I_2 = |E_2|^2 = \frac{1}{2}\left\{|E_s|^2 + |E_{LO}|^2 - 2\,\mathrm{Re}\{E_s E_{LO}^*\}\right\}$$

(7.15)

$$|E_s|^2 = |E_r|^2 + |n_o|^2 + 2\,\mathrm{Re}\{E_r n_o^*\}$$

(7.16)

$$|E_{LO}|^2 = I_{LO}(1 + I_{RIN}(t))$$

(7.17)

where I_{LO} and $I_{RIN}(t)$ are the average power and relative intensity noise (RIN) of the LO laser, and "Re" or "Im" denotes the real or imaginary part of a complex signal. For simplicity, the photo-detection responsivity is set as a unit, y. The three terms at the right hand of (7.16) represent signal-to-signal beat noise, signal-to-ASE beat noise, and ASE-to-ASE beat noise. Because of the balanced detection, using (7.14) and (7.15), the I component of the photocurrent becomes

$$I_I(t) = I_1 - I_2 = 2\operatorname{Re}\{E_s E_{LO}^*\} \tag{7.18}$$

Now the noise suppression mechanism becomes quite clear because the three noise terms in (7.16) and the RIN noise in (7.17) from a single detector are completely cancelled via balanced detection. Nevertheless, it is shown that coherent detection can be performed by using a single photo-detector, but with sacrifice of reduced dynamic range.[17]

In a similar fashion, the Q component from the other pair of balanced detectors can be derived as

$$I_Q(t) = I_3 - I_4 = 2\operatorname{Im}\{E_s E_{LO}^*\} \tag{7.19}$$

Using the results of (7.18) and (7.19), the complex photocurrent $\tilde{I}(t)$ that consists of both I and Q components becomes

$$\tilde{I}(t) = I_I(t) + jI_Q(t) = 2E_s E_{LO}^* \tag{7.20}$$

From (7.9), the linear down-conversion process via coherent detection becomes quite conspicuous: the complex photocurrent $\tilde{I}(t)$ is in essence a linear replica of the incoming complex signal that is frequency down-converted by a local oscillator frequency. This linear down-conversion is of critical importance for OFDM modulation which assumes linearity in every stage of signal processing.

7.2.7 Receiver Sensitivity for CO-OFDM

The channel model of Eq. (5.8) gives

$$r_{ki} = e^{\phi_i} H_{ki} c_{ki} + n_{ki} \tag{7.21}$$

where r_{ki} is the received information symbol, ϕ_i is the OFDM symbol phase or common phase error, H_{ki} is the frequency domain channel transfer function, and n_{ki} is the random noise. Temporally ignoring the index "i" and "k," we can see that each subcarrier channel is essentially a linear channel with an additive white Gaussian noise. Omitting the penalty from the channel and phase estimation errors that were discussed in Chapter 4, the ideal performance of the CO-OFDM system with quaternary phase-shift keying (QPSK) modulation is thus given by[19]

$$\mathrm{BER} = \frac{1}{2}\operatorname{erfc}\left(\sqrt{\frac{\gamma}{2}}\right) \tag{7.22}$$

$$\gamma = \frac{\sigma_r^2}{\sigma_n^2} \tag{7.23}$$

where γ is the signal-to-noise ratio; σ_r^2 and σ_n^2 are the variance for the received signal and the noise, respectively; and erfc is the error function.[19] It is quite instructive to explicitly write out the ideal coherent detection performance for CO-OFDM systems where the linewidths of the transmit/receive lasers are assumed to be zero. Because of the coherent detection, the noise is dominated by the signal amplified spontaneous emission (ASE) beat noise. The corresponding bit error ratio (BER), Q-factor, optical signal-to-noise ratio (OSNR), and γ for QPSK CO-OFDM in the ideal condition can be given by[20]

$$\gamma = 2 \cdot \text{OSNR} \cdot \frac{B_0}{R} \tag{7.24}$$

$$\text{BER} = \frac{1}{2}\text{erfc}\left(\sqrt{\text{OSNR} \cdot \frac{B_0}{R}}\right) \tag{7.25}$$

$$Q = 10\log_{10}\left(2 \cdot \text{OSNR} \cdot \frac{B_0}{R}\right) \tag{7.26}$$

where B_0 is the optical ASE noise bandwidth used for OSNR measurement ($\sim$12.5 GHz for 0.1 nm bandwidth); $R \equiv N_{sc} \cdot \Delta f$ is the total system symbol transmission rate or bandwidth of the OFDM signal; and N_{sc} and Δf are the number of the subcarriers and channel spacing of the subcarriers, respectively.

7.3 Direct Detection Optical OFDM

Direct detection optical OFDM (DDO-OFDM) has many more variants than the coherent counterpart. This mainly stems from the broader range of applications for direct detection OFDM due to its lower cost. For instance, the first report of DDO-OFDM[1] takes advantage of the fact that the OFDM signal is more immune to the impulse clipping noise in the cable access TV (CATV) network. Another example is the single-sideband (SSB)-OFDM, which has been proposed by Lowery et al.[21] and Djordjevic and Vasic[22] for long-haul transmission. Tang et al.[23,24] proposed an adaptively modulated optical OFDM (AMO-OFDM) that uses bit and power loading. This has shown promising results for both multimode fiber and short-reach SMF fiber link. The common feature for DDO-OFDM is, of course, the use of direct detection at the receiver, but we classify the DDO-OFDM into two categories according to how the optical OFDM signal is generated: (1) linearly mapped DDO-OFDM (LM-DDO-OFDM), where the optical OFDM spectrum is a replica of baseband OFDM, and (2) nonlinearly mapped DDO-OFDM (NLM-DDO-OFDM), where the optical OFDM spectrum does not display a replica of baseband OFDM. Next, we discuss the principles, design choices, and performance analysis for these two classes of direct detection OFDM systems.

7.3.1 Linearly Mapped DDO-OFDM

As shown in Figure 7.4, the optical spectrum of an LM-DDO-OFDM signal at the output of the O-OFDM transmitter is a linear copy of the RF OFDM spectrum plus an optical carrier that is usually 50% of the overall power. The position of the main optical carrier can be

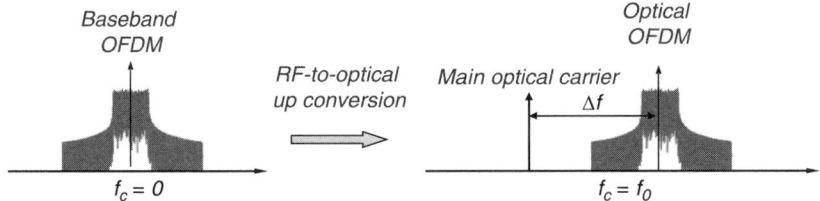

Figure 7.5: Illustration of LM-DDO-OFDM, where the optical OFDM spectrum is a replica of the baseband OFDM spectrum.

one OFDM spectrum bandwidth away[21,25] or at the end of the OFDM spectrum.[26,27] Formally, this type of DDO-OFDM can be described as

$$s(t) = e^{j2\pi f_0 t} + \alpha e^{j2\pi (f_0 + \Delta f)t} \cdot s_B(t) \tag{7.27}$$

where $s(t)$ is the optical OFDM signal, f_0 is the main optical carrier frequency, Δf is the guard band between the main optical carrier and the OFDM band (see Figure 7.5), and α is the scaling coefficient that describes the OFDM band strength related to the main carrier. $s_B(t)$ is the baseband OFDM signal given by

$$s_B = \sum_{k=-\frac{1}{2}N_{sc}+1}^{\frac{1}{2}N_{sc}} c_k e^{j2\pi f_k t} \tag{7.28}$$

where c_k and f_k are, respectively, the OFDM information symbol and the frequency for the kth subcarrier. For explanatory simplicity, only one OFDM symbol is shown in Eq. (7.28). After the signal passes through fiber link with chromatic dispersion, the OFDM signal can be approximated as

$$r(t) = e^{j(2\pi f_0 t + \Phi_D(-\Delta f) + \phi(t))} + \alpha e^{j(2\pi (f_0 + \Delta f)t + \phi(t))} \cdot \sum_{k=-\frac{1}{2}N_{sc}+1}^{\frac{1}{2}N_{sc}} c_{ik} e^{(j2\pi f_k t + \Phi_D(f_k))} \tag{7.29}$$

$$\Phi_D(f_k) = \pi \cdot c \cdot D_t \cdot f_k^2 / f_O^2 \tag{7.30}$$

where $\Phi_D(f_k)$ is the phase delay due to chromatic dispersion for the kth subcarrier, D_t is the accumulated chromatic dispersion in unit of picoseconds per picometer (ps/pm), f_0 is the center frequency of optical OFDM spectrum, and c is the speed of light. At the receiver, the photodetector can be modeled as the square law detector and the resultant photocurrent is

$$I(t) \propto |r(t)|^2 = 1 + 2\alpha \, \mathrm{Re} \left\{ e^{j2\pi \Delta f t} \sum_{k=-\frac{1}{2}N_{sc}+1}^{\frac{1}{2}N_{sc}} c_{ik} e^{(j2\pi f_k t + \Phi_D(f_k) - \Phi_D(-\Delta f))} \right\}$$

$$+ |\alpha^2| \sum_{k_1=-\frac{1}{2}N_{sc}+1}^{\frac{1}{2}N_{sc}} \sum_{k_2=-\frac{1}{2}N_{sc}+1}^{\frac{1}{2}N_{sc}} c_{k_2}^* c_{k_1} e^{(j2\pi (f_{k_1} - f_{k_2})t + \Phi_D(f_{k_1}) - \Phi_D(f_{k_2}))} \tag{7.31}$$

The first term is a DC component that can be easily filtered out. The second term is the fundamental term consisting of linear OFDM subcarriers that are to be retrieved. The third term is the second-order nonlinearity term that needs to be removed.

The following approaches can be used to minimize the penalty due to the second-order nonlinearity term:

Offset SSB-OFDM: Sufficient guard band is allocated such that the second-term and third-term RF spectra are nonoverlapping. As such, the third term in Eq. (7.31) can be easily removed using an RF filter, as proposed by Lowery et al.[21]

Baseband optical SSB-OFDM: The α coefficient is reduced as much as possible such that the distortion as a result of the third term is reduced to an acceptable level. This approach has been adopted by Djordjevic and Vasic[22] and Hewitt.[26]

Subcarrier interleaving: From Eq. (7.31), it follows that if only odd subcarriers are filled—that is, c_k is nonzero only for the odd subcarriers—the second-order intermodulation will be at even subcarriers, which are orthogonal to the original signal at the odd subcarrier frequencies. Subsequently, the third term does not produce any interference. This approach has been proposed by Peng et al.[28]

Iterated distortion reduction: The basic idea is to go through a number of iterations of estimation of the linear term and compute the second-order term using the estimated linear term, removing the second-order term from the right side of Eq. (7.31). This approach has been proposed by Peng et al.[27]

There are advantages and disadvantages to all four approaches. For instance, the second approach has the advantage of better spectral efficiency but receiver sensitivity is sacrificed. The fourth approach has both good spectral efficiency and receiver sensitivity, but with a burden of computational complexity. In the remainder of this section, we illustrate the four designs in detail.

7.3.1.1 Offset SSB-OFDM

Offset OFDM is proposed by Lowery et al.[21] They show that DDO-OFDM can mitigate enormous amounts of chromatic dispersion up to 5000 km SSMF fiber. The proof-of-concept experiment was demonstrated by Schmidt et al.[25] from the same group for 400 km CO-OFDM transmission at 20 Gb/s. Figure 7.6 shows the proposed system of offset OFDM in their simulation. The simulated system is 10 Gb/s with 4-QAM modulation with a bandwidth of approximately 5 GHz.[29] In the electrical OFDM transmitter, the OFDM signal is up-converted to an RF carrier at 7.5 GHz, generating an OFDM band spanning from 5 to 10 GHz. The RF OFDM signal is fed into an optical modulation. The output optical spectrum has the two side OFDM bands that are symmetric over the main optical subcarrier. An optical

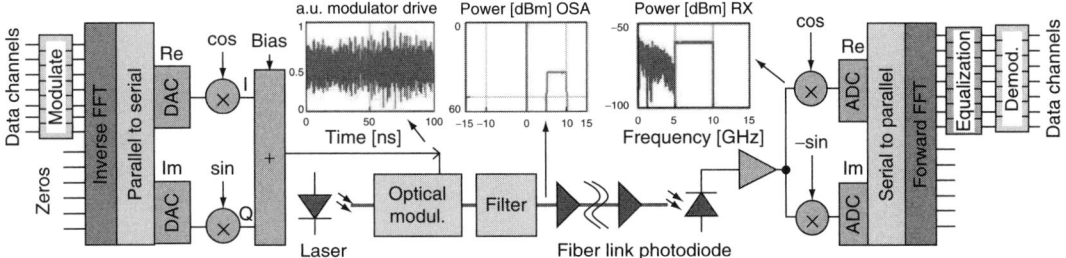

Figure 7.6: DDO-OFDM long-haul optical communication systems. After Lowery et al.[29]

filter is then used to filter out one OFDM sideband. This single sideband (SSB) is critical to ensure there is one-to-one mapping between the RF OFDM signal and the optical OFDM signal. The main optical subcarrier is also attenuated properly to maximize the sensitivity. At the receiver, only one photodetector is used. The RF spectrum of the photocurrent is depicted as an inset in Figure 7.6. It can be seen that the second-order intermodulation, the third-term in Eq. (7.31), is from DC to 5 GHz, whereas the OFDM spectrum, the second term in Eq. (7.31), spans from 5 to 10 GHz. As such, the RF spectrum of the intermodulation does not overlap with the OFDM signal, signifying that the intermodulation does not cause detrimental effects after proper electrical filtering.

Figure 7.7 shows the received Q-factor for a back-to-back system as a function of the carrier suppression varying modulation drive levels. The modulation depth is defined here as a standard deviation of the electrical drive as a percentage of the voltage required to turn the modulator from 0% to 100% transmission. Optical amplifier ASE noise is added to reach OSNR of 12.9 dB. Use of so-defined modulation depth, for drive levels above 20%, the OFDM waveform will be clipped at 0% and 100 % transmission, or the optical modulation index will be reached to 100%. This clipping reduces the maximum Q-factor to below the

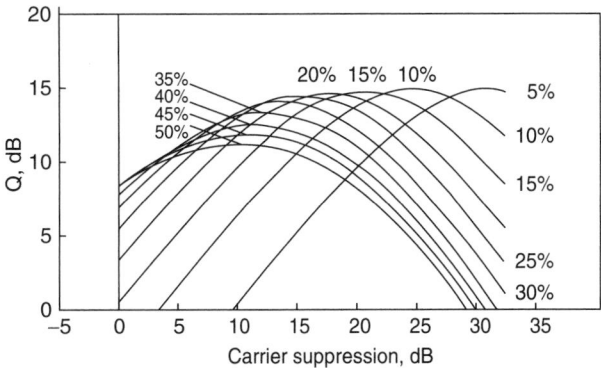

Figure 7.7: System performance against the carrier suppression for offset DDO-OFDM. After Lowery et al.[29]

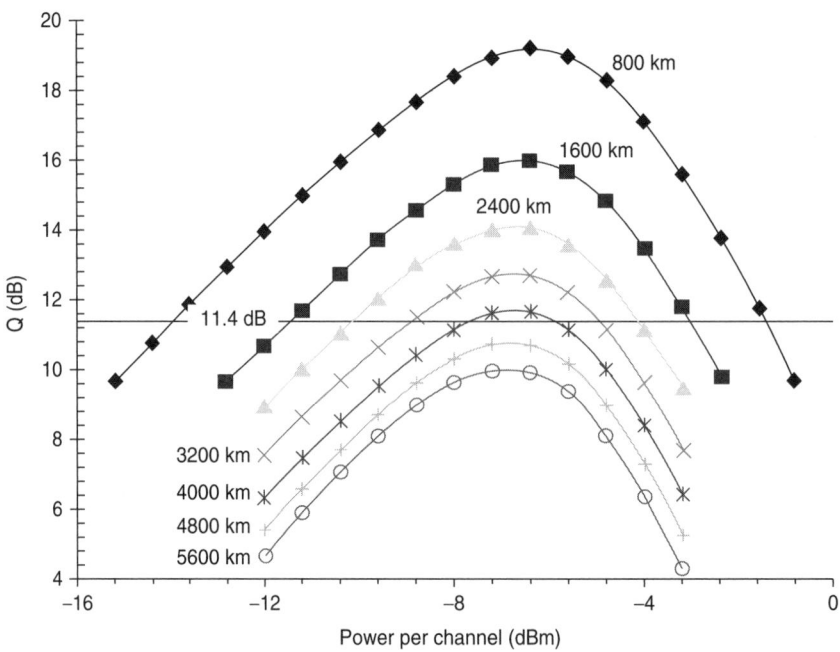

Figure 7.8: Q-factor versus the launch power per channel varying transmission distance. After Lowery et al.[29]

noise-limited level, which is not desirable. On the other hand, very low modulator drive levels (<10%) give very low optical sideband powers and so require strong attenuation of the optical carrier, which is also undesirable because a high gain optical amplifier will be required after the modulator to boost the signal into the fiber. It seems that the modulation depth of 10% and 20% gives optimal performance. Figure 7.8 shows the simulation results of eight WDM channels with a 15 GHz channel spacing as a function of the input power in such a DDO-OFDM system. For each reach (or transmission distance), there is an optimal launch power that maximizes the Q-factor. This optimization is a reflection of the accumulation of the optical amplifier ASE noise and fiber nonlinear effects. For lower input powers, the Q-factor is limited by optical amplifier ASE noise and increases decibel for decibel with the launch power per channel; for high input powers, the Q-factor is limited by fiber nonlinearity and decreases approximately 2 dB for every decibel increase in the launch power per channel.[29] The optimal Q-factor decreases with the system reach; for instance, the system longer than 4000 km cannot achieve a Q-factor of 11.4 dB.

7.3.1.2 Baseband Optical SSB-OFDM

Baseband optical SSB-OFDM is proposed by Hewitt[26] to improve the electrical and optical spectral efficiency. The transmitter setup is identical to that shown Figure 7.6. The difference is that for baseband SSB-OFDM, there is no need for up-conversion to the RF

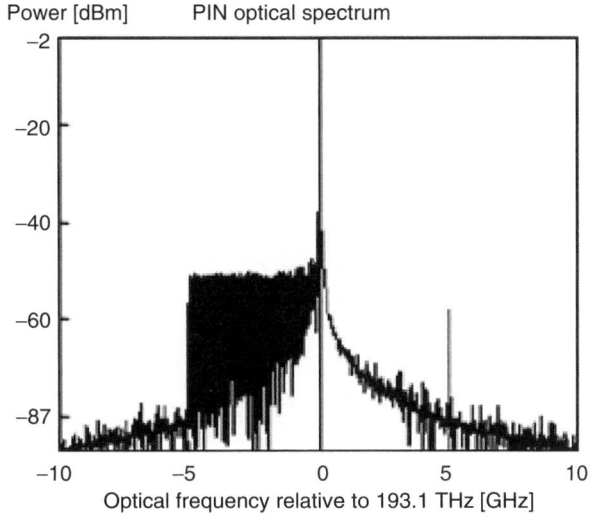

Figure 7.9: Optical spectrum of the baseband optical SSB-OFDM. After Hewitt.[26]

frequency. For the baseband SSB-OFDM, only the positive band is filled and the DC subcarrier has a large index to serve as the main subcarrier. Figure 7.9 shows the optical spectrum of baseband optical SSB-OFDM. As we analyzed in Eq. (7.31), such an approach is limited by the second-order nonlinearity. Djordjevic and Vasic[22] have also proposed a similar SSB-OFDM approach in combination with low-density parity-check coding. The performance of such a system was thoroughly discussed in Chapter 6.

7.3.1.3 RF Tone-Assisted OFDM

Offset OFDM has the disadvantage of spectral inefficiency and baseband SSB-OFDM has the disadvantage of second-order nonlinearity. RF tone-assisted OFDM (RFT-OFDM) has been proposed by Peng et al.[28] to address these two problems. Figure 7.10 shows the proposed RFT-OFDM system with two variations, OFDM-A and OFDM-B. For OFDM-A, a gap with the same width as the OFDM spectrum is used between the RF tone and the OFDM signal. For OFDM-B, odd subcarriers relative to the RF tone are filled for the data and the even subcarriers are left unfilled. For both schemes, the electrical OFDM signal after the inverse fast Fourier transform (IFFT) is complex and its real and imaginary parts are fed into the two arms of an optical I/Q modulator. Assume that the total OFDM bandwidth is B. For OFDM-A, the RF tone-induced main optical carrier is $B/2$ away from the OFDM band. This is similar to offset SSB-OFDM discussed in Section 7.3.1.1. However, OFDM-A does not need electrical RF up-conversion and requires smaller electrical bandwidth for the modulator. The spectra after the photodetection are shown in Figure 7.11. For OFDM-A, the second-order intermodulation will fall on the subcarriers

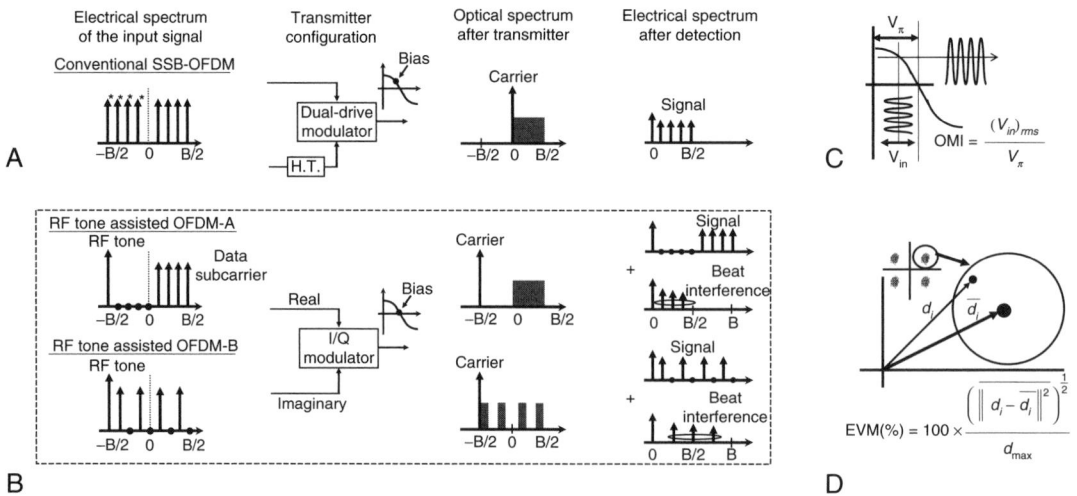

Figure 7.10: The principles for (a) baseband SSB-OFDM, (b) the proposed RF tone-assisted OFDM-A and OFDM-B, (c) definition of the optical modulation index (OMI), and (d) definition of the error vector magnitude (EVM). H.T., Hilbert transform. After Peng et al.[28]

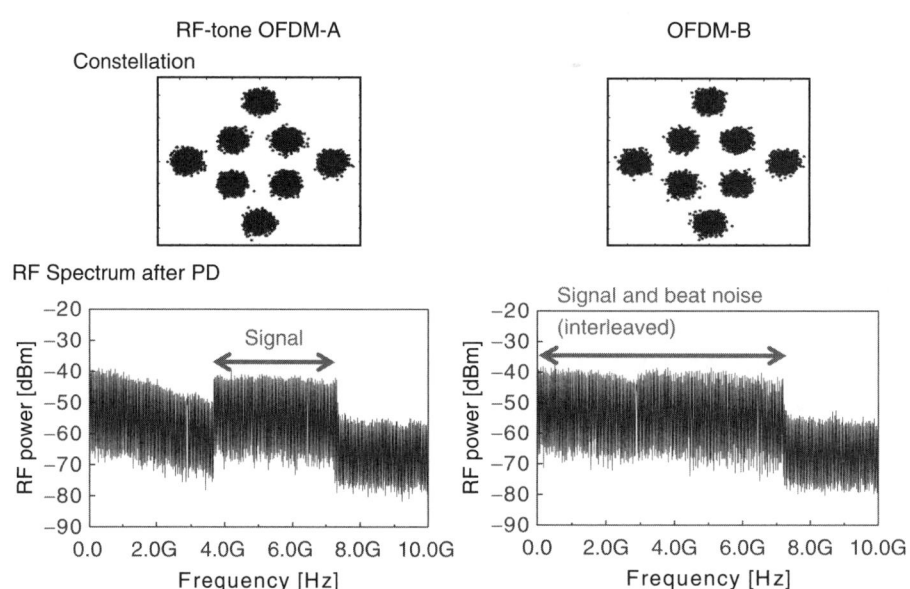

Figure 7.11: RF spectra for RFT-OFDM in both OFDM-A and OFDM-B configurations. After Peng et al.[28]

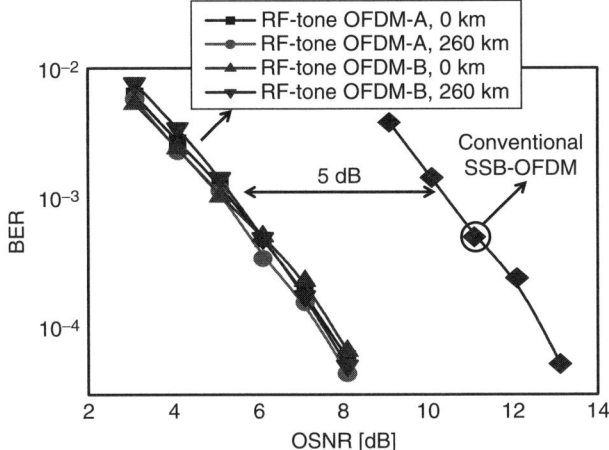

Figure 7.12: BER sensitivity performance for RFT-OFDM-A and OFDM-B. After Peng et al.[28]

within the gap, and for OFDM-B the intermodulation will fall on those subcarriers interleaved with the data subcarriers. Thus, the data of the two schemes can be extracted without suffering intermodulation distortion.

Figure 7.12 shows the BER performance for both RFT-OFDM schemes and the conventional baseband SSB-OFDM system[26] at 10 Gb/s with 8-QAM. For the RFT-OFDM scheme, the main RF tone consumes 50% of the total optical power, and for baseband SSB-OFDM the optimized OMI = 0.12 is used. The back-to-back sensitivities of the OFDM-A and OFDM-B are similar, and both are with a 5 dB better sensitivity than baseband SSB-OFDM. Following 260 km SSMF transmission, almost no penalties are observed for either of the RFT-OFDM systems.

7.3.1.4 Virtual SSB-OFDM

Virtually SSB-OFDM (VSSB-OFDM) is proposed by Peng et al.[27] to reduce both the modulation and the reception electrical bandwidth while maintaining high receiver sensitivity. Figure 7.13 shows the transmitter architecture of the proposed VSSB-OFDM.

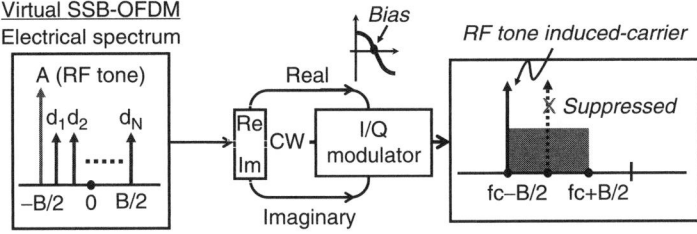

Figure 7.13: Transmitter architecture of virtual SSB-OFDM. After Peng et al.[27]

The RF tone as the main optical carrier is placed at the leftmost subcarrier. This RF subcarrier is treated as one of the regular subcarriers, and data subcarriers plus the main subcarrier are transferred into a complex baseband signal by IFFT. The carrier-to-signal power ratio (CSPR) is defined as the ratio between the main optical carrier and total power of OFDM subcarriers. For the up-conversion, the external modulator is biased at null that minimizes the modulation nonlinearity, as discussed in Section 7.2.2.4. Second-order nonlinearity from the beating among the data subcarriers at the photodiode will incur a penalty similar to the baseband SSB-OFDM. To remove the second-order nonlinearity, an iterative cancellation technique is performed by Peng et al.[27] as follows: (1) The signal with the distortion is processed and used to make the initial decision (potentially with errors); (2) the sliced data are used to reconstruct the second-order distortion; and (3) the distortion is subtracted from the original signal, and another iteration of the decision is made. The procedure can be iterated until the satisfactory result is reached. Obviously, the disadvantage of this VSSB-OFDM is that additional computation complexity is required. The BER performance for both the conventional baseband SSB-OFDM and the VSSB-OFDM is shown in Figure 7.14. It can be seen that the required OSNR at a BER of 10^{-3} for VSSB-OFDM is 5 dB better than that for the baseband SSB-OFDM, which has been optimized with OMI = 0.12 in back to back. The 5 dB gain is attributed to the optimum CSPR and the smaller constellation size used in VSSB-OFDM.

7.3.2 Nonlinearly Mapped DDO-OFDM

The second class of DDO-OFDM is nonlinearly mapped OFDM (NLM-DDO-OFDM), which means that there is no linear mapping between the electric field (baseband OFDM) and the optical field. Instead, NLM-DD-OFDM aims to obtain a linear mapping between

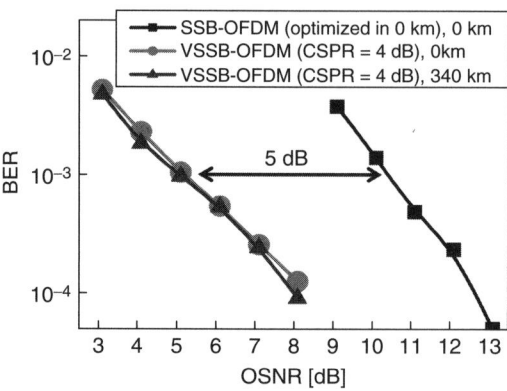

Figure 7.14: BER performance of VSSB-OFDM. After Peng et al.[27]

baseband OFDM and optical intensity. For exhibitory simplicity, we assume generation of NLM-DDO-OFDM using direct modulation of a DFB laser; the waveform after the direct modulation can be expressed as[30]

$$E(t) = e^{j2\pi f_o t} A(t)^{1+jC} \tag{7.32}$$

$$A(t) \equiv \sqrt{P(t)} = A_0 \sqrt{1 + \alpha \operatorname{Re}(e^{j(2\pi f_{\mathrm{IF}} t)} \cdot s_\mathrm{B}(t))} \tag{7.33}$$

$$s_\mathrm{B}(t) = \sum_{k=-\frac{1}{2}N_{\mathrm{sc}}+1}^{\frac{1}{2}N_{\mathrm{sc}}} c_k e^{j2\pi f_k t} \tag{7.34}$$

$$m \equiv \alpha \sqrt{\sum_{k=-\frac{1}{2}N_{\mathrm{sc}}+1}^{\frac{1}{2}N_{\mathrm{sc}}} |c_k|^2} \tag{7.35}$$

where $E(t)$ is the optical OFDM signal; $A(t)$ and $P(t)$ are the instantaneous amplitude and power of the optical OFDM signal, respectively; c_k is the transmitted information symbol for the kth subcarrier; C is the chirp constant for the direct modulated DFB laser[30]; f_{IF} is the IF frequency for the electrical OFDM signal for modulation; m is the optical modulation index; α is a scaling constant to set an appropriate modulation index m to minimize the clipping noise; and $s_\mathrm{B}(t)$ is the baseband OFDM signal. Assuming the chromatic dispersion is negligible, the detected current is

$$I(t) = |E(t)|^2 = |A|^2 = A_0 \left(1 + \alpha \operatorname{Re}\left(e^{j(2\pi f_{\mathrm{IF}} t)} \cdot s_\mathrm{B}(t)\right)\right) \tag{7.36}$$

Equation (7.36) shows that the photocurrent contains a perfect replica of the OFDM signal $s_\mathrm{B}(t)$ with a DC current. We also assume that modulation index m is small enough that clipping effect is not significant. Equation (7.36) shows that by using NLM-DDO-OFDM with no chromatic dispersion, the OFDM signal can be perfectly recovered. The fundamental difference between the NLM- and LM-DDO-OFDM can be discerned by studying their respective optical spectra. Figure 7.15 shows the optical spectra of NLM-DDO-OFDM using direct modulation of a DFB laser with the chirp coefficient C of 1 in Eq. (7.32), a modulation index m of 0.3 in Eq. (7.35), and offset SSB-OFDM as described in Section 7.3.1.1. It can be seen that in sharp contrast to SSB-OFDM, NLM-DDO-OFDM has multiple OFDM bands with significant spectral distortion. Therefore, there is no linear mapping from the baseband OFDM to the optical OFDM. The consequence of this nonlinear mapping is fundamental because when any type of dispersion—such as chromatic dispersion, polarization dispersion, or modal dispersion—occurs in the link, the detected photocurrent can no longer recover the linear baseband OFDM signal. Namely, any dispersion will cause the nonlinearity for NLM-DD-OFDM systems. In particular, unlike SSB-OFDM, the channel model for direct modulated OFDM is no longer linear under any form of optical dispersion. Subsequently, NLM-DD-OFDM is only fit for short-haul applications such as multimode fiber for LANs

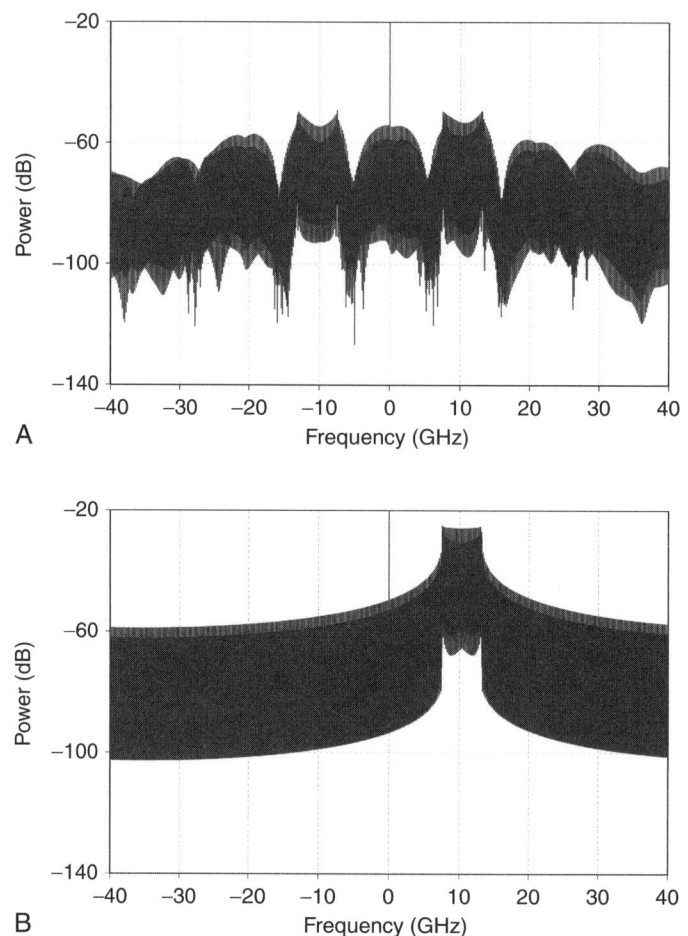

A

B

Figure 7.15: Comparison of optical spectra between (a) NLM-DDO-OFDM through direct modulation of a DFB laser and (b) externally modulated offset SSB-DDO-OFDM. The chirp constant C of 1 and the modulation index m of 0.3 are assumed for direct modulation in (a). Both OFDM spectrum bandwidths are 5 GHz comprising 256 subcarriers.

or short-reach SMF transmission. This class of optical OFDM has attracted significant attention due to its low cost. Next, we describe different types of NLM-DDO-OFDM with various benefits of using OFDM modulation. We would like to distinguish DDO-OFDM from radio-over-fiber applications, for which the fiber merely serves as a transmission medium for RF OFDM signal and OFDM is not used to mitigate the impairments in the optical domain.

7.3.2.1 Compatible OFDM

Similar to VSSB-OFDM, compatible SSB-OFDM (CompSSB) was proposed by Schuster et al.[31] to achieve higher spectral efficiency than offset SSB-OFDM. The idea behind CompSSB-OFDM is to apply the OFDM signal on the amplitude of the signal.[31] Ignoring the frequency up-conversion to the main optical carrier, the optical CompSSB-OFDM signal $E(t)$ is expressed as

$$E(t) = a(t)e^{j\phi(t)} \tag{7.37}$$

where $a(t)$ is a copy of the real-valued OFDM signal given by Eqs. (2.9) and (2.18) plus a DC component. The real-valued OFDM signal $a(t)$ is clipped to avoid being nonpositive.[31] $\phi(t)$ in Eq. (7.37) is not independent and is related to $a(t)$ by the following relationship:

$$\phi(t) = \mathcal{H}\{\ln(a(t))\} \tag{7.38}$$

where $\mathcal{H}\{\}$ stands for the Hilbert transform. We could rearrange Eq. (7.37) into

$$\ln(E(t)) = \ln a(t) + j\varphi(t) = \ln a(t) + j\mathcal{H}\{(\ln a(t))\} \tag{7.39}$$

It can be easily shown that $\ln(E(t))$ is an SSB signal, and so is $E(t)$. The photocurrent $I(t)$ after the square law photodetector is

$$I(t) = |a(t)|^2 \tag{7.40}$$

Therefore, the square root of the photocurrent $I(t)$ will recover the original OFDM signal $a(t)$. Figure 7.16 shows the optical and the detected (electrical) signal spectra of CompSSB-OFDM in comparison with the offset SSB-OFDM. The optical and electrical spectral requirement of compatible OFDM is reduced by approximately half from that of offset

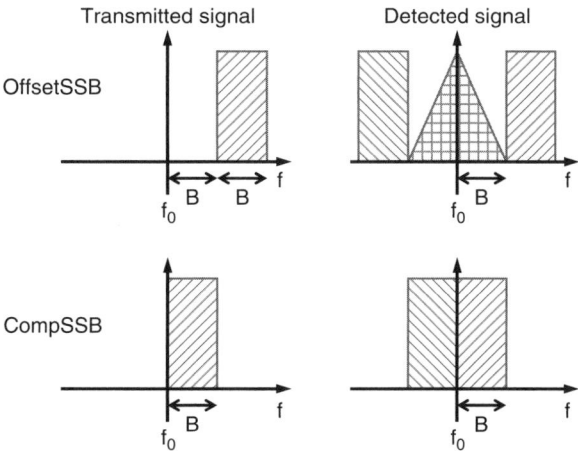

Figure 7.16: Optical and electrical spectra for offset SSB-OFDM and CompSSB-OFDM. After Schuster et al.[31]

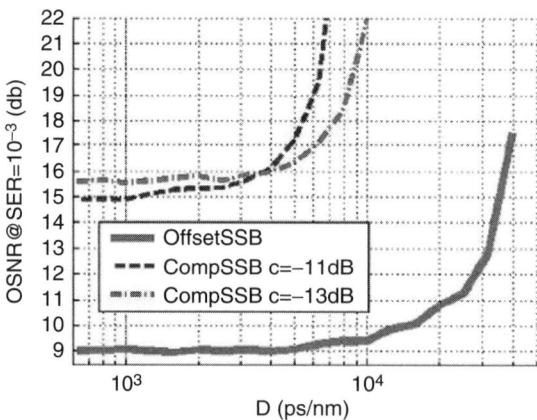

Figure 7.17: OSNR sensitivity against the chromatic dispersion for CompSSB-OFDM and offset SSB-OFDM. After Schuster et al.[31]

SSB-OFDM. Figure 7.17 shows the OSNR sensitivity performance as a function of the chromatic dispersion. The sensitivity of CompSSB-OFDM is approximately 6 dB worse than that of the offset SSB-OFDM. This is because the large DC component is needed to keep $a(t)$ positive, which reduces the effective power allocated to the OFDM data subcarriers. Second, CompSSB-OFDM is encoded in the amplitude, not the field; the optical OFDM signal is not a replica of the baseband OFDM signal, and subsequently it will not have the same capability of dispersion resilience as the offset SSB-OFDM. As shown in Figure 7.17, the chromatic dispersion tolerance for CompSSB-OFDM is limited to 7×10^3 ps/nm, or approximately 400 km of SMF transmission, whereas the offset SSB-OFDM can be retrieved beyond 2000 km of SMF transmission.[21,22]

7.3.2.2 OFDM for Mitigating Multipath Interference in Multimode Fiber

In 2001, Dixon et al.[33] proposed using OFDM to combat the multipath interference in multimode fiber. The motivation of that work was to investigate the feasibility of using multimode fiber as an inexpensive cell feed in broadband indoor picocellular systems. They found that in addition to its ability to mitigate a frequency-selective multipath environment in the wireless RF domain, OFDM can combat the frequency selectivity of a dispersive multimode fiber. They also concluded that rates in excess of 100 Mb/s over a multimode fiber channel are possible compared with 20–30 Mb/s using conventional ASK modulation. The breakthrough in DDO-OFDM took place in 2001 and had a profound impact on optical communications: Jolley et al.[32] from Nortel demonstrated the first 10 Gb/s OFDM (the highest OFDM data rate at the time) over 1000 m of multimode fiber using a directly modulated laser transmitter. Figure 7.18 shows the experimental setup used in their work, where the OFDM generation is performed using an arbitrary waveform generator and the

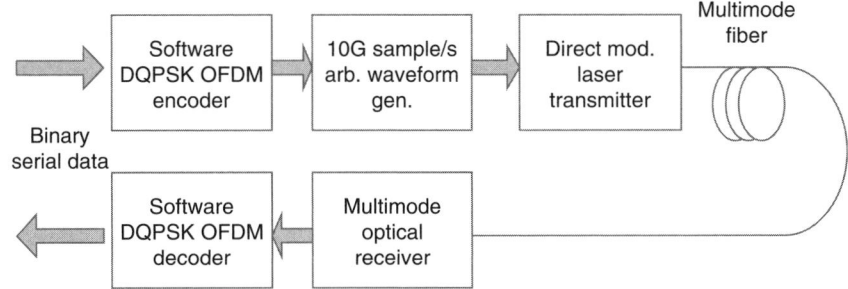

Figure 7.18: Experimental setup of OFDM over multimode fiber. After Jolley et al.[32]

demodulation is performed using a software decoder or offline signal processing. The transfer function of the optical receiver and the combined transfer functions of the receiver and the multimode fiber sections are plotted in Figure 7.19. The bandwidth of the multimode fiber is measured to be approximately 800 MHz/km. It can be observed that the multimode fiber is very frequency selective, with the amplitude change more than 6 dB for the interested frequency range of 0 to 5 GHz. This frequency selectivity can be equalized by using one tap equalizer within OFDM signal processing as illustrated in Eq. (4.8). Figure 7.20 shows the received spectra, constellations, and associated error rates for the signal. The back-to-back signal indicates the impairments from signal clipping and the transmitter and receiver frequency response. The spectrum is significantly modified after the transmission over the multimode fiber. The received error rates obtained in the range of 10^{-4} would result in error-free operation when combined with forward error correction.

Jolley et al.'s[32] work heralded subsequent intense research effort to apply OFDM technology in multimode fiber (MMF) systems.[34–37] There is a tremendous demand for higher bandwidth of 10 Gb/s and beyond in LANs and interconnects between the servers or data centers because of bandwidth-intensive applications such as IPTV and HDTV that have emerged as mainstream Internet traffic. There is a huge embedded base of MMF fiber in enterprise

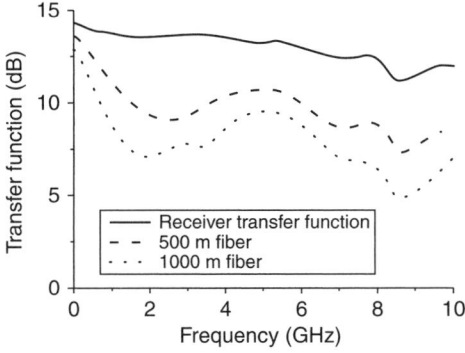

Figure 7.19: The receiver and fiber transfer function. After Jolley et al.[32]

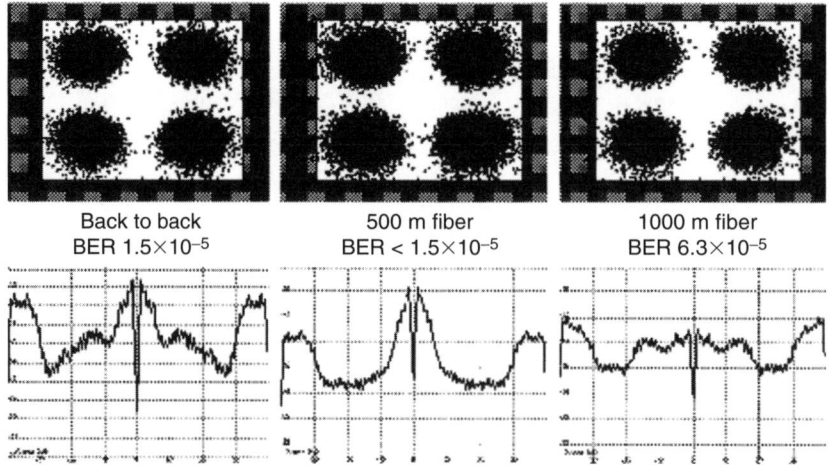

Figure 7.20: Received electrical spectra, the constellation diagrams, and associated error rates at back-to-back, 500 m and 100 m fiber. After Jolley et al.[32]

LANs. It is highly desirable to reuse this legacy system for the 10 Gb/s Ethernet. Tang et al.[34,35] proposed a variant of optical OFDM called adaptively modulated optical OFDM (AMO-OFDM) that takes advantage of bit and power loading in OFDM; Lowery and Armstrong[36] proposed a power-efficient OFDM to reduce the DC component required for direct modulated optical OFDM. Apart from the conventional silica MMF, the polymer optical fiber is gaining increasingly more interest for short-reach applications. Its main benefits are its 1 mm large-core diameter and numerical aperture of 0.5, leading to highly relaxed tolerance with regard to alignment and coupling with transceivers and robustness to mechanical stress.[38,39] Due to broad interest in optical OFDM for multimode link, we dedicate an entire chapter (Chapter 9) to this topic, including the detailed principle, design, and applications of OFDM in MMF systems.

7.3.2.3 OFDM for Short-Haul Single-Mode Fiber Application

We have argued that for the direct modulated or intensity modulation (IM) DDO-OFDM, the optical field of the OFDM signal does not produce a replica of the baseband OFDM, and therefore it is not fit for long-haul transmission. However, because of its simple implementation, it has become a very attractive option for the short-reach nonamplified link at 10 Gb/s and beyond. This application has been thoroughly investigated by Tang et al.[40,41] Figure 7.21 shows the system configuration studied by Tang et al.[40] The real-valued OFDM signal is used to directly drive a DFB laser. The link is one-span nonrepeated standard SMF link. The received optical signal is directly detected and converted to the real-valued OFDM signal. The uniqueness of the system is use of negotiation between the transmitter and receiver such that the different level of the modulation can be applied according to the quality

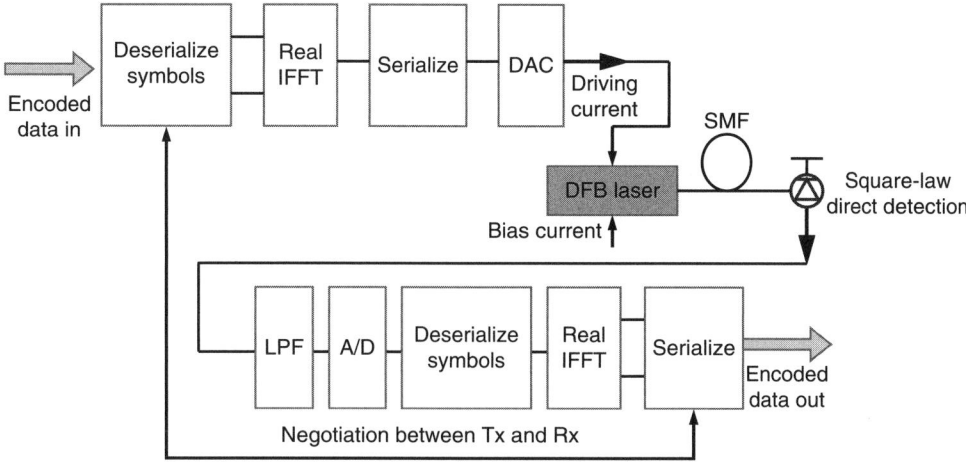

Figure 7.21: Experimental setup for IM-DDO-OFDM. After Tang and Shore.[40]

of the signal, which is the essence of the so-called AMO-OFDM. Figure 7.22 shows the modulation distribution for 80 km transmission over a nondispersion-shifted fiber (NDSF) link. It can be seen that the lower frequency subcarriers are loaded with 32-QAM and higher frequency subcarriers are loaded with lower modulation of 16-QAM. The lower order of the modulation at the high frequency is a result of the frequency chirp-induced spectral distortions at the high frequency. This adaptive bit loading ensures the optimized data rate can be achieved. It is shown that a 30 Gb/s transmission over a 40 km SMF can be achieved without involving optical amplification and chromatic dispersion compensation.[40]

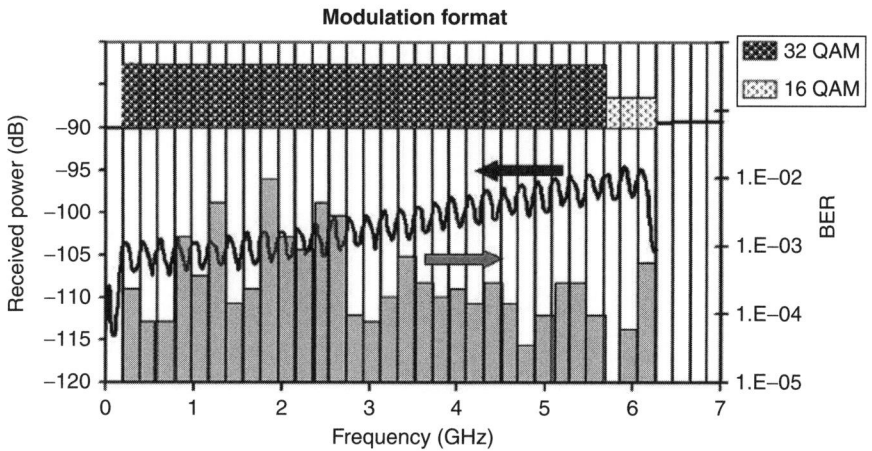

Figure 7.22: One example of use of different modulation formats. (Left) Normalized frequency response of an 80 km NDSF transmission link; (top) signal-modulation formats used on various subcarriers; and (right) their corresponding BER. After Tang and Shore.[40]

7.3.2.4 OFDM for Robustness against Clipping Impulse Noise

The first optical OFDM work was reported by Pan and Green[1] to combat clipping impulse noise in CATV networks. Because it is the first work on optical OFDM and the clipping noise is an important impairment for the direct modulation OFDM, it is of much interest to discuss their work in detail. At the time of their work, for CATV systems, digital modulation formats were proposed to overlay the existing analog subcarrier multiplexed system using amplitude-modulation vestigial-sideband modulation. The single-carrier M-QAM and OFDM M-QAM were both discussed as viable choices. Figure 7.23 shows the architecture of such hybrid SCM analog and digital systems. The electrical SCM signal is added with appropriate bias to directly modulate the laser. This biased signal will occasionally suffer from the clipping due to the large negative swing below the threshold where the laser output is clipped at zero regardless of the drive signal, as shown in Figure 7.24. This evidently will cause nonlinearity distortion to the signal, which is so-called impulse clipping noise. Formally, the noise of the output of the detected QAM optical systems comprises two noise components expressed as

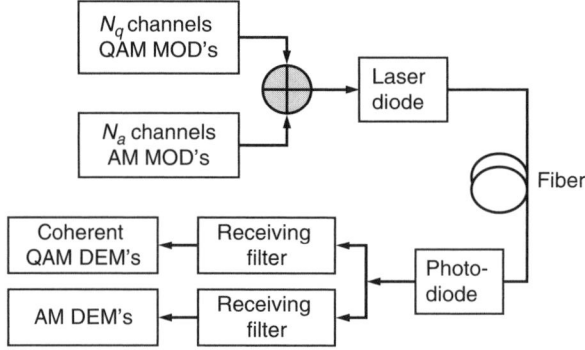

Figure 7.23: Schematic diagram of a hybrid AM/QAM system with N_a AM and N_q QAM channels. After Pan and Green.[1]

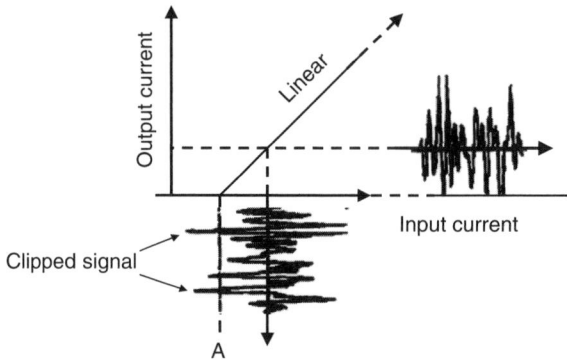

Figure 7.24: Principle of the clipping noise in direct modulation systems.

$$\sigma_t^2 = \sigma_c^2 + \sigma_g^2 \tag{7.41}$$

where σ_t^2, σ_c^2, and σ_g^2 are the variance for total noise, clipping noise, and Gaussian noise, respectively. The Gaussian noise σ_g^2 originates from the shot noise, thermal noise, and relative intensity noise (RIN) of the laser and is expressed in double-sideband spectral density as

$$\sigma_g^2 = \left[i_n^2 + 2qRP_0 + RIN(RP_0)^2 \right]/2 \tag{7.42}$$

where i_n is the receiver thermal current, q is the electron charge, R is the photodetector responsivity, and P_0 is the received DC optical power.[1] Mazo[42] developed an analytical model of the statistical properties of the clipping noise assuming that the SCM signal is a Gaussian process and the clipping noise is a rare event. In particular, the clipping event asymptotically approaches a Poisson process, the duration of the clipping events has Rayleigh probability density, and the clipping pulse assumes a parabola function. From this analytical model, Pan and Green[1] derived the probability density function (PDF) of the clipping noise $p_c(x)$ given by

$$p_c(x) = (1 - \upsilon T_0)\delta(x) + \frac{\upsilon T_0}{\pi} \int_0^{\frac{\pi}{2}} \frac{p_a(-|x|/\cos\lambda)}{\cos\lambda} d\lambda \tag{7.43}$$

where x is the noise amplitude for clipping noise, T_0 is the symbol duration of the (OFDM)-QAM signal, and υ is the average repetition rate of the clipping events. Because the QAM signal requires a much lower carrier-to-noise ratio than the analog signals, the optical modulation index (OMI) for each QAM channel, m_0, is much smaller than the OMI of an analog signal, m_a. Only the modulation from the analog signals is considered for the clipping noise. From Mazo,[42] υ can be given by

$$\upsilon = \left[\frac{f_b^3 - f_a^3}{3(f_b - f_a)} \right]^{1/2} \exp\left(-\frac{1}{2\sigma^2} \right) \tag{7.44}$$

where f_a and f_b are the lowest and highest frequencies of the AM signal band, respectively, and σ is the normalized effective modulation depth of the laser diode, equal to $\sigma = m_a \sqrt{N_a/2}$. $P_a(x)$ in Eq. (7.43) is the PDF for the clipping noise amplitude; a at the output of the matched filter and is given by

$$p_a(a) = \begin{cases} p_\tau\left(\sqrt[3]{\frac{6a\sqrt{2T_0}}{g}} \right) \left| \frac{1}{3a} \sqrt[3]{\frac{6a\sqrt{2T_0}}{g}} \right| & a \le 0, \\ 0 & a > 0 \end{cases} \tag{7.45}$$

where

$$p_\tau(\tau) = \frac{\pi}{2} \frac{\tau}{\tau_0^2} \exp\left[-\frac{\pi}{4}\left(\frac{\tau}{\tau_0} \right)^2 \right] \quad \tau \ge 0 \tag{7.46}$$

$$\tau_0 = \frac{1}{2}\frac{\mathrm{erfc}(\sigma/\sqrt{2})}{\upsilon}, g = (2\pi)^2 \frac{f_b^3 - f_a^3}{3(f_b - f_a)} \tag{7.47}$$

where A is the received DC current.

Clipping noise and Gaussian noise are independent of each other. The resultant composite PDF for the overall noise is the convolution of the two individual PDFs. The BER for (OFDM)-QAM can be worked out using the composite PDF, and the result is as follows[1]:

$$p(x) = \frac{2(1 - 1/\sqrt{M})}{\log_2 \sqrt{M}} \left\{ (1 - \upsilon T_0)\mathrm{erfc}\left(\frac{d_{\min}}{2\sqrt{2}\sigma_g}\right) + \frac{\upsilon T_0}{\pi} \times \right.$$

$$\left. \int_{-\infty}^{\infty} \left[\int_0^{\pi/2} \frac{P_a(-|y/\cos \lambda|)}{\cos \lambda} \, d\lambda \right] \times \mathrm{erfc}\left[\frac{(d_{\min}/2) - y}{\sqrt{2}\sigma_g}\right] dy \right\} \tag{7.48}$$

where $d_{\min}$ is the minimum distance for each M-QAM signal between the two neighboring constellation points, $d_{\min} = \sqrt{6P_{av}T_0/(M - 1)}$, and $P_{av} = (m_0RP_0)^2/2$ is the averaged received signal power for the M-QAM signal.

The parameters for the simulated system in Pan and Green[1] are 65 AM channels with 6 MHz spacing, with lowest and highest frequencies of 52.25 and 442.25 MHz, respectively. The comparison is made between single-carrier QAM and 1024-subcarrier QAM-OFDM. The single-carrier symbol period is T_q of 1/6 µs. The same data rate and the received optical power are assumed for OFDM-QAM. This implies that

$$m_0 = m_q/32 \quad \text{and} \quad T_0 = 1024T_q \tag{7.49}$$

where m_0 are T_0 are, respectively, the OMI and symbol period for each individual OFDM subcarrier, whereas m_q and T_q are for single-carrier QAM. The relevant optical parameters are as follows: received DC optical power of 0 dBm, receiver thermal current i_n of 10 pA/$\sqrt{Hz}$, responsivity of photodiode R of 0.9 W/A, and RIN of −155 dBc/Hz.

The first interesting aspect is the comparison of the PDF between single-carrier QAM and QAM-OFDM. Figure 7.25 shows the PDF for various noise sources as a function of the noise current at the output of the matched filter. The PDF for Gaussian noise is the same for the QAM-OFDM and single-carrier QAM. The fundamental difference lies in the clipping noise distribution. Compared with the single-carrier QAM, the QAM-OFDM clipping noise density is much reduced for the large current amplitude (larger than 2×10^{-11}). This means that the QAM-OFDM is unlikely to have a large distortion. The single carrier is more susceptible to the clipping-induced impulse noise because one complete bit or a few consecutive bits could be "ruined" by the impulse noise. In contrast, in the OFDM system, such impulse noise is spread across thousands of subcarriers, and the impulse noise changes into almost white noise

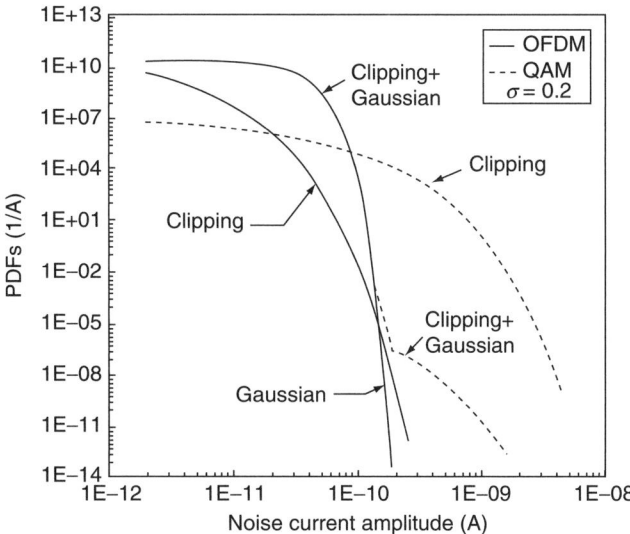

Figure 7.25: Probability density function for clipping noise, Gaussian noise, and the sum of the two noises at the output of the matched filter in OFDM and QAM systems, $\sigma = 0.2$. After Pan and Green.[1]

in the frequency domain. When combined with the Gaussian noise, the overall noise is almost solely dominated by Gaussian noise, as shown in Figure 7.25. Therefore, in the clipping-induced noise environment, OFDM systems have a large modulation dynamic range.

Figure 7.26 shows the BER performance as a function of the effective modulation depth σ. The BER floor at the lower σ is due to the dominance of the Gaussian noise. As the effective

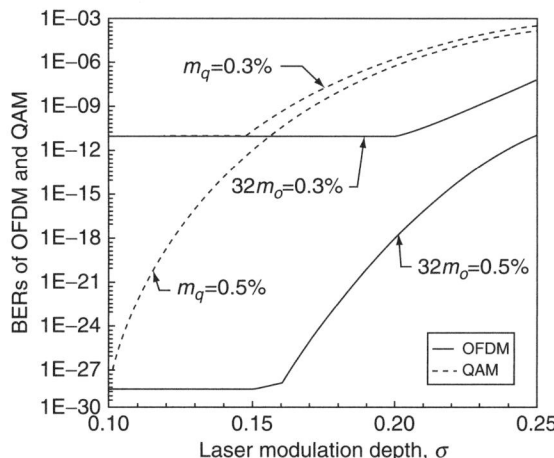

Figure 7.26: BERs of OFDM and QAM signals as a function of the laser modulation index, σ. 32 $m_0 = m_q$ is, respectively, 0.3 and 0.5%. After Pan and Green.[1]

modulation depth increases, the clipping noise increases above the Gaussian noise. It can be seen that the BERs of OFDMs for different modulation depths are much lower than those of QAM. The OFDM has a much wider region in which the noise is dominated by Gaussian noise. For instance, for an OMI m_0 of 0.5%, the corner modulation depth at which the clipping noise overtakes the Gaussian noise for OFDM and QAM is, respectively, 0.10 and 0.16 σ.

References

1. Pan Q, Green RJ. Bit-error-rate performance of lightwave hybrid AM/OFDM systems with comparison with AM/QAM systems in the presence of clipping impulse noise. *IEEE Photon Technol Lett* 1996;**8**:278–80.
2. Shieh W, Athaudage C. Coherent optical orthogonal frequency division multiplexing. *Electron Lett* 2006;**42**:587–9.
3. Shieh W, Yi X, Ma Y, Tang Y. Theoretical and experimental study on PMD-supported transmission using polarization diversity in coherent optical OFDM systems. *Opt Express* 2007;**15**:9936–47.
4. Shieh W, Yi X, Tang Y. Transmission experiment of multi-gigabit coherent optical OFDM systems over 1000 km SSMF fiber. *Electron Lett* 2007;**43**:183–5.
5. Jansen SL, Morita I, Takeda N, Tanaka H. 20-Gb/s OFDM transmission over 4160-km SSMF enabled by RF-Pilot tone phase noise compensation. In: *Opt. Fiber Commun. Conf.*, paper no. PDP15 Anaheim, CA; 2007.
6. Yamada E, Sano A, Masuda H, et al. Novel no-guard interval PDM CO-OFDM transmission in 4.1 Tb/s (50 88.8-Gb/s) DWDM link over 800 km SMF including 50-GHz spaced ROADM nodes. In: *Opt. Fiber Commun. Conf.*, paper no. PDP8. San Diego; 2008.
7. Tang Y, Shieh W, Yi X, Evans R. Optimum design for RF-to-optical up-converter in coherent optical OFDM systems. *IEEE Photon Technol Lett* 2007;**19**:483–5.
8. Ly-Gagnon DS, Tsukarnoto S, Katoh K, Kikuchi K. Coherent detection of optical quadrature phase-shift keying signals with carrier phase estimation. *J Lightwave Technol* 2006;**24**:12–21.
9. Savory SJ, Gavioli G, Killey RI, Bayvel P. Electronic compensation of chromatic dispersion using a digital coherent receiver. *Opt Express* 2007;**15**:2120–6.
10. Tang Y, Ho K, Shieh W. Coherent optical OFDM transmitter design employing predistortion. *IEEE Photon Technol Lett* 2008;**20**:954–6.
11. Barros DJF, Kahn JM. Optical modulator optimization for orthogonal frequency-division multiplexing. Accepted to publish in *J Lightwave Technol* 2009;**27**:2370–2378.
12. Auracher F, Keil R. Method for measuring the RF modulation characteristics of Mach–Zehnder-type modulators. *Appl Phys Lett* 1980;**36**:626–9.
13. Kolner BH, Dolfi DW. Intermodulation distortion and compression in an integrated electrooptic modulator. *Appl Optics* 1987;**26**:3676–80.
14. Mayrock M, Haunstein H. Impact of implementation impairments on the performance of an optical OFDM transmission system. In: *Eur. Conf. Opt. Commun.*, paper no. Th3.2.1. Cannes, France; 2006.
15. Cohn SB, Weinhouse NP. An automatic microwave phase measurement system. *Microwave Journal* 1964;**7**:49–56.
16. Hoer CA, Roe KC. Using an arbitrary six-port junction to measure complex voltage ratios. *IEEE Trans on MTT* 1975;**MTT-23**:978–84.
17. Tang Y, Chen W, Shieh W. Study of nonlinearity and dynamic range of coherent optical OFDM receivers. In: *Opt. Fiber Commun. Conf.*, paper JWA65. San Diego, CA; 2008.
18. Jackson JD. *Classical Electrodynamics*. 3rd ed. New York: Wiley; 1998.
19. Proakis J. *Digital Communications*. 3rd ed. New York: McGraw-Hill; 1995 [Chapter 5].

20. Shieh W, Tucker RS, Chen W, Yi X, Pendock G. Optical performance monitoring in coherent optical OFDM systems. *Opt Express* 2007;**15**:350–6.

21. Lowery AJ, Du L, Armstrong J. Orthogonal frequency division multiplexing for adaptive dispersion compensation in long haul WDM systems. In: *Opt. Fiber Commun. Conf.*, paper no. PDP 39. Anaheim, CA; 2006.

22. Djordjevic IB, Vasic B. Orthogonal frequency division multiplexing for high-speed optical transmission. *Opt Express* 2006;**14**:3767–75.

23. Tang JM, Shore KA. Maximizing the transmission performance of adaptively modulated optical OFDM signals in multimode-fiber links by optimizing analog-to-digital converters. *J Lightwave Technol* 2007;**25**:787–98.

24. Jin XQ, Tang JM, Spencer PS, Shore KA. Optimization of adaptively modulated optical OFDM modems for multimode fiber-based local area networks [Invited]. *J Opt Networking* 2008;**7**:198–214.

25. Schmidt BJC, Lowery AJ, Armstrong J. Experimental demonstrations of 20 Gbit/s direct-detection optical OFDM and 12 Gbit/s with a colorless transmitter. In: *Opt. Fiber Commun. Conf.*, paper no. PDP18. San Diego; 2007.

26. Hewitt DF. Orthogonal frequency division multiplexing using baseband optical single sideband for simpler adaptive dispersion compensation. In: *Opt. Fiber Commun. Conf.*, paper no. OME7. Anaheim, CA; 2007.

27. Peng W, Wu X, Arbab VR, et al. Experimental demonstration of 340 km SSMF transmission using a virtual single sideband OFDM signal that employs carrier suppressed and iterative detection techniques. In: *Opt. Fiber Commun. Conf.*, paper no. OMU1. San Diego; 2008.

28. Peng WR, Wu X, Arbab VR, et al. Experimental demonstration of a coherently modulated and directly detected optical OFDM system using an RF-tone insertion. In: *Opt. Fiber Commun. Conf.*, paper no. OMU2. San Diego; 2008.

29. Lowery AJ, Du LB, Armstrong J. Performance of optical OFDM in ultralong-haul WDM lightwave systems. *J Lightwave Technol* 2007;**25**:131–8.

30. Agrawal GP. *Fiber-Optic Communication Systems*. 3rd ed. New York: Wiley; 2002.

31. Schuster M, Randel S, Bunge CA, et al. Spectrally efficient compatible single-sideband modulation for OFDM transmission with direct detection. *IEEE Photon Technol Lett* 2008;**20**:670–2.

32. Jolley NE, Kee H, Pickard P, Tang J, Cordina K. Generation and propagation of a 1550 nm 10 Gbit/s optical orthogonal frequency division multiplexed signal over 1000 m of multimode fibre using a directly modulated DFB. In: *Opt. Fiber Commun. Conf.*, paper no. OFP3. Anaheim, CA; 2005.

33. Dixon BJ, Pollard RD, Iezekeil S. Orthogonal frequency-division multiplexing in wireless communication systems with multimode fiber feeds. *IEEE Trans Microwave Theory Tech* 2001;**49**:1404–9.

34. Tang JM, Shore KA. Maximizing the transmission performance of adaptively modulated optical OFDM signals in multimode-fiber links by optimizing analog-to-digital converters. *J Lightwave Technol* 2007;**25**:787–98.

35. Jin XQ, Tang JM, Spencer PS, Shore KA. Optimization of adaptively modulated optical OFDM modems for multimode fiber-based local area networks [Invited]. *J Opt Networking* 2008;**7**:198–214.

36. Lowery AJ, Armstrong J. 10Gbit/s multimode fiber link using power-efficient orthogonal-frequency-division multiplexing. *Opt Express* 2005;**13**:10003–9.

37. Armstrong J, Lowery AJ. Power efficient optical OFDM. *Electron Lett* 2006;**42**:370–2.

38. Lee SCJ, Breyer F, Randel S, et al. 24-Gb/s transmission over 730 m of multimode fiber by direct modulation of an 850-nm VCSEL using discrete multitone modulation. In: *Opt. Fiber Commun. Conf.*, paper no. PDP6. Anaheim, CA; 2007.

39. Lee SCJ, Breyer F, Randel S, van den Boom HPA, Koonen AMJ. High-speed transmission over multimode fiber using discrete multitone modulation [Invited]. *J Opt Networking* 2008;**7**:183–96.

40. Tang JM, Shore KA. 30 Gb/s signal transmission over 40-km directly modulated DFB-laser-based single-mode-fibre links without optical amplification and dispersion compensation. *J Lightwave Technol* 2006;**24**:2318–27.

41. Giacoumidis E, Wei JL, Jin XQ, Tang JM. Improved transmission performance of adaptively modulated optical OFDM signals over directly modulated DFB laser-based IMDD links using adaptive cyclic prefix. *Opt Express* 2008;**16**:9480–94.

42. Mazo JE. Asymptotic distortion spectrum of clipped, dc-biased Gaussian noise. *IEEE Trans Commun* 1992;**40**:1339–44.

CHAPTER 8

Spectrally Efficient High-Speed Coherent OFDM System

8.1 Introduction

Among all the variants of optical orthogonal frequency-division multiplexing (OFDM) systems, as we illustrated in Chapter 7, coherent optical OFDM (CO-OFDM) offers the ultimate performance in spectral efficiency, receiver sensitivity, and polarization or chromatic dispersion tolerance. The incorporation of the 2×2 multiple-input multiple-output OFDM (MIMO-OFDM) opens avenues for doubling the spectral efficiency while providing polarization dispersion and rotation robustness. This has become very attractive in light of the push for 100 Gb/s Ethernet, for which compliance to the 50 GHz ITU grid is critical. Since the conception of CO-OFDM, experimental and theoretical research has progressed rapidly. 100 Gb/s CO-OFDM transmission over 1000 km has been demonstrated by groups from the University of Melbourne,[1] KDDI,[2] and NTT.[3] Because CO-OFDM uses digital-to-analog converters (DACs) at the transmitter and also unique pilot subcarrier-based channel and phase estimation, it offers an extremely convenient way to achieve high spectral efficiency transmission through higher order modulation. 64-QAM in single polarization[4] and 16-QAM in dual polarization[5] have been reported, both of which represent record spectral efficiency for single polarization and dual polarization, respectively, using off-the-shelf commercial components. As such, CO-OFDM has emerged as the attractive modulation format for 100 Gb/s and beyond due to its high spectral efficiency and resilience to fiber dispersion. The superior computational efficiency of CO-OFDM that scales very well over speed and dispersion enables its smooth migration to a higher data rate of 400 Gb/s and beyond. More important, CO-OFDM is a rate-agnostic modulation format in which the hardware and software can be ported seamlessly from the current generation to the next one whenever transmission speed is upgraded. This is in contrast to other conventional single-carrier modulation formats, with which, as the network evolves and migrates, a complete change of the transceiver design, and at times even the link design, is mandated. In this chapter, we illustrate the application of CO-OFDM techniques to the optical long-haul transmission at 100 Gb/s and beyond, and we discuss the challenges and solutions for those high-performance systems.

Despite many niche applications for higher order modulation, due to the increased optical signal-to-noise ratio (OSNR) requirement and sensitivity to fiber nonlinearity, we anticipate that quaternary phase-shift keying (QPSK; 4-QAM) or 2 bit per symbol modulation will be the dominant modulation for long-haul transmission systems as a satisfactory compromise between spectral efficiency and system performance. In Section 8.2, we describe two experimental demonstrations of QPSK encoded 100 Gb/s transmission—orthogonal band multiplexing (OBM-OFDM) and no-guard-interval CO-OFDM. Specifically, OBM-OFDM offers a bandwidth scalable approach that can be implemented with the state-of-the-art silicon ASICs. No-guard-interval CO-OFDM is a novel alternative that takes advantage of the orthogonality principle—namely, overlapping two optical subcarrier spectra without incurring interference penalty. In Section 8.3, we investigate through simulation a host of issues in relation to high-speed transmission at 100 Gb/s and beyond. The issues are (1) performance comparison between dispersion-compensated link and nondispersion-compensated link, (2) the optical add–drop multiplexor (OADM) filtering impact, and (3) uniform filling and random filling for OBM-OFDM systems. The high spectral modulation format of over 2 bits per symbol has received significant attention.[4,5] It could potentially be very useful for some niche applications, such as adaptive date rate and ultra high capacity for metro networks.[6–8] In Section 8.4, we review some higher order modulation experiments and discuss the phase noise impact on the higher order QAM-modulated OFDM transmission systems.

8.2 Orthogonal Band Multiplexed OFDM

As IP traffic continues to increase at a rapid pace, the 100 Gb/s Ethernet is being considered as the next-generation transport standard for IP networks.[9] As the data rate approaches 100 Gb/s and beyond, the electrical bandwidth required for CO-OFDM would be at least 15 GHz,[10] which is not cost-effective to implement even with the best commercial DACs and analog-to-digital converters (ADCs) in silicon integrated circuit (IC).[11] To overcome this electrical bandwidth bottleneck, the concept of OBM-OFDM is proposed to divide the entire OFDM spectrum into multiple orthogonal bands.[10,12] Due to interband orthogonality, the multiple OFDM bands with zero or small guard bands can be multiplexed and demultiplexed without interband interference. With this scheme, transmission of a 107 Gb/s CO-OFDM signal over 1000 km (10 × 100 km) standard single-mode fiber (SSMF) has been realized using an erbium-doped fiber amplifier (EDFA) without the need for optical dispersion compensation.[10] Although several experiments have demonstrated transmission at 100 Gb/s and above at longer distances, relying on a dispersion compensation module and Raman amplification (RA) in each span,[13,14] Shieh et al.[10] achieved 1000 km transmission without optical dispersion compensation and without RA beyond 100 Gb/s. The 107 Gb/s OBM-OFDM can also be considered as 5 × 21.4 Gb/s wavelength-division multiplexing (WDM) channels without a frequency guard band, occupying 32 GHz optical bandwidth, implying a high spectral efficiency of 3.3 bits/Hz using only 4-QAM encoding.

By multiplexing and demulitpelxing multiple OFDM bands, OBM-OFDM has the following advantages: (1) High spectral efficiency can be achieved by allowing for zero or small guard band; (2) OBM-OFDM offers the flexibility of demodulating two OFDM sub-bands simultaneously with just one fast Fourier transform (FFT), whereas three inverse FFTs (IFFTs) would otherwise be needed for the same purpose; (3) OBM-OFDM can be readily partitioned with electrical anti-alias filters and subsequently processed with lower speed DAC/ADCs[6]; and (4) the required cyclic prefix length is shortened due to the sub-banding of the overall spectrum.

Without enforcing the band orthogonality, the band-multiplexed OFDM or multiband OFDM has also been proposed[15] for DDO-OFDM and demonstrated for CO-OFDM.[2] The multiband OFDM has been successfully implemented for 8-QAM OFDM at 122 Gb/s over 1000 km transmission,[2] and Jansen et al.[16] provide a detailed discussion of various multiband OFDM architectures. In this chapter, we illustrate the 100 Gb/s CO-OFDM performance with the configuration of OBM-OFDM where the orthogonality is enforced between the sub-bands.

8.2.1 Principle of OBM-OFDM

The basic principle of OBM-OFDM is to partition the OFDM into multiple sub-bands while maintaining their orthogonal property. As shown in Figure 8.1, the entire OFDM spectrum comprises N OFDM bands, each with the subcarrier spacing of Δf and band frequency guard spacing of Δf_G. The subcarrier spacing Δf is identical for each band because the same sampling clock within one circuit is used. The orthogonal condition between the different bands is given by

$$\Delta f_G = m \Delta f \qquad (8.1)$$

Namely, the guard band is a multiple (m times) of subcarrier spacing. In doing so, the orthogonality condition is satisfied for any two subcarriers inside the complete OFDM spectrum. For instance, the subcarrier f_i in band 1 is orthogonal to another subcarrier f_j in a different OFDM band (band 2). When $m = 1$ in Eq. (8.1), the OFDM bands can be multiplexed/demultiplexed even without the guard band, despite the fact that they originate

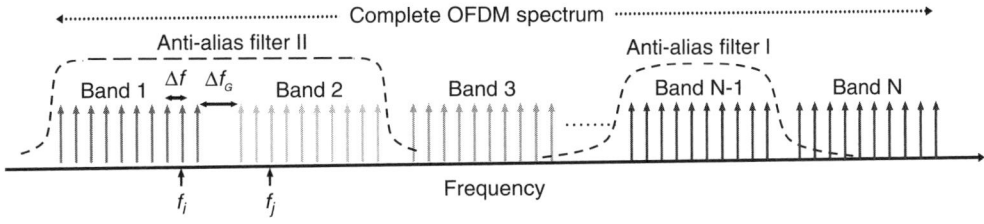

Figure 8.1: Conceptual diagram of OBM-OFDM. Anti-alias filters I and II correspond to two detection approaches described in Section 8.2.1.

from different bands. This method of subdividing OFDM spectrum into multiple orthogonal bands is called OBM-OFDM. An identical bandwidth-scalable and spectral-efficient multiplexing scheme for CO-OFDM was first proposed in Shieh et al.,[17] where it was called cross-channel OFDM (XC-OFDM). We adopt the term OBM-OFDM to stress the bandwidth reduction through sub-banding the OFDM spectrum.

Using such a scheme, each OFDM sub-band can be demultiplexed using an anti-alias filter slightly wider than the signal band. To detect OBM-OFDM, two approaches can be used. First, the receiver laser is tuned to the center of each band. Each band is detected separately by using an "anti-alias filter I" that low passes only one-band RF signals. Second, the local laser is tuned to the center of the guard band. Two bands are detected by using an "anti-alias filter II" that low passes two-band RF signals simultaneously. In either case, the interband interference can be avoided because of the orthogonality between the neighboring bands, despite the leakage of the subcarriers from neighboring bands.

8.2.2 Implementation of OBM-OFDM

OBM-OFDM can be implemented in the electrical domain, the optical domain, or a combination of both. As mentioned previously, OBM-OFDM is particularly suitable for realization with mixed-signal ICs to resolve the ADC/DAC bandwidth bottleneck, whereas the optical realization of OBM-OFDM serves as an alternative to the other spectrally efficient multiplexing schemes, including coherent WDM,[18] all-optical OFDM,[19] and electro-optically subcarrier-multiplexed OFDM.[20]

8.2.2.1 Electrical Implementation of OBM-OFDM

Figure 8.2 shows conceptual diagrams for implementing OBM-OFDM using mixed-signal circuits. In Figure 8.2a, each OFDM baseband transmitter is implemented using digital IC design (excluding DAC). The subsequent up-conversion, band filtering, and RF amplification can be implemented in RF IC design. The output of the OFDM baseband transmitter will be filtered through an anti-alias filter and up-converted to the appropriate RF band with the center frequency from f_1 to f_N using an IQ modulator or a complex multiplexer, the structure of which is shown in Figure 8.2c. The range of f_1 to f_N is centered around zero, given by

$$f_l = l \cdot \Delta f_b \quad l \in [-L, L] \tag{8.2}$$

where f_l is the center frequency of the lth OFDM band, Δf_b is the band spacing, and L is the maximum band number. The output of each IQ modulator is a complex value that has real and imaginary parts, as shown in Figure 8.2c. These complex signals are further summed up at the output, namely real and imaginary parts are added up in separate parallel paths. The combined complex OFDM signal will be used to drive an optical IQ modulator to be up-converted to the optical domain.[17,21] At the receive end as shown in Figure 8.2b, the incoming signal is split into multiple sub-bands and down-converted to baseband using IQ

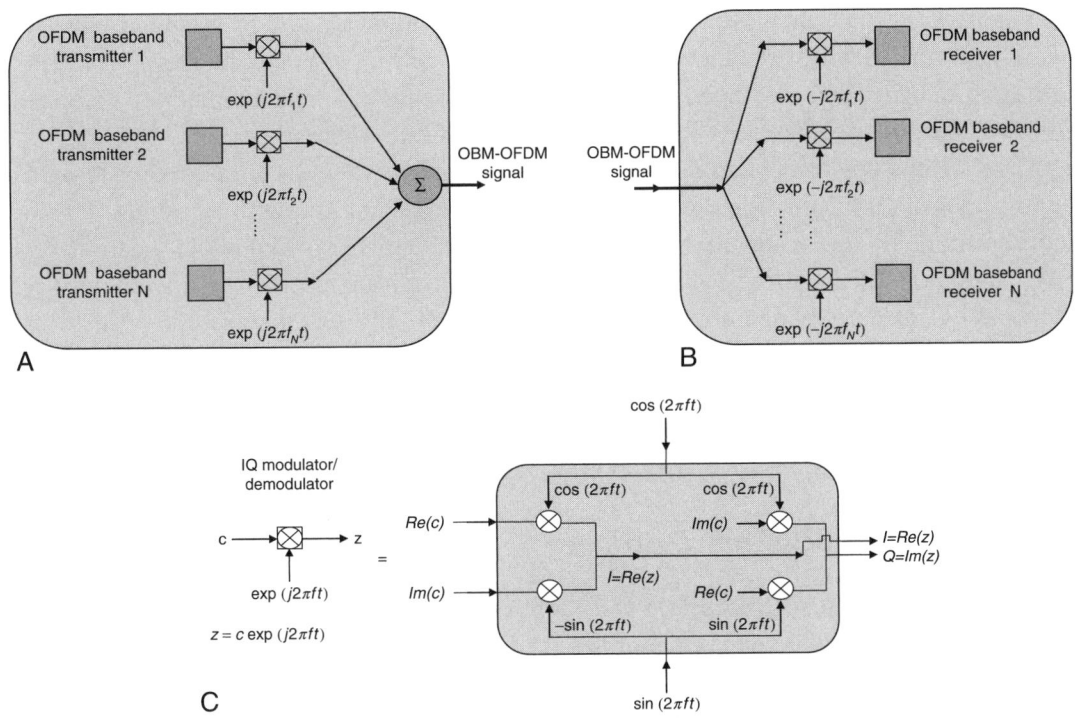

Figure 8.2: Schematic of OBM-OFDM implementation in mixed-signal circuits for (a) the transmitter, (b) the receiver, and (c) the IQ modulator/demodulator. Both the output from the transmitter in Figure 8.2a and the input to the receiver in Figure 8.2b are complex signals with real and imaginary components.

demodulators. Anti-alias filters should be used to remove unwanted high-frequency components at the output of the demodulators. In such a way, the DAC/ADC only needs to operate at the bandwidth of each OFDM band, which is approximately scaled down by a factor equal to the number of sub-bands from the original complete OFDM spectrum. For instance, the bandwidth of 107 Gb/s data rate with QPSK modulation and polarization multiplexing is approximately 35 GHz. If the number of sub-bands is five, each OFDM band will only need to cover approximately 7 GHz optical bandwidth. The electrical bandwidth required is 3.5 GHz, or half of the OFDM band spectrum if direct conversion is used at transmit and receive ends. The DAC/ADC with bandwidth of 3.5 GHz can be implemented in today's technology,[11] and using a wider bandwidth for each OFDM band will reduce the number of OFDM bands to two or three.

8.2.2.2 Optical Implementation of OBM-OFDM

The OBM-OFDM could be realized using either subcarrier multiplexing[3] or wavelength multiplexing to patch multiple orthogonal bands into a complete OFDM spectrum (see Figure 8.1). The OBM-OFDM can also be optically implemented by transmitting OFDM data through many WDM channels and locking all the lasers to the common optical standard such as an optical comb or directly using an optical comb.[22–24] The architecture of optical OBM-OFDM implementation is the same as the electronic implementation shown in Figure 8.2. However, the electronic components in Figure 8.2 will be replaced with optical counterparts; in particular, each RF IQ modulator will be replaced with an optical IQ modulator comprising two parallel Mach–Zehnder modulators (MZMs) with 90-degree phase shift,[21] and each RF local oscillator (LO) at either transmit or receive will be replaced with an optical LO, preferably demulitplexed from an optical comb. In doing so, the orthogonality condition is satisfied for all subcarriers across all WDM channels. This form of OFDM transmission is called XC-OFDM.[17] An optical filter with bandwidth slightly larger than the channel bandwidth can be used to select the desired channel. Consequently, no frequency guard band is necessary between neighboring WDM channels.

8.2.3 Experimental Setup and Description

Although the electronic OBM-OFDM is a more cost-effective solution, the research demonstration will involve expensive high-speed mixed-signal design, foundry run, and chip testing. We chose optical multiplexing to obtain OBM-OFDM for a proof-of-concept demonstration at 107 Gb/s. Figure 8.3 shows the experimental setup for 107 Gb/s CO-OFDM transmission. The details of the experimental setup are described next.

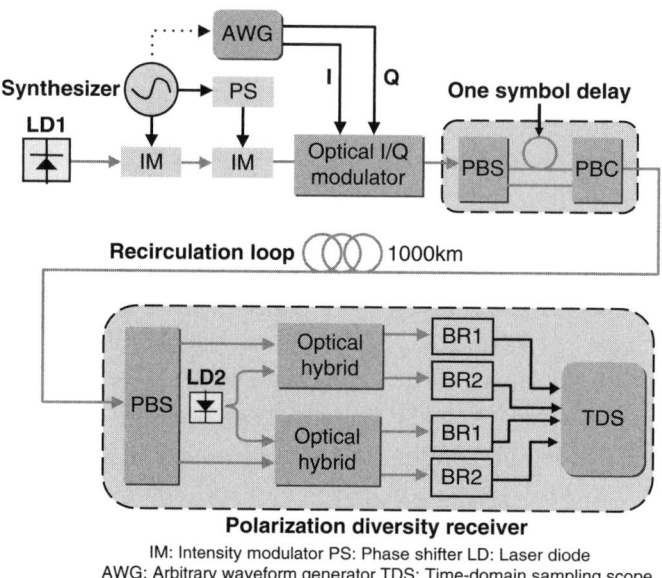

IM: Intensity modulator PS: Phase shifter LD: Laser diode
AWG: Arbitrary waveform generator TDS: Time-domain sampling scope
PBS/C: Polarization splitter/combiner BR: Balanced receiver

Figure 8.3: Experimental setup for 107 Gb/s OBM-OFDM systems.

8.2.3.1 CO-OFDM Transmitter

The 107 Gb/s OBM-OFDM signal is generated by multiplexing five OFDM (sub)bands. In each band, 21.4 Gb/s OFDM signals are transmitted in both polarizations. The multifrequency optical source spaced at 6406.25 MHz is generated by cascading two intensity modulators (IMs). The guard band equals just one subcarrier spacing ($m = 1$ in Eq. 8.1). Figure 8.4 shows the multiple tones generated by this cascaded architecture using two IMs, which is different from that using one IM and one phase modulator.[23,24] The two-modulator setup enables better flatness across the five bands and less leakage spectrum out of the intended five bands. Only the middle five tones with large and even power are used for performance evaluation. The transmitted signal is generated off-line by the Matlab program with a length of 2^{15}-1 PRBS and mapped to 4-QAM constellation. The digital time domain signal is formed after IFFT operation. The total number of OFDM subcarriers is 128, and the guard interval is one-eighth of the observation window. The middle 82 subcarriers out of 128 are filled, from which 4 pilot subcarriers are used for phase estimation. The I and Q components of the time domain signal are uploaded onto a Tektronix arbitrary waveform generator (AWG), which provides the analog signals at 10 GS/s for both I and Q components. Figure 8.5a shows the electrical spectrum of the I/Q channel at the output of AWG. It can be seen that the aliasing components of the OFDM signal are present above 6 GHz. When combining multiple OFDM sub-bands,

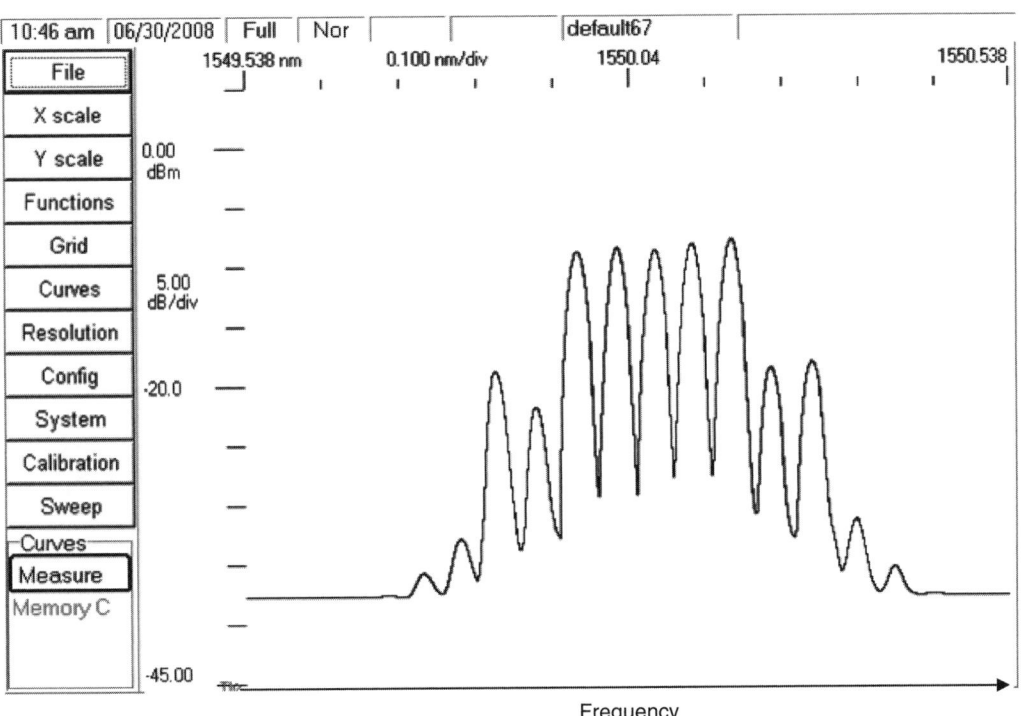

Figure 8.4: Multiple tones generated by two cascaded intensity modulators.

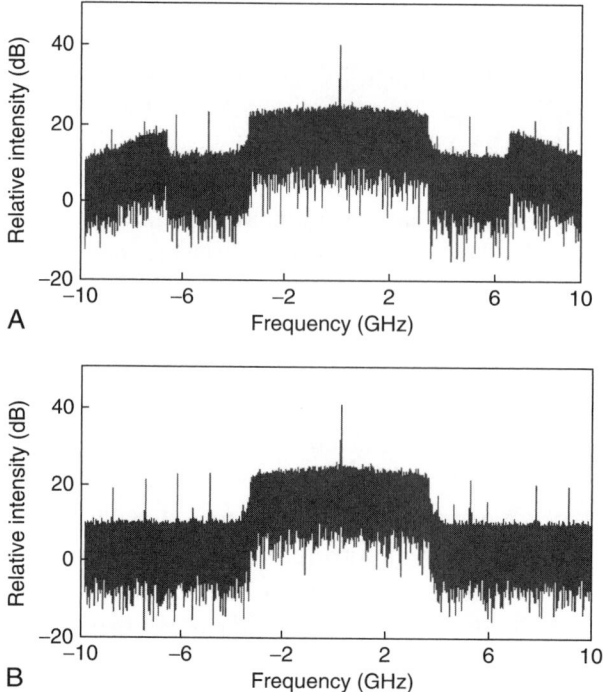

Figure 8.5: The electrical spectra (a) directly at the output of the AWG and (b) after 3 GHz anti-aliasing filters.

such aliasing frequency components will degrade the signals in the adjacent bands. A 3 GHz low-pass electrical filter is used to eliminate the aliasing OFDM components. Figure 8.5b shows the electrical spectrum after low-pass filtering, where the aliasing spectrum components are removed.

The AWG is phase locked to the synthesizer through 10 MHz reference. The optical I/Q modulator comprising two MZMs with 90-degree phase shift is used to directly impress the baseband OFDM signal onto five optical tones. The modulator is biased at the null point to completely suppress the optical carrier and perform linear RF-to-optical up-conversion.[17,21] The optical output of the I/Q modulator consists of five-band OBM-OFDM signals. Each band is filled with the same data at 10.7 Gb/s data rate and this is consequently called *uniform filling*. To improve the spectrum efficiency, 2×2 MIMO-OFDM is employed; that is, two OFDM transmitters are needed to send two independent data into each polarization, which are then detected by two OFDM receivers—one for each polarization.

A cost-effective method is adopted to emulate the two transmitters as follows: The single-polarization optical OFDM signal at the output of the I/Q modulator is first evenly split into two polarization branches with a polarization beam splitter, with one branch delayed by one OFDM symbol period—that is, 14.4 ns in this experiment, which equals one OFDM symbol length. The two polarization branches are subsequently combined, emulating two independent transmitters, one on each polarization, resulting in a composite data rate of 21.4 Gb/s. The two polarization components are completely independent due to the delay of 14.4 ns for each OFDM symbol.

8.2.3.2 Fiber Link

A multispan fiber link is emulated with a recirculation loop, which consists of 100 km SSMF fiber and an EDFA to compensate the link loss. Neither optical dispersion compensation nor RA is used for the transmission.

8.2.3.3 CO-OFDM Receiver

The signal is coupled out of the loop and received with a polarization diversity coherent receiver[25,26] comprising a polarization beam splitter, a local laser, and two hybrids, and four balanced receivers. The complete OFDM spectrum, composed of five sub-bands, is shown in Figure 8.6a. The entire bandwidth for 107 Gb/s OFDM signal is only 32 GHz.

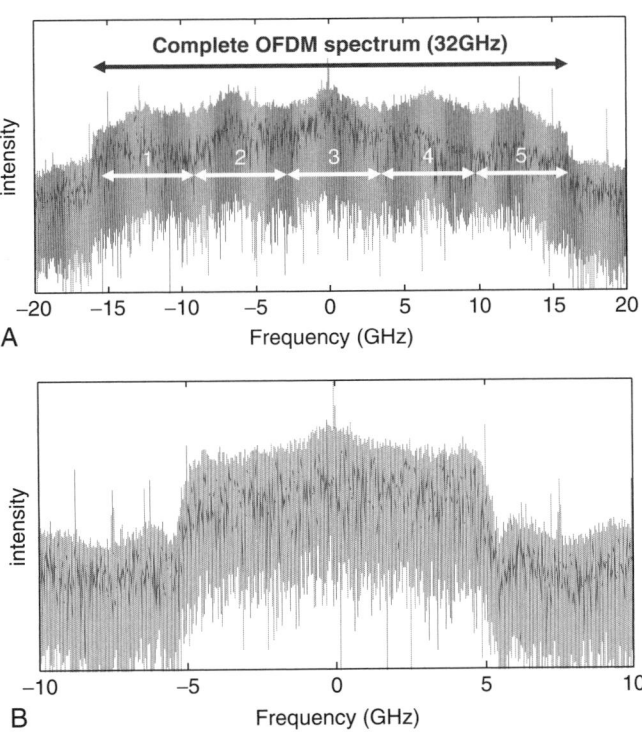

Figure 8.6: (a) Optical spectrum for the 107 Gb/s signal using a polarization diversity coherent receiver. The band numbers are depicted next to the corresponding bands. (b) The RF spectrum at the receiver after the 3.8 GHz anti-alias filter.

The local laser is tuned to the center of each band, and the RF signals from the four balanced detectors are first passed through the anti-aliasing low-pass filters with a bandwidth of 3.8 GHz, such that only a small portion of the frequency components from other bands is passed through, which can be easily removed during OFDM signal processing. The performance of each band is measured independently. The detected RF signals are then sampled with a Tektronix time domain sampling scope at 20 GS/s. The sampled data are processed with a Matlab program to perform 2×2 MIMO-OFDM processing. The receiver signal processing[25,26] involves (1) FFT window synchronization using Schmidl format to identify the start of the OFDM symbol, (2) software estimation and compensation of the frequency offset, (3) channel estimation in terms of the Jones matrix H, (4) phase estimation for each OFDM symbol, and (5) constellation construction for each subcarrier and bit error ratio (BER) computation.

The channel matrix H is estimated by sending 30 OFDM symbols using alternative polarization launch. Much fewer pilot symbols are needed if the spectral correlation of the channel matrix H is explored.[27] The total number of OFDM symbols evaluated is 500. Mathematically, the transmitter information symbols of the two polarizations in the form of a Jones vector are given by[28]

$$c = \begin{bmatrix} c_1 \\ c_2 \end{bmatrix} \tag{8.3}$$

where c_1 and c_2 are transmitted OFDM information symbols for two polarizations. Assume the fiber transmission Jones matrix H is

$$H = \begin{bmatrix} h_{11} & h_{12} \\ h_{21} & h_{22} \end{bmatrix} \tag{8.4}$$

Ignoring the additive noise, the received OFDM signal in the form of a Jones vector is thus given by

$$c' = \begin{bmatrix} c_1' \\ c_2' \end{bmatrix} = H \cdot c \tag{8.5}$$

or equivalently

$$\begin{cases} c_1' = h_{11}c_1 + h_{12}c_2 \\ c_2' = h_{21}c_1 + h_{22}c_2 \end{cases} \tag{8.6}$$

From Eq. (8.6), the transmitted information symbols can be recovered from the received signals by inverting H:

$$c = H \begin{bmatrix} c_1' \\ c_2' \end{bmatrix}, \quad H = \begin{bmatrix} h_{11} & h_{12} \\ h_{21} & h_{22} \end{bmatrix}^{-1} \tag{8.7}$$

The training symbols for channel estimation are generated by filling the odd symbols with known random data while nulling the even symbols. After the polarization multiplexing emulator, the training symbols form a pattern of alternative polarization launch for two consecutive OFDM symbols. Using odd training symbols, the associated channel estimation can be expressed as

$$\begin{bmatrix} c'_1 \\ c'_2 \end{bmatrix} = \begin{bmatrix} h_{11} & h_{12} \\ h_{21} & h_{22} \end{bmatrix} \begin{bmatrix} c_1 \\ 0 \end{bmatrix} \Rightarrow \begin{cases} h_{11} = c'_1/c_1 \\ h_{21} = c'_2/c_1 \end{cases} \tag{8.8}$$

and using even training symbols, it can be expressed as

$$\begin{cases} h_{12} = c'_1/c_2 \\ h_{22} = c'_2/c_2 \end{cases} \tag{8.9}$$

It can be seen from Eqs. (8.8) and (8.9) that by using the alternative polarization training symbol, the full channel estimation of H can be obtained. Then using the inverse of this matrix in Eq. (8.7) and the received information symbols, c', the transmitted symbols in the two polarizations can be estimated. The estimated transmitted symbols will be mapped to the closest constellation points to recover the transmitted digital bits.

8.2.4 Measurement and Discussion

Figure 8.6a shows the optical spectrum after 1000 km transmission measured with the polarization diversity coherent receiver shown. No frequency guard band ($m = 1$) is used in our transmission measurement. It can be seen that five OFDM bands, each with 6.4 GHz bandwidth, are closely patched together and the entire OFDM spectrum occupied is approximately 32 GHz. The out-of-band components are due to the multifrequency source generation not tightly bounded at five tones. This artifact will not exist in the real application using either subcarrier multiplexing or optical multiplexing OBM-OFDM. Figure 8.6b shows the detected electrical spectrum after using a 3.8 GHz electrical anti-alias filter for one-band detection. The anti-alias filter is critical for OBM-OFDM implementation. As shown in Figure 8.6a, without an electrical anti-alias filter, the electrical spectrum will be as broad as 16 GHz, indicating that at least 32 GS/s ADC has to be used. However, the filtered spectrum in Figure 8.6b can be easily sampled with 20 GS/s or even at a lower speed of 10 Gb/s. In addition, despite the fact that there are some spurious components from a neighboring band that is leaked at the edge of the 3.8 GHz filter, because they are orthogonal subcarriers to the interested OFDM subcarriers at the center, they do not contribute to the interference degradation.

Table 8.1 shows the performance of five bands at both back-to-back and 1000 km transmission. It can be seen that both polarizations in each band can be recovered successfully, and this is done without a need for a polarization controller at the receiver. The difference in BER for each entry is attributed to the tone power imbalance and instability as well as the receiver imbalance for two polarizations. The nonlinearity tolerance of OBM-OFDM transmission is first measured by varying the launch power into

Table 8.1: BER Distribution of Five Bands

BER Distribution	Band				
	1	2	3	4	5
Back-to-back at an OSNR of 17.0 dB					
x-polarization	4×10^{-4}	2.5×10^{-4}	4.63×10^{-4}	3.63×10^{-4}	2×10^{-4}
y-polarization	4×10^{-4}	2.38×10^{-4}	3.88×10^{-4}	3.00×10^{-4}	3×10^{-4}
After 1000 km transmission at an OSNR of 18.9 dB					
x-polarization	1.38×10^{-4}	1.0×10^{-4}	1.38×10^{-4}	1.5×10^{-4}	7.5×10^{-5}
y-polarization	1.63×10^{-4}	1.3×10^{-4}	3.13×10^{-4}	2.5×10^{-4}	1.2×10^{-4}

the optical links. Figure 8.7 shows the system Q performance of the 107 Gb/s CO-OFDM signal as a function of reach up to 1000 km. It can be seen that the Q reduces from 17.2 dB to 12.5 dB when reach increases from back-to-back to 1000 km transmission.

Figure 8.8 shows the BER sensitivity performance for the entire 107 Gb/s CO-OFDM signal at the back-to-back and 1000 km transmission with the launch power of -1 dBm. The BER is counted across all five bands and two polarizations. It can be seen that the OSNR required for a BER of 10^{-3} is, respectively, 15.8 and 16.8 dB for back-to-back and 1000 km transmission. To the best of our knowledge, the OSNR of 15.8 dB for a BER of 10^{-3} is the record sensitivity for 107 Gb/s transmission.

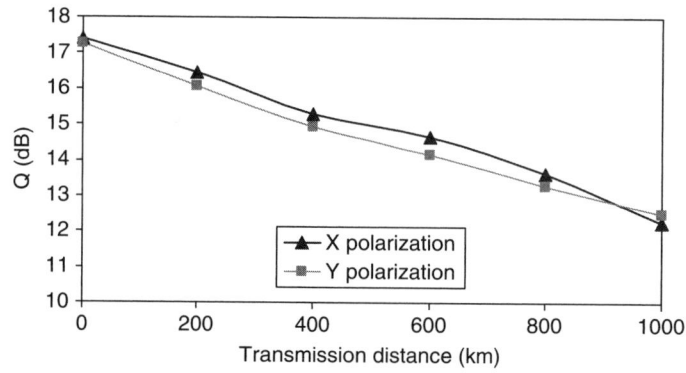

Figure 8.7: Q factor of 107 Gb/s CO-OFDM signal as a function of transmission distance. After Ref.[29]

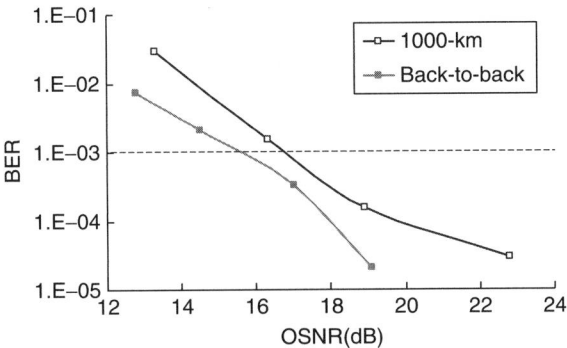

Figure 8.8: BER sensitivity of 107 Gb/s CO-OFDM signal at the back-to-back and 1000 km transmission. After Ref.[29]

8.3 111 Gb/s No-Guard Interval CO-OFDM Transmission

As illustrated in Chapter 7, no-guard interval CO-OFDM (NGI-CO-OFDM) is an alternative type of CO-OFDM that constructs the OFDM signal without FFT and cyclic prefix. It uses the same principle of optical OFDM through proper pulse shaping and phase locking the subcarrier to orthogonal frequency to achieve overlapping of the subcarrier spectrum without incurring interference penalty. It has the advantage over FFT-based CO-OFDM in that it can use the conventional transmitter setup, which is familiar to the general optical communication community. NGI-CO-OFDM was first proposed in references.[19,20,30] The first 110 Gb/s NGI-CO-OFDM transmission was shown by using the electro-optical subcarrier multiplexing technique.[3] By filling two subcarriers in each wavelength, 100 Gb/s WDM NGI-CO-OFDM transmission over 1300 km of G.652 SMFs and over 2100 km dispersion-shift fiber (DSF) was demonstrated.[3] In this section, we review the transmission experiment carried out by Yamada et al.[3] over DSF in which the fiber nonlinearity is a major issue due to small chromatic dispersion.

8.3.1 Experimental Configuration for 111 Gb/s NGI-CO-OFDM Transmission

Figure 8.9 shows the experimental setup for 10-channel 111 Gb/s NGI-CO-OFDM signals. The transmitter is composed of a two-subcarrier generator and an integrated OFDM QPSK subcarrier modulator. The two-subcarrier generator outputs two subcarriers spaced at the symbol rate of 13.9 GHz. The two subcarriers are split into two separate subcarriers with an interleaver, each passing through an IQ modulator for QPSK encoding at the rate of 13.9 Gbaud. The two QPSK subcarrier signals are combined with an optical coupler into a 55.5 Gb/s OFDM signal. The polarization multiplexing of the OFDM signal doubles the data rate to 111 Gb/s per wavelength.

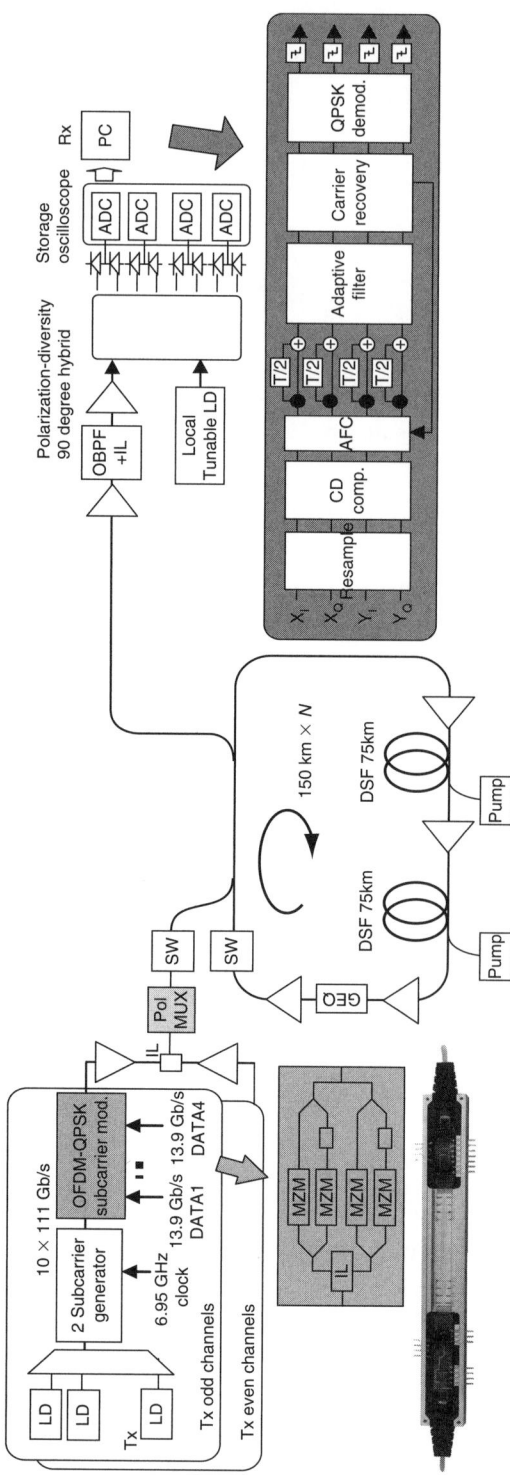

Figure 8.9: Experimental setup for multichannel 111 Gb/s NGI-CO-OFDM transmission. After Yamada et al.[3]

The transmission link is emulated with a 160 km recirculating loop containing two 75 km span DSFs. The DSF has an average chromatic dispersion at 1575 nm of 1.5 ps/nm/km; therefore, fiber nonlinearity is an important factor. At the receiver, each of the WDM signals is filtered and received by the polarization-diversity receiver. The four output signals from the balanced receiver are sampled at 50 GS/s using a real-time digital storage oscilloscope and converted to digital data for off-line processing. Similar to the single-carrier coherent systems,[31] the blind adaptive equalization with transversal filter is used to compensate the channel dispersion and thus the proposed scheme does not require the guard interval and training symbols. The subcarrier is extracted through a two-tap filter with the half symbol period delay. This is equivalent to the two-point FFT in conventional OFDM. The dynamic polarization rotation and polarization mode dispersion is compensated using 12-tap T/4-spaced finite impulse response adaptive filters based on the constant modulus algorithm.[31] The signal then goes through carrier recovery and chromatic compensation[31] to recover the transmitted QPSK data.

8.3.2 The NGI-CO-OFDM Transmission Experimental Results

Figure 8.10 shows the back-to-back OSNR sensitivity of the OFDM signal at 55.5 Gb/s for single polarization and 111 Gb/s for dual polarization. The required OSNR for a BER of 1×10^{-3}, corresponding to a Q-factor of 9.8 dB, is 13.1 and 16.4 dB for 55.5 and 111 Gb/s signals, respectively. The 3.3 dB OSNR sensitivity difference between single polarization and dual polarization shows insignificant degradation due to polarization multiplexing. Figure 8.11 shows the transmission performance and the optical spectrum at 2100 km. The Q-factors for all the WDM channels are better than 9.6 dB, which is 0.5 dB higher than the Q limit of 9.1 dB (dashed line) that yields BERs below 1×10^{-12} with the use of ITU-T G.975.1 enhanced FEC.

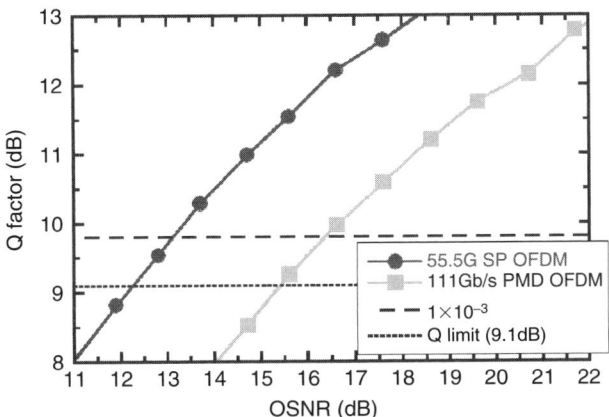

Figure 8.10: Q-factor against OSNR for NGI-CO-OFDM. After Yamada et al.[3]

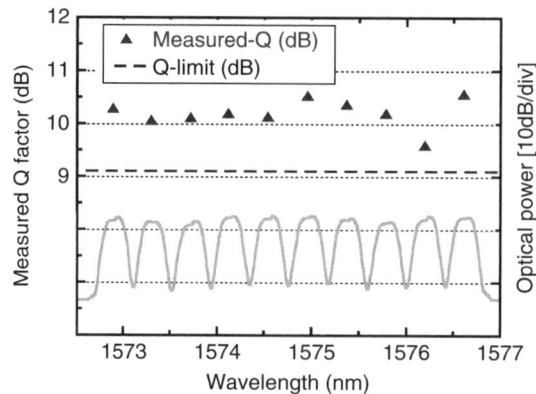

Figure 8.11: System performance of NGI-CO-OFDM after the 2100 km transmission. After Yamada et al.[3]

8.4 Simulation of 100 Gb/s CO-OFDM Transmission

Although the experiment as discussed in Sections 8.2 and 8.3 is more convincing and is the ultimate method of validation, the numerical simulation is more flexible and can carry out many more measurement scenarios for relatively low cost. Three aspects of the transmission issues are discussed in this section. First, in the experiment discussed in Section 8.2, five OFDM bands with the same data are multiplexed (uniform filling) to generate an OBM-OFDM signal. There is concern regarding whether this uniform filling will underestimate the nonlinearity such that the experimental demonstration may overestimate the CO-OFDM transmission performance. Therefore, it is important to compare the performance difference between the uniform filling used in the experiment and the scenario in which each band is filled with independent data (random filling). Second, the most deployed systems use dispersion compensation, although most of the experiments and simulations are focused on the noncompensated systems. It is instructive to determine whether the CO-OFDM can cope with the existing deployed fiber link and the difference in performance compared to that of the one without the dispersion compensation.[32] Finally, the influence of the OADM filtering effects on the 100 Gb/s signal has become an important issue, and the various OADM configurations are difficult to investigate experimentally. It is of great interest to identify through simulation the tolerance to OADM filtering for the CO-OFDM signals.

8.4.1 Comparison between Uniform Filling and Random Filling for 100 Gb/s OBM-OFDM

The basic CO-OFDM system with polarization multiplexed transmission is shown in Figure 8.12. The simulation is configured in such a way that it closely mirrors the experimental setup in Figure 8.13. Two optical OFDM signals are generated independently

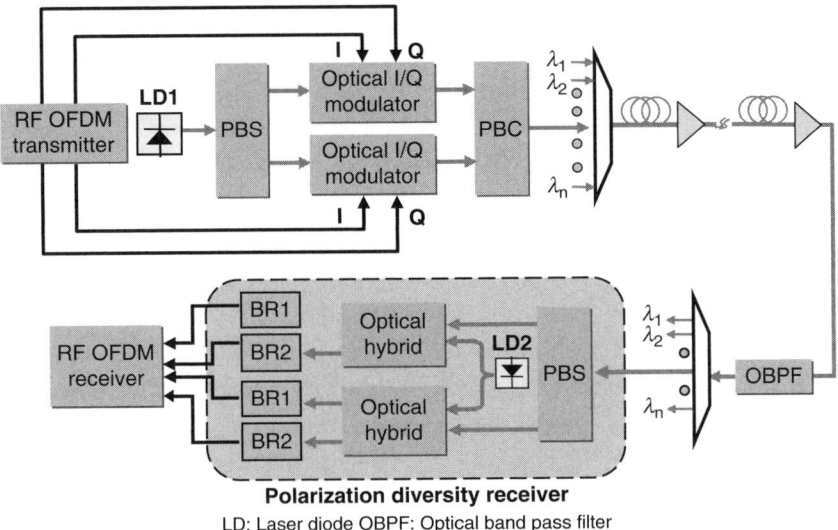

Figure 8.12: Conceptual diagram of a simulated CO-OFDM system.

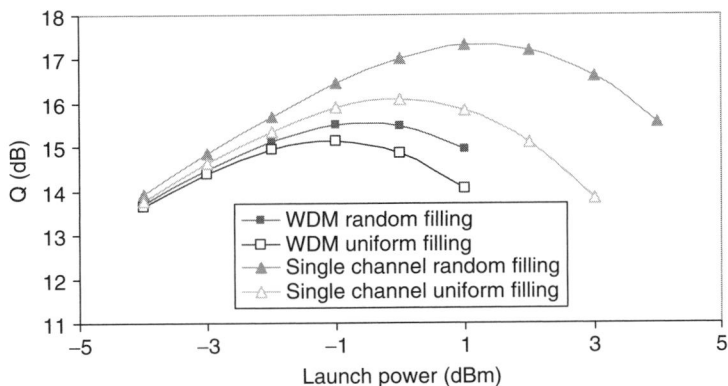

Figure 8.13: Simulated system Q performance as a function of fiber launch power after 1000 km for both single-channel and WDM CO-OFDM transmission. After Ref.[29]

and combined with a polarization beam combiner. Multiple WDM channels with CO-OFDM modulation format are launched into the optical link. The optical link consists of multispan SSMF fibers and an EDFA to compensate the loss. The output signal after transmission is detected using an optical coherent receiver that serves as an optical-to-RF OFDM direct down-converter. In the OFDM receiver, the signal is first sampled using an ADC and then demodulated by performing FFT to recover the data. The commonly used parameters are applied for the simulation: 100 km span distance, fiber chromatic dispersion of 16 ps/nm/km, 0.2 dB/km loss, and a nonlinear coefficient of 2.6×10^{-20} m²/W. The fiber span loss is compensated by an EDFA with a gain of 20 dB and noise figure of 6 dB. The linewidths of the transmit and receive lasers are assumed to be 100 kHz. Eight WDM channels are spaced at 50 GHz. The OFDM parameters are OFDM symbol period of 51.2 ns and 1280 subcarriers. The guard interval is set to one-fourth of the observation period, and 4-QAM encoding is used for each subcarrier. The bandwidth of each CO-OFDM signal is 31.25 GHz. The relative phase shift between subcarriers and channel estimation are calculated by using training sequences. The phase drift from the laser phase noise is also estimated and compensated using pilot subcarriers described in Yi et al.[33] For the uniform filling, the user data are mapped onto 256 subcarriers and the same data block is repeatedly used to fill out the remaining 1152 subcarriers. In this way, the transmitted OFDM signal similar to that used in the experiment (discussed in Section 8.2) can be simulated; that is, the five bands are filled with identical data. For the random filling, the data mapped onto each subcarrier has no correlation.

Figure 8.13 shows the simulated Q-factor as a function of launch power under uniform filling and random filling after 1000 km fiber for single-channel and WDM transmission. The single-channel optimal launch power is, respectively, 0 and 1 dBm for uniform filling and random filling, whereas the optimal launch power for WDM systems is 0 dBm for both uniform filling and random filling. It can be seen that if the power is increased beyond these optimal values, the nonlinearity of optical fiber will degrade the performance of the system, resulting in a reduced Q. On the other hand, if the input power is reduced from these optimal values, the OSNR of the received signal becomes lower, also resulting in a reduced Q. For both single-channel and WDM transmission, uniform filling shows worse nonlinearity performance because the uniform filling increases the peak-to-average power ratio (PAPR) due to the correlation between the bands, which will severely degrade the signal at the high launch power. The phase walk-off and decorrelation between neighboring bands arising from the fiber dispersion helps reduce the nonlinearity after long-distance fiber transmission. However, there is still a residual Q difference after 1000 km transmission between uniform filling and random filling.

A slight underestimation of system performance is observed by using the uniform filling in comparison with random filling due to enhanced nonlinearity in the uniform filling. For single-channel transmission, the optimum launch power difference between uniform filling and random filling is 1 dB and the optimum Q difference is approximately 1.2 dB. In the case of WDM, two filling schemes have the same optimum launch power of −1 dBm and the Q difference of 0.4 dB. It can be seen that at the fixed launch power of −1 dBm for the single-channel scenario, the Q difference between the random filling and uniform filling is only 0.5 dB, which is in agreement with the discussion in Takahashi et al.[5] Most important, the simulation signifies that for single-channel transmission, the true (random filled) 107 Gb/s

OBM-OFDM system will have better system performance than that observed in Section 8.2. The conclusion also applies to the WDM transmission. Therefore, OBM-OFDM with the uniform filling provides a conservative estimation of the true OBM-OFDM system. This observation is important because to completely decorrelate the neighboring band data, multiple AWGs and a band mixer have to be used, which in turn leads to a much more expensive proposition for constructing the experimental setup. It is observed that at the launch power of −1 dBm, the experimental Q of 12.5 dB (see Figure 8.7) is approximately 3.3 dB below the simulated Q for the uniformly filled single channel at the reach of 1000 km. This difference is attributed to the imperfection of CO-OFDM transmitter and receiver and to OFDM signal processing involving non-ideal phase estimation and channel estimation.

8.4.2 Dispersion Map Influence on 100 Gb/s CO-OFDM Transmission

The simulated inline dispersion-compensated system is the same as that in Figure 8.12 except that each span has the dispersion of SMF compensated by a dispersion compensating fiber (DCF) module that is sandwiched between two EDFAs (Figure 8.14). The simulated baseline dispersion map for SSMF and DCF is, respectively, 1600 and −1536 ps/nm, resulting in a 96% compensation ratio, which is similar to that used in most deployed systems.[34] The simulation parameters of DCF are set the same as those used in the experiment[35]: Dispersion is −1536 ps/nm per span, fiber loss is 8.47 dB per span, and the nonlinearity coefficient is 5.4 1/(W · km). To optimize performance, the power into the DCF is 6 dB less than the input power of SSMF.

Figure 8.15 shows the Q-factor as a function of launch power after 1000 km SSMF for both random filling and uniform filling under single-channel and 8-WDM transmission. As shown in Figure 8.15a, for random filling, the single-channel optimal launch power is, respectively, −1 and 1 dBm for the system with and without DCF and the maximum Q difference is 3.5 dB, whereas for uniform filling, the optimal launch power is 0 and −4 dBm with 4 dB maximum Q difference, which agrees well with the experiment in which the Q difference is approximately 3.5 dB.[35] For the case of WDM transmission in Figure 8.15b, the optimal launch power is −1 and −3 dBm for random filling under two dispersion maps. For both single-channel and WDM transmission, uniform filling shows worse nonlinearity performance. The reason is that the uniform filling increases the PAPR due to the correlation between the bands, which severely degrades the signal at the high launch power. For the transmission without DCF, the slight system performance underestimation is observed by

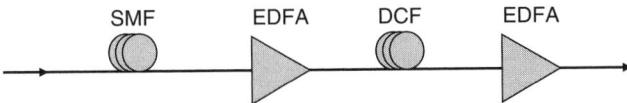

Figure 8.14: The schematic of one span of the inline dispersion map for the simulated transmission systems.

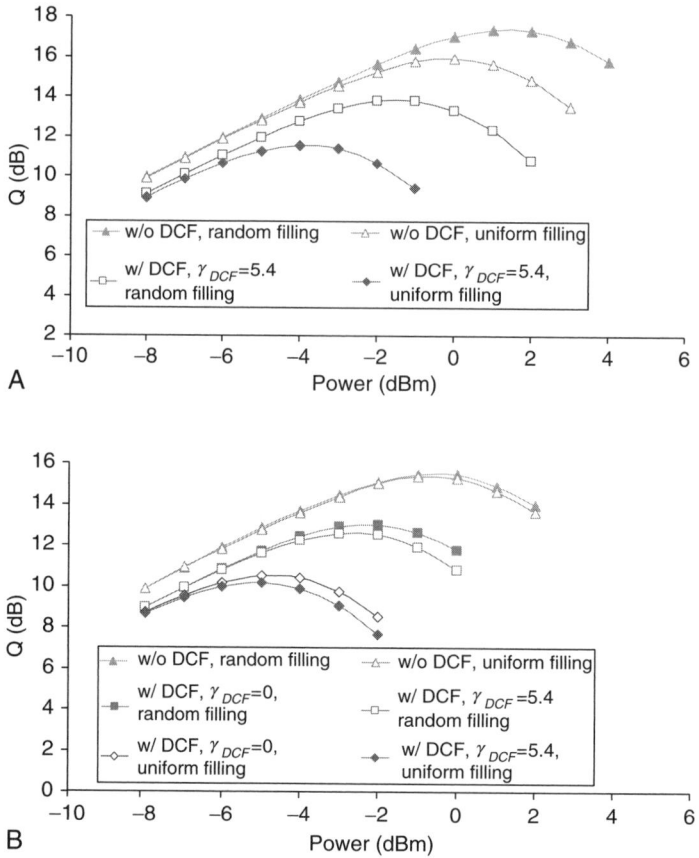

Figure 8.15: The simulated system performance as a function of the launch power for (a) 1000 km single channel and (b) 1000 km WDM transmission.

using the uniform filling in comparison to random filling due to enhanced nonlinearity in the uniform filling. For the single-channel transmission, the optimum launch power difference between uniform filling and random filling is 1dB and the optimum Q difference is approximately 1.4 dB. In the case of WDM, two filling schemes have almost the same Q performance. It is anticipated that the pre- and postcompensation do not influence the performance for CO-OFDM because the OFDM signal already exhibited a high PAPR at the beginning of the transmission.

We also intend to separate the contributing factors to the Q degradation from the dispersion compensation. At the optimal 6 dB input power ratio between SSMF and DCF, we set the nonlinearity coefficient of DCF to zero, and the Q performance with zero DCF nonlinear coefficient is shown in Figure 8.15b. For the WDM systems, the maximum Q is increased approximately 0.3 and 0.4 dB for uniform filling and random filling, respectively. We consider this as the degradation from the DCF nonlinearity. The second contribution is linear noise figure degradation due to the double-stage amplifier. The performance difference for the two dispersion maps at the lower launch powers reveals the linear noise figure degradation, which is approximately 0.8 dB. The third contributing factor is the dispersion map: For the inline compensated system, the OFDM signal almost resets itself at each span, whereas in the noncompensated system the OFDM signal continues to evolve along the transmission. The dispersion compensation enables the coherent addition of the nonlinearity-induced distortion in each span and therefore enhances the nonlinearity degradation.[36] Considering the aforementioned two contributors and overall Q degradation, it can be deduced that for the WDM systems (see Figure 8.15b), the dispersion map contributes 4.0 and 1.6 dB degradation for uniform filling and random filling, respectively.

8.4.3 100 Gb/s CO-OFDM Transmission with Cascaded ROADMs

In current optical fiber communication systems, dense wavelength division multiplexed (DWDM) transmission with 10 Gb/s wavelength channels at 50 GHz channel spacing is an International Telecommunication Union (ITU) standard for long-haul and metropolitan networks. Because 100 Gb/s technology has increasingly become a commercial reality, it is desirable to design future 100 Gb/s upgrades in compliance with the currently installed 10 Gb/s DWDM systems. One of the main limitations for the transparent networks with a large number of nodes is the signal degradation due to transmission through multiple reconfigurable OADMs (ROADMs). The degradation is mainly from (1) the spectral filtering due to narrowing of the overall filter passband and (2) the group delay introduced by the impact of filter phase response. This dispersion leads to pulse distortion, ultimately resulting in transmission performance degradation,[37] and it limits the bit rate. In this section, we study the feasibility of transmitting 100 Gb/s CO-OFDM signals over transparent DWDM systems with the impact of the cascaded ROADMs.

In general, a ROADM consists of a few filters, such as interleaver and blocker. The interleaver usually has an ITU standard channel grid (e.g., 50 GHz spacing for a 10 Gb/s DWDM system) and uniform passband bandwidth.[38] Figure 8.16a shows a transparent DWDM system with interleaver and wavelength blocker-based ROADMs.[39] The DWDM channel is separated into an "odd-channel" group and an "even-channel" group by the first interleaver, therefore relaxing the passband requirement on the wavelength blockers. The second interleaver combines the two groups of channels. The add–drop module consists of a passive coupler and a multiplexer/demultiplexer. The interleaver used in this study, shown in Figure 8.16b, is based on the optical bandpass filter (OBPF) proposed by Dingel and Aruga.[40] The OBPF is a modified Michelson interferometer in which one of its reflecting mirrors is replaced by a Gires–Tournois resonator.[37] A Michelson–GTE-based interleaver can provide a wide, flat-top, square-like frequency response and zero ripple factor. Figure 8.17 shows the

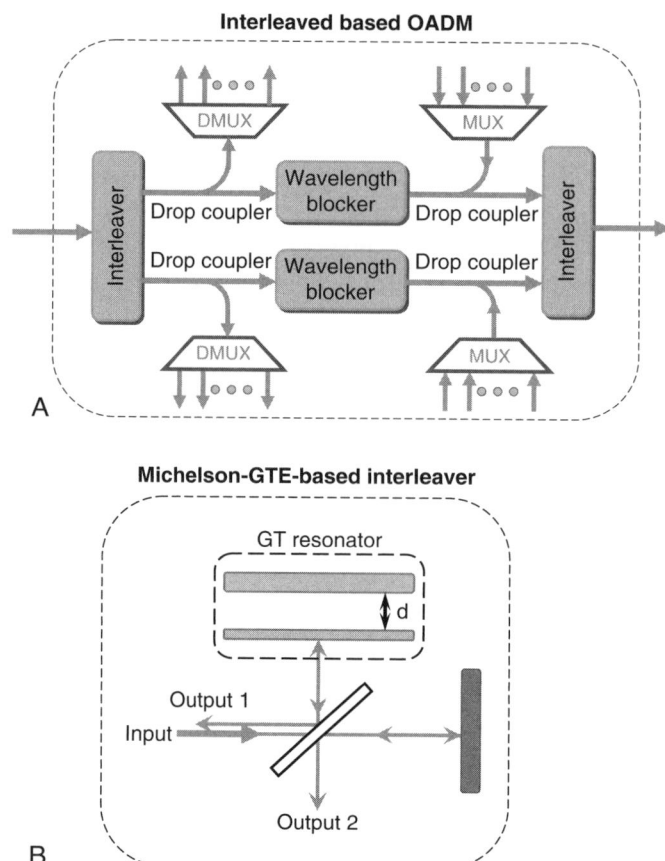

Figure 8.16: (a) Schematic of a transparent DWDM system with multiple interleaver-based ROADMs. (b) Michelson–GTE-based interleaver.

transmittance profile of 1, 4, and 20 cascaded interleavers and the group delay of a single interleaver when the power reflectance of the Michelson interferometer front mirror is 16%. It can be seen that the 1/3 dB bandwidth for a single interleaver filter is 44/50 GHz, reduced to 33/36.5 GHz after 20 stages. The peak-to-peak GDR within the 20 dB passband is 18 ps.

A Monte Carlo simulation is conducted to identify the filter concatenation impact on the performance of a 107 Gb/s CO-OFDM signal after 1000 km fiber transmission. The OFDM simulation parameters are shown in Table 8.2. The bandwidth of the 107 Gb/s CO-OFDM signal is 32 GHz. Eight WDM channels are transmitted. The curve with the open squares in Figure 8.18 represents the Q penalty as a function of the number of ROADM nodes when filters and laser are perfectly aligned. It can be seen that the Q penalty is almost zero even for 10 cascaded ROADMs. This is because the 107 Gb/s CO-OFDM signal only occupies 32 GHz of bandwidth, and the 1/3 dB bandwidth of 10 cascaded ROADMs is 33/36.5 GHz.

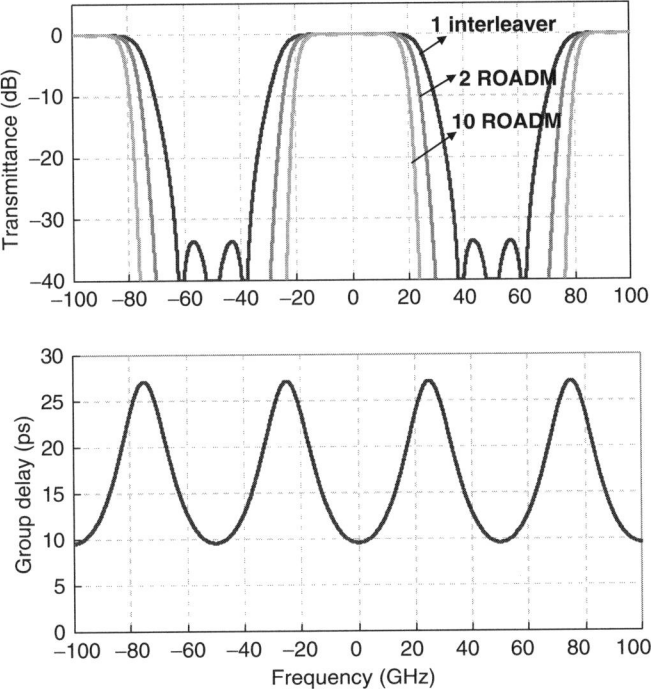

Figure 8.17: Transmittance and group-delay responses of a Michelson–GTE-based interleaver.

Table 8.2: OFDM Parameters

Data Rate (Gb/s)	Subcarrier No.	No. of Bands	Symbol Period (ns)
10.7	82	1	12.8
42.8	164	2	12.8
107	410	5	12.8
1070	4100	50	12.8

Therefore, most of the signal spectral will be confined in the filter passband. Second, because the guard interval $\Delta_G = 1.6$ ns is larger than the 0.36 ns delay induced by the filter phase response, the filter-induced dispersion impairment can be effectively eliminated.

To further investigate the filter concatenation performance, the tolerance to the filter misalignment was studied. The worst case for the filter misalignment was assumed, meaning the first interleaver was centered at $+10\%$ ($+5$ GHz) of the channel spacing, whereas the second was centered at -10% (-5 GHz). The Q penalty is indicated in Figure 8.18 (the curve with solid squares). The effect of detuning between the laser frequency and the filter center

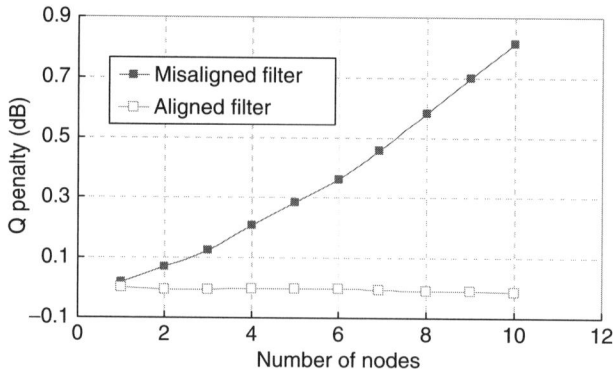

Figure 8.18: **Q penalty (dB) as a function of the number of nodes with and without filter misalignment.**

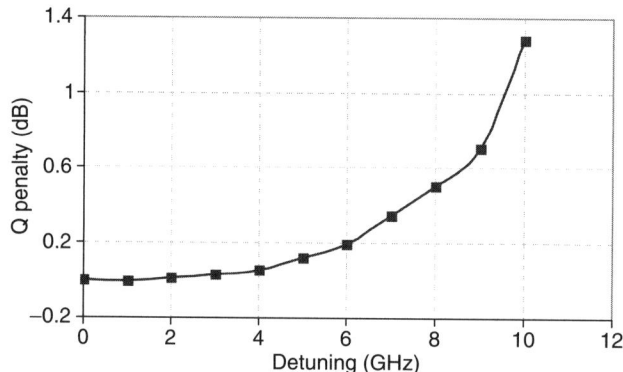

Figure 8.19: **Q penalty as a function of laser detuning.**

wavelength of 10 ROADMs is also examined (Figure 8.19). All of the filter frequencies are assumed to be aligned. It can be seen that a 1.3 dB Q penalty is observed for 10 GHz (20% of the channel spacing) detuning.

8.5 High Spectral Efficiency CO-OFDM Systems

As discussed in Chapter 1, one of the advantages of the optical OFDM is its compatibility of higher order modulation. This arises from the fact that the signal processing elements, including DSP and DAC, are available in the transmitter for optical OFDM systems. To migrate from conventional 4-QAM to higher QAM only requires a software modification without any optical hardware change, provided that the DAC resolution is sufficient. This compares advantageously to the single-carrier systems, in which the higher order modulation

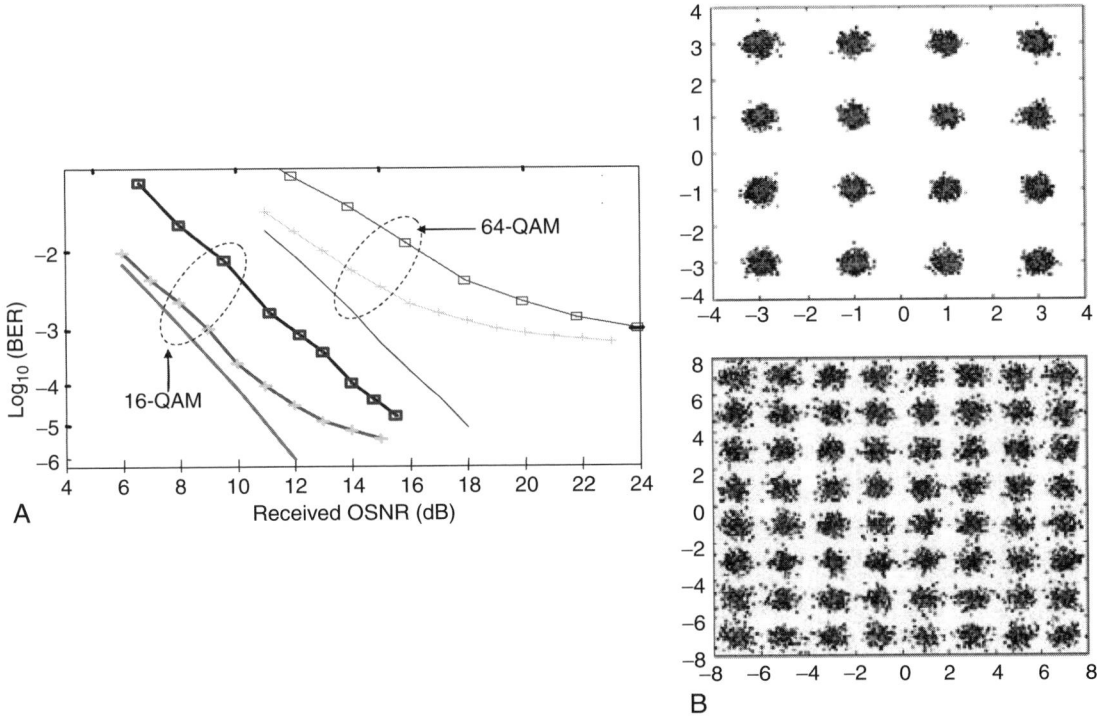

Figure 8.20: **(a) The BER performance of 16-QAM and 64-QAM CO-OFDM systems.**
(b) Constellation diagram for 16-QAM and 64-QAM CO-OFDM systems.

involves very complicated optical hardware configuration.[41] Another subtle advantage is the ease of the channel and phase estimation by using the pilot subcarriers in optical OFDM systems, which is independent of the modulation format that is placed on the data subcarriers. For the high-order modulation in single-carrier systems, both channel and phase estimation rely on the detailed modulation format. Some of the algorithms that are effective in PSK may not be applicable to QAM modulation.[31] It is no surprise that the CO-OFDM-based transmission systems surpass the single-carrier ones in demonstrating impressive high-speed, long-haul transmission up to 64-QAM in single-polarization and 16-QAM in dual-polarization configuration.[4,5] Figure 8.20 shows some experimental results from Yi et al.[4] Specifically, for 16-QAM and 64-QAM modulation, the experimental data (Figure 8.20, open squares), the theoretical data with 101 kHz linewidth (Figure 8.20, crosses), and theoretical data with zero laser linewidth (Figure 8.20, solid line) are shown. The theoretical curves have been shifted upward by 4.1 dB to account for the heterodyne detection and overhead power penalty.[4] The received OSNR is 11.1 dB at a BER of 10^{-3} for 16-QAM OFDM. However, 64-QAM OFDM is strongly limited by the laser phase noise and has an apparent BER floor, confirmed by both the simulation and the experiment. The error floor can be resolved by using

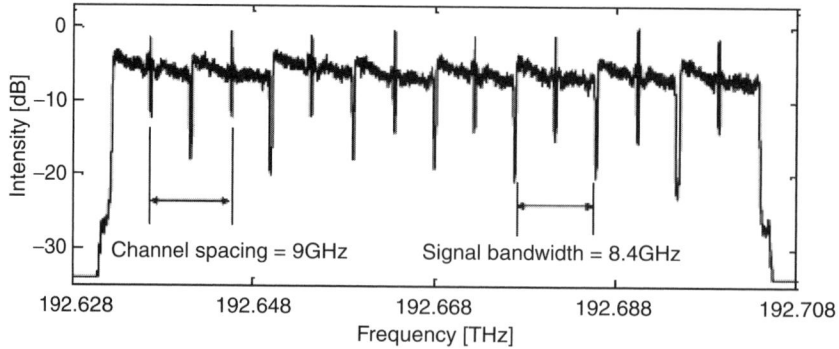

Figure 8.21: Optical spectrum of DWDM CO-OFDM signals. After Takahashi et al.[5]

powerful error-correction coding such as low-density parity-check codes.[42] Figure 8.20b shows the constellations from 14,400-QAM symbols. The SNR per symbol is approximately 25 dB for both constellations.

Figure 8.21 shows the optical spectrum for a another high spectral efficiency WDM experiment[5] in which eight wavelength channels are spaced by 9 GHz with the OFDM bandwidth of 8.4 GHz. The extremely small frequency guard band of 600 MHz is possible due to the intrinsic OFDM spectral tightness. Figures 8.22a and 8.22b, respectively, show the BER performance and the constellation for the 16-QAM CO-OFDM signal. The addition of neighboring channels introduces a penalty of 1.1 dB due to the narrow spacing of 9 GHz between WDM channels.

We now briefly discuss the challenges associated with higher order modulation. The fundamental origin of the difficulty in realizing higher order modulation is that it requires increasingly large SNR. This implies a much tighter requirement for laser

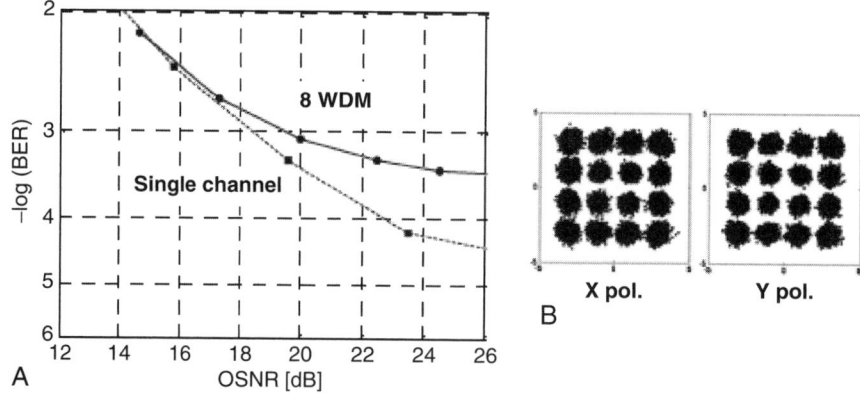

Figure 8.22: (a) The BER performance and (b) the constellation diagram for 16-QAM DWDM CO-OFDM systems. After Takahashi et al.[5]

linewidth, bit resolution, IQ modulator and demodulator imbalance, and fiber nonlinearity. For instance, we learned from Chapter 2 that the SNR penalty, $\Delta\gamma$, from the laser phase noise can be described as

$$\Delta\gamma(\text{dB}) = \frac{10}{\log(10)}\frac{11}{60}4\pi\beta T_s\gamma = 10\beta T_s\gamma \tag{8.10}$$

The γ, or SNR required to achieve the BER of 10^{-3} for 4-QAM and 64-QAM, is 9.8 and 22.8 dB, respectively. From Eq. (8.10), assuming the fixed symbol period, the 64-QAM will require approximately 20 times narrower laser linewidth than 16-QAM, which is simply the ratio between the corresponding required SNRs for the BER of 10^{-3}. A similar analysis applied to other specifications, such as the bit resolution and modulator/demodulator imbalance, will be more stringent when using higher order modulation.

References

1. Shieh W, Yang Q, Ma Y. 107 Gb/s coherent optical OFDM transmission over 1000-km SSMF fiber using orthogonal band multiplexing. *Opt Express* 2008;**16**:6378–86.
2. Jansen SL, Morita I, Tanaka H. 10 × 121.9-Gb/s PDM-OFDM transmission with 2-b/s/Hz spectral efficiency over 1000 km of SSMF. In: *Opt. Fiber Commun. Conf.*, paper no. PDP 2. San Diego; 2008.
3. Yamada E, Sano A, Masuda H, et al. 1 Tb/s (111Gb/s/ch × 10 ch) no-guard-interval CO-OFDM transmission over 2100 km. In: *Opto-Electronics Commun. Conf. and Aust. Conf. Opt. Fibre Technol.*, paper no. PDP 6. Sydney, Australia; 2008.
4. Yi X, Shieh W, Ma Y. Phase noise effects on high spectral efficiency coherent optical OFDM transmission. *J Lightwave Technol* 2008;**26**:1309–16.
5. Takahashi H, Amin AA, Jansen SL, Morita I, Tanaka H. 8 × 66.8-Gbit/s coherent PDM-OFDM transmission over 640 km of SSMF at 5.6-bit/s/Hz spectral efficiency. In: *Eur. Conf. Opt. Commun.*, paper no. Th.3.E.4. Brussels, Belgium; 2008.
6. Yang Q, Shieh W, Yiran M. Bit and power loading for coherent optical OFDM. *IEEE Photon Technol Lett* 2008;**20**:1305–7.
7. Bocoi A, Schuster M, Rambach F, et al. Cost comparison of networks using traditional 10 and 40 Gb/s transponders versus OFDM transponders. In: *Opt. Fiber Commun. Conf.*, paper no. OThB4. San Diego; 2008.
8. Qian D, Hu J, Yu J, et al. Experimental demonstration of a novel OFDM-A based 10Gb/s PON architecture. In: *Eur. Conf. Opt. Commun.*, paper no. 5.4.1. Berlin, Germany; 2007.
9. Duelk M. Next generation 100 Gb/s Ethernet. In: *Eur. Conf. Opt. Commun.*, paper no. Tu3.1.2. Glasgow, Scotland; 2005.
10. Shieh W, Yang Q, Ma Y. 107 Gb/s coherent optical OFDM transmission over 1000-km SSMF fiber using orthogonal band multiplexing. *Opt Express* 2008;**16**:6378–86.
11. Sun H, Wu K-T, Roberts K. Real-time measurements of a 40 Gb/s coherent system. *Opt Express* 2008;**16**:873–9.
12. Yang Q, Ma Y, Shieh W. 107 Gb/s coherent optical OFDM reception using orthogonal band multiplexing. In: *Opt. Fiber Commun. Conf.*, paper no. PDP 7. San Diego; 2008.
13. Winzer PJ, Raybon G, Duelk M. 107-Gb/s optical ETDM transmitter for 100 G Ethernet transport. In: *Eur. Conf. Op.t Commun.*, paper no. Th4.1.1. Glasgow, Scotland; 2005.
14. Fludger CRS, Duthel T, van den Borne D, et al. 10 × 111 Gbit/s, 50 GHz spaced, POLMUX-RZ-DQPSK transmission over 2375 km employing coherent equalisation. In: *Opt. Fiber Commun. Conf.*, paper no. PDP22. Anaheim, CA; 2007.

15. Djordjevic IB, Vasic B. Orthogonal frequency division multiplexing for high-speed optical transmission. *Opt Express* 2006;**14**:3767–75.

16. Jansen SL, Morita I, Schenk TCW, Tanaka H. 121.9-Gb/s PDM-OFDM transmission with 2-b/s/Hz spectral efficiency over 1000 km of SSMF. *J Lightwave Technol* 2009;**27**:177–88.

17. Shieh W, Bao H, Tang Y. Coherent optical OFDM: Theory and design. *Opt Express* 2008;**16**:841–59.

18. Ellis AD, Gunning FCG. Spectral density enhancement using coherent WDM. *IEEE Photon Technol Lett* 2005;**17**:504–6.

19. Sano A, Yoshida E, Masuda H, et al. 30 × 100-Gb/s all-optical OFDM transmission over 1300 km SMF with 10 ROADM nodes. In: *Eur. Conf. Opt. Commun.*, paper no. PD 1.7. Berlin, Germany; 2007.

20. Kobayashi T, Sano A, Yamada E, et al. Electro-optically subcarrier multiplexed 110 Gb/s OFDM signal transmission over 80 km SMF without dispersion compensation. *Electron Lett* 2008;**44**:225–6.

21. Tang Y, Shieh W, Yi X, Evans R. Optimum design for RF-to-optical up-converter in coherent optical OFDM systems. *IEEE Photon Technol Lett* 2007;**19**:483–5.

22. Washburn BR, Diddams SA, Newbury NR, et al. Phaselocked, erbium-fiber-laser-based frequency comb in the near infrared. *Opt Lett* 2004;**29**:250–2.

23. Sakamoto T, Kawanishi T, Izutsu M. Asymptotic formalism for ultraflat optical frequency comb generation using a Mach–Zehnder modulator. *Opt Lett* 2007;**32**:1515–17.

24. Wang Y, Pan Z, Yu C, et al. A multi-wavelength optical source based on supercontinuum generation using phase and intensity modulation at the line-spacing rate. In: *Eur. Conf. Opt. Commun.*, paper no. Th3.2.4. Rimini, Italy; 2003.

25. Shieh W. Coherent optical MIMO-OFDM for optical fibre communication systems. Berlin, Germany. In: *Eur. Conf. Opt. Commun.*, Berlin, Germany; 2007.

26. Jansen SL, Morita I, Tanaka H. 16 × 52.5-Gb/s, 50-GHz spaced, POLMUX-CO-OFDM transmission over 4160 km of SSMF enabled by MIMO processing KDDI R&D Laboratories. In: *Eur. Conf. Opt. Commun.*, paper no. PD1.3. Berlin, Germany; 2007.

27. Liu X, Buchali F. Intra-symbol frequency-domain averaging based channel estimation for coherent optical OFDM. *Opt Express* 2008;**16**:21944–57.

28. Shieh W, Yi X, Ma Y, Tang Y. Theoretical and experimental study on PMD-supported transmission using polarization diversity in coherent optical OFDM systems. *Opt Express* 2007;**15**:9936–47.

29. Yang Q, Tang Y, Ma Y, Shieh W. Experimental demonstration and numerical simulation of 107-Gb/s high spectral efficiency coherent optical OFDM. *J Lightwave Technol* **27**:168–176.

30. Yamada E, Sano A, Masuda H, et al. Novel no-guard interval PDM CO-OFDM transmission in 4.1 Tb/s (50 88.8-Gb/s) DWDM link over 800 km SMF including 50-GHz spaced ROADM nodes. In: *Opt. Fiber Commun. Conf.*, paper no. PDP8. San Diego; 2008.

31. Savory SJ. Digital filters for coherent optical receivers. *Opt Express* 2008;**16**:804–17.

32. Forozesh K, Jansen SL, Randel S, Morita I, Tanaka H. The influence of the dispersion map in coherent optical OFDM transmission systems. *IEEE/LEOS Summer Topical Meetings, Technical Digest* 2008;135–6.

33. Yi X, Shieh W, Ma Y. Phase noise effects on high spectral efficiency coherent optical OFDM transmission. *J Lightwave Technol* 2008;**26**:1309–16.

34. Li MJ. Recent progress in fiber dispersion compensators. In: *Eur. Conf. Opt. Commun.*, paper no. Th.M.1.1; 2001.

35. Tang Y, Ma Y, Shieh W. 107 Gb/s CO-OFDM transmission with inline chromatic dispersion compensation. In: *Opt. Fiber Commun. Conf.*, paper no. OWW3. San Diego; 2009.

36. Nazarathy M, Khurgin J, Weidenfeld R. Phased-array cancellation of nonlinear FWM in coherent OFDM dispersive multi-span links. *Opt Express* 2008;**16**:15777–810.

37. Nykolak G, Lenz G, Eggleton BJ, Strasse TA. Impact of fiber grating dispersion on WDM system performance. In: *Opt. Fiber Commun. Conf.*, paper no. TuA3. San Jose, CA; 1998.

38. Keyworth BP. ROADM subsystems and technologies. In: *Opt. Fiber Commun. Conf.*, paper no. OWB3. Anaheim, CA; 2005.

39. Xiang L. Can 40-gb/s duobinary signals be carried over transparent DWDM systems with 50-GHz channel spacing? *IEEE Photon Technol Lett* 2005;**17**:1328–30.

40. Dingel BB, Aruga T. Properties of a novel noncascaded type, easy-to-design, ripple-free optical bandpass filter. *J Lightwave Technol* 1999;**17**:1461–9.

41. Yu J, Zhou X, Huang M, et al. 17 Tb/s (161 × 114 Gb/s) PolMux-RZ-8PSK transmission over 662 km of ultra-low loss fiber using C-band EDFA amplification and digital coherent detection. In: *Eur. Conf. Opt. Commun.*, paper no. Th.3.E.2. Brussels, Belgium; 2008.

42. Djordjevic IB, Vasic B. LDPC-coded OFDM in fiber-optics communication systems. *J Opt Networking* 2008;**7**:217–26.

CHAPTER 9

OFDM for Multimode Fiber Systems

Multimode fibers (MMFs) are widely used in local area networks (LANs) and optical interconnects. Many of the installed Ethernet backbones are based on MMFs and operate at data rates of approximately 1 Gb/s. A great deal of research is devoted to upgrading 1 Gb/s Ethernet backbones to 10 Gb/s. Because orthogonal frequency-division multiplexing (OFDM) has shown robustness in a variety of wireless channels, it has been adopted in many communication standards[1,2] including wireless LAN (Wi-Fi), digital video and audio broadcasting standards, digital subscribed loop (DSL), and Worldwide Interoperability for Microwave Access (WiMAX, or IEEE 802.16). It has been shown in previous chapters that OFDM can successfully be applied in long-haul optical transmission systems to deal with chromatic dispersion and polarization mode dispersion. Because the high data rate signal in OFDM is split into many lower data rate substreams, it demonstrates high robustness against different dispersion effects, including multimode dispersion present in MMF links, which is well documented in a number of publications.[3–16]

In this chapter, we describe OFDM use in MMF links to extend transmission distance and to increase the aggregate data rate. In Section 9.1, we discuss the basic characteristics of MMFs. In Section 9.2, we describe the use of MMFs for short-reach systems. The possible use of OFDM in MMF links for beyond short-reach applications is discussed in Section 9.3. Finally, in Section 9.4, we describe the use of the multiple-input multiple-output (MIMO) OFDM concept as a means to increase the spectral efficiency and system reach.

9.1 Multimode Fibers

MMFs transfer light through a collection of spatial transversal modes, with each mode being defined through a specified combination of electrical and magnetic components. Each mode occupies a different cross section of the optical fiber core and takes a slightly distinguished path along the optical fiber. The difference in mode path lengths in MMFs results in different arrival times at the receiver side. This effect is known as multimode (intermodal) dispersion and causes signal distortion and imposes the limitations in signal bandwidth. The effect of intermodal dispersion is illustrated in Figure 9.1 for meridional rays, for both step-index

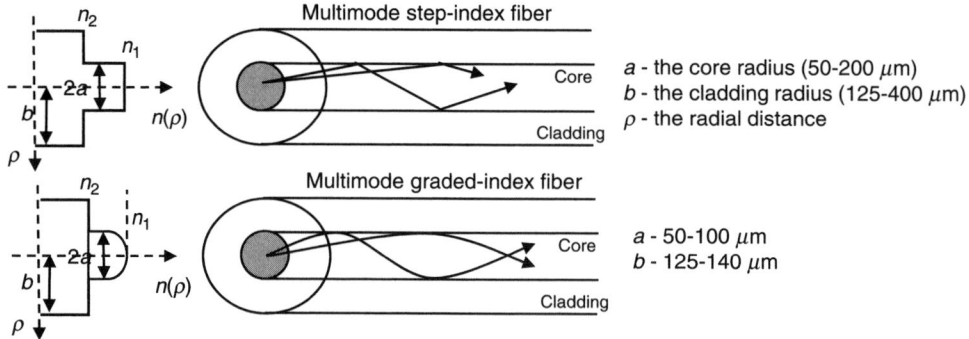

Figure 9.1: Intermodal dispersion in MMFs: (a) multimode step-index fiber and (b) multimode graded-index fiber.

MMF (Figure 9.1a) and graded-index MMF (Figure 9.1b). The steeper the angle of the propagation of ray of congruence, the higher the mode number and the slower the axial group velocity.[17] This variation in group velocities of different modes results in group delay spread, which is the root cause of intermodal dispersion. The maximum pulse broadening can be obtained as the difference in travel time of the highest order mode T_{max} and that of the fundamental mode T_{min} as follows:

$$\Delta T_{mod} = T_{max} - T_{min} = \frac{Ln_1^2}{cn_2} - \frac{Ln_1}{c} = \frac{Ln_1^2}{cn_2}\Delta \approx \frac{L}{2cn_2}(NA)^2, \quad (\Delta \ll 1) \tag{9.1}$$

where L is the MMF length; n_1 and n_2 are refractive indices related to the fiber core and cladding, respectively; and NA is numerical aperture defined as the light-gathering capacity of MMF:

$$NA = \sqrt{n_1^2 - n_2^2} \approx n_1\sqrt{2\Delta} \quad (\Delta \ll 1)$$

$\Delta = (n_1 - n_2)/n_1$ is the normalized index difference. The delay spread is larger for fibers with larger numerical aperture or larger normalized index difference. A rough measure of the delay spread that can be tolerated is half of the bit duration, which results in the following bit maximum bit rate × distance product:

$$BL < \frac{n_2c}{2n_1^2\Delta} \approx \frac{n_2c}{(NA)^2} \tag{9.2}$$

For example, for typical MMF parameters, $n_1 \approx n_2 \approx 1.5$ and $\Delta = 0.01$, we obtain $BL < 10$ (Mb/s) km for non-return-to-zero systems. To improve BL product, graded-index MMFs are used, whose index profile can be described by[17]

$$n(\rho) = \begin{cases} n_1[1 - 2\Delta(\rho/a)^\alpha]^{1/2}, & \rho \leq a \\ n_1(1 - 2\Delta)^{1/2} \square n_1(1 - \Delta) = n_2, & \rho > a \end{cases} \quad (\alpha - profile) \tag{9.3}$$

A step-index fiber is approached in the limit for large α, and for $\alpha = 2$ the parabolic-index profile is obtained. Because the index of refraction is lower at the outer edges of the core, the light rays will travel faster in that region than in the center, where the refractive index is higher. Therefore, the ray congruence characterizing the higher order mode will travel farther than the fundamental ray congruence but at a faster rate, resulting in similar arrival times at the receiver side. Multimode dispersion in graded-index fibers is reduced compared to that of step-index fibers.

The trajectory of a light ray is obtained by solving the following differential equation:

$$\frac{d}{ds}\left[n(r)\frac{dr}{ds}\right] = \nabla n(r) \tag{9.4}$$

where r is the point along the trajectory, s is the path along the trajectory, and $n(r)$ is the index of refraction. By observing meridional paraxial rays $r = (\rho, z)$ and noticing that $n(r) = n(\rho)$, the differential equation (9.4) becomes

$$\frac{d^2\rho}{dz^2} = \frac{1}{n}\frac{dn}{d\rho} \tag{9.5}$$

which can easily be solved for the parabolic profile to get

$$\rho(z) = \rho_0 \cos(pz) + \frac{\rho'_0}{p}\sin(pz), \quad p = \sqrt{\frac{2\Delta}{a^2}} \tag{9.6}$$

where ρ_0 and ρ'_0 are the slope and slope radius at $z = 0$, respectively. Therefore, the trajectory of meridional rays is periodic with a period $2\pi/p$.

The root mean square (RMS) pulse-broadening σ in a graded-index fiber can be obtained by

$$\sigma = \sqrt{\sigma^2_{\text{intermodal}} + \sigma^2_{\text{intramodal}}} \tag{9.7}$$

where $\sigma_{\text{intermodal}}$ is the RMS pulse spread resulting from intermodal dispersion, and $\sigma_{\text{intramodal}}$ is the RMS pulse spread resulting from intramodal dispersion. The intermodal delay distortion can be found by

$$\sigma_{\text{intermodal}} = \sqrt{\langle\tau^2_g\rangle - \langle\tau_g\rangle^2} \tag{9.8}$$

where τ_g is the group delay of the mode, which is a function of the order (v, m) of the mode

$$\tau_g = \frac{L}{c}\frac{d\beta_{vm}}{dk} \tag{9.9}$$

with L being the fiber length, β_{vm} being the propagation constant, and $k = 2\pi/\lambda$ being the wave number. The quantities $<\tau_g>$ and $<\tau^2_g>$ can be obtained from mode distribution P_{vm} by[17]

$$\langle\tau_g\rangle = \sum_{v,m}\frac{P_{vm}\tau_g(v,m)}{M} \qquad \langle\tau^2_g\rangle = \sum_{v,m}\frac{P_{vm}\tau^2_g(v,m)}{M} \tag{9.10}$$

where M is the total number of modes. To determine the group delay, we use the following expression for propagation constant[17]:

$$\beta = k n_1 \left[1 - 2\Delta \left(\frac{m}{M} \right)^{\alpha/(\alpha+2)} \right]^{1/2} \tag{9.11}$$

where m is the number of modes with propagation constant between $k n_1$ and β. By substituting Eq. (9.11) into Eq. (9.9) and assuming $\Delta \ll 1$, we obtain

$$\tau_g = \frac{N_1 L}{c} \left[1 + \frac{\alpha - 2 - \varepsilon}{\alpha + 2} \Delta \left(\frac{m}{M} \right)^{\alpha/(\alpha+1)} + \frac{3\alpha - 2 - 2\varepsilon}{2(\alpha + 2)} \Delta^2 \left(\frac{m}{M} \right)^{\alpha/(\alpha+1)} + O(\Delta^3) \right] \tag{9.12}$$

where $N_1 = n_1 + k \partial n_1 / \partial k$, and $\varepsilon = [2 n_1 k/(N_1 \Delta)] \partial \Delta / \partial k$. The previous equation indicates that to the first order in Δ, the group delay difference between the modes is zero if $\alpha = 2 + \varepsilon$. If all modes are equally excited, $P_{vm} = P$, and the number of modes is large, the summations in Eq. (9.10) become integrals, and upon substitution of Eq. (9.12) into Eq. (9.10) we obtain the following expression for intermodal pulse spread[17]:

$$\sigma_{\text{intermodal}} = \frac{L N_1}{2c} \frac{\alpha}{\alpha + 1} \left(\frac{\alpha + 2}{3\alpha + 2} \right)^{1/2} \left[c_1^2 + \frac{4 c_1 c_2 (\alpha + 1)\Delta}{2\alpha + 1} + \frac{16 \Delta^2 c_2^2 (\alpha + 1)^2}{(5\alpha + 2)(3\alpha + 2)} \right]^{1/2} \tag{9.13}$$

where $c_1 = (\alpha - 2 - \varepsilon)/(\alpha + 2)$ and $c_2 = (3\alpha - 2 - 2\varepsilon)/(\alpha + 2)$. The optimum index profile can be found by minimizing Eq. (9.13) to get

$$\alpha_{\text{opt}} = 2 + \varepsilon - \Delta \frac{(4 + \varepsilon)(3 + \varepsilon)}{5 + 2\varepsilon} \tag{9.14}$$

By setting $\varepsilon = 0$ and substituting Eq. (9.14) into Eq. (9.13), we obtain

$$\sigma_{\text{opt}} = \frac{n_1 \Delta^2 L}{20\sqrt{3}c} \tag{9.15}$$

The corresponding expression for step-index fiber is obtained by setting $\alpha \to \infty$ (and also assuming $\varepsilon = 0$) as follows:

$$\sigma_{\text{step}} = \frac{n_1 \Delta L}{c} \frac{1}{2\sqrt{3}} \left(1 + 3\Delta + \frac{12\Delta^2}{5} \right)^{1/2} \approx \frac{n_1 \Delta L}{2\sqrt{3}c} \tag{9.16}$$

The ratio of Eqs. (9.16) and (9.15) gives $\sigma_{\text{step}}/\sigma_{\text{opt}} = 10/\Delta$. Because typically $\Delta = 0.01$, graded-index fiber capacity is three orders of magnitude larger than step-index fiber capacity. The corresponding bandwidth $\times$ length product is then $BL < 10$ (Gb/s) km.

The impulse response of graded-index MMFs can be written as follows[18]:

$$h(t) = \begin{cases} \frac{\alpha + 2}{\alpha} \left| \frac{\alpha + 2}{\Delta(\alpha - 2)} \right|^{(2/\alpha)+1} |t|^{2/\alpha}, & \alpha \neq 2 \\ \frac{2}{\Delta^2}, & \alpha \approx 2 \end{cases}, \quad 0 \leq t < T; \quad T = \begin{cases} \frac{\alpha - 2}{\alpha + 2}\Delta, & \alpha \neq 2 \\ \frac{\Delta^2}{2}, & \alpha \approx 2 \end{cases} \tag{9.17}$$

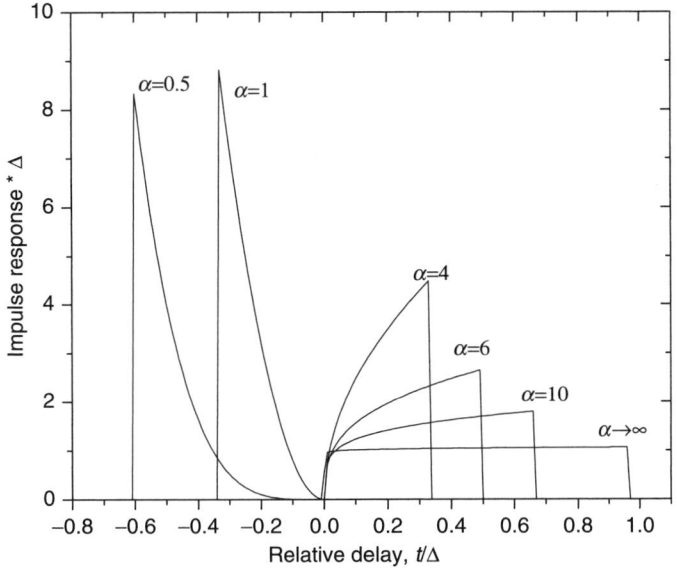

Figure 9.2: Impulse responses of MMFs with $\alpha = 0.5$, 1, 2, 4, 6, 10, and ∞ (step index) for the index profile given by Eq. (9.3).

where the total delay of the channel is normalized with delay of the central mode. Notice that for $\alpha < 2$, the delay spread is negative, meaning that higher modes arrive earlier than the central mode. The impulse responses for different α-profile MMFs are shown in Figure 9.2. For $\alpha \to \infty$, the fiber becomes a step index.

9.2 Optical OFDM in MMF Links

The growth of bandwidth-intensive applications such as HDTV, IPTV, and data processing for medical applications has resulted in a high demand for capacity of LAN backbones and server interconnects. Research on 10 Gb/s transmission and beyond over MMF links has increased.[19–22] These approaches, however, require the use of high-speed components, external modulators, and single-mode components.

On the other hand, by using multicarrier transmission, the transmitted bitstream can be divided into many different substreams (commonly orthogonal to each other) and send these over many different subchannels, avoiding the need for high-speed components. The number of substreams is chosen in such a way to make the symbol time on each substream much greater than the delay spread of the channel, reducing the intersymbol interference (ISI). In particular, OFDM deals with time-varying channel conditions such as MMF links through adaptive modulation and coding. One particular scheme, suitable for high-speed transmission over plastic optical fibers, has been described[8] (see also Chapter 11, Section 11.4). To reduce the system cost, this scheme is based on direct detection. However, it still requires the use of

RF up- and down-converters. To further reduce system cost, which is of high importance for LAN applications, Lee et al.[15] advocate the use of discrete multitone (DMT) modulation, a version of OFDM. On the other hand, the same sequence is transmitted twice, resulting in reduced data rate. In addition to system cost, DMT has two advantages compared to traditional approaches[19–22]: (1) the ability to maximize the information rate by tailoring the information-bearing distribution across the channel according to the channel conditions and (2) adaptivity to time-varying channel conditions, which is achieved by virtue of the data stream being divided among many subcarriers, and the subcarriers affected by the channel can be avoided through adaptive modulation.

The block diagram of the DMT system for transmission over MMF links is shown in Figure 9.3. The demultiplexer converts the incoming information data stream into parallel form. The constellation mapper maps parallel data into N subcarriers using quadrature amplitude modulation (QAM). The modulation is performed by applying the inverse fast Fourier transform (IFFT), which transforms frequency domain parallel data into time domain parallel data. In OFDM, the time domain sequence upon IFFT, for frequency domain QAM sequence C_k ($k = 0, 1, \ldots, N - 1$), can be written as follows:

$$s_k = \frac{1}{\sqrt{N}} \sum_{n=0}^{N-1} C_n \, \exp\left(j 2\pi k \frac{n}{N}\right); \quad k = 0, 1, \ldots, N - 1 \tag{9.18}$$

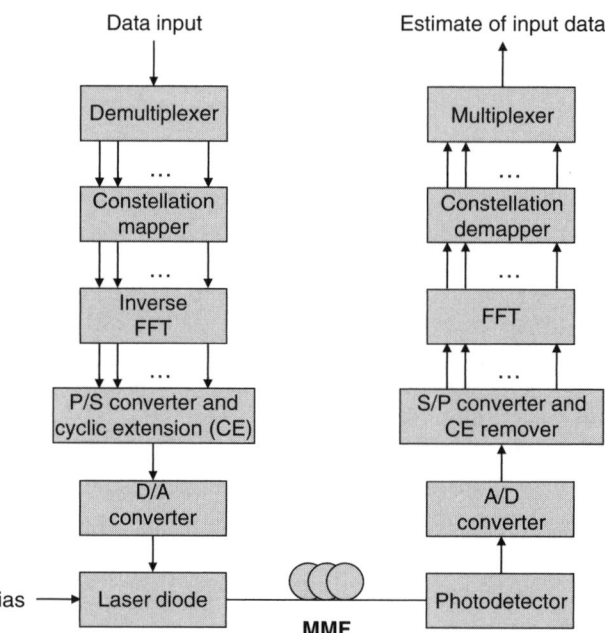

Figure 9.3: Block diagram of the DMT system for transmission over MMF. P/S, parallel-to-serial; S/P, serial-to-parallel.

where s_k is the complex-valued time domain sequence of length N. In DMT, the time domain sequence is real-valued, which is achieved by employing $2N$ IFFT instead, with input values satisfying the Hermitian symmetry property:

$$C_{2N-n} = C_n^*; \quad n = 1, 2, \ldots, N - 1; \quad \text{Im}\{C_0\} = \text{Im}\{C_N\} = 0 \qquad (9.19)$$

meaning that the second half of the input sequence (to the IFFT block) is a complex conjugate of the first half. In addition, the 0th and Nth subcarriers must be real-valued as given in Eq. (9.19). Therefore, the $2N$ point IFFT in DMT is obtained as follows:

$$s_k = \frac{1}{\sqrt{2N}} \sum_{n=0}^{2N-1} C_n \exp\left(j2\pi k \frac{n}{2N}\right); \quad k = 0, 1, \ldots, 2N - 1 \qquad (9.20)$$

where s_k is now a real-valued sequence of length $2N$. The corresponding, discrete time signal upon parallel-to-serial conversion can be written as

$$s\left(k\frac{T}{2N}\right) = \frac{1}{\sqrt{2N}} \sum_{n=0}^{2N-1} C_n \exp\left(j2\pi n \frac{kT}{2N}\right); \quad k = 0, 1, \ldots, 2N - 1 \qquad (9.21)$$

where T is the time duration of the DMT frame. In parallel-to-serial block (see Figure 9.3), the cyclic extension is performed in a similar manner as done in previous chapters. The cyclic extension should be longer than the maximum pulse spread due to multimode dispersion described previously. The digital-to-analog (D/A) converter performs the conversion from the digital to the analog domain. Typically, the D/A converter contains a transmit filter. As shown in Figure 9.3, to reduce the system cost, direct modulation of the laser diode is used to impose the DMT signal. Because the negative signals cannot be transmitted over an intensity modulation with direct detection, the bias voltage is used to convert the negative portion of the DMT signal to positive. Since this approach is power inefficient, the clipping can be used.

At the receiver side, upon optical-to-electrical conversion by the photodetector, DC bias blocking, analog-to-digital (A/D) conversion, cyclic removal, and serial-to-parallel conversion, the demodulation is performed by $2N$ point FFT:

$$\hat{C}_n = \frac{1}{\sqrt{2N}} \sum_{k=0}^{2N-1} c_k \exp\left(-j2\pi k \frac{n}{2N}\right); \quad n = 0, 1, \ldots, 2N - 1 \qquad (9.22)$$

where $\hat{C}_n$ ($n = 0, 1, \ldots, N - 1$) represents the estimation of the transmitted sequence.

The cyclic extension, in addition to being used to deal with multimode dispersion, can be used for frequency offset correction due to common phase error and for synchronization (Figure 9.4). When the cyclic extension is performed by repeating the last $N_G/2$ samples of the effective FFT part as prefix and the first $N_G/2$ samples as suffix, on a receiver side (by ignoring distortion effects) the first $N_G/2$ samples of the effective DMT frame are identical to the suffix, spaced T_{FFT} (the duration of FFT portion) seconds apart. By calculating the

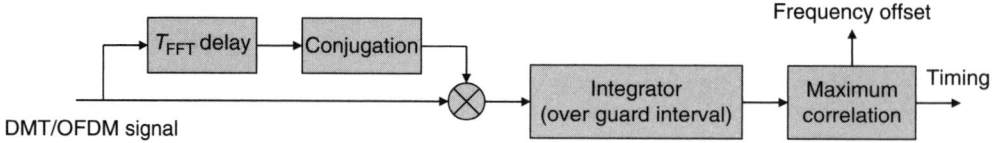

Figure 9.4: Frequency offset correction and timing using the cyclic prefix.

autocorrelation of these two parts, we are able to estimate the frequency offset due to common phase error and remove it. At the same time, correlation peaks obtained after every DMT symbol can be used for timing. Once the rough estimate of the beginning of the DMT frame is found, the fine timing and frequency synchronization is performed by using the preamble sent periodically.

Because in DMT/OFDM different subcarriers are modulated independently, some of the subcarriers can add constructively, leading to high peak-to-average power ratio (PAPR), defined as

$$PAPR = 10 \log_{10} \left(\frac{\max\limits_{0 \leq t < T_{DMT}} \left| s_{DMT}(t) \right|^2}{E\left[\left| s_{DMT}(t) \right|^2 \right]} \right) \quad [dB] \tag{9.23}$$

where $s_{DMT}(t)$ represents a DMT frame of duration T_{DMT}, and $E[]$ is the averaging operator. High PAPR can introduce nonlinearities in driver amplifier and D/A and A/D converters. Although the DMT frames can have high PAPR values, they occur with certain probability and can be characterized using the complementary cumulative distribution function Pr(PAPR > PAPR$_{ref}$), where PAPR$_{ref}$ is the referent PAPR (usually expressed in decibels). It has been shown that Pr(PAPR > 15 dB) (when 512 subcarriers are used) is 10^{-4}, indicating that Pr(PAPR > 27 dB) (when all subcarriers add constructively) would be even smaller.[15] Therefore, there is no need to accommodate for full dynamic range. By limiting the dynamic range of D/A and A/D converters to an appropriately chosen range, the optimum performance can be obtained, as has been shown in Jin et al.[14] The use of different numbers of subcarriers leads to different probability density functions of PAPR. Lee et al.[15] showed that the use of a larger number of subcarriers provides better robustness to poor channel response by allowing for longer guard intervals for the same overhead.

The adjustment of the dynamic range of D/A and A/D converters can simply be achieved by clipping, as shown here:

$$s_{clipped}(t) = \begin{cases} s_{DMT}(t), & \left| s_{DMT}(t) \right| \leq A \\ 0, & \text{otherwise} \end{cases} \tag{9.24}$$

where $s_{clipped}(t)$ denotes the clipped DMT signal, and A denotes the maximum tolerable amplitude before clipping. By measuring the bit error ratio (BER) against the clipping level C_L, defined as

$$C_{\mathrm{L}} = 10 \log_{10} \left(\frac{A^2}{E\left[\left|s_{\mathrm{DMT}}(t)\right|^2\right]} \right) \quad [\mathrm{dB}] \tag{9.25}$$

we can determine the optimum clipping level minimizing the BER. For example, it has been shown[15] that for 256 subcarriers and 64-QAM, the optimum clipping ratio for different A/D (D/A) converter resolutions is between 8.5 and 9.5 dB.

The clipping introduces the distortion of DMT signals. It is possible to reduce PAPR by using distortionless PAPR reduction.[23–25] Another approach to reduce PAPR is through coding.[26] The complexity of some of these distortionless methods is significantly higher than that of clipping. This is the reason why medium-level complexity methods have been advocated,[15] such as selective mapping,[23,24] which is based on dual use of real and imaginary parts of a complex FFT. Let us observe the complex sequence D_n ($n = 0, 1, \ldots, N - 1$). The sequence of complex numbers of length $2N$ satisfying the symmetry property

$$D_{2N-n} = -D_n^*; \ n = 1, 2, \ldots, N - 1; \ \mathrm{Im}\{D_0\} = \mathrm{Im}\{D_N\} = 0 \tag{9.26a}$$

has for the $2N$ point IFFT a purely imaginary-valued sequence. The IFFT of sequence $X_n = C_n + jD_n$ follows:

$$
\begin{aligned}
s_k &= \frac{1}{\sqrt{2N}} \sum_{n=0}^{2N-1} X_n \exp\left(j2\pi k \frac{n}{2N}\right); \quad k = 0, 1, \ldots, 2N - 1 \\
&= \frac{1}{\sqrt{2N}} \left[\sum_{n=0}^{2N-1} C_n \exp\left(j2\pi k \frac{n}{2N}\right) + j \sum_{n=0}^{2N-1} D_n \exp\left(j2\pi k \frac{n}{2N}\right) \right] \\
&= \mathrm{Re}\{s_k\} + j\mathrm{Im}\{s_k\}
\end{aligned}
\tag{9.26b}
$$

We have just created a two-channel input–output modulator with a single $2N$ point IFFT. By using a gray mapping for C_n and anti-gray mapping for D_n and applying the same input sequence $D_n = C_n$, we get two DMT frames with different PAPR values. We can then transmit the DMT frame with a smaller PAPR value. Of course, the receiver has to know which sequence was used on a transmitter side. It can be shown that the same $\mathrm{Pr}(\mathrm{PAPR} > \mathrm{PAPR}_{\mathrm{ref}})$ of 10^{-4} is now obtained at approximately 13.2 dB (for 512 subcarriers and 64-QAM).[15]

The capacity of the DMT system with N independent subcarriers of bandwidth B_N and subcarrier gain g_i ($i = 0, \ldots, N - 1$) can be evaluated by

$$C = \max_{P_i: \sum P_i = P} \sum_{i=0}^{N-1} B_N \log_2\left(1 + \frac{g_i(P_i + P_{bias})}{N_0 B_N}\right) \tag{9.27}$$

where P_i is the optical power allocated to the ith subcarrier. The signal bias is used to convert negative values of DMT signal to positive, and P_{bias} represents the corresponding power used to transmit the bias. N_0 is the power spectral density of the transimpedance amplifier. The subcarrier gain g_i can be calculated by $g_i = R^2|H_i|^2$, where R is the photodiode responsivity and H_i is the MMF transfer function amplitude of the ith subcarrier (the FFT of Eq. 9.17). Because direct detection is used, Eq. (9.27) represents essentially a lower bound on channel capacity. Nevertheless, it can be used as an initial figure of merit. It can be shown by using the Lagrangian method that the optimum power allocation policy is the water-filling over frequency[21]:

$$\frac{P_i + P_{\text{bias}}}{P} = \begin{cases} 1/\gamma_c - 1/\gamma_i, & \gamma_i \geq \gamma_c \\ 0, & \text{otherwise} \end{cases}, \; \gamma_i = g_i(P + P_{\text{bias}})/N_0 B_N \tag{9.28}$$

where γ_i is the signal-to-noise ratio (SNR) of the ith subcarrier, and γ_c is the threshold SNR. By substituting Eq. (9.28) into Eq. (9.27), the following channel capacity expression is obtained:

$$C = \sum_{i:\, \gamma_i > \gamma_c} B_N \log_2\left(\frac{\gamma_i}{\gamma_c}\right) \tag{9.29}$$

The ith subcarrier is used when the corresponding SNR is above threshold. The number of bits per ith subcarrier is determined by $m_i = \lfloor B_N \log_2(\gamma_i/\gamma_c) \rfloor$, where $\lfloor \; \rfloor$ denotes the largest integer smaller than the enclosed number. Therefore, signal constellation size and power per subcarrier are determined based on MMF channel coefficients. When the subcarrier SNR is high, larger constellation sizes are used and power per subcarrier is chosen as given by Eq. (9.28). Smaller constellation sizes are used when the subcarrier SNR is low, and nothing is transmitted when the subcarrier SNR falls below a certain threshold value. An adaptive QAM example is shown in Figure 9.5.

Lee et al.[15] have shown that the transmission of 24 Gb/s (in aggregate rate) is possible over 730 m of MMF with a bandwidth of only approximately 3 GHz. In a typical baseband

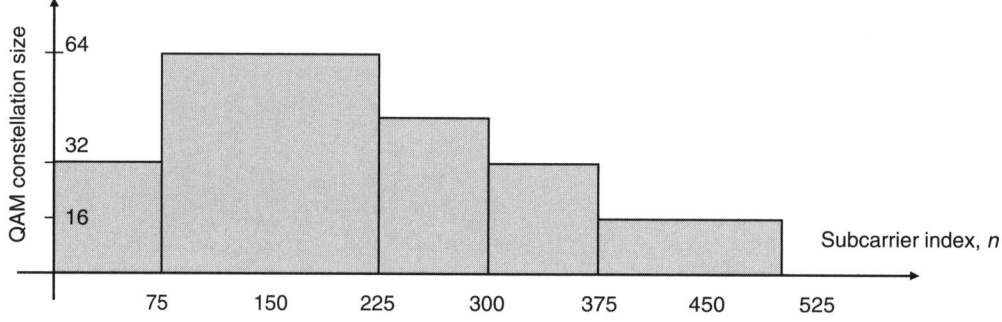

Figure 9.5: Illustration of adaptive QAM constellation mapping per subcarrier.

single-carrier transmission system using on–off keying (OOK), it would be impossible to transmit 24 Gb/s signal over MMF of bandwidth 3 GHz.

9.2.1 Power-Efficient OFDM

In Chapter 6, we described the clipped OFDM (C-OFDM) and unclipped OFDM (U-OFDM) as two possible alternatives to power-inefficient biased OFDM (B-OFDM). Here, we describe a power-efficient scheme by Armstrong and Lowery,[27] which is suitable for use in DMT systems, and it is called here asymmetrically clipped optical OFDM (ACO-OFDM) (according to the original authors).[27,28] Namely, it has been shown[27] that when only odd subcarriers are modulated, all of the intermodulation caused by clipping falls on even subcarriers so that the clipping does not affect the data-carrying subcarriers. On the other hand, the odd frequency clipping requirement and DMT Hermitian constraint mean that only $N/4$ complex values out of N used in FFT/IFFT are independent, which reduces the spectral efficiency. Nevertheless, the spectral efficiency of this scheme is still much better than that of pulse position modulation (PPM),[28] which is a well-known power-efficient modulation scheme.

Let C_n denote the complex number representing the signal constellation point transmitted on the nth subcarrier of a given DMT symbol; the corresponding time domain IFFT sequence is given by Eq. (9.18). The clipping process is a memoryless operation on time domain sequence s_k, which introduces the unwanted (clipping) noise component d_k as follows from Bussgang's theorem (see Corollary 1 of Theorem 3 in Dardari et al.)[29]:

$$s_k^c = Ks_k + d_k \tag{9.30a}$$

where K is the constant and d_k is the random process uncorrelated with time domain sequence s_k, $E[x_k d^*_k] = 0$ ($E[]$ denotes the expectation operator), and s_k^c denotes the time domain clipped sequence. The constant K can be expressed from Eq. (9.30a) as follows:

$$K = E\left[s_k^c s_k^*\right]/E\left[s_k s_k^*\right] \tag{9.30b}$$

Because s_k is real ($s^*_k = s_k$) and s_k is symmetrically distributed around zero, $E[s_k^c s_k] = E[s_k s_k]/2$ so that $K = 1/2$ and $s_k^c = s_k/2 + d_k$. Therefore, the clipping of the signal at zero level reduces the amplitude of wanted signal by half, whereas the mean of optical power is reduced to[27]

$$E\left[s_k^c\right] = \frac{1}{\sqrt{2\pi}} \int\limits_0^\infty z e^{-z^2/2} dz = \frac{1}{\sqrt{2\pi}} \tag{9.31}$$

The overall improvement in electrical SNR can be found as $20 \log_{10}(0.5 \cdot 2 \cdot \sqrt{(2\pi)}) = 7.98$ dB (0.5 derives from a reduction in amplitude of each subcarrier due to clipping, and the factor 2 derives from the fact that DC bias is not used). The effect of clipping on both even and odd

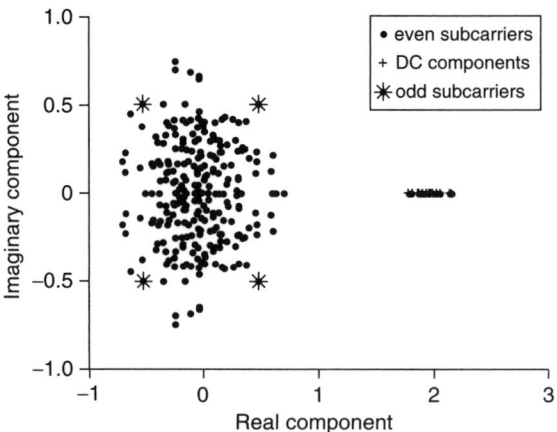

Figure 9.6: The QPSK signal constellation after clipping. After Armstrong and Lowery.[27]

subcarriers is illustrated in Figure 9.6, in which quaternary phase-shift keying (QPSK) with $N = 32$ is used, with only odd subcarriers carrying the data. We see that odd subcarriers are not affected at all, whereas even subcarriers have a random-like distribution coming from clipping noise.

Figure 9.7 provides a comparison of ACO-OFDM and DC-biased optical OFDM (DCO-OFDM), OOK, and PPM in terms of normalized bandwidth and required optical energy per

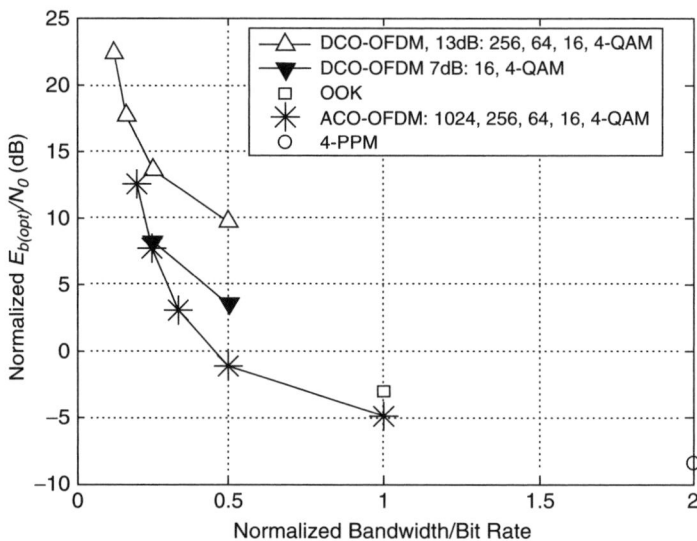

Figure 9.7: Normalized optical SNR $E_{b(opt)}/N_0$ at BER of 10^{-3} versus normalized bandwidth for ACO-OFDM against DCO-OFDM with 7 and 13 dB DC bias, OOK, and PPM. After Armstrong and Schmidt.[28]

bit to single-sided noise power spectral density $E_{b(opt)}/N_0$ to achieve BER of 10^{-3}. The bandwidth was defined as the first null bandwidth ($1 + 2/N$ for ACO-OFDM).[28] For each modulation, the largest constellation corresponds to the leftmost point on the corresponding curves. Significant improvement in ACO-OFDM was found compared to DCO-OFDM. For example, ACO-OFDM using 16-QAM of normalized bandwidth comparable to 4-QAM DCO-OFDM with 7 dB bias has better power efficiency for 4.7 dB. The PPM has slightly better power efficiency than that of ACO-OFDM but much lower spectral efficiency.

9.3 The Use of Optical OFDM in MMF Links for beyond Short-Reach Applications

Traditionally, as indicated previously, MMF is considered as the medium for short-reach applications. Until recently, the longest transmission distance at 10 Gb/s was 5 km. Nevertheless, it is important to note that MMF has several advantages over single-mode fiber (SMF): (1) easy installation, maintenance, and handling, which leads to low cost; (2) the effective cross-sectional area of MMF is much larger than that of SMF, indicating much better tolerance to fiber nonlinearities; and (3) from the information theory standpoint, the existence of many modes can be exploited in MIMO fashion, already in use in wireless communications,[21] to improve the spectral efficiency. In this section, we describe the experiment by Shieh's group[16] in which the 21.4 Gb/s polarization multiplexed coherent OFDM transmission over 200 km of MMF was demonstrated.

Figure 9.8 shows one experimental setup with MMF recirculation loop, suitable for studying coherent OFDM transmission over MMF links, that was used by Shieh et al.[16] The OFDM signal is generated by means of MATLAB and uploaded into an arbitrary waveform generator (AWG), which generates the analog signals at 10 GS/s. The total number of OFDM

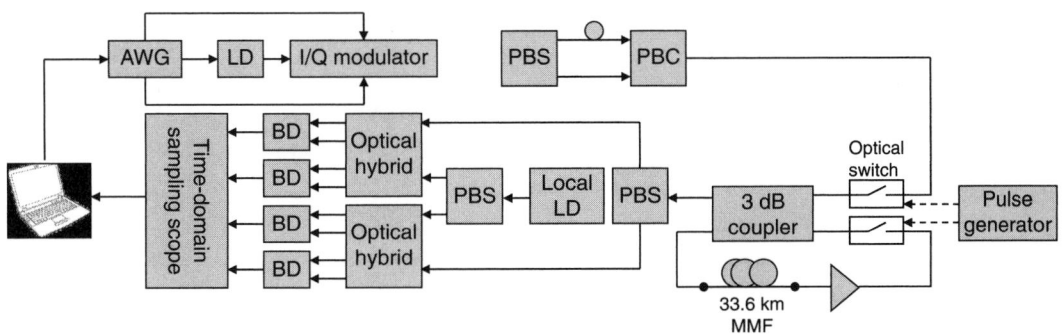

Figure 9.8: Experimental setup for the study of polarization multiplexed coherent OFDM transmission over MMF recirculating loop. BD, balanced detector; LD, laser diode; PBC, polarization beam combiner; PBS, polarization beam splitter.

subcarriers being used is 128, and the guard interval is set to 1/8 of the observation window. Out of 128 subcarriers, only 82 are filled with data, and 4 of them are used for phase estimation. The aggregate data rate resulting from the uniform loading of 4-QAM is 10.7 Gb/s. The real and imaginary parts of the OFDM signal are uploaded into the AWG, which performs D/A conversion. Then the parts are fed into I- and Q-ports of an optical I/Q modulator, composed of two Mach–Zehnder modulators with $\pi/2$ phase shift between them, which is used to convert the OFDM baseband signals into the optical domain.

The optical OFDM signal from the optical I/Q modulator is then split by using the polarization beam splitter (PBS), whose outputs are decorrelated by one OFDM symbol period (of duration 14.4 ns), and then recombined by the polarization beam combiner (PBC). The two polarization components are therefore completely independent of each other. The purpose of this manipulation is to emulate the polarization multiplexing transmitter with the aggregate data rate of 21.4 Gb/s. The signal is further used as input into a recirculation loop consisting of 33.6 km 50/125 μm MMF and an erbium-doped fiber amplifier to compensate for the MMF loss of 16 dB. The optical signal is launched from the 9-μm core-diameter SMF into the 33.6 km of 50-μm core diameter of graded-index MMF. SMF-to-MMF launch is used to perform the spatial mode filtering effect, resulting in only a small number of lower order modes excited in the MMF link. The 33.6-km MMF link consists of two 16.8-km fiber spools spliced with a piece of SMF in between, with each end of the loop being directly connected with SMF. This approach provides that the light is center launched into MMF, offering a twofold benefit: (1) At the beginning of the fiber loop, most of the light launched into MMF is transmitted in the axial area so that higher order modes are reduced, and (2) at the end of MMF, the signal is received with SMF pigtailed photodetectors, where the mode filtering gives a further reduction on the higher order modes. The remaining higher mode ISI due to multimode dispersion is mitigated with coherent optical OFDM-based signal processing. The OFDM signal is then coupled out from the recirculation loop and fed into a polarization multiplexing coherent receiver composed of a local laser, a PBS, two hybrids, and four balanced detectors (BDs). The local laser is tuned to the center of the signal, and the RF signals from the four balanced detectors are used as inputs to the Tektronix time domain sampling scope and are further processed by a MATLAB program using 2×2 MIMO-OFDM models. The following are the 2×2 MIMO-OFDM signal processing steps: (1) FFT window synchronization using Schmidl format to identify the start of the OFDM symbol, (2) software-performed estimation and compensation of the frequency offset, (3) channel estimation in terms of Jones matrix H, (4) phase estimation for each OFDM symbol, and (5) constellation construction for each carrier and BER computation. The channel matrix H is estimated by sending 30 OFDM symbols using alternative polarization launch. A total of 500 OFDM symbols are evaluated.

As mentioned previously, an important advantage of MMF compared to SMF is that it can be driven with a large optical power without inducing fiber nonlinearity. Figure 9.9a shows the

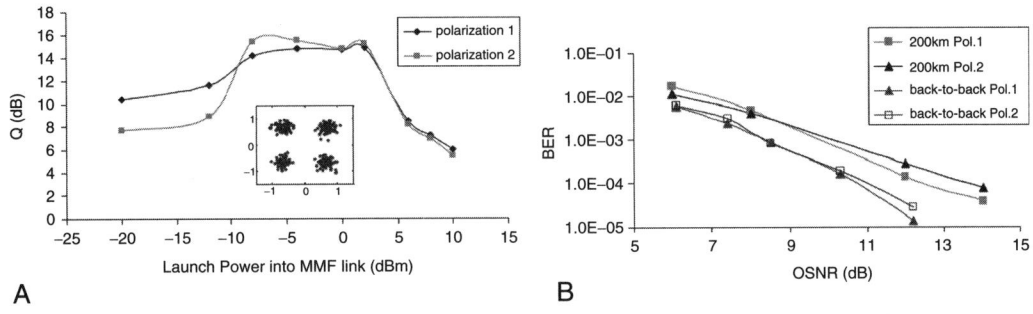

Figure 9.9: (a) Q-factor against MMF launch power for both polarizations and (b) BERs against optical SNR after 200 km of MMF.

Q-factor against the launch power into MMF after 200 km of transmission (or equivalently after six loop circulations). The maximum Q-factor is obtained for a wide range of launch powers from −8 to 3 dBm. In contrast to SMF transmission, in which launch power has to be kept below a relatively small threshold, a large input power into fiber up to 3 dBm used in MMF-based transmission did not introduce the significant nonlinearity impairments. The inset in Figure 9.9 shows the constellation diagram after 200 km of MMF transmission.

Figure 9.9b shows the BER performance in the back-to-back configuration and after 200 km of MMF transmission at a launch power of 2.5 dBm. The penalty after 200 km of MMF transmission at BER of 10^{-3} is only approximately 1 dB.

9.4 Optical OFDM in Broadcast MIMO Signaling over MMF Links

ISI represents the main limiting factor for high-speed transmission due to multimode dispersion. Namely, in MMFs the light propagates in modes, each with different propagation constant and group velocity. The modes being excited depend on the launch conditions and mode coupling within the MMF. Because the pulse of light that excites many modes in MMF arrives as multiple pulses at the fiber output, the multimode dispersion is similar to the multipath effect in wireless communications.[21] Therefore, different methods already in use in wireless communications are also applicable here, including equalization (in either the electrical or the optical domain), multicarrier modulation and OFDM, spread spectrum, diversity, and MIMO. Some of these methods have been studied for possible use in both point-to-point links and broadcasting.[30–35] For example, as an alternative to electrical equalization, the use of adaptive optical compensation has been advocated.[35,36] The main idea behind this proposal is to use the spatial shaping of the electric field at the MMF input, such as a spatial light modulator (a two-dimensional image), to excite only desired principal modes (PMs), which represent a complete set of orthogonal modes. The pulse emitted in a PM at the MMF input emerges as the single pulse at the MMF output, even in the presence of coupling.[37] By using the basic principles of PMs, Panicker et al.[36] described several algorithms suitable for equalization of ISI due to multimode dispersion by adaptive optics. This method requires a low-rate adaptive feedback channel, and it was studied for single-channel transmission in point-to-point links. The next natural step would be to use the MMFs in MIMO fashion for broadcasting applications, which has been proposed.[30–34] Different MIMO signaling scenarios over MMF can be classified as either with coherent detection[30–32] or with direct detection.[33,34]

An example of the MIMO-OFDM signaling scheme over MMF with coherent detection, employing both polarizations, is shown in Figure 9.10. In Figure 9.10a, a 2×2 MIMO system is given along with propagation paths of two modes, which occupy a different cross section of the optical fiber core and take a slightly different path along the optical fiber. In Figure 9.10b, an example of a MIMO system with M_T transmitters and M_R receivers is given; for simplicity, we set $M_T = M_R = N$. Because this system employs both polarizations, it is able to carry two M_T independent data streams. The laser diode continuous-wave output signal is split among M_R branches, each containing a transmitter, whose configuration is given in Figure 9.10c. Two independent data streams are used as inputs to OFDM transmitter x, whose signal is transmitted over x-polarization, and OFDM transmitter y, whose signal is transmitted over y-polarization. To convert OFDM signals in the optical domain, I/Q modulators are used. The optical OFDM signals from both polarizations are combined by a PBC. On the receiver side, N different outputs from MMF are used as inputs to N coherent receivers, whose configuration is shown in Figure 9.10d. The receiver configuration is typical

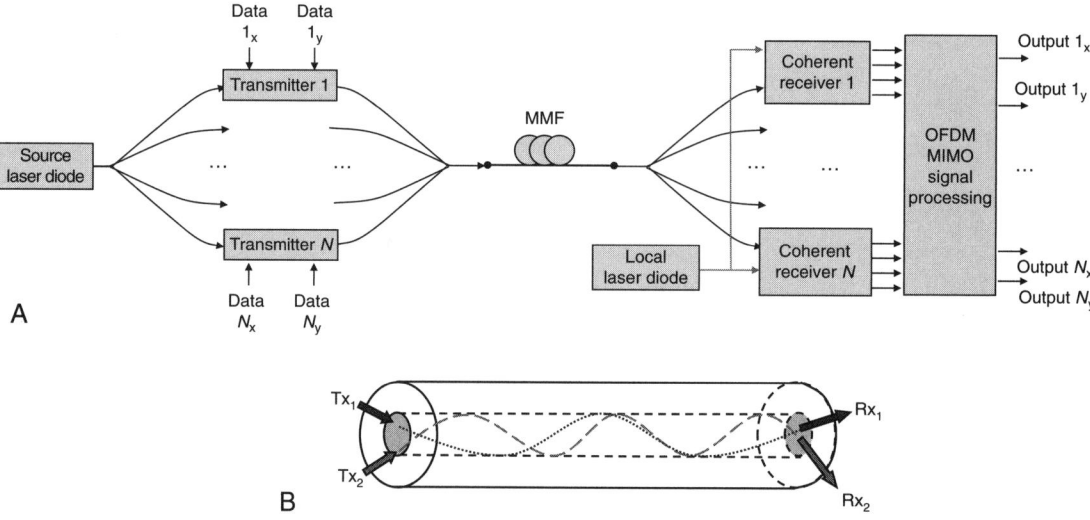

Figure 9.10: Coherent optical MIMO-OFDM over MMF: (a) 2×2 MIMO example, (b) a coherent optical MIMO MMF link,

(continued)

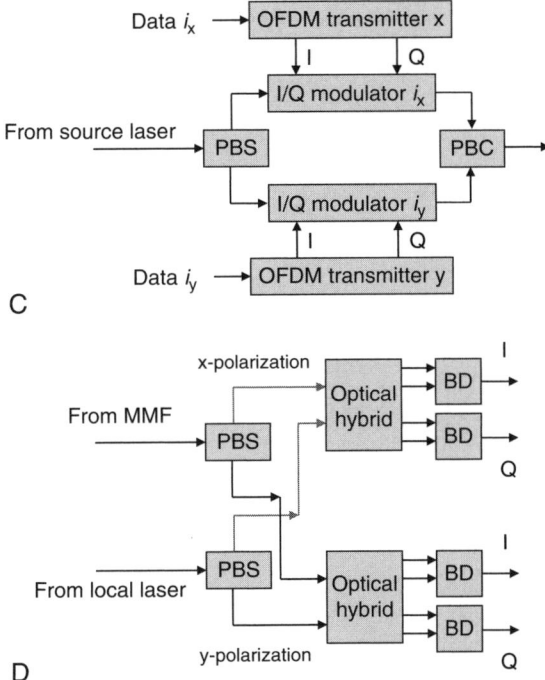

C

D

**Figure 9.10—cont'd (c) *i*th transmitter configuration, and (d) coherent receiver configuration.
BD, balanced detector; PBC, polarization beam combiner; PBS, polarization beam splitter.**

for coherent OFDM systems, as explained in previous chapters. One local laser is used for all coherent receivers to reduce the system cost. The in-phase (I) and quadrature (Q) channel outputs from both polarizations are used as inputs to MIMO-OFDM signal processing block. The use of OFDM facilitates multimode dispersion compensation. The scheme shown in Figure 9.10 can be used in both broadcasting and multicasting applications.

Let us now consider a MIMO system over MMF with M_T transmitter and M_R receivers. The corresponding input–output relation can be established as

$$y_i(t) = \sum_{j=1}^{M_T}\sum_{k=1}^{P} h_{ji}(k)e^{j\omega_c\left(t-\tau_{p,k}\right)}x_j\left(t-\tau_{g,k}\right) + w_i(t) \tag{9.32}$$

where $y_i(t)$ is the signal received by the ith receiver ($i = 1, 2, \ldots, M_R$), whereas $h_{ji}(k)$ is the channel gain from the jth transmitter through the kth mode. $\tau_{p,k}$ and $\tau_{g,k}$ denote the phase and group delay associated with the kth mode, respectively. ω_c is the carrier frequency, and w_i is the noise signal. $x_j(t)$ denotes the transmitted signal originating from the jth transmitter, and P is the number of modes. Notice that Eq. (9.32) observes only one polarization. When the group delay spread $\Delta\tau_g = \tau_{g,P} - \tau_{g,1}$ is small compared to symbol duration, the following is valid: $x(t - \tau_{g,k}) \approx x(t - \tau_g)$. This means that all paths arrive at approximately the same time, which is valid up the certain MMF length. Then the sampled baseband equivalent of Eq. (9.32) can be written in matrix form as follows:

$$y(n) = Hx(n) + w(n) \tag{9.33a}$$

where

$$y(n) = \begin{bmatrix} y_1(nT_s) \\ \cdots \\ y_{M_R}(nT_s) \end{bmatrix}, \quad H_{ij} = \sum_{k=1}^{P} h_{ij}(k)e^{-j\omega_c\tau_{p,k}}, \quad x(n) = \begin{bmatrix} x_1(nT_s) \\ \cdots \\ x_{M_T}(nT_s) \end{bmatrix}, \quad w(n) = \begin{bmatrix} w_1(nT_s) \\ \cdots \\ w_{M_R}(nT_s) \end{bmatrix} \tag{9.33b}$$

When the number of modes is large so that each Tx and Rx launch and sample from sufficiently different groups of modes, and when the product of carrier frequency and phase-delay spread $\omega_c\tau_{p,k} \gg 2\pi$ so that each element of H can be considered as uniformly distributed, then the elements of H will have complex Gaussian distribution with Rayleigh distribution for amplitudes.[21] The simulations performed in Hsu et al.[31] confirm the validity of this assumption.

Because the MMF channel varies at a relatively slower rate compared to data being transmitted, sending the estimated channel state information (CSI) back to the transmitter is feasible. The channel capacity can be evaluated for two different scenarios: (1) when CSI is available at the transmitter side (CSIT) and (2) when CSI is not available at the transmitter side.

The channel capacity for the CSIT case can be obtained by averaging the capacities associated with each channel realization[21]:

$$C = E_H\left\{ \max_{R_x:\, \text{Tr}(R_x)=\rho} B\log_2\left[\det\left(I_{M_R} + HR_xH^\dagger\right)\right]\right\}, \quad \rho = \sum_{i=1}^{M_T} E\left[x_i x_i^*\right] \tag{9.34}$$

where R_x is the covariance matrix of transmitted data, and I is the identity matrix. We use Tr() to denote the trace of matrix (the sum of diagonal elements), $E[\,]$ to denote the expectation operator, and $\dagger$ to denote Hermitian operation (transposition and complex conjugate). By performing the singular value decomposition of channel matrix

$$H = U\Sigma V^\dagger; \quad U^\dagger U = I_{M_R}; \quad V^\dagger V = I_{M_T}; \quad \Sigma = \text{diag}(\sigma_i), \quad \sigma_i = \sqrt{\lambda_i}, \quad \lambda_i = \text{eigenvalues}\left(HH^\dagger\right) \tag{9.35}$$

the MIMO channel capacity can be written as

$$C = E_H\left\{ \max_{P_i:\, \sum_i P_i \le \bar{P}} \sum_{i=1}^{R_H} B\log_2\left(1 + \frac{P_i\gamma_i}{\bar{P}}\right)\right\}, \quad \gamma_i = \frac{\sigma_i^2\bar{P}}{\sigma^2}, \quad \sigma^2 = E\left[w_i^2\right] \tag{9.36}$$

where $R_H = \text{rank}(H) \le \min(M_T, M_R)$ is the number of nonzero singular values of H.

When the CSI is not known on the transmitter side, the transmitter assumes the zero-mean spatially white model for H so that corresponding ergodic channel capacity can be determined by[21]

$$C = \max_{R_x:\ Tr(R_x)=\rho} E_H\left\{ B\log_2\left[\det\left(I_{M_T} + HR_xH^\dagger\right)\right]\right\} \tag{9.37}$$

where the expectation is performed with respect to the distribution on the channel matrix. The optimum input covariance that maximizes ergodic channel capacity is the scaled identity matrix $R_x = (\rho/M_T)I_{M_T}$, so ergodic channel capacity can be written as

$$C = E_H\left\{ B\log_2\left[\det\left(I_{M_R} + \frac{\rho}{M_T}HH^\dagger\right)\right]\right\} \tag{9.38}$$

The channel capacity of a 2 × 2 MIMO over MMF for both scenarios is shown in Figure 9.11. The results are obtained by propagating two signals over MMF link modeled using Rsoft's LinkSIM. For comparison purposes, the ideal Gaussian MIMO case is provided as well. For high SNRs, we can see degradation of optical MIMO over MMF in comparison with ideal Gaussian MIMO. The significant spectral efficiency improvement was found for optical MIMO over optical single-input single-output. Notice that those numerical results are obtained assuming that all modes arrive at the same time, and only one polarization is

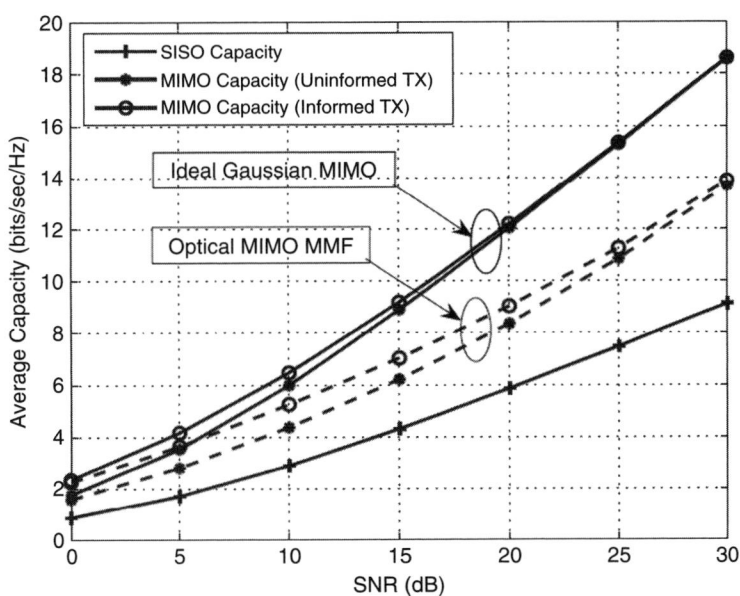

Figure 9.11: Ergodic channel capacity against SNR for 2 × 2 MIMO over MMF. The number of modes was set to 500. SISO, single-input single-output. After Hsu et al.[31]

observed. These results can be observed as the upper bound for the channel capacity of optical MIMO over MMF.

The channel capacity discussed previously is related to single-carrier coherent MIMO over MMF. The corresponding channel capacity expressions given previously can be generalized to MIMO-OFDM systems with N_c subcarriers as follows[38,39]:

$$C = E_H\left\{\frac{1}{N_c}\max_{\mathrm{Tr}(\Sigma)\leq P} B\log_2\left[\det\left(I_{M_RN_c} + \frac{1}{\sigma_n^2}H\Sigma H^\dagger\right)\right]\right\} \qquad (9.39)$$

where Σ is the covariance matrix of Gaussian input vector, defined as

$$\Sigma = \mathrm{diag}\{\Sigma_k\}_{k=0}^{N_c-1}, \ \Sigma_k = \frac{P}{M_RN_c}I_{M_R}; \ k = 0, 1, \ldots, N_c - 1 \qquad (9.40)$$

with P being the maximum overall transmit power, and σ_n being the standard deviation of noise process. The channel matrix in Eq. (9.39) is now a block diagonal matrix with ith block diagonal element corresponding to $M_T \times M_R$ MIMO over MMF channel:

$$H = \begin{bmatrix} H(0) & \cdots & 0 \\ \cdots & \cdots & \cdots \\ 0 & \cdots & H(N_c - 1) \end{bmatrix} \qquad (9.41)$$

With uniform power allocation, the OFDM MIMO over MMF channel capacity is given by

$$C = E_H\left\{\frac{1}{N_c}\sum_{k=0}^{N_c-1} B\log_2\left[\det\left(I_{M_R} + \rho H(k)H^\dagger(k)\right)\right]\right\} \qquad (9.42)$$

where $\rho = P/(M_RN_c\sigma_n^2)$.

One possible implementation of the MIMO-OFDM system with direct detection is shown in Figure 9.12a, with the corresponding transmitter configuration shown in Figure 9.12b. We use a particular OFDM version, DMT described in Section 9.1, so that the resulting OFDM signal is real. This approach assumes that the MIMO channel matrix estimate is available on the transmitter side. The precoder performs the following operation:

$$s = \xi^{1/2}H^{-1}s_{\mathrm{DMT}}^{1/2}$$

where H is the channel matrix, $s_{\mathrm{DMT}} = [s_1 \ldots s_{N_T}]^T$ is the sequence of DMT symbols originating from different transmitters, and ξ is the constant that is used to provide appropriate transmission power. This approach is also known as channel inversion (CI)[33] or beamforming in wireless literature,[21] and it is also applicable for coherent MIMO-OFDM over MMFs. There are several problems related to CI: (1) invertibility of the channel matrix; (2) if the channel matrix is not square, we must use pseudo-inverse instead; and (3) through CI, most of the power is allocated to the weakest channel eigenmodes. To solve these problems, we can employ truncated channel inversion[21] or regularized channel inversion

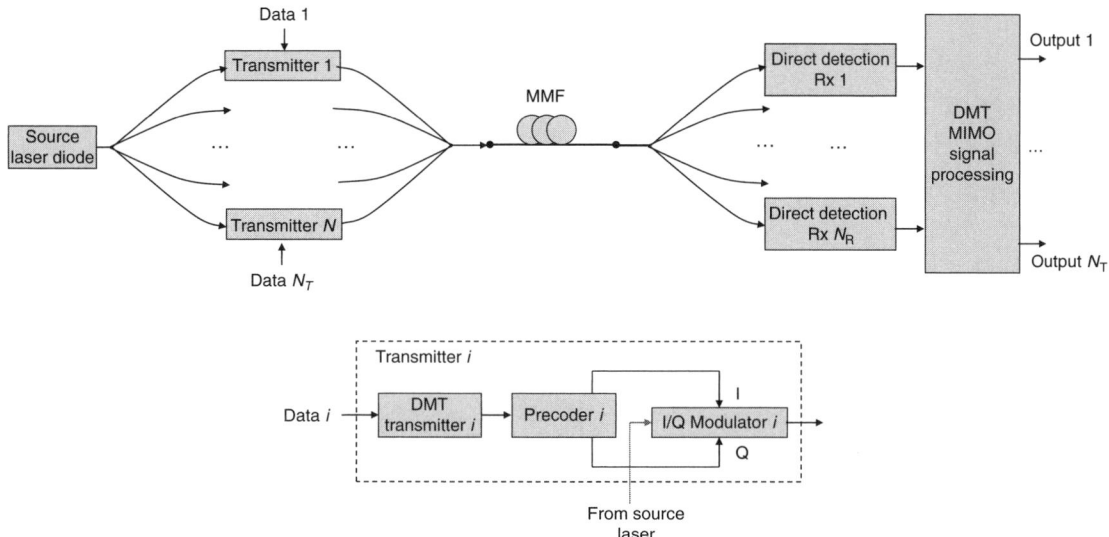

Figure 9.12: Direct detection optical DMT-MIMO over MMF: (a) a direct detection optical MIMO MMF system and (b) the *i*th transmitter configuration.

(RCI).[33,40] The key idea behind RCI is to perform the CI (precoding) in an iterative manner, $s = G_{RCI}s_{DMT}^{1/2}$, so that the overall data rate is maximized. This method, however, does not lead to interference-free transmission. Namely, for total transmit power of P and unit variance noise at each receiver, the corresponding effective signal-to-interference plus noise ratio of the *m*th user can be determined as follows[33]:

$$\rho_m = \frac{\left|h^{(m)}g_{RCI,m}\right|^2}{\mathrm{Tr}\left(G_{RCI}G_{RCI}^\dagger\right)/P + \sum_{k \neq m}\left|h^{(k)}g_{RCI,k}\right|^2} \tag{9.43}$$

where $g_{RCI,m}$ is the *m*th column of G_{RCI}.

The proper implementation of linear transmit precoding (beamforming) requires knowledge of the MMF channel matrix H. The conventional training methods in both single-carrier systems and OFDM require the coherent detection. Unfortunately, in direct detection the dependence between photocurrent and photodetector input electrical field is quadratic, as shown here:

$$i = |Hs|^2 + w \tag{9.44}$$

where $i = [i_1 \ldots i_{N_R}]^T$ is the received current vector, and w is the corresponding noise vector (to keep the exposition simple, we ignore the influence of photodiode responsivity). To determine the channel matrix when direct detection is used, novel training methods need to be developed. One such method was proposed in Bikhazi et al.[33] and will be described here (see also[34]).

Consider the sequence of N_s transmitted vectors organized in the form of $N_T \times N_s$ matrix $S = [s^{(1)} \ldots s^{(N_S)}]$. The corresponding received matrix can be written as follows:

$$Y = |HS|^2 + W \tag{9.45}$$

where Y is the $N_R \times N_s$ matrix of received currents, and W denotes the corresponding noise matrix. The magnitude of the channel matrix is relatively easy to estimate by transmitting N_b blocks independently, each of which excites one transmitter individually $S = \alpha^{1/2}[II \ldots I]$, where I is the identity matrix and α is the corresponding laser diode intensity. The maximum likelihood (ML) estimate of $|H_{mn}|^2$ can be determined by[33]

$$\left|H_{mn}\right|^2 = \frac{1}{\alpha N_b} \sum_{l=1}^{N_b} Y_{mn}(l) \tag{9.46}$$

where $Y(l)$ represents the matrix of received currents given by Eq. (9.45) for the lth block.

Because of the square-low photodetection process, the mth photocurrent output can be written as

$$i_m = \left|h^{(m)}s\right|^2 + w_m = \left|e^{j\phi_m}h^{(m)}s\right|^2 + w_m \tag{9.47}$$

which means that in order to perform proper precoding, we need to know only the relative phase along each row of H. This information can be obtained by exciting two sources that are simultaneously in phase for an even index of training blocks and simultaneously in quadrature for an odd index of training blocks. The training sequence for an even number of blocks N_b is given by[33]

$$S = \alpha^{1/2}\left[D_1\ D_1\ \ldots\ D_{N_b}\right] \tag{9.48}$$

where

$$D_l = \begin{bmatrix} 1 & 1 & \ldots & 1 \\ e^{j\phi_l} & 0 & \ldots & 0 \\ 0 & e^{j\phi_l} & \ldots & 0 \\ \ldots & \ldots & & \\ 0 & 0 & \ldots & e^{j\phi_l} \end{bmatrix} \tag{9.49}$$

with ϕ_l being set to 0, and $\pi/2$ for l being odd and even, respectively. The phase difference between H_{m1} and H_{mn} can be found by[33]

$$\theta_{mn} = \arg\min_{\theta_d} \frac{1}{\alpha} \sum_{l=1}^{N_b} Y_{mn}(l) - \left||H_{m1}| + |H_{mn}|e^{j(\theta_d + \phi_l)}\right|; \quad n = 2, \ldots, N_T \tag{9.50}$$

The accuracy of phase estimate is inversely proportional to SNR and training sequence length, whereas the accuracy of magnitude square estimate is determined by the Cramer–Rao bound of $\sigma^2/\alpha N_b$,[33] where σ^2 is the variance of noise elements in Eq. (9.45). Notice that this method is valid for MMF lengths for which all modes arrive approximately at the same time. For longer MMFs, we have to use coherent detection in a manner similar to that described in Section 9.3.

Because the coherent MIMO MMF channel has the input–output relationship $\boldsymbol{y} = \boldsymbol{Hx} + \boldsymbol{w}$, the symbol transmitted over the MMF channel at every symbol interval is rather a vector than a scalar. When the signal design extends over both the space coordinate (by multiple modes in MMF) and the time coordinate (by multiple symbol intervals), we can refer to it as the space–time code because of the similarities between multimode dispersion and multipath fading in wireless communications. Most of the space–time codes designed for wireless communications are straightforwardly applicable here. Under the assumption that the MMF channel is static for the duration of N_s symbols, the MMF channel input and output become matrices, with dimensions corresponding to the space coordinate (MMF modes) and time coordinate (the symbol intervals), given by

$$\boldsymbol{Y} = \boldsymbol{HX} + \boldsymbol{W} \tag{9.51}$$

where

$$\boldsymbol{Y} = \left[\boldsymbol{y}_1, \boldsymbol{y}_2, \ldots, \boldsymbol{y}_{N_s}\right] = \left(Y_{ij}\right)_{M_R \times N_s}, \quad \boldsymbol{X} = \left[\boldsymbol{x}_1, \boldsymbol{x}_2, \ldots, \boldsymbol{x}_{N_s}\right] = \left(X_{ij}\right)_{M_T \times N_s},$$

$$\boldsymbol{W} = \left[\boldsymbol{w}_1, \boldsymbol{w}_2, \ldots, \boldsymbol{w}_{N_s}\right] = \left(W_{ij}\right)_{M_R \times N_s} \tag{9.52}$$

with N_T being the number of transmitters, and N_R being the number of receivers. Let us observe a space–time code in which the receiver has the knowledge of the channel matrix $\boldsymbol{H}$. Under ML detection, the optimum transmit matrix is obtained by the following minimization:

$$\hat{\boldsymbol{X}} = \arg \min_{X \in X^{M_T \times N_s}} \sum_{i=1}^{N_s} \left\| \boldsymbol{y}_i - \boldsymbol{Hx}_i \right\|^2 \tag{9.53}$$

where the minimization is performed over all possible space–time input matrices. For example, the code matrix of Alamouti code is given by[41]

$$\boldsymbol{X} = \begin{bmatrix} x_1 & -x_2^* \\ x_2 & x_1^* \end{bmatrix} \tag{9.54}$$

in which, in the first time interval, the symbol x_1 is transmitted using propagation mode 1, and symbol x_2 is transmitted using mode 2. In the second time interval, the symbol $-x_2^*$ is transmitted using propagation mode 1, whereas the symbol x_1^* is transmitted using mode 2. The corresponding MMF channel matrix for the 2×2 MIMO example shown in Figure 9.10a can be written as

$$\boldsymbol{H} = \begin{bmatrix} h_{11} & h_{12} \\ h_{21} & h_{22} \end{bmatrix} \tag{9.55}$$

The received currents can be organized in matrix form as follows:

$$\begin{bmatrix} y_{11} & y_{12} \\ y_{21} & y_{22} \end{bmatrix} = \begin{bmatrix} h_{11} & h_{12} \\ h_{21} & h_{22} \end{bmatrix} \begin{bmatrix} x_1 & -x_2^* \\ x_2 & x_1^* \end{bmatrix} + \begin{bmatrix} w_{11} & w_{12} \\ w_{21} & w_{22} \end{bmatrix} = \begin{bmatrix} h_{11}x_1 + h_{12}x_2 + w_{11} & -h_{11}x_2^* + h_{12}x_1^* + w_{12} \\ h_{21}x_1 + h_{22}x_2 + w_{21} & h_{21}x_2^* + h_{22}x_1^* + w_{22} \end{bmatrix}$$

(9.56)

The combiner outputs, described in Alamouti,[41] can be obtained as follows:

$$\tilde{x}_1 = h_{11}^* y_{11} + h_{12} y_{12}^* + h_{21}^* y_{21} + h_{22} y_{22}^* = \left(|h_{11}|^2 + |h_{12}|^2 + |h_{21}|^2 + |h_{22}|^2 \right) x_1 + \text{noise}$$

$$\tilde{x}_2 = h_{12}^* y_{11} - h_{11} y_{12}^* + h_{22}^* y_{21} - h_{21} y_{22}^* = \left(|h_{11}|^2 + |h_{12}|^2 + |h_{21}|^2 + |h_{22}|^2 \right) x_2 + \text{noise}$$

(9.57)

The Alamouti code belongs to the class of linear space–time block codes.[42] In linear space–time codes L, symbols $x_1, x_2, \ldots, x_L$ are transmitted using M_T transmitters in T time intervals, and their code matrix has the form

$$X = \sum_{l=1}^{L} (\alpha_l A_l + j\beta_l B_l), \quad \alpha_l = \text{Re} - \{x_l\}, \quad \beta_l = \text{Im}\{x_l\}$$

(9.58)

where A_l and B_l are the complex matrices of dimension $M_T \times T$. For example, the Alamouti code can be put in the form

$$X = \begin{bmatrix} x_1 & -x_2^* \\ x_2 & x_1^* \end{bmatrix} = \begin{bmatrix} \alpha_1 + j\beta_1 & -\alpha_2 + j\beta_2 \\ \alpha_2 + j\beta_2 & \alpha_1 - j\beta_1 \end{bmatrix} = \alpha_1 \begin{bmatrix} 1 & 0 \\ 0 & 1 \end{bmatrix} + j\beta_1 \begin{bmatrix} 1 & 0 \\ 0 & -1 \end{bmatrix}$$

$$+ \alpha_2 \begin{bmatrix} 0 & -1 \\ 1 & 0 \end{bmatrix} + j\beta_2 \begin{bmatrix} 0 & 1 \\ 1 & 0 \end{bmatrix}$$

(9.59)

Another interesting class of space–time codes are trellis space–time codes.[42] Trellis space–time codes are trellis-coded modulation schemes in which transition among states, described by a trellis branch, is labeled by M_T signals, each associated with one transmitter. At each time t, depending on the state of the encoder and the input bits, a transition branch is chosen. If the label of this branch is $q_t^1 q_t^2 \ldots q_t^{M_T}$, then transmitter i is used to send constellation symbol q_t^i, and all these transmissions are simultaneous. A QPSK trellis space–time coding scheme with $M_T = 2$ is given in Figure 9.13.

Assuming the ideal CSI, and that the path gains h_{ij} ($I = 1, \ldots, N_T; j = 1, \ldots, N_R$) are known to the detector, and assuming that r_t^j is the received signal at the jth receiver at time instance t, the branch metric for a transition labeled $q_t^1 q_t^2 \ldots q_t^{N_T}$ is given by

$$\sum_{j=1}^{M_R} \left| r_t^j - \sum_{i=1}^{M_T} h_{ij} q_t^i \right|^2$$

(9.60)

The Viterbi algorithm is further used to compute the path with the lowest accumulated metric.

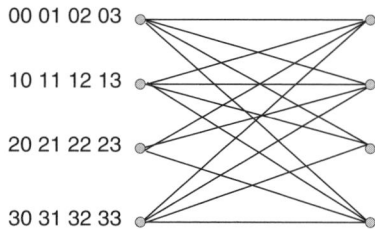

Figure 9.13: QPSK trellis space–time code.

Other approaches currently in use in MIMO wireless communications that can also be adopted in coherent MIMO over MMF systems include both linear interfaces (zero-forcing and minimum mean-square error) and nonlinear interfaces, such as Bell Labs Layered Space–Time architectures.[43]

9.5 Summary

In this chapter, we described the use of optical OFDM in MMF links to increase the spectral efficiency and system reach. In Section 9.1, we provided some basic properties of MMFs, including the impulse response. In Section 9.2, we described the use of discrete multitone, a particular version of OFDM, for transmission over MMFs. We described DMT modulation and demodulation, timing and synchronization, PAPR reduction, the clipping process, adaptive loading, and channel capacity calculation.

It was shown in Section 9.3 that 21.4 Gb/s transmission over 200 km of MMF is possible by using the polarization multiplexed coherent optical OFDM transmission. The launch power up to 3 dBm does not introduce significant nonlinearity degradation. The RF spectrum of the OFDM signal after 200 km of MMF was significantly deteriorated due to multimode dispersion. Nevertheless, most of the degradation was compensated by coherent OFDM signal processing, and the penalty after 200 km of MMF was found to be within 1 dB at BER of 10^{-3}.

In Section 9.4, we described the use of optical DMT-OFDM over MMFs for broadcasting and multicasting applications. Both coherent detection and direct detection versions were described. For coherent detection, the channel capacity calculation was provided. For the direct detection version, the method to estimate the MMF channel matrix was provided as well as a particular precoding method suitable for channel inversion. We also described the potential use of space–time coding for coherent MIMO over MMF systems.

References

1. Van Nee R, Prasad R. *OFDM wireless multimedia communications*. Boston: Artech House; 2000.
2. Prasad R. *OFDM for wireless communications systems*. Boston: Artech House; 2004.

3. Dixon BJ, Pollard RD, Iezekiel S. Orthogonal frequency-division multiplexing in wireless communication systems with multimode fiber feeds. *IEEE Trans Microwave Theory Tech* 2001;**49**(8):1404–9.

4. Lowery AJ, Armstrong J. 10 Gbit/s multimode fiber link using power-efficient orthogonal-frequency-division multiplexing. *Opt Express* 2005;**13**(25):10003–9.

5. Tang JM, Lane PM, Shore KA. Transmission performance of adaptively modulated optical OFDM signals in multimode fiber links. *IEEE Photon Technol Lett* 2006;**18**:205–7.

6. Kurt T, Yongaçoğlu A, Choinard J-Y. OFDM and externally modulated multi-mode fibers in radio over fiber systems. *IEEE Trans Wireless Commun* 2006;**5**(10):2669–74.

7. Randel S, et al. 1 Gbit/s transmission with 6.3 bits/s/Hz spectral efficiency in a 100 m standard 1 mm step-index plastic optical fibre link using adaptive multiple sub-carrier modulation. *Proc Post-Deadline Papers ECOC 2006*, paper no. Th 4.4.1, 2006.

8. Djordjevic IB. LDPC-coded OFDM transmission over graded-index plastic optical fiber links. *IEEE Photon Technol Lett* 2007;**19**(12):871–3.

9. Wei X, Guijun H, Qing D. Implementation of OFDM multimode fiber communication system in the simulink environment. *Information Optic and Photonics Technologies II, Proc. SPIE* 2007;**6837**:68371M-1– 68371M-7.

10. Wei C, Guijun H, Qing D. Application of turbo codes in optical OFDM multimode fiber communication system. *Opt Commun* 2008;**281**:1118–22.

11. Xu L, Qian D, Hu J, Wei W, Wang T. OFDMA-based passive optical networks (PON). *2008 Digest of the IEEE/LEOS Summer Topical Meetings* 2008;159–60, 21–3.

12. Qian D, Hu J, Ji P, Wang T, Cvijetic M. 10-Gb/s OFDMA-PON for delivery of heterogeneous services. *Opt Fiber Commun Conf and Natl Fiber Opt Engineers Conf*, OSA Technical Digest (CD) (Optical Society of America), paper no. OWH4, 2008.

13. Wei W, et al. Resource provisioning for orthogonal frequency division multiple access (OFDMA)-based virtual passive optical networks (VPON), In: *Opt Fiber Commun Conf and Natl Fiber Opt Engineers Conf*, OSA Technical Digest (CD) (Optical Society of America), paper no. OTUI1, 2008.

14. Jin XQ, Tang JM, Spencer PS, Shore KA. Optimization of adaptively modulated optical OFDM modems for multimode fiber-based local area networks [Invited]. *J Opt Networking* 2008;**7**(3):198–214.

15. Lee SCJ, Breyer F, Randel S, van den Boom HPA, Koonen AMJ. High-speed transmission over multimode fiber using discrete multitone modulation [Invited]. *J Opt Networking* 2008;**7**(2):183–96.

16. Tong Z, Yang Q, Ma Y, Shieh W. 21.4 Gb/s coherent optical OFDM transmission over multimode fiber, *Post-Deadline Papers Technical Digest*. 13th Optoelectronics and Communications Conference (OECC) and 33rd Australian Conference on Optical Fibre Technology (ACOFT), Paper No. PDP-5. 2008.

17. Keiser G. *Optical fiber communications*. Boston: McGraw-Hill; 2000.

18. Gloge D, Marcatili EAJ. Multimode theory of graded-core fibers. *Bell Systems Tech J* 1973;1563–78.

19. Pepeljugoski PK, Kuchta DM. Design of optical communications data links. *IBM J Res Dev* 2003;**47**:223–37.

20. Matthijsse P, Kuyt G, Gooijer F, et al. Multimode fiber enabling 40 Gbit/s multi-mode transmission over distances >400 m. In: *Proc. Opt. Fiber Commun. Conf.* paper no. OW113 Anaheim, CA; 2006.

21. Goldsmith A. *Wireless communications*. Cambridge, UK: Cambridge University Press; 2005.

22. Yam SSH, Achten F. Single wavelength 40 Gbit/s transmission over 3.4 km broad wavelength window multimode fibre. *Electron Lett* 2006;**42**:592–4.

23. Mestdagh DJG, Spruyt PMP. A method to reduce the probability of clipping in DMT-based transceivers. *IEEE Trans Commun* 1996;**44**:1234–8.

24. Eetvelt PV, Wade G, Thompson M. Peak to average power reduction for OFDM schemes by selected scrambling. *IEEE Electron Lett* 1996;**32**:1963–4.

25. Friese M. Multicarrier modulation with low peak-to-average power ratio. *IEEE Electron Lett* 1996;**32**:712–13.

26. Davis JA, Jedwab J. Peak-to-mean power control in OFDM, Golay complementary sequences, and Reed–Muller codes. *IEEE Trans Inf Theory* 1999;**45**:2397–417.

27. Armstrong J, Lowery AJ. Power efficient optical OFDM. *IEEE Electron Lett* 2006;**42**(6).

28. Armstrong J, Schmidt BJC. Comparison of asymmetrically clipped optical OFDM and DC-biased optical OFDM in AWGN. *IEEE Commun Lett* 2008;**12**(5):343–5.

29. Dardari D, Tralli V, Vaccari A. A theoretical characterization of nonlinear distortion effects in OFDM systems. *IEEE Trans Commun* 2000;**48**(10):1755–64.

30. Shah AR, Hsu RCJ, Sayed AH, Jalali B. Coherent optical MIMO (COMIMO). *J Lightwave Technol* 2005;**23**(8):2410–19.

31. Hsu RCJ, Tarighat A, Shah A, Sayed AH, Jalali B. Capacity enhancement in coherent optical MIMO (COMIMO) multimode fiber links. *J Lightwave Technol* 2005;**23**(8):2410–19.

32. Tarighat A, Hsu RCJ, Sayed AH, Jalali B. Fundamentals and challenges of optical multiple-input multiple output multimode fiber links. *IEEE Commun Mag* 2007;**45**:57–63.

33. Bikhazi NW, Jensen MA, Anderson AL. MIMO signaling over the MMF optical broadcast channel with square-law detection. *IEEE Trans Commun* 2009;**57**(3):614–17.

34. Agmon A, Nazarathy M. Broadcast MIMO over multimode optical interconnects by modal beamforming. *Opt Express* 2007;**15**(20):13123–8.

35. Alon E, Stojanovic V, Kahn JM, Boyd SP, Horowitz M. Equalization of modal dispersion in multimode fibers using spatial light modulators. *Proc IEEE Global Telecommun Conf*, Dallas, TX, Nov 29–Dec 2004;3.

36. Panicker RA, Kahn JM, Boyd SP. Compensation of multimode fiber dispersion using adaptive optics via convex optimization. *J Lightwave Technol* 2008;**26**(10):1295–303.

37. Fan S, Kahn JM. Principal modes in multi-mode waveguides. *Opt Lett* 2005;**30**(2):135–7.

38. Bölcskei H, Gesbert D, Paulraj AJ. On the capacity of wireless systems employing OFDM-based spatial multiplexing. *IEEE Trans Commun* 2002;**50**:225–34.

39. Wang J, Zhu S, Wang L. On the channel capacity of MIMO-OFDM systems. *Proc Int Symp Commun Information Technol 2005* 2005; 1325–1328, Beijing, China.

40. Stojnic M, Vikalo H, Hassibi B. Rate maximization in multi-antenna broadcast channels with linear preprocessing. *IEEE Trans Commun* 2006;**5**:2338–42.

41. Alamouti S. A simple transmit diversity technique for wireless communications. *IEEE J Sel Areas Commun* 1998;**16**:1451–8.

42. Biglieri E, Calderbank R, Constantinides A, et al. *MIMO wireless communications*. Cambridge, UK: Cambridge University Press; 2007.

43. Foschini GJ. Layered space–time architecture for wireless communication in a fading environment when using multi-element antennas. *Bell Labs Tech J* 1996;**1**:41–59.

CHAPTER 10

OFDM in Free-Space Optical Communication Systems

10.1 Introduction

High bandwidth demands in metropolitan area networks (MANs) and requirements for flexible and cost-effective service cause the imbalance known as the "last mile bottleneck." Fiber optics, RF, and copper/coaxial lines are the main state-of-the-art technologies used to address the high bandwidth requirements. The incompatibility of RF/microwave and optical communication technologies due to a large bandwidth mismatch between RF and optical channels is now widely believed to be the limiting factor in efforts to further increase transport capabilities. For this reason, RF/microwave–optical interface solutions that will enable aggregating multiple RF/microwave channels into an optical channel are becoming increasingly important. Free-space optical (FSO) communication is the technology that can address any connectivity needed in optical networks, such as core, edge, or access networks.[1–9] In MANs, the FSO can be used to extend the existing MAN rings; in enterprise, the FSO can be used to enable local area network (LAN)-to-LAN connectivity and intercampus connectivity; and the FSO is an excellent candidate for the last-mile connectivity. However, an optical wave propagating through the air experiences fluctuations in amplitude and phase due to atmospheric turbulence. The intensity fluctuation, also known as the scintillation, is one of the most important factors that degrade the performance of an FSO communication link, even under the clear sky condition.

FSO links are considered a viable solution for the last mile bottleneck problem because they have the following properties[1–9]:

1. High directivity of the optical beam provides high power efficiency and spatial isolation from other potential interferers, a property not inherent in RF/microwave communications.

2. The FSO transmission is unlicensed.

3. Large fractional-bandwidth coupled with high optical gain using moderate powers permits very high data rate transmission.

4. The state-of-the-art fiber-optics communications employ intensity modulation with direct detection (IM/DD), and the components for IM/DD are widely available.

5. The FSO links are relatively easy to install and easily accessible for repositioning when necessary.

The FSO communication may also be considered as an enabling technology to bring different technologies together and easy to integrate with a variety of interfaces and network elements.

In this chapter, we show that low-density parity-check (LDPC)-coded orthogonal frequency-division multiplexing (OFDM) is able to significantly outperform LDPC-coded on–off keying (OOK) over the atmospheric turbulence channel in terms of both coding gain and spectral efficiency. The key idea of this approach is to lower the symbol rate by using OFDM and, in combination with interleaving and LDPC codes, obtain high tolerance to the deep fades that characterize transmission through the atmospheric turbulent channel.

The incompatibility between RF/microwave and optical channels is widely accepted as an important limiting factor in efforts to increase future transport capabilities in mixed RF/optical networks. For this reason, RF/optical interface solutions that enable aggregating multiple RF channels into a single optical channel are of utmost importance. One such interface can be implemented based on coded OFDM.

The use of coherent detection can offer the potential of up to 24 dB improvement over the uncoded direct detection counterpart. One portion of improvement (10–13 dB) occurs from the fact that coherent detection can approach the quantum detection limit easier than can direct detection. The second portion (approximately 11 dB) occurs from the use of large-girth LDPC codes, which were described in Chapter 6.

Due to the high complexity associated with coherent detection, current FSO communication systems[1–9] employ IM/DD. Such systems use point-to-point communication between two optical transceivers along a line of sight. For example, an 8×10 Gb/s terrestrial FSO transmission over 3.4 km using an optical repeater has been demonstrated.[8] The IM/DD technique is also used in state-of-the-art fiber-optic communications, and the availability of optical components used in fiber-based systems makes FSO communication a cost-effective solution for high-rate image, voice, and data transmission.[1–9]

However, an optical wave propagating through the air experiences fluctuations in amplitude and phase due to atmospheric turbulence.[1–9] The atmospheric turbulence is caused by variations in the refractive index of the transmission medium due to inhomogeneities in temperature and pressure caused by solar heating and wind. The atmospheric turbulence optical channel has been intensively studied, and various models have been proposed to describe turbulence-induced performance degradation and intensity fluctuations.[1–7] The intensity fluctuation, also known as scintillation, is one of the most important factors that degrade the performance of an FSO communication link. Due to constraints on the receiver size, it is not always possible to ensure that the receiver aperture is significantly larger than the turbulent correlation length. In such a case, aperture averaging becomes ineffective, and alternative techniques to mitigate the intensity fluctuations are required.[1] These techniques can be classified into two broad categories. Spatial domain techniques[1] involve diversity detection using multiple receivers, and time domain techniques[7] adaptively optimize the decision threshold according to the maximum likelihood criterion. When the receiver has knowledge of the joint temporal distribution of intensity fluctuations, maximum-likelihood sequence detection (MLSD) can be employed. MLSD has high computational complexity, and suboptimal implementations of MLSD such as those based on suboptimal per-survivor processing (PSP)[7] are more likely to be implemented in practice. At bit error ratio (BER) below 10^{-6}, both MLSD and PSP require an electrical signal-to-noise ratio larger than 20 dB even in the weak turbulence regime. Such signal powers are unacceptably high for many applications, and novel modulation techniques for IM/DD FSO systems are needed. In this chapter, we show that OFDM combined with error control coding is a very good modulation format for FSO IM/DD systems.

OFDM[10–13] is a special case of a multicarrier transmission in which a single information-bearing stream is transmitted over many lower rate subchannels. It has been used for digital audio broadcasting[10] and high-definition television terrestrial broadcasting,[11] in digital subscriber line systems,[10] in IEEE 802.11, in high-performance LAN type 2 and multimedia mobile access communication wireless LANs,[10] and has been studied for use in lightwave hybrid AM/OFDM cable systems[12] and in radio over fiber-based networks.[13,14] It is interesting to note that modern digital TV broadcasting is based on OFDM,[11] and that OFDM is intensively studied for wireless applications.[10] Because the FSO link is a cost-effective solution for transmission of high-speed signals, the study of OFDM transmission over the FSO link is becoming increasingly important.

As discussed in previous chapters, OFDM uses the fast Fourier transform (FFT) algorithm for modulation and demodulation, and it requires no equalization. At the same time, it provides high spectral efficiency. These features, together with its immunity to burst errors due to intensity fluctuations, make OFDM an interesting candidate for FSO transmission. We have shown that significant performance improvement can be obtained by using LDPC codes and iterative decoding based on a sum-product algorithm that does not require knowledge of the joint temporal probability distribution functions.[9,15,16] LDPC codes have been shown to achieve impressive coding gains for a variety of channels.[15,16] They perform significantly better than turbo product and Reed–Solomon (RS) codes in bursty error channels such as the fiber-optics communication channel at 40 Gb/s or higher[15,16] and the FSO channel,[9] making them an excellent error-control coding scheme to combine with OFDM.

In this chapter, we describe an LDPC-coded FSO-OFDM system built using standard optical and RF components. The key idea is to lower the symbol rate by using OFDM and, in combination with interleaving and LDPC codes, to obtain high tolerance to the deep fades that are inherent to a turbulent channel. Note that the arbitrary forward error correction (FEC) scheme can be used; however, the use of LDPC codes leads to channel capacity achieving performance. On the other hand, OFDM is more sensitive to phase noise and has a relatively large peak-to-average power ratio,[10] so a careful design is needed to fully exploit the advantages and minimize the disadvantages of OFDM. In particular, the use of OFDM in the IM/DD system is not very power efficient if implemented as suggested in You and Kahn.[17] To improve the power efficiency, we introduce single-sideband clipped- and unclipped-OFDM schemes.

Future Internet should be able to support a wide range of services containing large amounts of multimedia over different network types at high speed. The future optical networks will therefore be hybrid, composed of different single-mode fiber (SMF), multimode fiber (MMF), and FSO links. In these networks, novel modulation and coding techniques are needed to deal with different channel impairments in SMF, MMF, and FSO links. We describe a coded modulation scheme suitable for use in hybrid FSO–fiber-optics networks, which is based on polarization multiplexing and coded OFDM with large-girth quasi-cyclic LDPC codes as channel codes.

The chapter is organized as follows: The concept of FSO-OFDM transmission is introduced in Section 10.2. The simulation model and error control coding scheme are described in Sections 10.3 and 10.4, respectively. The numerical results related to LDPC-coded OFDM with direct detection are presented in Section 10.5. In Section 10.6, we describe the use of coded OFDM in hybrid optical networks. Section 10.7 summarizes the chapter.

10.2 FSO-OFDM Transmission System

FSO-OFDM systems support high data rates by splitting a high-rate data stream into a number of low-rate data streams and transmitting these over a number of narrowband subcarriers. The narrowband subcarrier data streams experience smaller distortions than high-speed ones and require no equalization. Moreover, most of the required signal processing is performed in the RF domain. This is advantageous because microwave devices are much more mature than their optical counterparts and because the frequency selectivity of microwave filters and the frequency stability of microwave oscillators are significantly better than those of corresponding optical devices. Furthermore, the phase noise levels of microwave oscillators are significantly lower than those of distributed feedback (DFB) laser diodes, which means that RF coherent detection is easier to implement than optical coherent detection. This, in turn, allows a system architect to directly apply the most advanced coherent modulation formats already developed for wireless communication.

The basic FSO-OFDM transmitter and receiver configurations are shown in Figures 10.1a and 10.1b, respectively. The corresponding FSO link is shown in Figure 10.1c. A 10 Gb/s

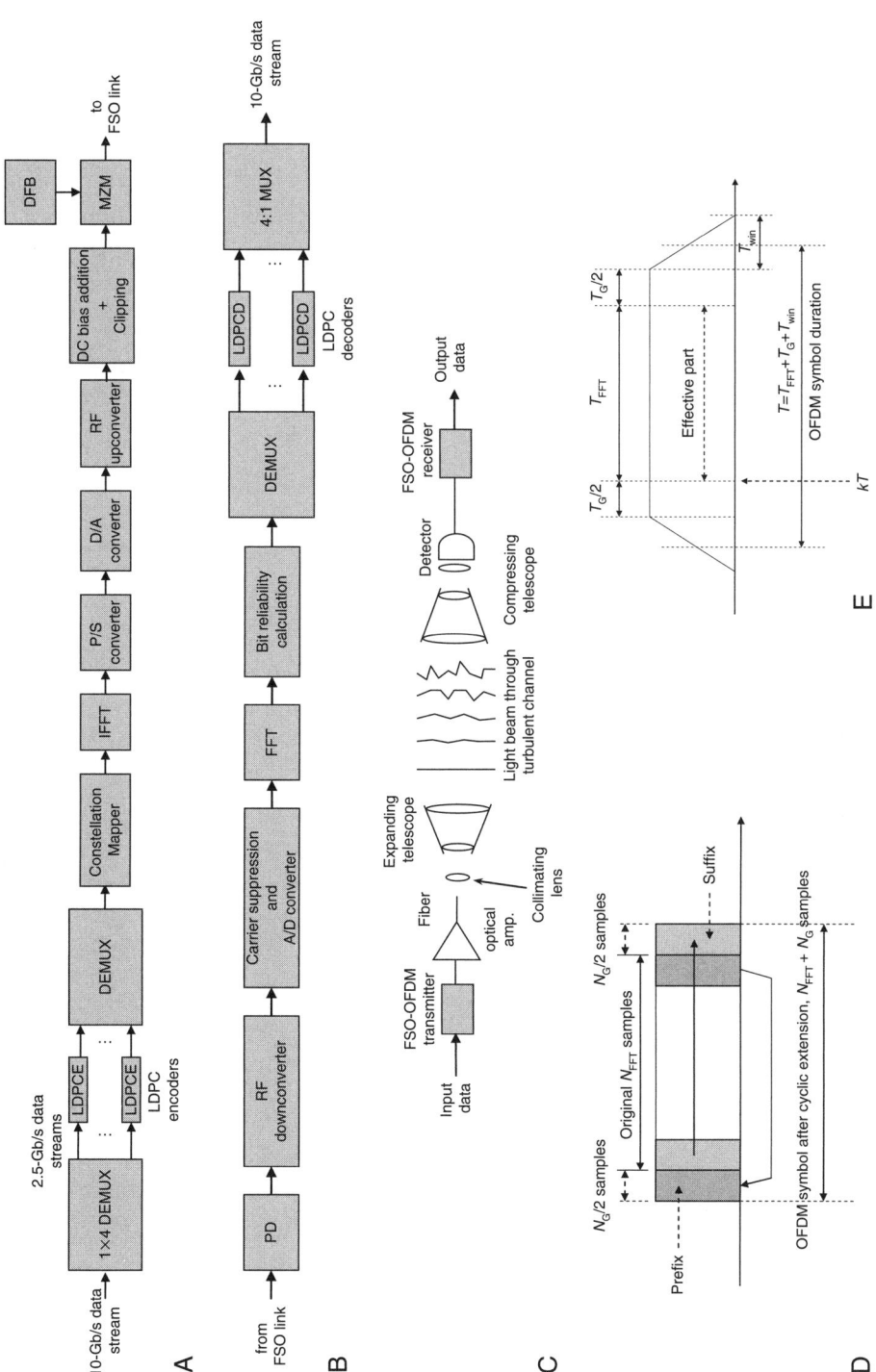

Figure 10.1: FSO-OFDM system: (a) transmitter configuration, (b) receiver configuration, (c) FSO link, (d) OFDM symbol cyclic extension, and (e) OFDM symbol after windowing. DFB, distributed feedback laser; LDPCD, LDPC decoder; LDPCE, LDPC encoder; MZM, dual-drive Mach–Zehnder modulator; P/S, parallel-to-serial.

information-bearing stream is demultiplexed into four 2.5 Gb/s streams, each encoded by identical LDPC encoders (for examples of currently available LDPC chips, see Lin et al.).[18] The LDPC-encoded outputs are further demultiplexed and parsed into groups of B bits. The B bits in each group (frame) are subdivided into K subgroups, with the ith subgroup containing b_i bits. The b_i bits from the ith subgroup are mapped into a complex-valued signal from a 2^{b_i} point signal constellation such as quadrature amplitude modulation (QAM). The complex-valued signal points from K subchannels are considered to be the values of the discrete Fourier transform (DFT) of a multicarrier OFDM signal. Therefore, the symbol length (the time between two consecutive OFDM symbols) in an OFDM system is $T = KT_s$, where T_s is the symbol interval length in an equivalent single-carrier system. By selecting K, the number of subchannels, sufficiently large, the OFDM symbol interval can be made significantly larger than the dispersed pulse width in a single-carrier system, resulting in an arbitrarily small intersymbol interference. Following the description given in Van Nee and Prasad,[10] the transmitted OFDM signal can be written as

$$s(t) = s_{OFDM}(t) + D \tag{10.1}$$

where

$$s_{OFDM}(t) = \text{Re}\left\{ \sum_{k=-\infty}^{\infty} w(t - kT) \sum_{i=-N_{FFT}/2}^{N_{FFT}/2 - 1} X_{i,k} \cdot e^{j2\pi \frac{i}{T_{FFT}} \cdot (t - kT)} e^{j2\pi f_{RF} t} \right\} \tag{10.2}$$

is defined for

$$kT - T_G/2 - T_{win} \leq t \leq kT + T_{FFT} + T_G/2 + T_{win}$$

In the previous expression, $X_{i,k}$ denotes the kth OFDM symbol in the ith subcarrier, $w(t)$ is the window function, and f_{RF} is the RF carrier frequency. The duration of the OFDM symbol is denoted by T, whereas T_{FFT} is the FFT sequence duration, T_G is the guard interval duration (the duration of cyclic extension), and T_{win} is the length of the windowing interval. The details of the resulting OFDM symbol are shown in Figures 10.1d and 10.1e. The symbols are generated as follows: $N_{QAM}(= K)$ consecutive input QAM symbols are zero-padded to obtain $N_{FFT} (= 2^m, m > 1)$ input samples for inverse FFT, then N_G samples are inserted to create the guard interval T_G, and, finally, the OFDM symbol is multiplied by the window function (raised cosine function is used in Van Nee and Prasad,[10] but the Kaiser, Blackman–Harris, and other window functions are also applicable).

The purpose of the cyclic extension is to preserve the orthogonality among subcarriers when the neighboring OFDM symbols partially overlap, and the purpose of the windowing is to reduce the out-of-band spectrum. The cyclic extension, illustrated in Figure 10.1d, is performed by repeating the last $N_G/2$ samples of the FFT frame (of duration T_{FFT} with N_{FFT} samples) as the prefix and repeating the first $N_G/2$ samples (out of N_{FFT}) as the suffix. (Notice that windowing is more effective for smaller numbers of subcarriers.) After a D/A conversion and RF up-conversion, we convert the RF signal to the optical domain using one of two options: (1) for symbol rates up to 10 Gsymbols/s the OFDM signal directly modulates the DFB laser, and (2) for symbol rates above 10 Gsymbols/s the OFDM signal drives the dual-drive Mach–Zehnder modulator (MZM). The DC component (D in Eq. 10.1) is inserted to enable noncoherent recovery of the QAM symbols. The following three OFDM schemes are presented here:

A. *Biased-OFDM single-sideband scheme*: This scheme is based on intensity modulation and is referred to as the "biased-OFDM" (B-OFDM) scheme. Because bipolar signals cannot be transmitted over an IM/DD link, it is assumed that the bias component D is sufficiently large so that when added to $s_{OFDM}(t)$, the resulting sum is non-negative. For illustrative purposes, the DFB laser driving signal (which is identical to the MZM RF input signal of schemes B and C) is shown in Figure 10.2a. The main disadvantage of the B-OFDM scheme is the poor power efficiency.

B. *Clipped-OFDM single-sideband scheme*: To improve the power efficiency, we present two alternative schemes. The first of these, which we refer to as the "clipped-OFDM" (C-OFDM) scheme, is based on single-sideband (SSB) transmission, with clipping of the negative portion of the OFDM signal after bias addition. The bias is varied to find the optimum one for fixed optical launched power. It was found that the optimum case is one in which approximately 50% of the total electrical signal energy before clipping is allocated for transmission of a carrier. To convert the signal from double-sideband (DSB) to SSB, we can (1) use Hilbert transformation of in-phase signal as the quadrature signal in the electrical domain or (2) perform DSB-to-SSB transformation by an optical filter. The MZM RF input signal for the C-OFDM scheme is shown in Figure 10.2b.

C. *Unclipped-OFDM single-sideband scheme*: The second power-efficient scheme, which we refer to as the "unclipped-OFDM" (U-OFDM) scheme, is based on SSB transmission employing LiNbO$_3$ MZM. To avoid distortion due to clipping, the information-bearing signal is transmitted by modulating the electrical field (instead of intensity modulation employed in the B-OFDM and C-OFDM schemes) so that the negative part of the OFDM signal is transmitted to the photodetector. Distortion introduced by the photodetector, caused by squaring, is successfully eliminated by proper filtering, and the recovered signal distortion is insignificant. Notice that U-OFDM is less power efficient than C-OFDM because the negative portion of the OFDM signal is transmitted and then discarded (see Figure 10.2c). For U-OFDM, the detector nonlinearity is compensated by postdetection filters that reject (potentially useful) signal energy and compromise power efficiency. Despite this drawback, U-OFDM is still significantly more power efficient than B-OFDM. Note that the DC bias shifts the average of the C-OFDM signal toward positive values, whereas in the case of B-OFDM, a much larger bias is needed to completely eliminate the negative portion of the signal. The MZM RF input signal for

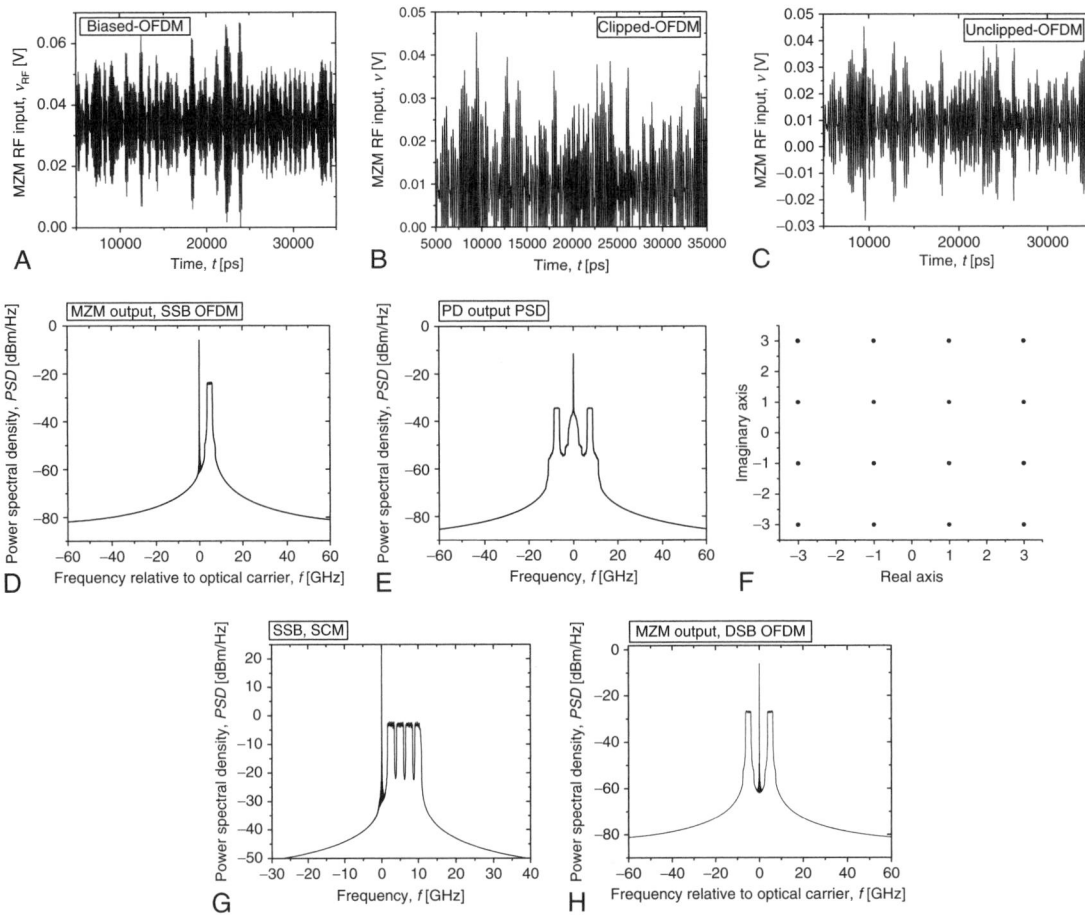

Figure 10.2: Waveforms and power spectral densities of the SSB-OFDM signal with 64 subcarriers at different points during transmission of an OFDM signal in a back-to-back configuration: (a) MZM RF input for B-OFDM, (b) MZM RF input for C-OFDM, (c) MZM RF input for U-OFDM, (d) PSD after MZM (U-OFDM), (e) photodetector output PSD (U-OFDM), (f) receiver constellation diagram for 16-QAM (U-OFDM), (g) PSD of SCM signal with four OFDM channels (U-OFDM), and (h) PSD of double-sideband OFDM signal after MZM (U-OFDM).

U-OFDM is shown in Figure 10.2c, and the recovered constellation diagram for 16-QAM SSB is shown in Figure 10.2g. The transmitted signal is recovered with negligible distortion.

The point-to-point FSO system considered here, shown in Figure 10.1c, consists of an FSO-OFDM transmitter, propagation medium, and an FSO-OFDM receiver. The modulated beam is projected toward the receiver using the expanding telescope. At the receiver, an optical system collects the light and focuses it onto a detector, which delivers an electrical signal proportional to the incoming optical power. The receiver commonly employs the transimpedance design, which is a good compromise between noise and bandwidth. A PIN photodiode plus preamplifier or an avalanche photodiode is typically used as an optical detector. During propagation through the air, the optical beam experiences amplitude and phase variations caused by scattering, refraction caused by atmospheric turbulence, absorption, and building sway. The photodiode output current can be written as

$$i(t) = R_{PD}\left|a(t)s_{OFDM}(t) + a(t)D\right|^2 = R_{PD}\left[\left|a(t)s_{OFDM}(t)\right|^2 + \left|a(t)D\right|^2 + 2R_e\left\{a(t)s_{OFDM}(t)a^*(t)D\right\}\right]$$

(10.3)

where $|a(t)|^2$ denotes the intensity fluctuation due to atmospheric turbulence, and R_{PD} denotes the photodiode responsivity.

The signal after RF down-conversion and appropriate filtering can be written as

$$r(t) = \left[i(t)k_{RF}\cos\left(\omega_{RF}t\right)\right] * h_e(\tau) + n(t)$$

(10.4)

where $h_e(t)$ is the impulse response of the low-pass filter (having the transfer function $H_e(j\omega)$); $n(t)$ is electronic noise in the receiver, commonly modeled as a Gaussian process; k_{RF} denotes the RF down-conversion factor; and the asterisk is the convolution operator. Finally, after the A/D conversion and cyclic extension removal, the transmitted signal is demodulated by the FFT algorithm. The soft outputs of the FFT demodulator are used to estimate the bit reliabilities that are fed to four identical LDPC iterative decoders based on the sum-product algorithm.[19] The parameters of the overall FSO-OFDM system must be carefully chosen, as explained later in this section, so that the reconstructed sequence constellation diagram suffers minimal distortion in a back-to-back configuration.

For the sake of illustration, consider the signal waveforms and power spectral densities (PSDs) at various points in the OFDM system given in Figure 10.2. These examples were generated using SSB transmission in a back-to-back configuration. The bandwidth of the OFDM signal is set to 2.5 GHz and the RF carrier to 7.5 GHz. The number of OFDM subchannels is set to 64. The OFDM sequence is zero-padded, and the FFT is calculated using 128 points. The guard interval is obtained by a cyclic extension of 2×16 samples as explained previously. The windowing (2×16 samples) is based on the Blackman–Harris windowing function. The average transmitted launched power is set to 0 dBm. The RF driver amplifier and MZM operate in linear regime (see Figures 10.2a–10.2c). The PSD for an SSB-OFDM MZM output signal is shown in Figure 10.2d, and the photodetector output signal (for SSB-OFDM transmission) is shown in Figure 10.2e. The OFDM term after beating in the photodetector (PD) (the third term in Eq. 10.2), the low-pass term, and the squared OFDM term (the first term in Eq. 10.2) can be easily identified.

If a 16-QAM OFDM system employing 64 subcarriers is used in combination with 39 Mb/s subchannels, the OFDM system proposed here allows transmitting a 10 Gb/s signal over a 2.5 GHz bandwidth, thereby increasing the spectral efficiency of OOK. To facilitate the implementation at higher speeds, OFDM may be combined with subcarrier multiplexing (SCM) in a similar manner as proposed for fiber-optic communication.[20] This approach is known as multiband OFDM in wireless communications literature. In this case (PSD shown in Figure 10.2g), the spectral efficiency of 4×10 Gb/s/11.25 GHz = 3.55 bits/s/Hz is achieved, which is significantly better than that for OOK transmission over an FSO link. (For illustrative purposes, the PSD of DSB-OFDM signal is also provided; see Figure 10.2h.)

After this high-level description of the system, we describe a statistical model of atmospheric turbulence based on gamma–gamma distribution (see Section 10.3) and an efficient LDPC error correction scheme based on quasi-cyclic LDPC codes suitable for combining with OFDM (see Section 10.4). The numerical results are reported in Section 10.5. Before we turn our attention to the FSO channel model, let us briefly explain how to interface RF/microwave and optical channels based on OFDM.

10.2.1 Aggregation of RF/Microwave Channels Using OFDM

In this section, we describe one possible interface between RF/microwave and optical channels, based on coded OFDM. The block diagrams of the possible transmitter and receiver configurations are shown in Figure 10.3. The data streams from L different RF channels are combined using OFDM and encoded using an LDPC encoder. The LDPC-encoded data stream is then parsed into groups of B bits, in a manner similar to that explained previously. The b_i bits

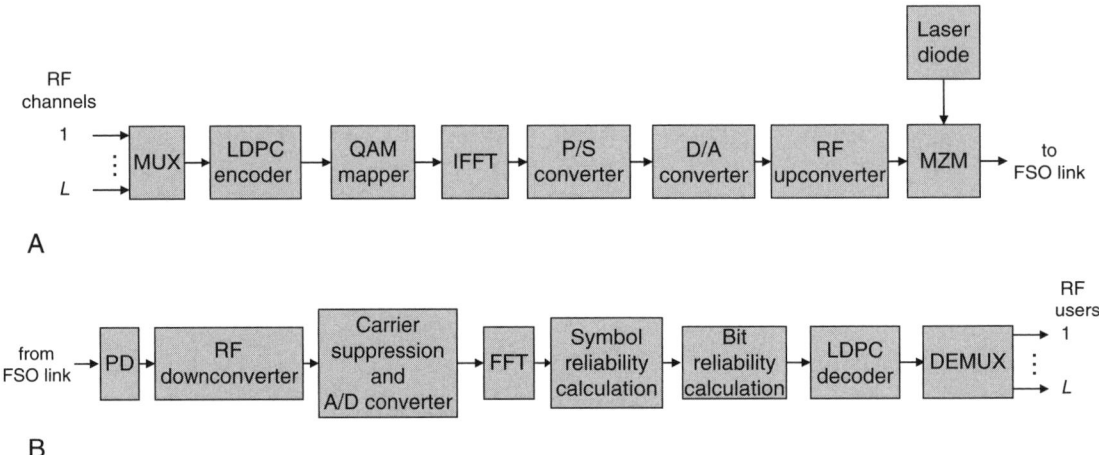

Figure 10.3: The aggregation of RF/microwave channel by coded OFDM: (a) transmitter configuration and (b) receiver configuration.

from the ith subgroup are mapped into a complex-valued signal from a 2^{b_i} point signal constellation such as QAM. The complex-valued signal points from all K subchannels are considered as the values of the DFT of a multicarrier OFDM signal. After D/A conversion and RF up-conversion, the OFDM signal drives an MZM for transmission over the FSO link. The DC component facilitates recovering the QAM symbols incoherently. At the receiver, an optical system collects the light and focuses it onto a detector, which delivers an electrical signal proportional to the incoming optical power. After the RF down-conversion, carrier suppression, A/D conversion, and cyclic extension removal, the transmitted signal is demodulated using the FFT algorithm. The soft outputs of the FFT demodulator are used to estimate the symbol reliabilities, which are converted to bit reliabilities and provided as input to an LDPC iterative decoder. Therefore, the configuration of this scheme is similar to that of Figure 10.1, from QAM block to bit reliability calculation block. In this scheme, we used one LDPC code for all RF channels. Another option would be to use the LDPC encoder/decoder for every RF channel.

10.3 Atmospheric Turbulence Channel Modeling

A commonly used turbulence model assumes that the variations of the medium can be understood as individual cells of air or eddies of different diameters and refractive indices. In the context of geometrical optics, these eddies may be observed as lenses that randomly refract the optical wavefront, generating a distorted intensity profile at the receiver of a communication system. The intensity fluctuation is known as scintillation, and it represents one of the most important factors that limit the performance of an atmospheric FSO communication link. The most widely accepted theory of turbulence is due to Kolmogorov.[21–24] This theory assumes that kinetic energy from large turbulent eddies, characterized by the parameter known as outer scale L_0, is transferred without loss to the eddies of decreasing size down to sizes of a few millimeters characterized by the inner scale parameter l_0. The inner scale represents the cell size at which energy is dissipated by viscosity. The refractive index varies randomly across the different turbulent eddies and causes phase and amplitude variations to the wavefront. Turbulence can also cause the random drifts of optical beams—a phenomenon usually referred to as wandering—and can induce beam focusing.

Outer scale is assumed to be infinite in this chapter. We consider zero and nonzero inner scale conditions. Understanding the turbulence effects under zero inner scale is important because it represents a physical bound for the optical atmospheric channel and as such has been of interest to researchers.[21]

To account for the strength of the turbulence, we use the unitless Rytov variance, given by[4,21–23]

$$\sigma_R^2 = 1.23 \, C_n^2 \, k^{7/6} L^{11/6} \tag{10.5}$$

where $k = 2\pi/\lambda$ is the wave number, λ is the wavelength, L is the propagation distance, and C_n^2 denotes the refractive index structure parameter, which is constant for horizontal paths. Although the Rytov variance has been used as an estimate of the intensity variance in weak turbulence, we also use it here as an intuitive metric that brings together all the physical operating conditions.

To characterize the FSO channel from a communication theory standpoint, it is useful to give a statistical representation of the scintillation. The reliability of the communication link can be determined if we use a good probabilistic model for the turbulence. Several probability density functions (PDFs) have been proposed for the intensity variations at the receiver of an optical link.[25–30] Al-Habash et al.[6] proposed a statistical model that factorizes the irradiance as the product of two independent random processes, each with a gamma PDF. The PDF of the intensity fluctuation is therefore[6]

$$f(I) = \frac{2(\alpha\beta)^{(\alpha+\beta)/2}}{\Gamma(\alpha)\Gamma(\beta)} I^{(\alpha+\beta)/2-1} K_{\alpha-\beta}\left(2\sqrt{\alpha\beta I}\right), \quad I > 0 \tag{10.6}$$

where I is the signal intensity, α and β are parameters of the PDF, Γ is the gamma function, and $K_{\alpha-\beta}$ is the modified Bessel function of the second kind of order $\alpha-\beta$.

10.3.1 Zero Inner Scale

The parameters α and β of the PDF that predicts the scintillation experienced by plane waves in the case of $l_0 = 0$ are given by the expressions[23,24]

$$\alpha = \left(\exp\left[\frac{0.49\sigma_R^2}{(1 + 1.11\sigma_R^{12/5})^{7/6}}\right] - 1\right)^{-1} \qquad \beta = \left(\exp\left[\frac{0.51\sigma_R^2}{(1 + 0.69\sigma_R^{12/5})^{5/6}}\right] - 1\right)^{-1} \tag{10.7}$$

where σ_R^2 is the Rytov variance as given in Eq. (10.5). This is a very interesting expression because the PDF of the intensity fluctuations at the receiver can be predicted from the physical turbulence conditions. The predicted distribution matches very well the distributions obtained from numerical propagation simulations.[21,25]

10.3.2 Nonzero Inner Scale

In the presence of a nonzero inner scale, the model must be modified to account for the change in the power spectrum of the refractive index variations. The PDF model is again a gamma–gamma distribution, but its parameters are now given by[21,25]

$$\alpha = \left\{\exp\left[\sigma_{\ln X}^2\right] - 1\right\}^{-1} \qquad \beta = \left\{\exp\left[\frac{0.51\sigma_P^2}{(1 + 0.69\sigma_P^{12/5})^{5/6}}\right] - 1\right\}^{-1} \tag{10.8}$$

where $\sigma_{\ln X}^2(l_0)$ is given by

$$\sigma_{\ln X}^2 = 0.16\sigma_1^2 \left(\frac{\eta_x Q}{\eta_x + Q}\right)^{7/6} \left[1 + 1.75\left(\frac{\eta_x}{\eta_x + Q}\right)^{1/2} - 0.25\left(\frac{\eta_x}{\eta_x + Q}\right)^{7/12}\right] \quad (10.9)$$

and

$$\eta_x = \frac{2.61}{1 + 0.45\sigma_1^2 Q^{1/6}}, \qquad Q = \frac{10.89L}{kl_0^2} \quad (10.10)$$

The variance σ_P^2 parameter, from the second term in Eq. (10.8), is given by

$$\sigma_P^2 = 3.86\sigma_1^2 \left\{(1 + 1/Q^2)^{11/12}\left[\sin\left(\frac{11}{6}\tan^{-1}Q\right) + \frac{1.51}{(1 + Q^2)^{1/4}}\sin\left(\frac{4}{3}\tan^{-1}Q\right)\right.\right.$$
$$\left.\left. - \frac{0.27}{(1 + Q^2)^{7/24}}\sin\left(\frac{5}{4}\tan^{-1}Q\right)\right] - 3.5Q^{-5/6}\right\} \quad (10.11)$$

The influence of both the atmospheric turbulence and the electronic noise on quaternary phase-shift keying (QPSK) and 16-QAM SSB FSO-OFDM systems (and zero inner scale) is illustrated in Figure 10.4. Results for an SSB-OFDM system with 64 subcarriers are shown. The average launched power is set to 0 dBm, the electrical signal-to-noise ratio at the PD is set to 18 dB, and the received signal constellation diagrams are obtained assuming weak atmospheric turbulence ($\sigma_R = 0.6$) and zero inner scale. The atmospheric turbulence changes the symmetry of clusters from circular for the additive white Gaussian noise (AWGN) channel to elliptic (see Figure 10.4). Both C-OFDM and U-OFDM schemes are more immune to the atmospheric turbulence than the B-OFDM scheme. The U-OFDM system is only slightly more immune to the atmospheric turbulence than the C-OFDM scheme. It appears that the better power efficiency of C-OFDM compensates the distortion introduced by clipping. The reason is simple. The average launched power is fixed for all three OFDM schemes, meaning that more energy per bit is allocated in the C-OFDM scheme (because the power in DC bias is lower) and, as a consequence, the scheme is more immune to electrical noise. Higher immunity to electrical noise may result in slightly better BER performance of the C-OFDM scheme compared to the U-OFDM scheme.

10.3.3 Temporal Correlation FSO Channel Model

The channel model discussed previously ignores the temporal correlation. In the presence of temporal correlation, the consecutive bits experience similar channel conditions. Because of the lack of literature on the temporal statistics in the FSO channel and the complexity of multidimensional joint distributions, we restrict our discussion to the independent and uncorrelated case. In many OFDM systems, this approach is reasonable

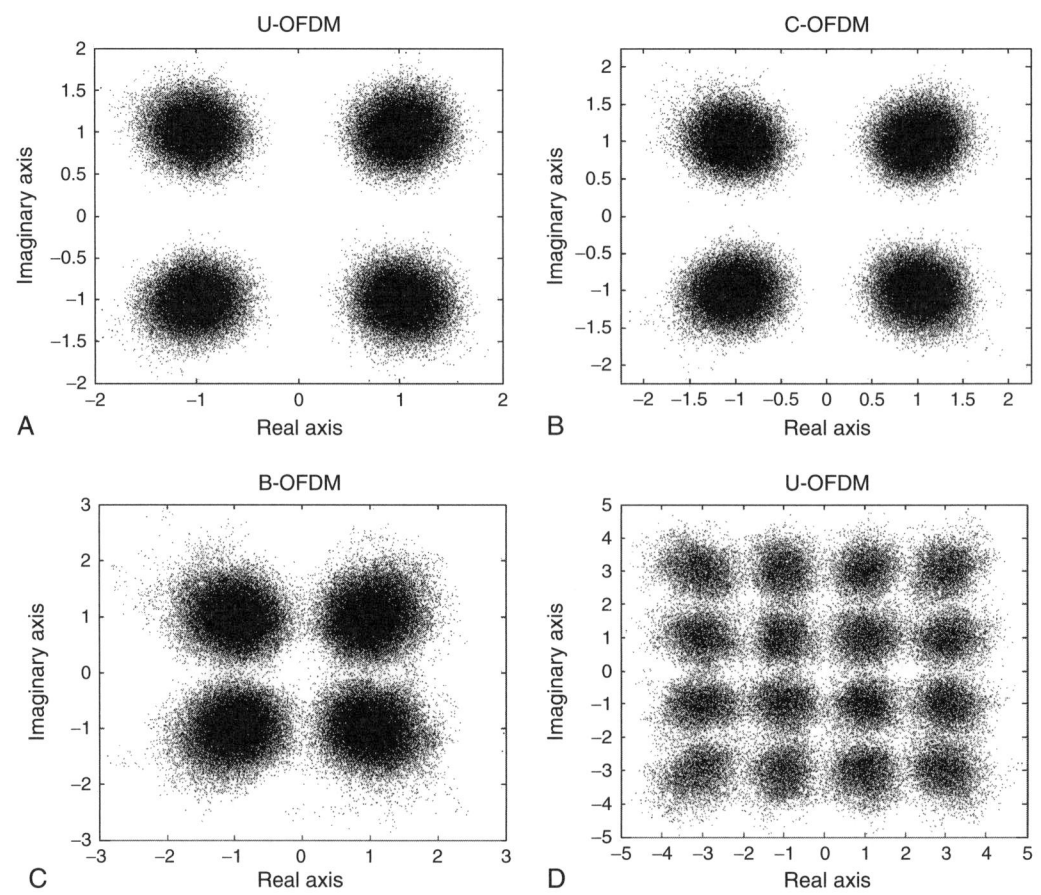

Figure 10.4: Received constellation diagrams of QPSK (a–c) and 16-QAM (d) SSB FSO-OFDM systems with electrical signal-to-noise per bit of 18 dB under the weak turbulence ($\sigma_R = 0.6$) for (a, d) U-OFDM scheme, (b) C-OFDM scheme, and (c) B-OFDM scheme.

for the following reasons: (1) When the channel conditions do not vary, a simple channel estimation technique based on pilot signals (see Van Nee and Prasad[10] for details) can be used to overcome the temporal correlation, and (2) the immunity to temporal correlation can further be improved by using interleaving. The interleaving can be visualized as the forming of an $L \times N$ (N is the codeword length) array of L LDPC codewords (the parameter L is known as interleaving degree) written row by row and the transmitting of the array entries column by column. If the original code can correct a single error burst of length l or less, then the interleaved code can correct a single error burst of length lL. Therefore, interleaved OFDM can successfully eliminate temporal correlation introduced by the FSO channel.

To illustrate the applicability of LDPC-coded OFDM in the presence of temporal correlation, we performed simulations (see Section 10.5) by employing the joint temporal correlative distribution model from Zhu and Kahn,[5] which describes the fading in an FSO channel at a single point of space at multiple instances in time. This method is based on the Rytov method to derive the normalized log-amplitude covariance function for two positions in a receiving plane perpendicular to the direction of propagation[5]:

$$b_X(d_{ij}) = \frac{B_X(P_i,P_j)}{B_X(P_i,P_i)} \tag{10.12}$$

where d_{ij} is the distance between points P_i and P_j. B_X denotes the log-amplitude covariance function

$$B_X(P_i,P_j) = E\left[X(P_i)X(P_j)\right] - E\left[X(P_i)\right]E\left[X(P_j)\right] \tag{10.13}$$

and X is the log-amplitude fluctuation.

In the weak turbulence regime, the Rytov method is commonly used to represent the field of electromagnetic wave as follows[21] (using cylindrical coordinates, $R = (r, L)$, where L is the transmission distance):

$$U(\mathbf{R}) \equiv U(\mathbf{r},L) = U_0(\mathbf{r},L)\exp\left[\psi(\mathbf{r},L)\right] \tag{10.14}$$

where $U_0(\mathbf{r}, L)$ is the electromagnetic field in the absence of turbulence, and ψ is the complex phase perturbation due to turbulence. The complex phase perturbation can be expressed as follows:

$$\psi(\mathbf{r},L) = \log\left[\frac{U(\mathbf{r},L)}{U_0(\mathbf{r},L)}\right] = X + jY \tag{10.15}$$

where X is the log-amplitude fluctuation, and Y is the corresponding phase fluctuation. In the weak turbulence regime, it is reasonable to assume X and Y to be Gaussian random processes. To deal with phase fluctuations, which is important in coherent FSO communication systems, one may use active modal compensation of wavefront phase distortion.[31] The residual phase variance after modal compensation can be described in Zernike terms by[31]

$$\sigma_Y^2 = Z_J\left(\frac{D}{d_0}\right)^{5/3} \tag{10.16}$$

where D is the aperture diameter, d_0 is the correlation length, and Z_J denotes the Jth Zernike term not being compensated (commonly $J = 3, 6, 10, 20$).

The joint temporal distribution of n intensity samples $(I_1, I_2, \ldots, I_n)$ is given by[5]

$$f_{\vec{I}}(I_1,I_2,\ldots,I_n) = \frac{1}{2\pi\prod\limits_{i=1}^{n} I_i \, (2\pi)^{n/2}\left|C_X\right|^{1/2}}\exp\left[-\frac{1}{8}\left(\ln\frac{I_1}{I_0}\cdots\ln\frac{I_n}{I_0}\right)\right] \tag{10.17}$$

where C_X is the covariance matrix of intensity samples:

$$C_X = \begin{bmatrix} \sigma_X^2 & \sigma_X^2 b_X\left(\dfrac{T}{\tau_0}d_0\right) & \cdots & \sigma_X^2 b_X\left(\dfrac{(n-1)T}{\tau_0}d_0\right) \\ \sigma_X^2 b_X\left(\dfrac{T}{\tau_0}d_0\right) & \sigma_X^2 & \cdots & \sigma_X^2 b_X\left(\dfrac{(n-2)T}{\tau_0}d_0\right) \\ \cdots & \cdots & \cdots & \cdots \\ \sigma_X^2 b_X\left(\dfrac{(n-1)T}{\tau_0}d_0\right) & \sigma_X^2 b_X\left(\dfrac{(n-2)T}{\tau_0}d_0\right) & \cdots & \sigma_X^2 \end{bmatrix} \tag{10.18}$$

σ_X^2 denotes the variance of the log-normally distributed amplitude, which for the plane wave can be approximated as[21]

$$\sigma_X^2 \cong 0.56 k^{7/6} \int_0^L C_n^2(x)(L-x)^{5/6} dx \tag{10.19}$$

where the wave number k, propagation length L, and the refractive index structure parameter C_n were introduced previously. T is the time interval between observations, which corresponds to the OFDM symbol period, whereas τ_0 is the coherence time. Notice that Eqs. (10.14) to (10.16) are valid in the weak turbulence regime. In the same regime, the covariance function (Eq. 10.15) is found to be exponential for both plane and spherical waves[21]:

$$b_X(\tau) = \exp\left(-\left(\frac{|\tau|}{\tau_0}\right)^{5/3}\right) \tag{10.20}$$

The typical values of coherence time τ_0 are in the range of 10 µs to 10 ms.

10.4 Soft Iterative Decoding

In this section, we give a brief description of the algorithm for calculating the required bit likelihoods in the iterative decoder. LDPC codes have been shown to significantly outperform turbo product codes in bursty error-prone channels such as the fiber-optics channel in the presence of intrachannel nonlinear effects.[15,16] The quasi-cyclic LDPC codes similar to those described in Chapter 6 are employed here. In FSO communications, the receiver electronics noise is commonly modeled as a Gaussian noise.[1-5,7,9] If r_I is the in-phase demodulator sample, and r_Q is the quadrature demodulator sample, then the symbol log-likelihood ratio (LLR) is calculated as

$$\lambda\left(s = \left(s_I, s_Q\right)\right) = -\frac{\left(r_I - s_I\right)^2}{2\sigma^2} - \frac{\left(r_Q - s_Q\right)^2}{2\sigma^2} \tag{10.21}$$

where s_I and s_Q are the coordinates of a transmitted signal constellation point, and AWGN variance (σ^2) is determined from the required electrical signal-to-noise ratio (SNR) per bit E_b/N_o:

$$\frac{E_b}{N_o} = \frac{E\{s_{i,k}\}}{\log_2 M} \frac{P_o}{\sigma^2} \tag{10.22}$$

where P_o is the normalized received power,[10] and $s_{i,k}$ denotes the QAM symbol in the kth subcarrier channel of the ith OFDM frame. (M denotes the number of points in the corresponding constellation diagram.) Notice that the definition of electrical SNR per bit, common in digital communications,[10,32] is different from that used in references.[5,7,9]

The initial bit likelihoods, provided to the iterative decoder, are calculated from the symbol LLRs, $\lambda(s)$, as

$$L\left(s_j\right) = \log\frac{\sum_{s:s_j=1} \exp\left[\lambda(\mathbf{s})\right]}{\sum_{s:s_j=0} \exp\left[\lambda(\mathbf{s})\right]}. \tag{10.23}$$

The Gaussian assumption in Eq. (10.21) may lead to BER performance degradation because the joint distribution is actually a convolution of the Gaussian and gamma–gamma PDFs. To reduce complexity, we use the Gaussian approximation in the calculation of symbol reliabilities. Nevertheless, dramatic performance improvement of an LDPC-coded FSO-OFDM system over an LDPC-coded FSO OOK system is obtained, as shown in Section 10.5. In calculating bit reliabilities from symbol reliabilities (Eq. 10.23; see also Chapter 6), the following "max-star" operator, defined as $\max^*(x, y) = \log(e^x + e^y)$, is applied recursively: $\max^*(x, y) = \max(x, y) + \log(1 + e^{-|x-y|})$.

10.5 Performance Assessment of Coded FSO-OFDM Systems with Direct Detection

Simulation results of an LDPC-coded SSB U-OFDM system for two different turbulence strengths and zero inner scale are shown in Figure 10.5. The influence of the atmospheric turbulence channel is included by perturbing the intensities of the OFDM samples (before photodiode) according to the PDF in Eq. (10.6). For binary phase-shift keying (BPSK) and QPSK, the coding gain improvement of an LDPC-coded FSO-OFDM system over an LDPC-coded FSO OOK system increases as the turbulence strength increases. However, the 16-QAM FSO-OFDM system is not able to operate in the regime of strong turbulence. For weak turbulence ($\sigma_R = 0.6$) (see Figure 10.5a), the coding gain improvement of the LDPC-coded FSO-OFDM system with 64 subcarriers over the LDPC-coded FSO OOK system is 8.47 dB for QPSK and 9.66 dB for BPSK at the BER of 10^{-5}. For strong turbulence ($\sigma_R = 3.0$) (see Figure 10.5b), the coding gain improvement of the LDPC-coded FSO-OFDM

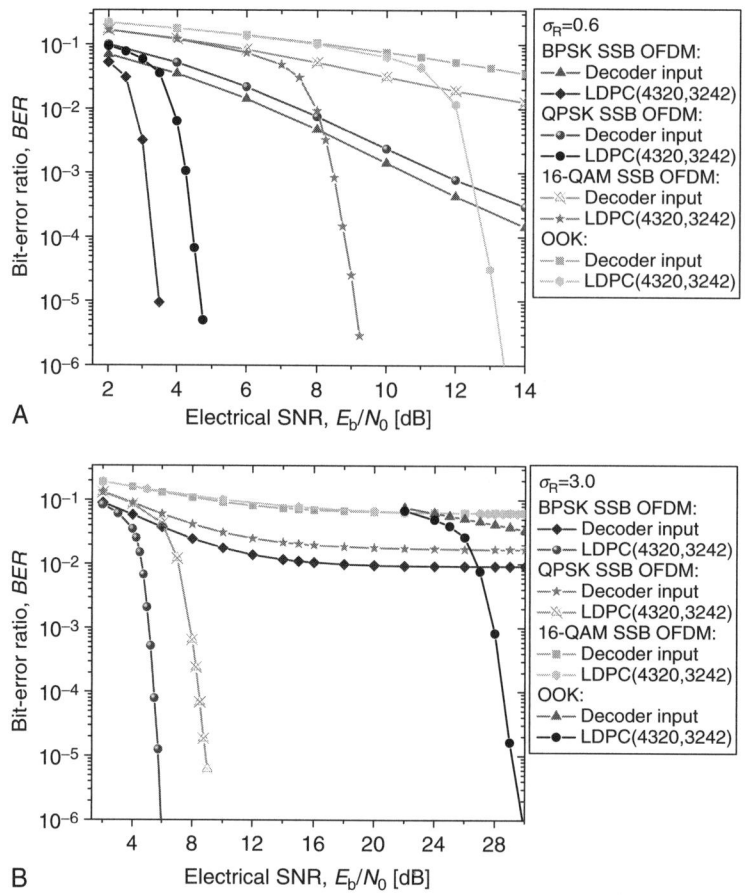

Figure 10.5: BER performance of LDPC-coded SSB U-OFDM system with 64 subcarriers under (a) weak turbulence ($\sigma_R = 0.6$) and (b) strong turbulence ($\sigma_R = 3.0$).

system over the LDPC-coded FSO OOK system is 20.24 dB for QPSK and 23.38 dB for BPSK. In both cases, the block-circulant[33] LDPC code (4320, 3242) of rate 0.75 is employed.

The comparison of different LDPC-coded SSB-OFDM schemes, under weak turbulence ($\sigma_R = 0.6$), is given in Figure 10.6. The C-OFDM scheme slightly outperforms the U-OFDM scheme. Both C-OFDM and U-OFDM schemes outperform the B-OFDM scheme by approximately 1.5 dB at BER of 10^{-5}. For the results shown in Figures 10.5 and 10.6, the received intensity samples are considered to be independent and uncorrelated, similarly as in references.[6,9,17] The results of simulations using the model described by Eqs. (10.17) and (10.18) are shown in Figure 10.7. The standard deviation σ_X is set to 0.6 (notice that σ_X is different from the Rytov standard deviation σ_R used previously, and for horizontal paths $\sigma_X \sim 0.498\sigma_R$). It is clear from Figure 10.7 that LDPC-coded OFDM with or without interleaver provides excellent performance improvement even in the presence of temporal

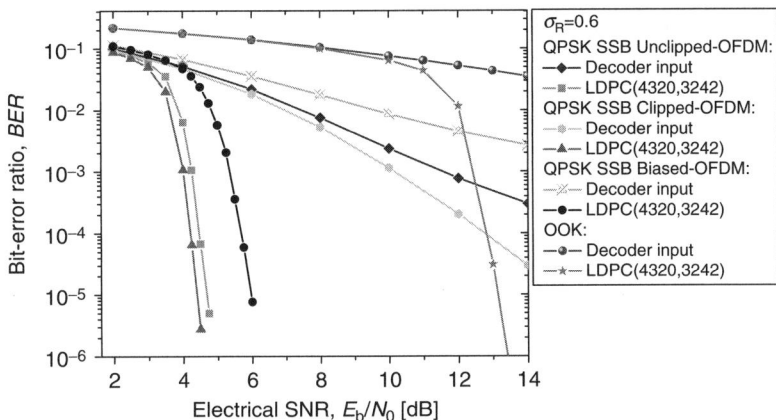

Figure 10.6: Comparison of different LDPC-coded SSB FSO-OFDM systems with 64 subcarriers under weak turbulence ($\sigma_R = 0.6$).

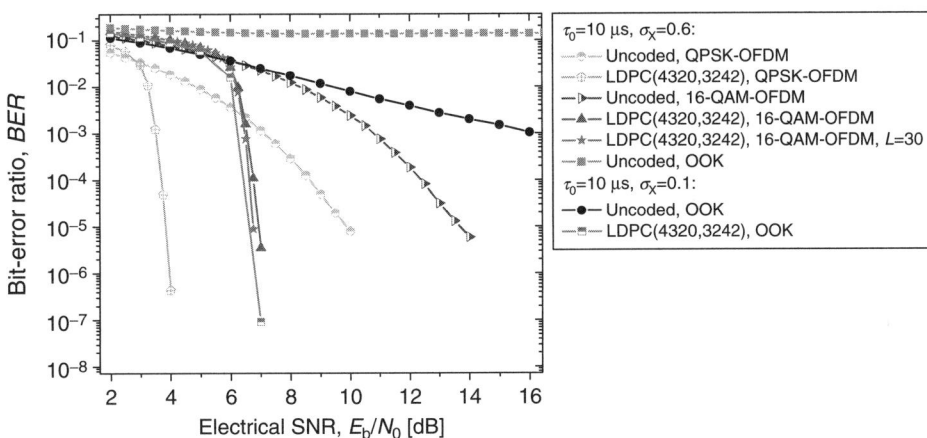

Figure 10.7: BER performance of LDPC-coded OFDM in the presence of temporal correlation.

correlation. The BER performance can further be improved by using the interleaver with larger interleaving degree than that used in Figure 10.7 (the star curve) at the expense of increasing encoder/decoder complexity. Notice that the OOK scheme enters the BER floor for this value of standard deviation ($\sigma_X = 0.6$), and even advanced FEC is not able to help much. However, LDPC-coded OOK is able to operate properly at lower standard deviations σ_X. To generate temporally correlated samples, we used two different methods: The first was based on the Levinson–Durbin algorithm[34,35] and the second was based on an algorithm from Wood and Chan.[36]

10.6 OFDM in Hybrid Optical Networks

Future optical networks will allow the integration of fiber optics and FSO technologies and will therefore have different portions of network composed of fiber (either SMF or MMF) and FSO sections. These hybrid optical networks might have a significant impact for both military and commercial applications, when pulling ground fiber is expensive and deployment takes a long time. Given the fact that hybrid optical networks will contain both FSO and fiber-optic sections, someone has to study the influence of not only the atmospheric turbulence present in the FSO portion of the network but also the influence of fiber nonlinearities, polarization mode dispersion (PMD), and chromatic dispersion in the fiber-optic portion of the network.

In this section, which is based on Djordjevic,[37] we describe a coded modulation scheme that is able to simultaneously deal with atmospheric turbulence, chromatic dispersion, and PMD in future hybrid optical networks. Moreover, the presented scheme supports 100 Gb/s per dense wavelength division multiplexing channel transmission and 100 Gb/s Ethernet while employing the mature 10 Gb/s fiber-optics technology.[38] The described hybrid optical network scheme employs the OFDM as a multiplexing and modulation technique, and it uses the LDPC codes as channel codes. With a proper design for 16-QAM-based polarization multiplexed coded OFDM, the aggregate data rate of 100 Gb/s can be achieved for the OFDM signal bandwidth of only 12.5 GHz, which represents a scheme suitable for 100 Gb/s Ethernet. Note that an arbitrary FEC scheme can be used in the proposed hybrid optical network. However, the use of large-girth LDPC codes leads to the channel capacity achieving performance.

We describe two scenarios: (1) the FSO channel characteristics are known on the transmitter side and (2) the FSO channel characteristics are not known on the transmitter side. In both scenarios, we assume that fiber-optic channel properties are known on the receiver side, obtained by pilot-aided channel estimation. Given the fact that transmitter and receiver nodes might be connected through several FSO and fiber-optic links, and that FSO link properties can vary significantly during the day, it is reasonable to assume that FSO link channel conditions are not known on the receiver side. In the presence of rain, snow, and fog, we assume that an RF feedback channel is used to transmit the channel coefficients to the transmitter, which adapts the transmitted power and data rate according to the channel conditions.

The described scheme has many unique advantages, including the following:

1. Demodulation, equalization, and decoding are jointly performed.

2. It is able to operate in the presence of channel impairments over different optical links, in SMF, MMF, and FSO.

3. It has high bandwidth efficiency (up to 10 bits/s/Hz).

4. It is compatible with future 100 Gb/s Ethernet technologies.

5. The employed coded modulation provides excellent coding gains.

We also describe how to determine the symbol reliabilities in the presence of laser phase noise and describe a particular channel inversion technique suitable for dealing with PMD effects.

10.6.1 Hybrid Optical Networks

An example of a hybrid FSO–fiber-optic network is shown in Figure 10.8a. This example includes intersatellite links and connection to aircrafts. The fiber-optic portion of the network could be a part of an already installed MAN or a wide area network. The FSO network portion should be used whenever pulling ground fiber is expensive and/or takes too long for deployment, such as in urban and rural areas where the optical fiber links are not already installed. The corresponding hybrid optical networking architecture is shown in Figure 10.8b. We can identify three ellipses representing the core network, the edge network, and the access network. The FSO links can be used in both edge and access networks. The hybrid optical network imposes a major challenge to engineers because novel signal processing techniques must be developed that simultaneously deal with atmospheric turbulence in FSO links and with chromatic dispersion, PMD, and fiber nonlinearities in fiber-optic links. One such coded modulation technique is described next. By using retroreflectors, FSO systems can be applied even when there is no line of sight between transmitter and receiver.

The described coded modulation scheme employs the coded OFDM with coherent detection for the reasons given at the beginning of the chapter. The transmitter and receiver shown in Figure 10.9, to be used in the hybrid optical network from Figure 10.8, are able to simultaneously deal with atmospheric turbulence, residual chromatic dispersion, and PMD. The bit streams originating from m different information sources are encoded using different (n, k_i) LDPC codes of code rate $r_i = k_i/n$. k_i denotes the number of information bits of ith $(i = 1, 2, \ldots, m)$ component LDPC code, and n denotes the codeword length, which is the same for all LDPC codes. The use of different LDPC codes allows us to optimally allocate the code rates. If all component LDPC codes are identical, the corresponding scheme is commonly referred to as the bit-interleaved coded modulation. The outputs of m LDPC encoders are written row-wise into a block-interleaver block. The mapper accepts m bits at time instance i from the $(m \times n)$ interleaver columnwise and determines the corresponding M-ary ($M = 2^m$) signal constellation point $(\phi_{I,i}, \phi_{Q,i})$ in two-dimensional constellation diagrams such as M-ary PSK or M-ary QAM. (The coordinates correspond to in-phase and quadrature components of M-ary two-dimensional constellation.) The operations of most of

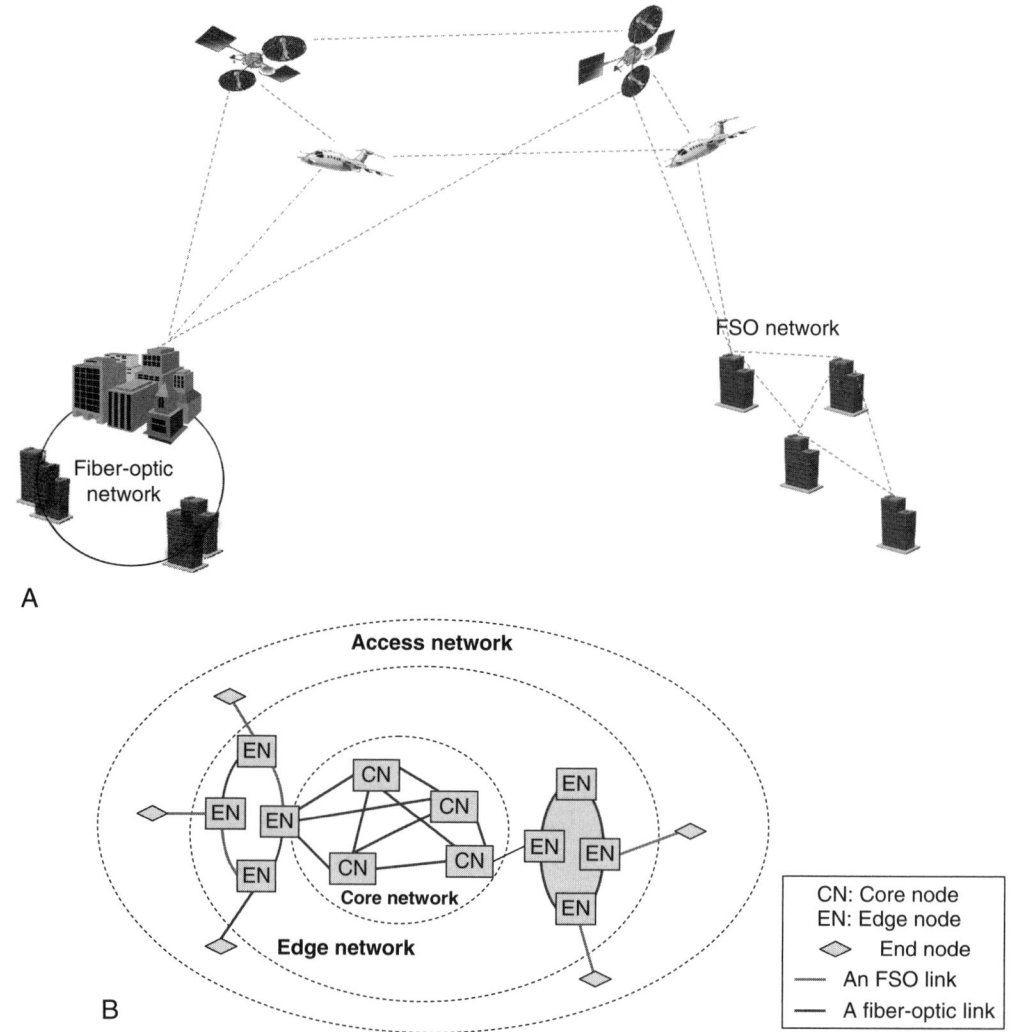

Figure 10.8: (a) A hybrid FSO–fiber-optic network example and (b) a hybrid optical networking architecture.

the blocks in the transmitter and receiver are similar to those presented in Chapter 6. Here, we provide only details relevant to hybrid optical networks.

The main motivation for using OFDM in hybrid optical networks is the fact that the OFDM symbols with a large number of subcarriers (on the order of thousands) have duration on the order of microseconds, and by means of interleavers on the order of thousands we are able to overcome the atmospheric turbulence with temporal correlation on the order of 10 ms. For the OFDM scheme to be capable of simultaneously compensating for chromatic dispersion and PMD, in addition to the atmospheric turbulence, the cyclic extension guard

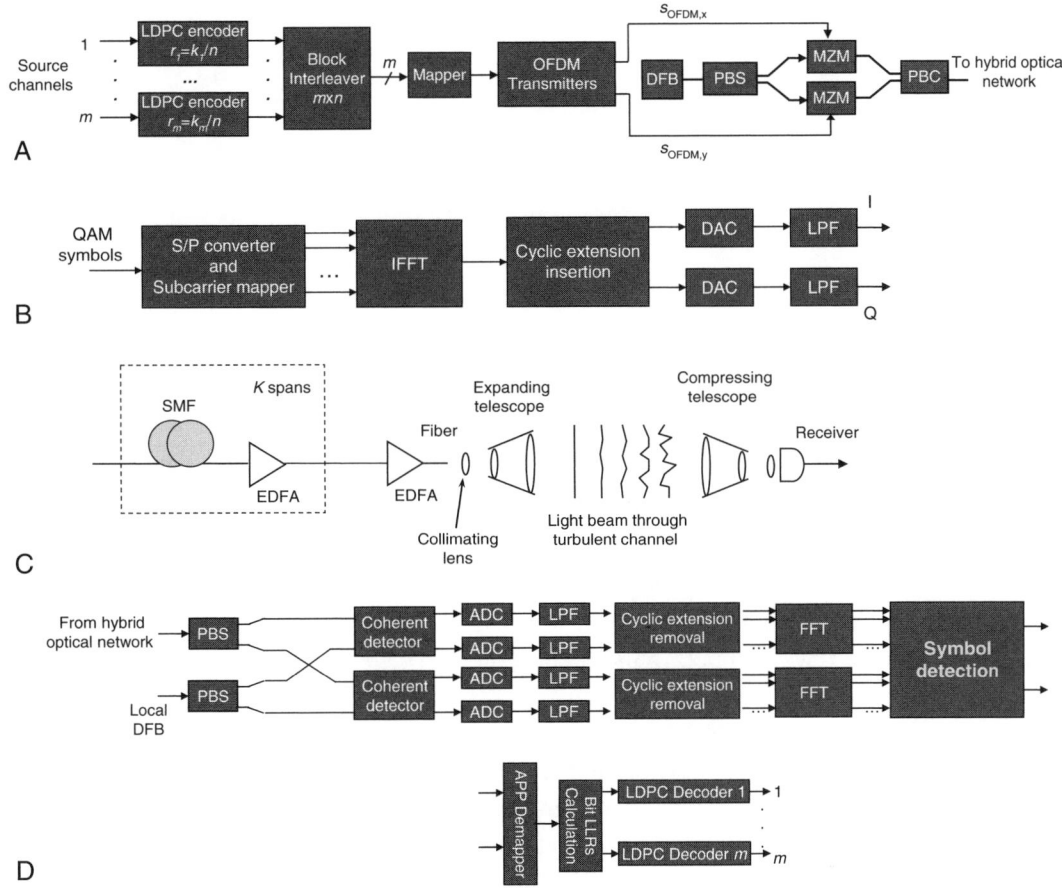

Figure 10.9: **The transmitter and receiver configurations for an LDPC-coded OFDM hybrid optical system with polarization multiplexing and coherent detection: (a) transmitter architecture, (b) OFDM transmitter architecture, (c) a hybrid optical link example, and (d) receiver architecture.**

interval should be longer than the total delay spread due to chromatic dispersion and differential group delay (DGD).

Because for high-speed signals a longer sequence of bits is affected by the deep fade in the milliseconds range due to atmospheric turbulence, we propose to employ the polarization multiplexing and large QAM constellations in order to achieve the aggregate data rate of $R_D = 100$ Gb/s while keeping the OFDM signal bandwidth on the order of 10 GHz. For example, by using the polarization multiplexing and 16-QAM we can achieve $R_D = 100$ Gb/s for an OFDM signal bandwidth of 12.5 GHz, resulting in bandwidth efficiency of 8 bits/s/Hz. Similarly, by using polarization multiplexing and 32-QAM, we can achieve the same data rate ($R_D = 100$ Gb/s) for OFDM signal bandwidth of 10 GHz, with bandwidth efficiency of 10 bits/s/Hz.

The receiver description requires knowledge of the channel. In the following, we assume that fiber-optic channel characteristics are known on the receiver side because the fiber-optics channel coefficients can easily be determined by pilot-aided channel estimation. On the other hand, the hybrid optical network may contain different FSO and fiber-optic sections, whereas the channel characteristics of FSO link can change rapidly even during the day, so it is reasonable to assume that FSO channel characteristics are not known on the receiver side. The FSO transmitter can use a retroreflector and a training sequence to sense the FSO channel. Next, we describe two scenarios: (1) The transmitter does not have any knowledge about the FSO link and (2) the transmitter knows the FSO link properties. When the transmitter knows the FSO link properties, we can employ the transmitter diversity concept.

10.6.2 Description of Receiver and Transmission Diversity Scheme

In this section, we describe the operation of the receiver by observing two different transmission scenarios. In the first scenario, we assume that the transmitter does not have any knowledge about the FSO channel. In the second scenario, we assume that the transmitter has knowledge about the FSO link, which is obtained by using the short training sequence transmitted toward the retroreflector. In both scenarios, we assume that the receiver knows the properties of the fiber-optic portion of the network obtained by pilot-aided channel estimation. This can be achieved by organizing the OFDM symbols in OFDM packets with several initial OFDM symbols being used for channel estimation. Note that this approach is also effective in estimating the FSO channel properties in the regime of weak turbulence. The immunity to atmospheric turbulence can be improved by employing the diversity approaches. To maximize the receiver diversity, multiple receivers should be separated enough so that the independence condition is satisfied. Given the fact that the laser beam is getting expanded during propagation, it might not be possible to always separate the receivers sufficiently that the independence condition is satisfied. On the other hand, by using transmission diversity instead, the independence condition is easier to satisfy. Moreover, it has been shown[39] that transmitter diversity performs comparably to the maximum ratio combining receiver diversity. In transmission diversity, the signal to be transmitted from the ith transmitter, characterized by path gain $r_i \exp[-j\theta_i]$, is premultiplied by complex gain $\alpha_i = a_i \exp[-j\theta_i]$ ($0 \le a_i \le 1$). On the receiver side, the weight a_i that maximizes the SNR is chosen by[39]

$$a_i = r_i \Big/ \sqrt{\sum_{i=1}^{L} r_i^2}$$

where L is the number of transmitter branches. When the channel is not known on the transmitter side, we have to set up a_i to 1, and $\theta_I = 0$, and use an Alamouti-type scheme instead.[40] Note, however, that use of an Alamouti-type receiver requires knowledge of the FSO channel and as such is not considered here.

The received complex symbols in x- and y-polarization, in the presence of PMD and scintillation, can be written by

$$r_{x,i,k} = a_i(k)\alpha_i(k)e^{j\phi_Y}\left[h_{xx}(k)s_{x,i,k} + h_{xy}(k)s_{y,i,k}\right] + n_{x,i,k} \tag{10.24}$$

$$r_{y,i,k} = a_i(k)\alpha_i(k)e^{j\phi_Y}\left[h_{yx}(k)s_{x,i,k} + h_{yy}(k)s_{y,i,k}\right] + n_{y,i,k} \tag{10.25}$$

where we use the index k to denote the kth subcarrier; index i to denote the ith OFDM symbol; $h_{ij}(k)$ ($i, j \in \{x, y\}$) are the channel coefficients due to PMD introduced in Chapter 6; $s_{x,i,k}$ and $s_{y,i,k}$ denote the transmitted symbols in x- and y-polarization, respectively; and corresponding received symbols are denoted by $r_{x,i,k}$ and $r_{y,i,k}$. The weight a_i is chosen in such a way to maximize the SNR, as explained previously. In Eqs. (10.24) and (10.25), $n_{x,i,k}$ and $n_{y,i,k}$ denote the amplified spontaneous emission (ASE) noise processes in x- and y-polarization. In the absence of ASE noise, Eqs. (10.24) and (10.25) represent the system of linear equations with two unknowns $s_{x,i,k}$ and $s_{y,i,k}$, and upon solving we obtain

$$\tilde{s}_{x,i,k} = \frac{\dfrac{h_{xx}^*(k)}{\left|h_{xx}(k)\right|^2}\left[r_{x,i,k} - \dfrac{h_{xy}(k)h_{yy}^*(k)}{\left|h_{yy}(k)\right|^2}r_{y,i,k}\right]}{1 - \dfrac{h_{xx}^*(k)h_{xy}(k)}{\left|h_{xx}(k)\right|^2}\dfrac{h_{yx}(k)h_{yy}^*(k)}{\left|h_{yy}(k)\right|^2}} \tag{10.26}$$

$$\tilde{s}_{y,i,k} = \frac{h_{yy}^*(k)}{\left|h_{yy}(k)\right|^2}r_{y,i,k} - \frac{h_{yx}(k)h_{yy}^*(k)}{\left|h_{yy}(k)\right|^2}\tilde{s}_{x,i,k} \tag{10.27}$$

where $\tilde{s}_{x,i,k}$ and $\tilde{s}_{y,i,k}$ denote the detector estimates of symbols $s_{x,i,k}$ and $s_{y,i,k}$ transmitted on the kth subcarrier of the ith OFDM symbol. Notice that the OFDM scheme with polarization diversity (described in Chapter 6), assuming that both polarizations are used on a transmitter side and equal gain combining on a receiver side, is the special case of symbol detector described by Eqs. (10.26) and (10.27). By setting $s_{x,i,k} = s_{y,i,k} = s_{i,k}$ and using the symmetry of channel coefficients, the transmitted symbol can be estimated by

$$\tilde{s}_{i,k} = \frac{h_{xx}^*(k)r_{x,i,k} + h_{xy}^*(k)r_{y,i,k}}{\left|h_{xx}(k)\right|^2 + \left|h_{xy}(k)\right|^2}$$

In the presence of laser phase noise, the symbol detector estimates are a function of the laser phase noise process:

$$\tilde{s}_{x,i,k} = \frac{\dfrac{h_{xx}^*(k)}{\left|h_{xx}(k)\right|^2}e^{-j\phi_{PN}}\left[r_{x,i,k} - \dfrac{h_{xy}(k)h_{yy}^*(k)}{\left|h_{yy}(k)\right|^2}r_{y,i,k}\right]}{1 - \dfrac{h_{xx}^*(k)h_{xy}(k)}{\left|h_{xx}(k)\right|^2}\dfrac{h_{yx}(k)h_{yy}^*(k)}{\left|h_{yy}(k)\right|^2}} \tag{10.28}$$

$$\tilde{s}_{y,i,k} = \frac{h_{yy}^*(k)e^{-j\phi_{PN}}}{|h_{yy}(k)|^2}r_{y,i,k} - \frac{h_{yx}(k)h_{yy}^*(k)}{|h_{yy}(k)|^2}\tilde{s}_{x,i,k} \tag{10.29}$$

The detector soft estimates of symbols carried by the kth subcarrier in the ith OFDM symbol, $\tilde{s}_{x(y)i,k}$, are forwarded to the a posteriori probability demapper, which determines the symbol LLRs $\lambda_{x(y)}(s)$ of x- (y-) polarization by

$$\lambda_{x(y)}(s \mid \phi_{PN}) = -\frac{\left(\text{Re}\left[\tilde{s}_{i,k,x(y)}(\phi_{PN})\right] - \text{Re}[\text{QAM}(\text{map}(s))]\right)^2}{2\sigma^2}$$

$$-\frac{\left(\text{Im}\left[\tilde{s}_{i,k,x(y)}(\phi_{PN})\right] - \text{Im}[\text{QAM}(\text{map}(s))]\right)^2}{2\sigma^2}; \quad s = 0,1,\ldots,2^{n_b} - 1 \tag{10.30}$$

where Re[] and Im[] denote the real and imaginary part of a complex number, QAM denotes the QAM constellation diagram, σ^2 denotes the variance of an equivalent Gaussian noise process originating from ASE noise, and map(s) denotes a corresponding mapping rule (Gray mapping rule is applied here). (n_b denotes the number of bits carried by symbol.) Note that symbol LLRs in Eq. (10.30) are conditioned on the laser phase noise sample $\phi_{PN} = \phi_T - \phi_{LO}$, which is a zero-mean Gaussian process (the Wiener–Lévy process) with variance $\sigma^2_{PN} = 2\pi(\Delta\nu_T + \Delta\nu_{LO})|t|$ ($\Delta\nu_T$ and $\Delta\nu_{LO}$ are the corresponding laser linewidths introduced previously). This derives from the fact that estimated symbols $\tilde{s}_{x(y)i,k}$ are functions of ϕ_{PN}. To remove the dependence on ϕ_{PN}, we have to average the likelihood function (similarly as we did in Chapter 6) over all possible values of ϕ_{PN}:

$$\lambda_{x(y)}(s) = \log\left\{\int_{-\infty}^{\infty} \exp\left[\lambda_{x(y)}(s \mid \phi_{PN})\right]\frac{1}{\sigma_{PN}\sqrt{2\pi}}\exp\left(-\frac{\phi_{PN}^2}{2\sigma_{PN}^2}\right)d\phi_{PN}\right\} \tag{10.31}$$

The calculation of LLRs in Eq. (10.31) can be performed by numerical integration. For the laser linewidths on the order of tens of kilohertz it is sufficient to use the trapezoidal rule, with samples of ϕ_{PN} obtained by pilot-aided channel estimation. Let us denote by $b_{j,x(y)}$ the jth bit in an observed symbol s binary representation $\boldsymbol{b} = (b_1, b_2, \ldots, b_{nb})$ for x- (y-) polarization. The bit LLRs required for LDPC decoding are calculated from symbol LLRs by

$$L\left(\hat{b}_{j,x(y)}\right) = \log\frac{\sum_{s:b_j=0}\exp\left[\lambda_{x(y)}(s)\right]}{\sum_{s:b_j=1}\exp\left[\lambda_{x(y)}(s)\right]} \tag{10.32}$$

Therefore, the jth bit LLR in Eq. (10.32) is calculated as the logarithm of the ratio of a probability that $b_j=0$ and probability that $b_j=1$, in a similar manner as described in Chapter 6.

10.6.3 Performance Evaluation of Hybrid Optical Networks

We now discuss the evaluation of the described hybrid optical network. In simulation results shown in Figures 10.10 and 10.11, we assume that PMD channel coefficients are known at

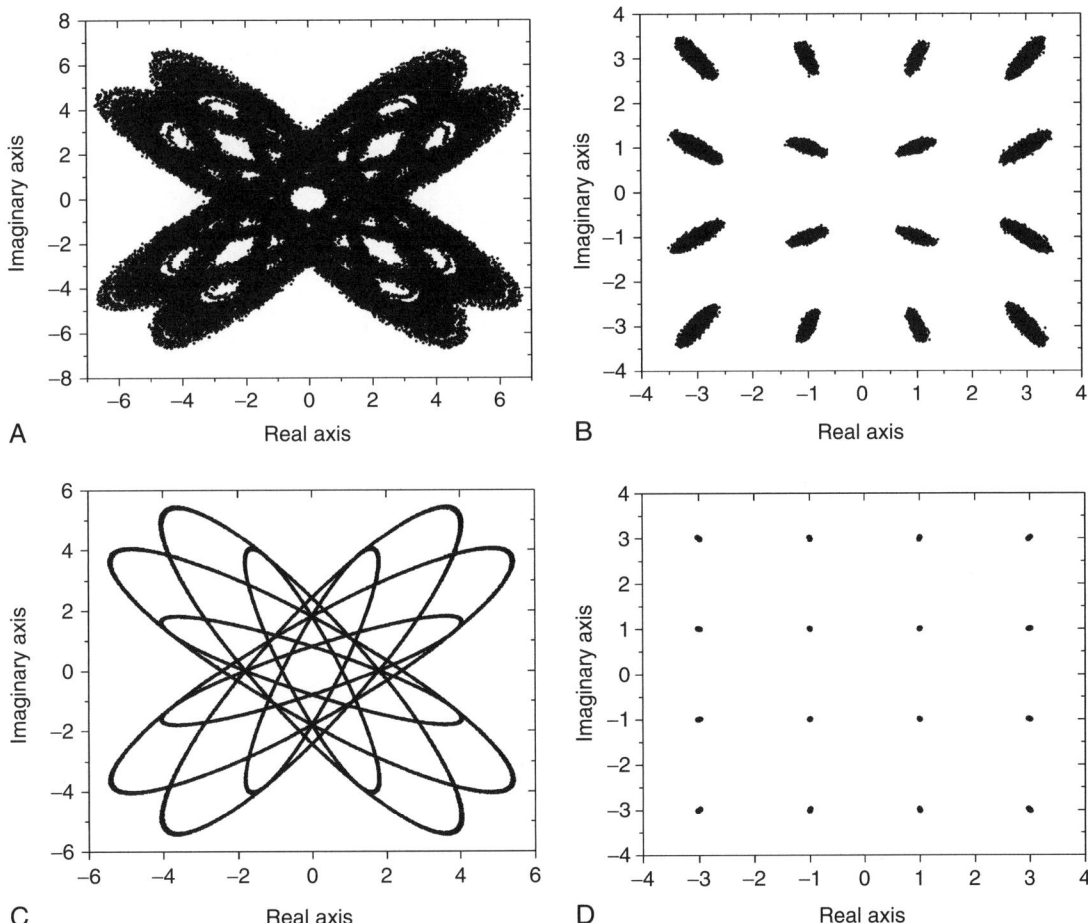

Figure 10.10: The constellation diagrams for polarization multiplexed 16-QAM (the aggregate data rate is 100 Gb/s) after 500 ps of DGD for $\sigma_X = 0.1$, $\sigma_Y = 0.05$, and optical SNR = 50 dB observing the worst-case scenario ($\theta = \pi/2$ and $\varepsilon = 0$) without transmission diversity (a) before PMD compensation and (b) after PMD compensation. The corresponding constellation diagrams in the presence of PMD only (for DGD of 500 ps) (c) before PMD compensation and (d) after PMD compensation.

the receiver because they can easily be determined by pilot-aided channel estimation. On the other hand, the FSO channel may change significantly during the daytime, and as such it is difficult to estimate. To illustrate the efficiency of this scheme, Figures 10.10a and 10.10b show the constellation diagrams for aggregate rate of 100 Gb/s, corresponding to the $M = 16$-QAM and the OFDM signal bandwidth of 12.5 GHz in the presence of atmospheric turbulence ($\sigma_X = 0.1$ and $\sigma_Y = 0.05$) before (Figure 10.10a) and after (Figure 10.10b) PMD compensation, assuming the worst-case scenario ($\theta = \pi/2$ and $\varepsilon = 0$). The corresponding constellation diagrams in the presence of PMD only are shown in Figures 10.10c and 10.10d.

The proposed coded modulation scheme is able to compensate for the PMD with DGD of up to 500 ps in the presence of atmospheric turbulence characterized by $\sigma_X = 0.1$ and $\sigma_Y = 0.01$.

Figure 10.11 shows the BER performance of the described scheme for both the uncoded case (Figure 10.11a) and LDPC-coded case (Figure 10.11b). The OFDM system parameters were

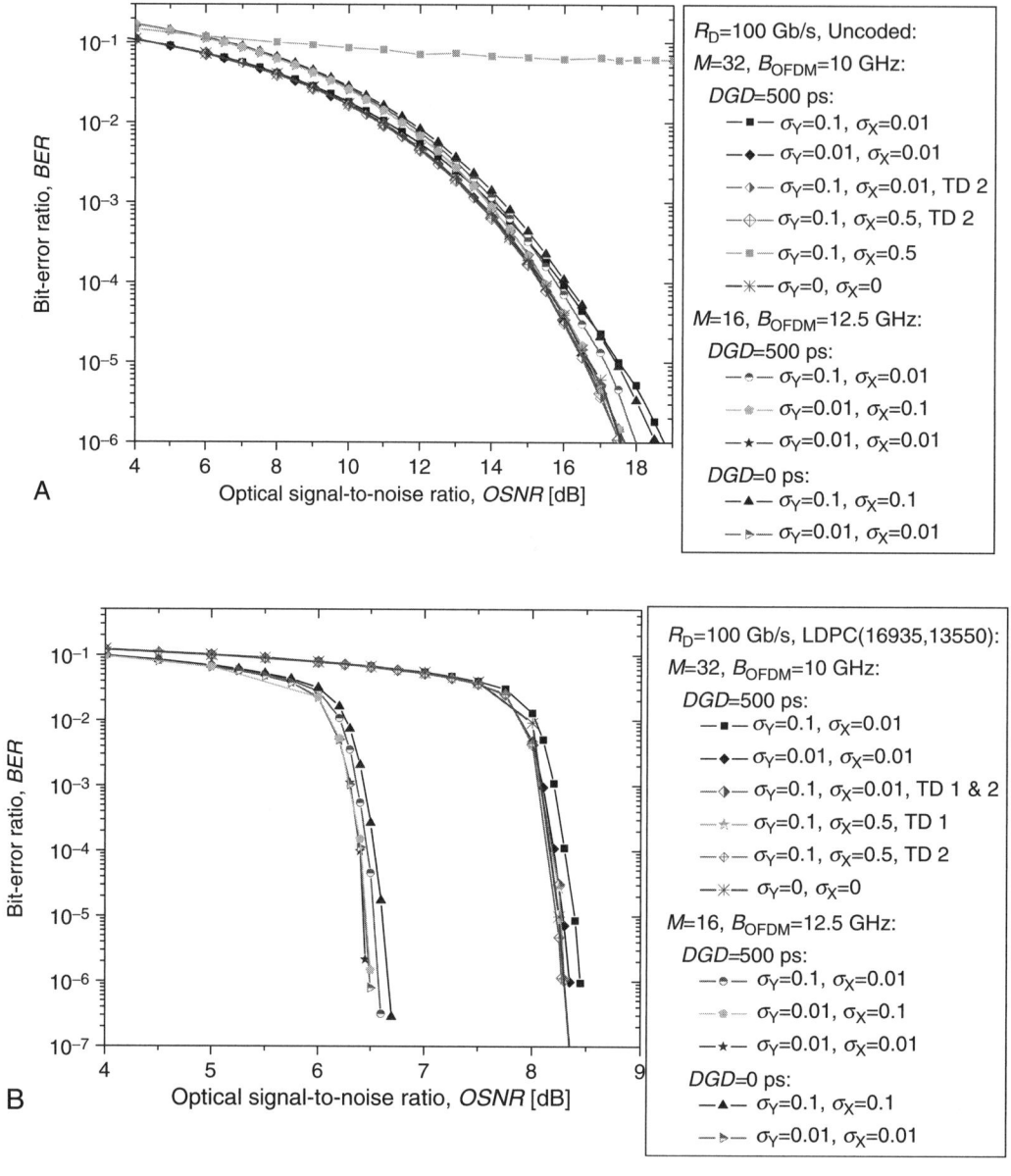

Figure 10.11: BER performance of the proposed hybrid optical network scheme: (a) uncoded BER curves and (b) LDPC-coded BERs. R_D denotes the aggregate data rate, and B_{OFDM} is the OFDM signal bandwidth. TD i, transmission diversity of order i.

chosen as follows: The number of QAM symbols $N_{QAM} = 4096$, the oversampling is two times, OFDM signal bandwidth is set to either 10 GHz ($M = 32$) or 12.5 GHz ($M = 16$), and the number of samples used in cyclic extension $N_G = 64$. For the fair comparison of different M-ary schemes, the optical signal-to-noise ratio (OSNR) on the x-axis is given per information bit, which is also consistent with the digital communication literature. The code rate influence is included in Figure 10.11 so that the corresponding coding gains are net effective coding gains. To generate the temporally correlated samples according to Eq. (10.18), we used the Levinson–Durbin algorithm.[34,35] From Figures 10.10 and 10.11 it can be concluded that PMD can be successfully compensated even in the presence of atmospheric turbulence. Most of the degradation is coming from the FSO channel, as shown in Figures 10.10 and 10.11. The 32-QAM case with aggregate data rate $R_D = 100$ Gb/s performs 1.9 dB (at BER $= 10^{-6}$) worse than 16-QAM (with the same aggregate rate), although the occupied bandwidth is smaller.

The net coded gain improvement (at BER of 10^{-6}) of LDPC-coded OFDM over uncoded-OFDM is between 11.05 dB ($M = 16$, $\sigma_X = 0.01$, $\sigma_Y = 0.01$, corresponding to the weak turbulence regime) and 11.19 dB ($M = 16$, $\sigma_X = 0.1$, $\sigma_Y = 0.01$, corresponding to the medium turbulence regime). The additional coding gain improvement due to transmission diversity with two lasers is 0.19 dB for 32-QAM-based OFDM ($\sigma_X = 0.01$ and $\sigma_Y = 0.1$) at a BER of 10^{-6}. The improvement due to transmission diversity for the uncoded case (at the same BER) is 1.26 dB. Therefore, in the regime of weak atmospheric turbulence, the improvements due to transmission diversity are moderate. On the other hand, in the moderate turbulence regime the use of transmission diversity is unavoidable. Otherwise, the uncoded BER error floor is so high (see the $\sigma_X = 0.5$, $\sigma_Y = 0.1$ curve in Figure 10.11a) that even the best LDPC codes are not able to handle the turbulence, if the complexity is to be kept reasonably low. With transmission diversity, in the moderate turbulence regime, we obtain BER performance comparable to the case in the absence of turbulence regime, as shown in Figure 10.11. The strong turbulence regime is not considered here due to the lack of an appropriate temporal correlation model. (The atmospheric turbulence model with temporal correlation described in Section 10.3 is not a valid model in the strong turbulence regime.)

The laser linewidths of transmitting and local laser were set to 10 kHz so that the atmospheric turbulence, PMD, and ASE noise are predominant effects. Note that the BER threshold required to achieve BER $= 10^{-6}$ at the output of LDPC decoder is 0.0196, and in this region BER values for different laser linewidths are comparable.

10.7 Summary

In this chapter, we described an LDPC-coded IM/DD OFDM system as a modulation/coding scheme for FSO communication over atmospheric turbulence channels that provides a number of advantages: (1) It has excellent coding gains (defined at BER of 10^{-5}) ranging from 8.47 dB in the regime of weak turbulence (for QPSK) to 23.38 dB in the regime of strong turbulence (for BPSK) compared to LDPC-coded OOK, (2) there is significant spectral efficiency improvement, (3) simple equalization is required, and (4) a simple FFT is used for modulating and demodulating. To further improve spectral efficiency, the FSO-OFDM SSB transmission scheme may be combined with subcarrier multiplexing. We also explained how to use OFDM in hybrid wireless–optical communications and in hybrid optical networks. We described a particular polarization multiplexed coded OFDM scheme with coherent detection suitable for use in hybrid FSO–fiber-optics networks. This scheme is able to simultaneously deal with atmospheric turbulence, chromatic dispersion, and PMD. We showed that PMD can be compensated even in the presence of atmospheric turbulence. This coded modulation scheme supports 100 Gb/s per wavelength transmission and 100 Gb/s Ethernet, and it provides high bandwidth efficiency.

References

1. Djordjevic B, Vasic B, Neifeld MA. LDPC coded orthogonal frequency division multiplexing over the atmospheric turbulence channel [Invited]. In: *Proc. CLEO/QELS 2006, paper no. CMDD5*, Long Beach, CA; May 21–26, 2006.
2. Djordjevic IB, Vasic B, Neifeld MA. LDPC coded OFDM over the atmospheric turbulence channel. *Opt Express* 2007;**15**:6332–46.
3. Djordjevic IB, Vasic B, Neifeld MA. LDPC-coded OFDM for optical communication systems with direct detection. *J Sel Topics Quantum Electron* 2007;**13**:1446–54.
4. Anguita JA, Djordjevic IB, Neifeld MA, Vasic BV. Shannon capacities and error-correction codes for optical atmospheric turbulent channels. *J Opt Networking* 2005;**4**:586–601.
5. Zhu X, Kahn JM. Free-space optical communication through atmospheric turbulence channels. *IEEE Trans Commun* 2002;**50**:1293–300.
6. Al-Habash MA, Andrews LC, Phillips RL. Mathematical model for the irradiance probability density function of a laser beam propagating through turbulent media. *Opt Eng* 2001;**40**:1554–62.
7. Zhu X, Kahn JM. Markov chain model in maximum-likelihood sequence detection for free-space optical communication through atmospheric turbulence channels. *J Lightwave Technol* 2003;**51**:509–16.
8. Jeong M-C, Lee J-S, Kim S-Y, et al. 8 × 10 Gb/s terrestrial optical free-space transmission over 3.4 km using an optical repeater. *IEEE Photon Technol Lett* 2003;**15**:171–3.
9. Anguita JA, Djordjevic IB, Neifeld MA, Vasic BV. Shannon capacities and error-correction codes for optical atmospheric turbulent channels. *J Opt Networking* 2005;**4**:586–601.
10. Van Nee R, Prasad R. *OFDM wireless multimedia communications*. Boston: Artech House; 2000.
11. Wu Y, Caron B. Digital television terrestrial broadcasting. *IEEE Commun Mag* 1994;**32**:46–52.
12. Pan Q, Green RJ. Bit-error-rate performance of lightwave hybrid AM/OFDM systems with comparison with AM/QAM systems in the presence of clipping impulse noise. *IEEE Photon Technol Lett* 1996;**8**:278–80.
13. Kim A, Hun Joo Y, Kim Y. 60 GHz wireless communication systems with radio-over-fiber links for indoor wireless LANs. *IEEE Trans Commun Electron* 2004;**50**:517–20.
14. Dixon BJ, Pollard RD, Iezekiel S. Orthogonal frequency-division multiplexing in wireless communication systems with multimode fiber feeds. *IEEE Trans Microwave Theory Tech* 2001;**49**:1404–9.
15. Djordjevic IB, Milenkovic O, Vasic B. Generalized low-density parity-check codes for optical communication systems. *J Lightwave Technol* 2005;**23**:1939–46.
16. Djordjevic IB, Vasic B. Nonbinary LDPC codes for optical communication systems. *IEEE Photon Technol Lett* 2005;**17**:2224–6.

17. You R, Kahn JM. Average power reduction techniques for multiple-subcarrier intensity-modulated optical signals. *IEEE Trans Commun* 2001;**49**:2164–71.

18. Lin C-C, Lin K-L, Chang HC, Lee C-Y. A 3.33 Gb/s (1200, 720) low-density parity check code decoder. *Proc ESSCIRC* 2005;211–14.

19. Xiao-Yu H, Eleftheriou E, Arnold D-M, Dholakia A. Efficient implementations of the sum-product algorithm for decoding of LDPC codes. *Proc IEEE Globecom* 2001;**2**:1036–1036E.

20. Hui R, Zhu B, Huang R, et al. Subcarrier multiplexing for high-speed optical transmission. *J Lightwave Technol* 2002;**20**:417–27.

21. Andrews LC, Philips RL. *Laser beam propagation through random media.* Bellingham, WA: SPIE Optical Engineering Press; 1998.

22. Ishimaru A. *Wave propagation and scattering in random media.* New York: Academic Press; 1978.

23. Andrews LC, Phillips RL, Hopen CY. *Laser beam scintillation with applications.* Bellingham, WA: SPIE Press; 2001.

24. Karp S, Gagliardi R, Moran SE, Stotts LB. *Optical channels.* New York: Plenum; 1988.

25. Flatté SM, Bracher C, Wang G-Y. Probability-density functions of irradiance for waves in atmospheric turbulence calculated by numerical simulation. *J Opt Soc Am A* 1994;**11**:2080–92.

26. Andrews LC, Phillips RL. I-K distribution as a universal propagation model of laser beams in atmospheric turbulence. *J Opt Soc Am* 1985;**2**:160–3.

27. Andrews LC, Phillips RL. Mathematical genesis of the I-K distribution for random optical fields. *J Opt Soc Am* 1986;**3**:1912–19.

28. Churnside JH, Hill RJ. Probability density of irradiance scintillations for strong path-integrated refractive turbulence. *J Opt Soc Am A* 1987;**4**:727–33.

29. Hill RJ, Frehlich RG. Probability distribution of irradiance for the onset of strong scintillation. *J Opt Soc Am A* 1997;**14**:1530–40.

30. Churnside JH, Frehlich RG. Experimental evaluation of log-normally modulated Rician and IK models of optical scintillation in the atmosphere. *J Opt Soc Am A* 1989;**6**:1760–6.

31. Noll RJ. Zernike polynomials and atmospheric turbulence. *J Opt Soc Am* 1976;**66**:207–11.

32. Proakis JG. *Digital communications.* Boston: McGraw-Hill; 2001.

33. Milenkovic O, Djordjevic IB, Vasic B. Block-circulant low-density parity-check codes for optical communication systems. *J Sel Topics Quantum Electron* 2004;**10**:294–9.

34. Levinson N. The Wiener RMS error criterion in filter design and prediction. *J Math Phys* 1947;**25**:261–78.

35. Durbin J. Efficient estimation of parameters in moving-average models. *Biometrica* 1959;**46**:306–16.

36. Wood ATA, Chan G. Simulation of stationary Gaussian processes in [0,1]^d. *J Comp Graph Stat* 1994;**3**:409–32.

37. Djordjevic IB. Coded-OFDM in hybrid optical networks. *IET Optoelectronics*, accepted for publication.

38. Djordjevic IB, Xu L, Wang T. Beyond 100 Gb/s optical transmission based on polarization multiplexed coded-OFDM with coherent detection. *IEEE J Sel Areas Commun Opt Commun Networking* 2009;**27**(3).

39. Goldsmith A. *Wireless communications.* Cambridge, UK: Cambridge University Press; 2005.

CHAPTER 11

OFDM Applications in Access Optical Networks

Orthogonal frequency-division multiplexing (OFDM) as a technology that enables robust transmission over both wireless and optical channels has already been employed in many communication standards, including wireless local area network (LAN; also known as Wi-Fi), digital video and audio broadcasting standards, digital subscribed loop (DSL), and Worldwide Interoperability for Microwave Access (WiMAX, or IEEE 802.16).

In this chapter, we describe the use of OFDM in radio-over-fiber (RoF) systems, passive optical networks (PON), WiMAX, ultra wideband (UWB) communications, and for indoor wireless optical applications.

11.1 OFDM in Radio-over-Fiber Systems

WiMAX is a wireless technology that provides wireless last mile broadband access with fast connection speeds comparable to T1, cable, and DSL data rates.[1] WiMAX provides up to 75 Mb/s over ranges up to 30 miles. WiMAX is the technology that brings the network to the building, whereas end users can be connected using wireless LAN, Ethernet, or plastic optical fibers (POFs). The working group responsible for WiMAX is IEEE 802.16, which initially developed the standard for the 10- to 66-GHz portion of the spectrum. This region of frequencies requires line-of-sight (LOS) between the transmitting and receiving antennas. Because of the need for high antennas, focus has shifted to the 2- to 11-GHz portion of the spectrum, which does not require LOS (IEEE 802.16a standard). In 2004, the IEEE 802.16-2004 standard was released, which completed the fixed wireless access. The corresponding mobile WiMAX amendment is known as IEEE 802.16e.

To reduce the deployment and maintenance cost of WiMAX and other wireless networks while providing low power consumption and large bandwidth, RoF technology is considered as a promising candidate.[2–9] In RoF systems, the fiber is used to distribute the RF signal from a central station (CS) to remote antenna units. Different types of fibers, such as single-mode fibers (SMFs), multimode fibers (MMFs), and POFs, can be used.[4–14] Possible applications include (1) in cellular systems to establish the connection between the mobile telephone switching office (MTSO) and base stations (BSs), (2) in WiMAX to extend the coverage and reliability by connecting WiMAX BSs and remote antenna units (RAUs), and (3) in UWB communications[1] to extend the wireless coverage range.[14] RoF can also be used to eliminate so-called dead zones (in tunnels, mountain areas, etc.), in hybrid fiber coaxial systems, and in fiber to the home applications. For example, to reduce system installation and maintenance costs for indoor applications, the POFs or MMFs can be used from residential gateway to either fixed or mobile wireless units inside a building. The RoF technologies offer many advantages with respect to wireless, such as low attenuation loss, large bandwidth, improved security, immunity to electromagnetic interference, reduced power consumption, and easy installation and maintenance.

One typical example of RoF systems is shown in Figure 11.1a. The data for a given end user are generated in a CS, imposed on a set of OFDM subcarriers assigned to that particular user, transmitted over optical fiber upon modulation in a Mach–Zehnder modulator (MZM), and converted into electrical domain by the optical receiver in the BS. From the BS, the signal is transmitted over a wireless channel to the end user. An example of downlink transmission is shown in Figure 11.1b.

RoF systems can be used to improve the reliability of WiMAX networks through transmission diversity by deploying several RAUs as shown in Figure 11.2. To satisfy ever-increasing needs for higher data rates, 60-GHz systems are receiving increased attention. These systems can provide services with data rates on the order of gigabytes per second. However, before the migration to such radio systems, the following problems must be solved: (1) Because of huge propagation loss, this band can only be used for short-range wireless access in indoor environments, and (2) because the millimeter radio wave is strongly affected by the attenuation, the radio cell can span only one room. To extend the coverage, RoF

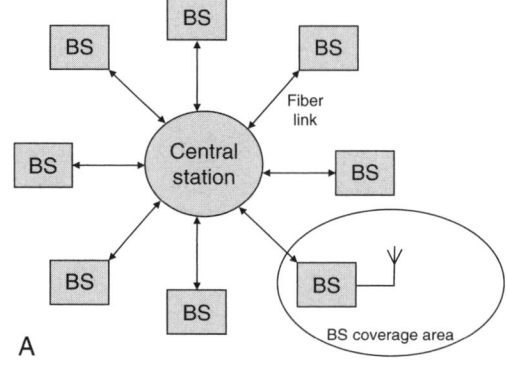

A

Figure 11.1: Examples of (a) an RoF system and

(continued)

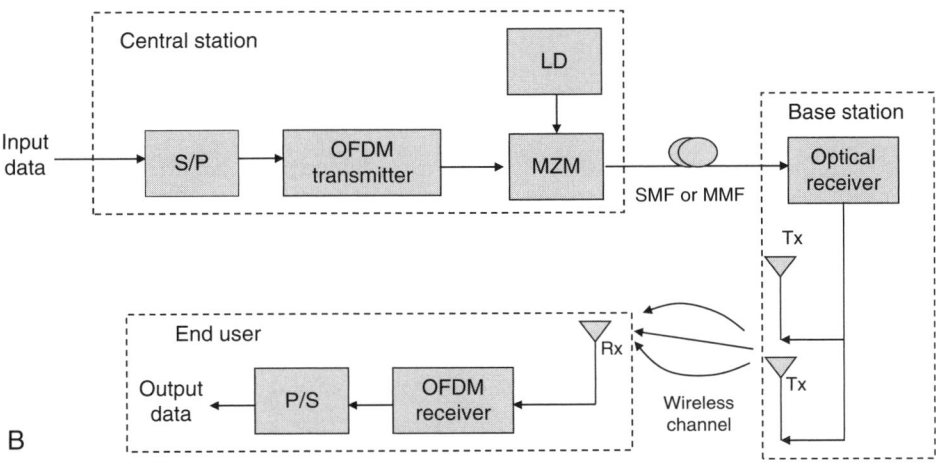

Figure 11.1—cont'd (b) a downlink transmission. BS, base station; LD, laser diode; MZM, Mach–Zehnder modulator; Rx, wireless receiver; Tx, wireless transmitter.

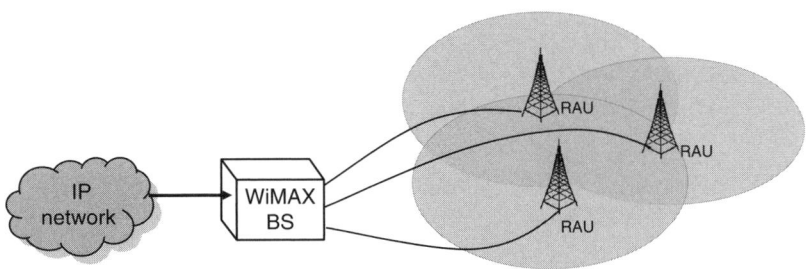

Figure 11.2: Improvement of WiMAX coverage through deployment of RAUs.

systems seem to provide an excellent alternative, with one example shown in Figure 11.3. The CS provides consolidation of radio access control and signal processing, and it delivers the radio signal transparently to antenna units. However, the direct transmission of high-frequency radio signal requires the use of high-speed electro-optical devices. To avoid this problem, commonly the RF signal is down-converted at an intermediate frequency and transmitted over fiber link (such as POF in Figure 11.3) by employing intensity modulation with direct detection (IM/DD).

The RoF system must be capable of generating wireless radio signals and allow reliable radio signal transmission over the optical links. Among different techniques, optical frequency multiplication (OFM) is commonly used. The OFM principle is based on harmonics generation by frequency modulation–intensity modulation (IM) conversion through a periodic bandpass filter (BPF), such as the Mach–Zehnder interferometer (MZI), and a BPF to select the desired harmonic at the remote node (RN). An example of a point-to-multipoint

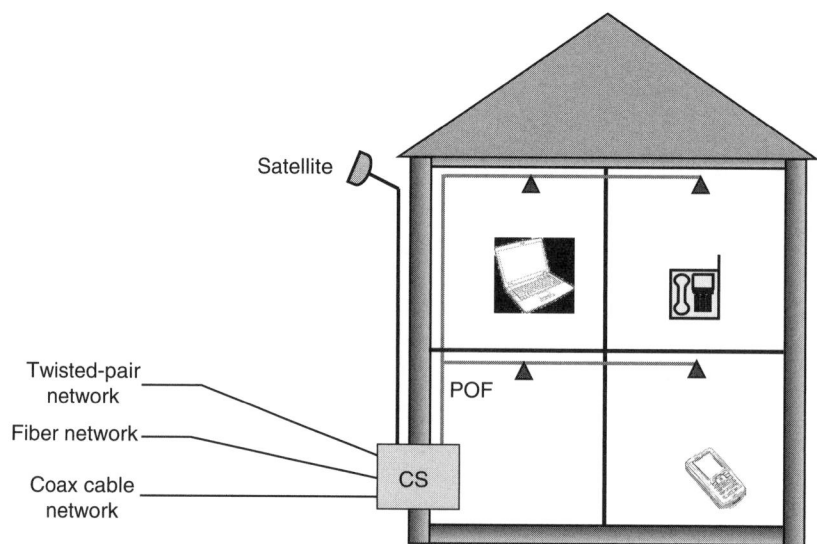

Figure 11.3: RoF distribution antenna system.

OFM system is shown in Figure 11.4. The fiber links in the optical ring are bidirectional, and each RN employs two wavelengths for uplink and downlink signals. The downlink signal at RN is filtered with BPF_1 and radiated, whereas one of the sidebands of the f_ith channel is selected by BPF_2 and used as a local oscillator for the uplink signal. If the data signal is transmitted using OFDM, with proper design of OFDM system parameters, we are able to deal simultaneously with dispersion effects present in fiber links and multipath fading present in wireless links.

The signal in RoF systems is impaired by multimode dispersion (when MMFs are used) or chromatic dispersion (when SMFs are used), imperfection of components at the BS, and multipath wireless fading. In RoF systems, system designers have to deal not only with multipath fading present in wireless links but also with dispersion effects present in fiber links. In previous chapters, we showed that chromatic dispersion and polarization mode dispersion can be compensated for by employing optical OFDM, provided that the guard interval is longer than the total delay spread due to chromatic dispersion and maximum differential group delay. In wireless networks, the number of subcarriers in OFDM is chosen in such a way that the bandwidth of the signal per subcarrier is smaller than the coherence bandwidth of the wireless channel so that each subcarrier experiences the flat fading. Therefore, for simultaneous compensation of multipath fading in wireless channels and dispersion effects in optical fiber links, we have to provide that the cyclic prefix is longer than the total delay spread due to multipath spreading and dispersion effects. Kurt et al.[15] studied the influence of the cyclic prefix to the multimode dispersion for wireless LAN systems operating at both 5 and 60 GHz. They observed the worst-case scenario—the

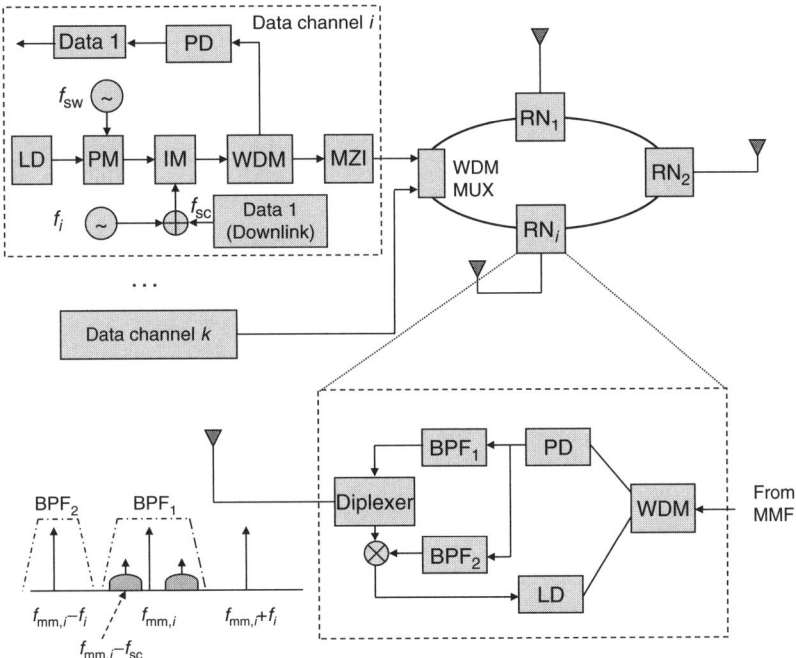

Figure 11.4: Full-duplex point-to-multipoint RoF system. BPF, bandpass filter; IM, intensity modulator; LD, laser diode; PD, photodetector; PM, phase modulator; RN, remote node; WDM, wavelength-division multiplexing coupler.

Rayleigh fading wireless channel (without LOS) with 250-ns delay spread—and set the cycle prefix to standard 800 ns. As expected, it was found that 60-GHz systems are more sensitive to multimode dispersion than 5-GHz systems. As long as the cyclic prefix is longer than the overall delay spread due to both multimode dispersion and wireless channel multipath spread, the performance of both systems is unaffected.

11.2 OFDM in Passive Optical Networks

The OFDM is also an excellent candidate to be used in passive optical network (PON) applications.[7–10] Currently, PON is being deployed to substitute conventional cable-based access networks. With the optical fiber used as the transmission media, PON can offer much higher bandwidth while supporting various communication services. PON has attracted much attention in both industry and academia. For example, the Verizon Fiber Optic Service (FiOS) is a PON-based system.[11] In academia, researchers are working on the development of next-generation PON. A number of different PONs have been standardized to provide broadband access services, including BPON (ITU-T G.983), EPON (IEEE 802.3ah), and GPON (ITU-T G.984).

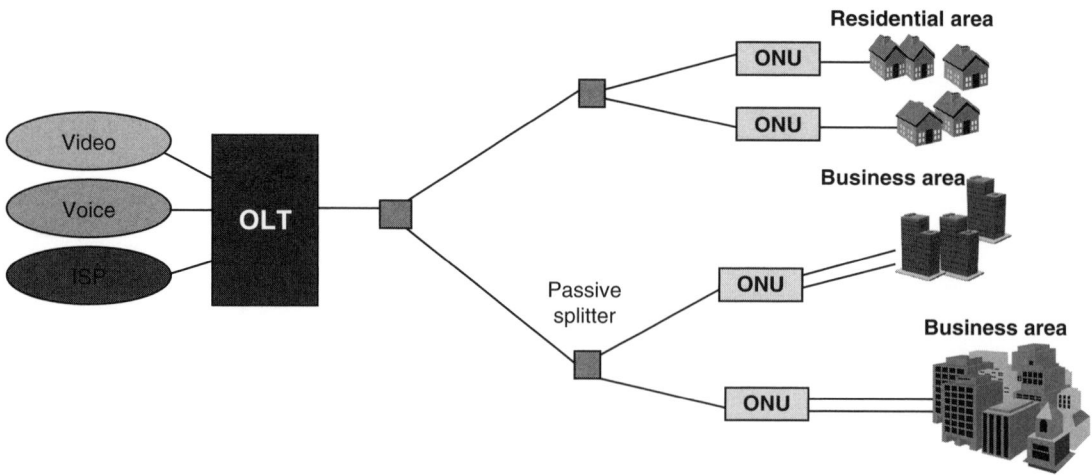

Figure 11.5: A passive optical network. OLT, optical line terminal; ONU, optical network unit.

An example of PON is shown in Figure 11.5. Basically, PON is a point-to-multipoint topology. To minimize the system cost, all the components used between the optical line terminal (OLT) and optical network units (ONUs) are passive. Different services, such as video, voice, and data traffic, can be delivered to end users. The links in Figure 11.5 are bidirectional. For example, the FiOS system uses three different wavelengths to support the three services: (1) 1310 nm for upstream data at 155 Mb/s or 1.2 Gb/s when GPON is used, (2) 1490 nm for downstream data at 622 Mb/s or 2.4 Gb/s with GPON, and (3) 1550 nm for RF video with 870 MHz of available bandwidth. One OLT can support up to 32 subscribers.

Different PON technologies have been proposed, including time-division multiplexing (TDM)-PON, wavelength-division multiplexing (WDM)-PON (Figure 11.6), subcarrier multiplexing (SCM)-PON, and optical code-division multiplexing (OCDM)-PON. An example of broadcast-and-select WDM-PON is shown in Figure 11.6. A WDM array of transmitters is used at the OLT side. Each ONU receiver operates at its own wavelength, and the type of electronics is dependent on the data rate at each particular ONU. For the upstream, different ONUs use the same wavelength channel (at 1330 nm), and TDM is used to share the wavelength channel. ONUs are synchronized to OLT by the ranging process (each ONU measures the delay from OLT and adjusts its clock accordingly).

In references,[7–9] the orthogonal frequency-division multiple access (OFDMA)-PON has been advocated. TDM-PON assigns different time slots to ONUs and uses complex scheduling algorithms and framing technologies to support services of various kinds. In SCM-PON, one dedicated electrical subcarrier for each ONU is used, and it allows multiple users to share the same optical channel and its corresponding components. In OCDM-POM, different CDM codes are assigned to different users. This technology provides asynchronous

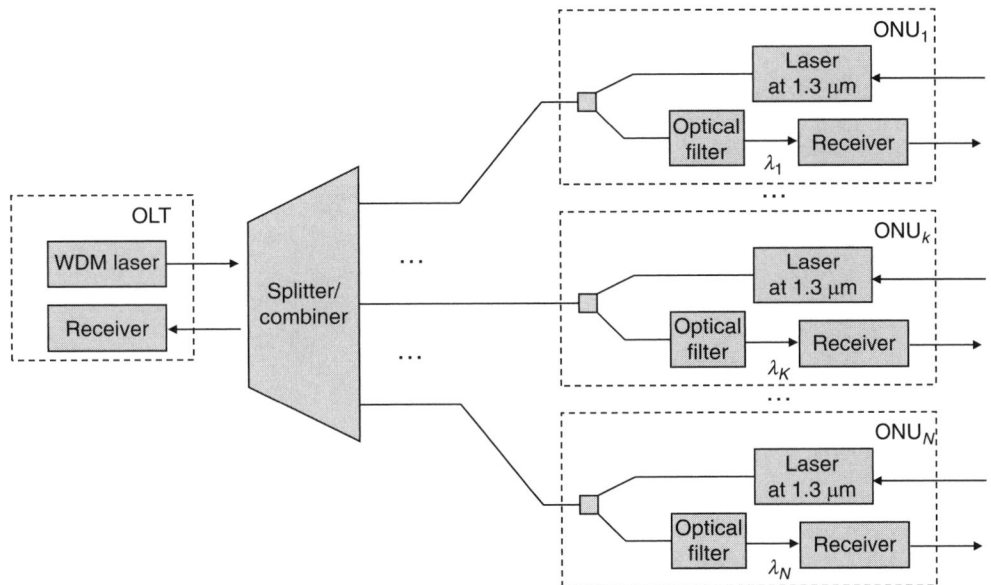

Figure 11.6: A broadcast-and-select WDM-PON. OLT, optical line terminal; ONU, optical network unit.

communications and security against unauthorized users. However, the encoders and decoders are still expensive, and because the CDM sequences are nonorthogonal (in IM/DD systems), multiaccess interference and noise are still important issues. On the other hand, OFDMA is a multiple-access technology that allows assigning different subcarriers to multiple users in a dynamic manner. It simultaneously enables time and frequency domain resource partitioning.[7–9] Compared to TDM-PONs, OFDMA-PONs can be combined with TDM to offer one additional dimension for resource management. For example, in time domain the PON can accommodate burst traffic, and in frequency domain the PON can offer fine granularity and channel-dependent scheduling.[7–9] The OFDMA-PON principle is illustrated in Figure 11.7. OLT assigns to every particular user (ONU) a subset of subcarriers. Legacy users (denoted as white (E1/T1) and black (wireless cell) subcarrier frequency slots in Figure 11.7) can coexist with Internet users. In the upstream direction, each ONU modulates the data over the assigned subcarrier set, whereas all the other subcarriers belonging to other ONUs are set to zero.

OFDMA-PON systems have many advantages compared to other PON technologies[7]: (1) improved bandwidth efficiency (e.g., 4 bit/s/Hz for 16-QAM modulation, and with 2.5 GHz bandwidth it can support 10 Gb/s aggregate data rate); (2) unique flexibility in dealing with bandwidth resource sharing and virtualization; (3) protocol independence and service transparency (subsets of subcarriers, similarly as for transparent pipes, can support both digital and analog signals with a variety of quality of service requirements);

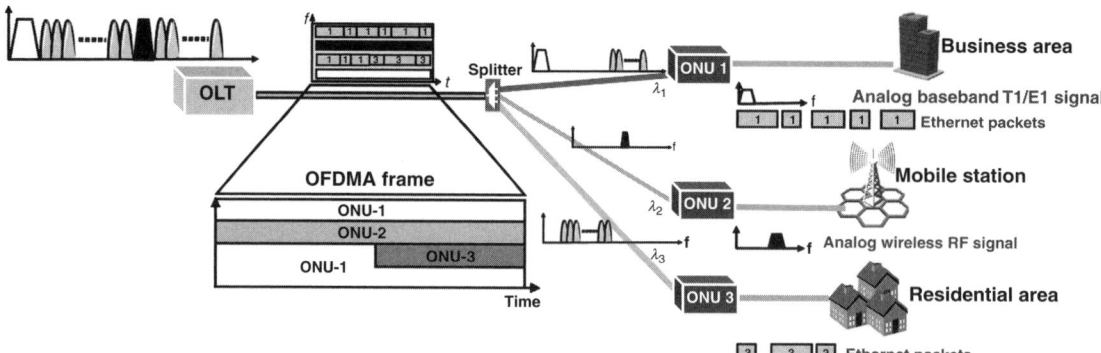

Figure 11.7: An example of OFDMA-PON. OLT, optical line terminal; ONU, optical network unit. Adapted from references.[7-9]

(4) OFDMA-PON is a scalable architecture (namely, it can coexist with TDM-PON and WDM-PON); (5) it is a cost-effective solution (fewer receivers are needed in OLT compared to traditional WDM-PONs); and (6) it can operate with simple medium access control (MAC) with low overhead.

The research group from NEC Laboratories America introduced an OFDMA-based virtual PON (VPON) architecture.[7-9] The VPON can be implemented as either a user-defined or a customized PON system. The VPON design supports the coexistence of EPON, BPON, and GPON, which is illustrated in Figure 11.8. It utilizes the OFDM subcarriers for bandwidth virtualization and data isolation. The set of subcarriers is dynamically allocated to different types of services. To enable high flexibility, the use of digital signaling processing and field programmable gate array has been advocated. The subset of subcarriers can employ different modulation formats. The same authors performed an experiment in which two data channels at 10 Gb/s and one RF WiMAX channel (of bandwidth 3.5 MHz at 3.4 GHz) were transmitted over the PON. They found no throughput degradation when the WiMAX signal was transmitted over the OFDMA-PON (20 km of SMF and 1:32 splitter) together with the other two ONU signals carrying the PON traffic.

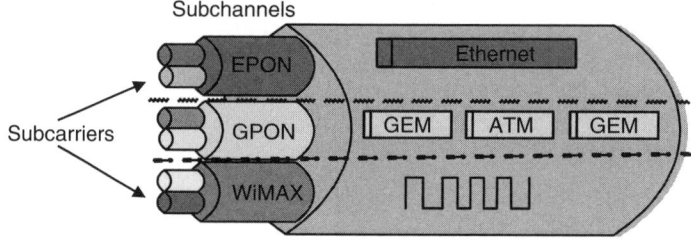

Figure 11.8: Coexistence of different VPONs. ATM, asynchronous transfer mode; GEM, GPON encapsulation method. Adapted from references 7-9.

Two types of resource partitioning for VPON have been studied by Wei et al.[9]: (1) dedicated resource provisioning (resources are explicitly preassigned to each VPON based on the average demands of traffic flow) and (2) shared resource provisioning (the resources are shared among different VPONs). The authors performed the comparison of these two schemes with respect to the average packet delay and the packet jitter. It was found that the average packet delay increases dramatically as the number of different VPONs increases, with the shared resource partitioning always outperforming dedicated partitioning. The jitter performance comparison, on the other hand, shows a quite different trend, with the dedicated model outperforming the shared model.

11.3 Ultra Wideband Signals and Optical OFDM

UWB communication is a fast-emerging radio technology that can be used at very low energy levels for short-range, high-bandwidth communications by using a large radio spectrum.[1] UWB complements other longer range wireless networking technologies, such as Wi-Fi, WiMAX, and cellular networks. UWB has found important applications in relaying data from a host device to other devices in the immediate area (within 10 m). In the United States, UWB radio transmissions can legally operate in the range from 3.1 to 10.6 GHz, with transmitting power compliance. Consequently, UWB provides dramatic channel capacity at short range that limits interference. UWB is considered a complementary communication solution within future 4G systems.

The transmission of UWB signal over fiber/free-space optical links is a cost-effective solution to extend the coverage of UWB radios to hundreds of meters. Such a scheme has several advantages: (1) It supports a transparent electro-optical signal conversion regardless of the UWB's modulation methods, (2) it avoids using some expensive high-frequency electronic components required during signal transmission, and (3) optoelectronic integration technologies can potentially reduce the size and power consumption of the optical UWB transceivers.

Two major UWB technologies are impulse radio UWB (IR-UWB) and multiband OFDM. Two popular pulse shapes for IR-UWB are Gaussian monocycle and doublet, obtained as the first and second derivative of Gaussian pulse $y_{g1}(t) = K_1 \exp[(-t/\tau)^2]$ as follows[1]:

$$y_{g2}(t) = K_2 \frac{-2t}{\tau^2} \exp\left[-(t/\tau)^2\right] \qquad y_{g3}(t) = K_3 \frac{-2}{\tau^2} \left(1 - \frac{2t^2}{\tau}\right) \exp\left[-(t/\tau)^2\right] \qquad (11.1)$$

To cover the whole UWB spectrum, from 3.1 to 10.6 GHz, the pulse duration should be on the order of 200 ps. Different modulation formats suitable for use in IR-UWB communications, given in Figure 11.9, can be classified into two categories: (1) time-based techniques (pulse position modulation (PPM) and pulse duration modulation) and (2) shape-based techniques (biphase modulation (BPM), pulse amplitude modulation, on–off keying (OOK), orthogonal pulse modulation (OPM), etc.).

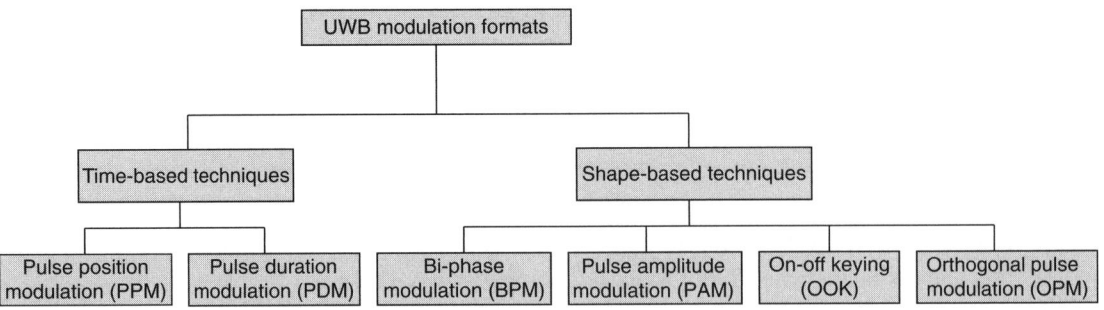

Figure 11.9: Classification of different modulation formats for IR-UWB communications.

The signal constellation for M-ary PPM (M is the signal constellation size) is represented by $\{s_i = p(t - \tau_i)\}$ $(i = 1, \ldots, M)$, where τ_i is the delay parameter associated with the ith signal constellation point, and $p(t)$ is the pulse shape. The corresponding signal constellation for BPM is represented by $\{s_i = d_i p(t)\}$ $(i = 1, 2; d_i \in \{-1, 1\})$, whereas the signal constellation for OOK is represented by $\{s_i = d_i p(t)\}$ $(i = 1, 2; d_i \in \{0, 1\})$. Finally, the signal constellation for M-ary OPM is represented by $\{p_i(t)\}$ $(i = 1, \ldots, M)$, where $p_i(t)$'s are orthogonal to each other. For example, modified Hermite polynomials are orthogonal to each other and can be used as symbols for M-OPM as follows[1]:

$$h_n(t) = k_n e^{-t^2/4\tau^2} \tau^2 \frac{d^n}{dt^n}\left(e^{-t^2/2\tau^2}\right) = k_n e^{-t^2/4\tau^2} \tau^2 \sum_{i=0}^{[n/2]} \left(-\frac{1}{2}\right)^i \frac{(t/\tau)^{n-2i}}{(n-2i)!i!} \tag{11.2}$$

where the parameter k_n is related to the energy E_n by[1]

$$k_n = \sqrt{\frac{E_n}{\tau n! \sqrt{2\pi}}}$$

A UWB signal is typically composed of a sequence of sub-nanosecond pulses of low duty cycle (approximately 1%), with the bandwidth on the order of gigahertz. Because the total power is spread over a wide range of frequencies, its power spectral density (PSD) is low, and interference to other existing wireless systems is extremely low. Similarly as in code division multiple access, each user is assigned a unique pseudo-random noise (PN) sequence that is used to encode the pulses in either position (PPM) or polarity (BPM), respectively, as[1]

$$s_{\text{PPM}}(t) = A \sum_{j=-\infty}^{\infty} \sum_{i=0}^{N_c - 1} p\left(t - jN_c T_f - iT_f - c_i T_c - \tau b_j\right) \tag{11.3}$$

and

$$s_{\text{BPM}}(t) = A \sum_{j=-\infty}^{\infty} \sum_{i=0}^{N_c - 1} b_j c_i p\left(t - jN_c T_f - iT_f\right) \tag{11.4}$$

where $b_j \in \{-1, 1\}$ is the jth information bit, c_i ($i = 0, 1, \ldots, N_c-1$) is the PN sequence of length N_c, N_cT_f is the bit duration, and A is the amplitude of pulse $p(t)$. For PPM, the PN sequence modulates the positions in increments/decrements of multiple T_c, whereas the information bit b_j introduces additional time shift $b_j\tau$. For BPM (also known as binary phase-shift keying), the PN sequence modulates the polarity of pulse within each frame T_f, whereas the information bit b_j changes the polarity of the block of N_c pulses. Typical receiver architecture is shown in Figure 11.10a, whereas the optimum demodulator with matched filter (correlator) is shown in Figure 11.10b. The processing gain (PG; defined as the improvement in effective signal-to-noise ratio) can be evaluated by PG [dB] $= 10\log_{10}(N_c) + 10\log_{10}[1/(\text{duty cycle})]$. $r(t)$ denotes the received signal, and for PPM $s_0(t)$ and $s_1(t)$ can be obtained from Eq. (11.3) by

$$s_0(t) = \sum_{i=0}^{N_c-1} \text{rect}\left(t - iT_f - c_iT_c + \tau\right)$$

and

$$s_1(t) = \sum_{i=0}^{N_c-1} \text{rect}\left(t - iT_f - c_iT_c - \tau\right)$$

respectively, where rect(t) is a unit rectangular pulse. For BPM, $s_1(t)$ and $s_0(t)$ can be obtained from Eq. (11.4) by

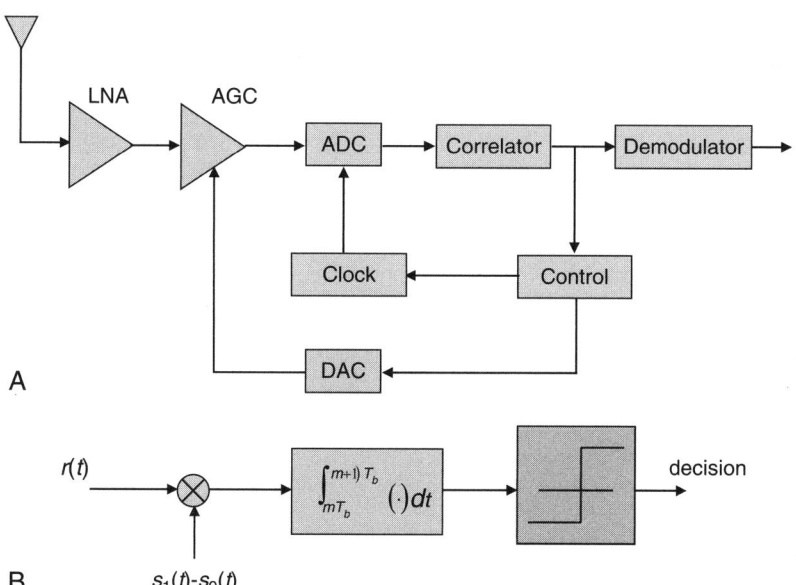

Figure 11.10: Block diagram of UWB receiver (a) and demodulator of UWB signals (b). ADC, A/D converter; AGC, automatic gain control; DAC, D/A converter; LNA, low noise amplifier.

$$s_1(t) = \sum_{i=0}^{N_c-1} c_i \operatorname{rect}(t - iT_f)$$

and

$$s_0(t) = -s_1(t)$$

respectively.

Multiband (MB) modulation is another approach to modulate information in UWB systems, in which the 7.5-GHz band is split into multiple smaller frequency bands. The key idea is to efficiently utilize UWB spectrum by transmitting multiple UWB signals at the same time. The MB-OFDM system can be considered as a combination of OFDM and frequency hopping.[1,14] With this approach, the information bits are "spread" across the entire UWB spectrum, thus exploiting frequency diversity and providing robustness against multipath fading and narrowband interference. The 7.5-GHz UWB spectrum is divided into 14 bands, each with a bandwidth of $\Delta f = 528$ MHz. The center frequency of nth band $f_{c,n}$, the lower boundary frequency $f_{l,n}$ and upper boundary frequency $f_{h,n}$ can be determined by

$$f_{c,n} = f_0 + n\Delta f \quad f_{l,n} = f_{c,n} - \Delta f/2 \quad f_{h,n} = f_{c,n} + \Delta f/2 \quad f_0 = 2904 \quad \text{MHz}$$

The frequency band allocation is shown in Figure 11.11, based on references.[1,16] The MB-OFDM signal is obtained by frequency hopping of the OFDM signal of bandwidth 528 MHz over various carrier frequencies $f_{c,n}$ so that the entire UWB spectrum is occupied. The time-frequency codes (TFCs) are used to determine which band is occupied by the OFDM signal of bandwidth 528 MHz.[1,14] Table 11.1, based on reference,[16] illustrates the TFC patterns for band group 1. For band groups 2–4, TFCs can be formed in a similar manner as described in Table 11.1. For band group 5, TFCs are based only on TFC numbers 5 and 6.

The MB-OFDM transmitter and receiver configurations are shown in Figure 11.12. The binary data bits are scrambled and then encoded employing a convolutional code of rate 1/3, with constraint length $K = 7$, and generating polynomials $g_0 = 133_8$, $g_1 = 165_8$, and

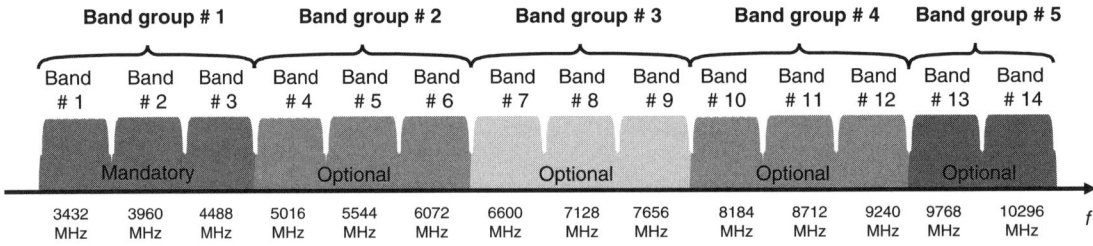

Figure 11.11: Frequency band allocation for MB-OFDM proposal with center band frequencies shown.

Table 11.1: Time-Frequency Codes for Band Group 1

TFC No.	TFC of length 6
1	1 2 3 1 2 3
2	1 3 2 1 3 2
3	1 1 2 2 3 3
4	1 1 3 3 2 2
5	1 2 1 2 1 2
6	1 1 1 2 2 2

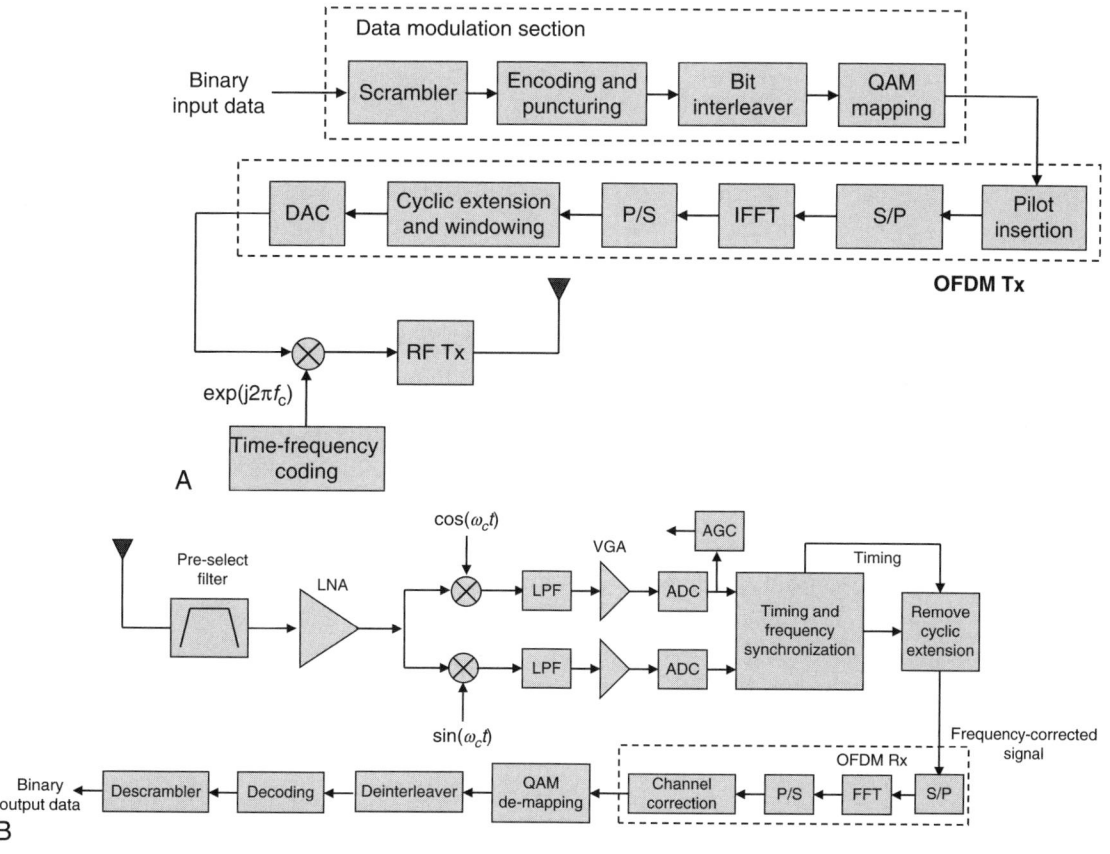

Figure 11.12: Block diagram of MB-OFDM system: (a) transmitter configuration and (b) receiver configuration. ADC, A/D conversion; AGC, automatic gain control; DAC, D/A conversion; LNA, low noise amplifier; LPF, low-pass filter; P/S, parallel-to-serial conversion; RF Tx, radio frequency transmitter; S/P, serial-to-parallel conversion; VGA, variable gain amplifier.

$g_2 = 171_8$ (the subscript denotes the octal number system). Higher rate convolutional codes (of rate 11/32, 1/2, 5/8, and 3/4) are obtained by puncturing. The number of subcarriers used in MB-OFDM is 128, 100 of which are used to transmit quaternary phase-shift keying (QPSK) data, 12 are pilots, and 10 are guard tones (the remained subcarriers are set to zero). The pilots are used for channel estimation and carrier-phase tracking on the receiver side. To improve performance of the MB-OFDM scheme, low-density parity-check (LDPC)-coded MB-OFDM can be used instead.[17,18] It has been shown that the transmission distance over the wireless channel of LDPC-coded MB-OFDM can be increased 29–73% compared to that of MB-OFDM with convolutional codes.[18] After pilot insertion, serial-to-parallel conversion (S/P), inverse fast Fourier transform (IFFT) operation, and parallel-to-serial (P/S) conversion, cyclic extension and windowing are performed. After D/A conversion and time-frequency coding, the RF signal is transmitted over the transmit (Tx) antenna.

The transmitted MB-OFDM signal can be written as[16]

$$s_{\text{MB-OFDM}}(t) = \text{Re}\left\{\sum_{k=0}^{N-1} s_{\text{OFDM},k}(t - kT_{\text{OFDM}})\exp\left[j2\pi f_{(k \bmod 6)}t\right]\right\} \tag{11.5}$$

where $s_{\text{OFDM},k}(t)$ is the OFDM symbol of duration T_{OFDM}, and N is the number of OFDM symbols transmitted. f_k is the carrier frequency over which the kth OFDM symbol is transmitted. The carrier frequency range over three frequencies assigned to the band group is organized in sequences of length 6 (TFCs), as illustrated in Table 11.1. The kth OFDM symbol is generated in a similar manner as described in previous chapters. Receiver configuration, shown in Figure 11.12b, is very similar to that of a conventional OFDM receiver, except a preselect filter is used to select an OFDM signal of 528-MHz bandwidth at a given carrier frequency.

The RoF UWB system with external modulation is shown in Figure 11.13a. The data modulation section and OFDM Tx are the same as those in Figure 11.12a. The RoF UWB system with direct modulation is shown in Figure 11.13b, with the MB-OFDM transmitter (Tx) block shown in Figure 11.13a. The optical fiber links could be SMFs, MMFs, or POFs. For both systems, downlink receivers are essentially wireless MB-OFDM UWB receivers shown in Figure 11.12b. The RoF approach is used to extend the transmission distance to several hundred meters. System designers have to consider not only frequency-selective fading in wireless channels but also multimode dispersion in MMF links. For example, the cyclic extension interval should be longer than overall spread due to multipath wireless fading and multimode dispersion spread in MMF. Moreover, because direct detection is used, MZM behaves as a nonlinear device whose input–output characteristic is given by

$$P_{\text{out}}(t) = P_{\text{in}}(t)\cos^2\left(\pi\frac{s_{\text{MB-OFDM}}(t) + V_{\text{bias}}}{V_\pi}\right) \tag{11.6}$$

where $s_{\text{MB-OFDM}}(t)$ is given by Eq. (11.5), V_{bias} is bias voltage shown in Figure 11.13a, and V_π is the half-wave voltage; we also have to take into account the nonlinear characteristics of

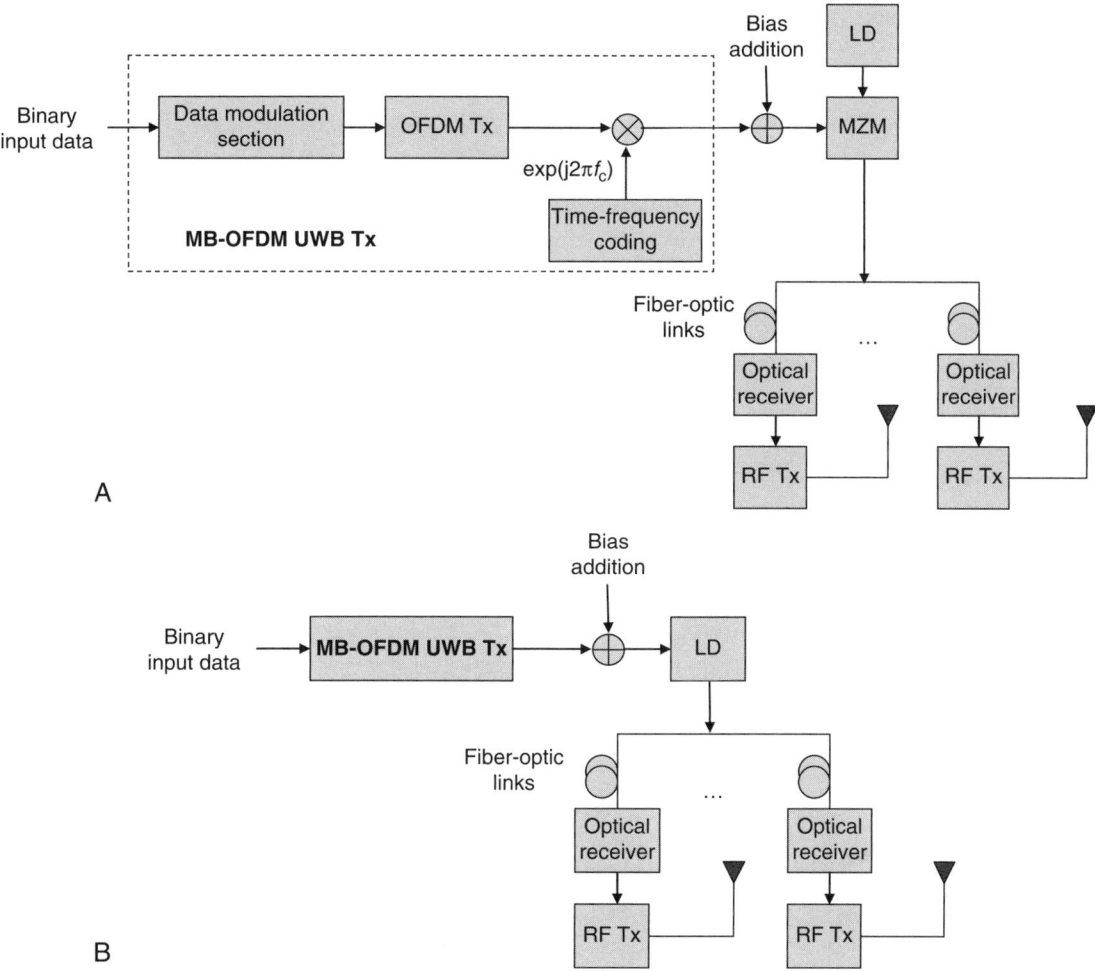

Figure 11.13: MB-OFDM UWB signal transmission over fiber-optic links with (a) external modulation and (b) direct modulation. Data modulation section and OFDM transmitter (Tx) are the same as shown in Figure 11.12. Fiber-optic link could be SMF, MMF, or POF. LD, laser diode; MZM, dual-drive Mach–Zehnder modulator; RF Tx, radio frequency transmitter.

MZM. We might need to redesign the OFDM frame but keep the 528-MHz bandwidth fixed to be compatible with the MB-OFDM UWB proposal, which represents an open area for research.

11.4 Coded-OFDM over Plastic Optical Fibers

The graded-index plastic optical fiber (GI-POF)[19] is a robust, low-cost, easy-to-install medium that provides relatively high bandwidth.[19,20] Possible applications of POFs include gigabit Ethernet and uncompressed HDTV.[21] One possible scenario was shown in Figure 11.3. The signals from different sources (satellite TV signal, cable HDTV signal, OFDMA-PON Internet traffic, MB-OFDM UWB signal over fiber, etc.) are all collected in the CS and then transmitted by POFs to different offices and rooms. However, due to limited bandwidth-length product of GI-POF (approximately 3.45 GHz × 100 m[19]), the POFs impose serious bandwidth limitations for applications at 2.5 Gb/s and above. To counter the bandwidth problem, an adaptive multiple subcarrier operating at 1 Gb/s with spectral efficiency of 6.3 bits/s/Hz and employing either 256-QAM or 64-QAM has been proposed.[21]

To improve the spectral efficiency and intersymbol interference (ISI) tolerance, we describe an LDPC-coded OFDM scheme adopted from Djordjevic.[13] Through the concept of cyclic extension, the overlapping of neighboring OFDM symbols is allowed, and by using a sufficient number of subcarriers the ISI can be significantly reduced. To improve poor power efficiency of multiple subcarrier systems, we employ single-sideband (SSB) transmission and clipped- and unclipped-OFDM. Due to severe distortion of signals at 2.5 Gb/s and above transmitted over GI-POFs, a power forward error correction (FEC) scheme would be required. Several such FEC schemes based on girth-8 LDPC codes are employed here. However, through the use of adaptive loading and LDPC coding, in combination with OFDM, it is even possible to transmit 40 Gb/s signal and beyond over 100 m of GI-POFs.[22]

The transmitter configuration, receiver configuration, and the OFDM symbol after cyclic extension are shown in Figure 10.1. The only difference is that the OFDM signal is transmitted over POF instead of the free-space optical link. An information-bearing stream at 10 Gb/s is demultiplexed into four 2.5 Gb/s streams, which are further encoded using identical LDPC codes. This step is determined by currently existing LDPC chips.[23] The outputs of LDPC decoders are demultiplexed and parsed into groups of B bits corresponding to the OFDM frame. The B bits in each OFDM frame are subdivided into K subchannels with the ith subcarrier carrying b_i bits, $B = \sum b_i$. The b_i bits from the ith subchannel are mapped into a complex-valued signal from a 2^{b_i} point signal constellation such as QAM, which is considered here. For example, $b_i = 2$ for QPSK and $b_i = 4$ for 16-QAM. The complex-valued signal points from subchannels are considered to be the values of the discrete Fourier transform of a multicarrier OFDM signal. By selecting the number of subchannels K sufficiently large, the OFDM symbol interval can be made significantly larger than the dispersed pulse width of an equivalent single-carrier system, resulting in significantly reduced ISI due to dispersion. The OFDM symbol, shown in Figure 10.1c, is generated as follows: $N_{QAM}(= K)$ input QAM symbols are zero-padded to obtain N_{FFT} input samples for IFFT, the N_G samples are inserted to create the guard interval, and the OFDM symbol is multiplied by the window function. The purpose of cyclic extension is to preserve the orthogonality among subcarriers even when the neighboring OFDM symbols partially overlap due to dispersion, and the purpose of the windowing is to reduce the out-of-band spectrum. The cyclic extension, illustrated in Figure 10.1c, is done by repeating the last $N_G/2$

samples of the effective OFDM symbol part (N_{FFT} samples) as the prefix and repeating the first $N_G/2$ samples as the suffix.

After a D/A conversion and RF up-conversion, the RF signal is converted to the optical domain using one of two options: (1) The OFDM signal directly modulates the DFB laser or (2) the OFDM signal is used as RF input of the MZM. The DC bias component is inserted to enable recovering the QAM symbols incoherently. Next, three different OFDM schemes, described in Chapter 10, are discussed. The first scheme is based on direct modulation and shall be referred to as the "biased-OFDM" (B-OFDM) scheme. Because bipolar signals cannot be transmitted over an IM/DD link, it is assumed that the bias component is sufficiently large so that when added to the OFDM signal, the resulting sum is non-negative. The main disadvantage of the B-OFDM scheme is poor power efficiency.

To improve the power efficiency, in Chapter 10 we described two alternative schemes. The first scheme, which we refer to as the "clipped-OFDM" (C-OFDM) scheme, is based on SSB transmission and clipping of the OFDM signal after bias addition. The bias is varied to find the optimum one for fixed optical launched power. It was found that the optimum case is one in which approximately 50% of the total electrical signal energy before clipping is allocated for transmission of a carrier. The second power-efficient scheme, which we refer to as the "unclipped-OFDM" (U-OFDM) scheme, is based on SSB transmission and employs LiNbO$_3$ MZM. To avoid distortion due to clipping, the information is imposed by modulating the electrical field (instead of intensity modulation employed in the B-OFDM and C-OFDM schemes) so that the negative part of the OFDM signal is transmitted to the photodetector. Distortion introduced by the photodetector, caused by squaring, is successfully eliminated by proper filtering, and recovered signal does not exhibit significant distortion. Note, however, that the U-OFDM scheme is less power efficient that the C-OFDM scheme.

The receiver commonly employs the transimpedance design, which is a good compromise between noise and bandwidth. A preamplified PIN photodiode or an avalanche photodiode is typically used as an optical detector. The PIN photodiode output current can be written as

$$i(t) = R|(s_{\text{OFDM}}(t) + b) * h(t)|^2 = R\left[|s_{\text{OFDM}}(t) * h(t)|^2 + |b * h(t)|^2 + 2R_e\{(s_{\text{OFDM}}(t) * h(t))(b * h(t))\}\right]$$

(11.7)

where $s_{\text{OFDM}}(t)$ denotes the transmitted OFDM signal, b is the DC bias component, and R denotes the photodiode responsivity. With $h(t)$ we denote the impulse response of the GI-POF, which was found to be Gaussian in references.[19,20] (The asterisk operator is the convolution operator.) The complex envelope of the OFDM signal in Eq. (11.7) can be written as

$$s_{\text{OFDM}}(t) = \sum_{k=-\infty}^{\infty} w(t - kT) \sum_{i=-N_{\text{FFT}}/2}^{N_{\text{FFT}}/2 - 1} X_{i,k} \cdot e^{j2\pi \frac{i}{T_{\text{FFT}}} \cdot (t - kT)} e^{j2\pi f_{\text{RF}} t}$$

(11.8)

which is defined for

$$kT - T_G/2 - T_{win} \leq t \leq kT + T_{FFT} + T_G/2 + T_{win}$$

In Eq. (11.8), $X_{i,k}$ denotes the QAM symbol in the ith subcarrier of the kth OFDM symbol, $w(t)$ is the window function, and f_{RF} is the RF carrier frequency. The duration of the OFDM symbol is denoted by T, whereas T_{FFT} is the FFT sequence duration, T_G is the guard interval duration (the duration of cyclic extension), and T_{win} is the length of the windowing interval. The signal after RF down-conversion and appropriate filtering can be written as

$$r(t) = \left[i(t)k_{RF}\cos\left(\omega_{RF}t\right)\right] * h_e(\tau) + n(t) \tag{11.9}$$

where $h_e(t)$ is the impulse response of the low-pass filter (having the transfer function $H_e(j\omega)$); $n(t)$ is electronic noise in the receiver, commonly modeled as Gaussian; and k_{RF} denotes the RF down-conversion factor. Finally, after the A/D conversion and cyclic extension removal, the transmitted signal is demodulated by FFT algorithm. The soft outputs of FFT demodulator are used to estimate the bit reliabilities that are fed to identical LDPC iterative decoders based on the sum-product algorithm.

For the sake of illustration, let us consider the signal waveforms and PSDs at various points in the OFDM system, as shown in Figure 11.14. The bandwidth of the OFDM signal is set

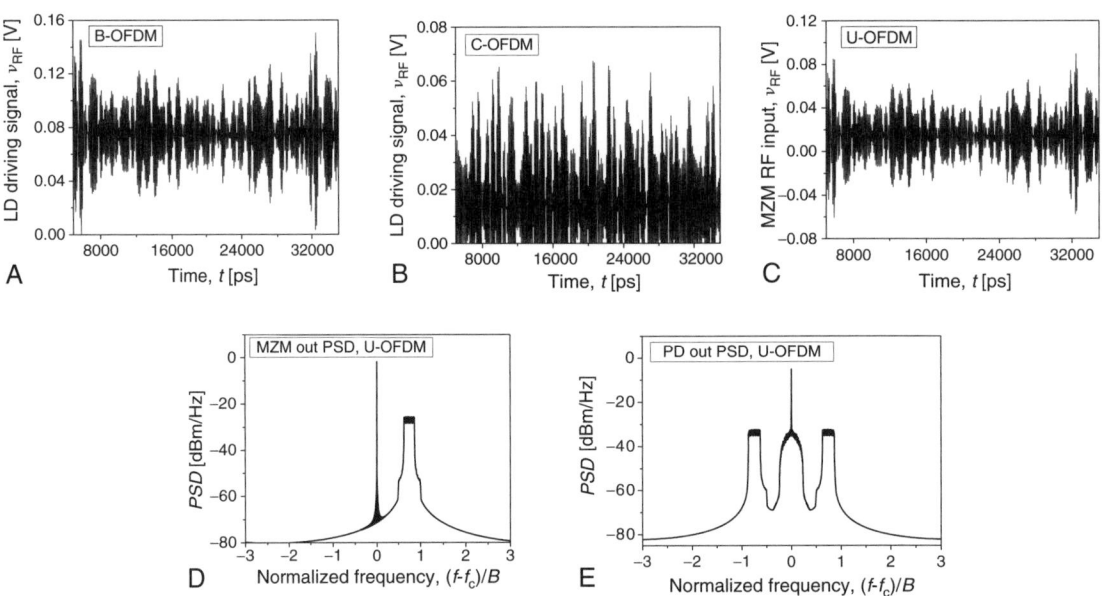

Figure 11.14: Waveforms and PSDs of SSB QPSK-OFDM signal at different points during transmission for electrical signal-to-noise ratio (per bit) of 6 dB. f_c, optical carrier frequency; LD, laser diode.

to B GHz and the RF carrier to $0.75\ B$. With B, we denote the aggregate data rate. The number of OFDM subchannels is set to 64, the OFDM sequence is zero-padded, and the FFT is calculated using 128 points. The guard interval is obtained by a cyclic extension of 2×16 samples, whereas the windowing (2×16 samples) is based on the Blackman–Harris windowing function. The average transmitted launched power is set to 0 dBm. The OFDM system parameters are carefully chosen such that the RF driver amplifier and MZM operate in linear regime (see Figures 11.14a–11.14c). The laser driving signal for B-OFDM and the MZM RF input signals for U-OFDM and C-OFDM are given in Figures 11.14a–11.14c, respectively. The PSD for an SSB QPSK-OFDM MZM output signal is shown in Figure 11.14d, and photodetector output signal is shown in Figure 11.14e.

The transmitted QAM symbols are estimated by

$$\hat{X}_{i,k} = \left(h_i^* / \left| h_i \right|^2 \right) e^{-j\theta_k} Y_{i,k} \tag{11.10}$$

where h_i is channel distortion introduced by POF, and θ_k is the corresponding phase shift of the kth OFDM symbol. With $Y_{i,k}$ we denoted the received QAM symbol in the ith subcarrier of the kth OFDM symbol. To determine the channel coefficients, it is enough just to pretransmit a short training OFDM sequence.

Given this high-level description, we turn our attention to the performance evaluation of LDPC-coded OFDM transmission over GI-POF links.

11.4.1 Performance Analysis of LDPC-Coded OFDM over POFs

The results of simulations for LDPC-coded OFDM are shown in Figure 11.15 for different values of GI-POF normalized bandwidth B (the bandwidth is normalized with respect to the aggregate bit rate, which was set to 10 Gb/s).

For normalized bandwidths $B \geq 1$, the OFDM system parameters are selected as follows. The OFDM signal bandwidth is set to $0.25\ B$, the number of subcarriers is set to $N_{QAM} = 64$, FFT/IFFT is calculated in $N_{FFT} = 128$ points, RF carrier frequency is set to $0.75\ B$, the bandwidth of optical filter for SSB transmission is set to $2\ B$, and the total averaged launched power is set to 0 dBm. The guard interval is obtained by cyclic extension of $N_G = 2 \times 16$ samples as explained previously. For normalized bandwidth $B < 1$, all parameters are the same except for the RF carrier frequency set to $0.3\ B$. The results reported in Figure 11.15 are given in terms of normalized OFDM signal bandwidth so that they are applicable for a variety of OFDM systems having the same normalized bandwidth. Three classes of LDPC codes are considered in simulations. The first class is the girth-8 regular LDPC code (8547, 6922) of rate 0.81. The second class is irregular girth-8 LDPC code (6419, 4794) of rate 0.75. The third class is girth-8 regular quasi-circular LDPC code (4320, 3242) of rate 0.75.

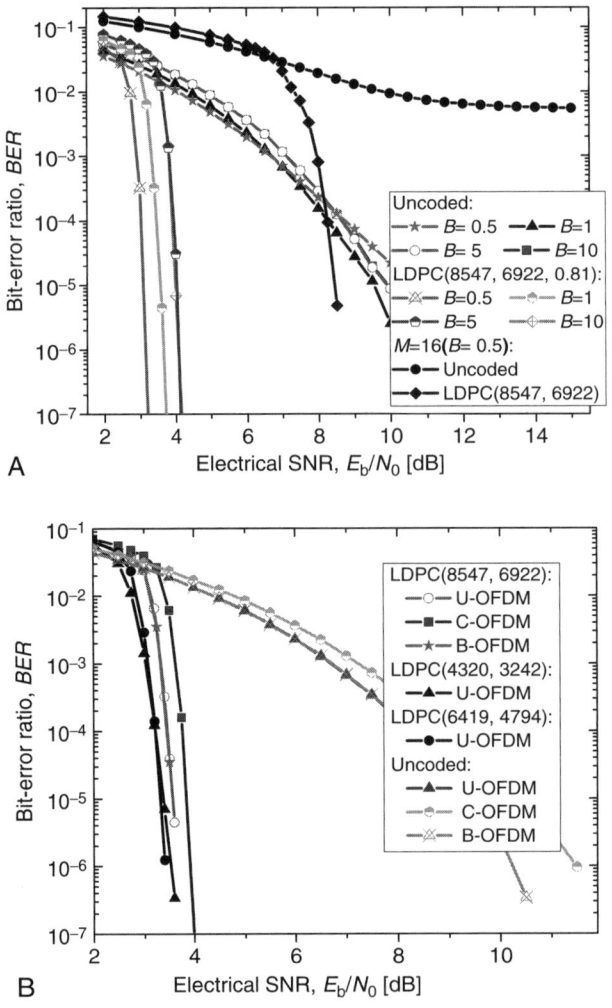

Figure 11.15: BERs of LDPC-coded OFDM over GI-POF links (4:1 multiplexer output in Figure 11.1b): (a) comparison for different normalized bandwidths and (b) comparison of different LDPC-coded OFDM schemes.

In simulations shown in Figure 11.15a, the LDPC code of rate 0.81 is employed, and bit error ratio (BER) results for different GI-POF normalized bandwidths are reported for U-OFDM scheme. For normalized bandwidth $B = 0.5$ and QPSK U-OFDM, the LDPC code of rate 0.81 provides the coding gain of approximately 9 dB at BER of 10^{-6}; a much larger coding gain is expected for lower BERs. The uncoded 16-QAM U-OFDM exhibits the BER floor, whereas the LDPC-coded OFDM scheme is able to operate error-free. Note that other FEC schemes based on RS codes (e.g., RS(255, 223)) or concatenated RS codes are not able to operate at all for 16-QAM OFDM because the error floor of uncoded signal is too high for their error correction capabilities. In Figure 11.15b, different classes of LDPC codes are compared for QPSK U-OFDM, and it was found that the irregular LDPC code of rate 0.75 outperforms the other two classes of LDPC codes. In the same figure, LDPC-coded B-OFDM, U-OFDM, and C-OFDM are compared in terms of BER, and it was found that B-OFDM and U-OFDM perform comparably, whereas the C-OFDM is approximately 0.5 dB worse. Note, however, that C-OFDM is the most efficient in terms of power efficiency, and B-OFDM is the worst.

11.5 Indoor Optical Wireless Communications and OFDM

Optical wireless communications (OWCs) represent a viable and promising supplemental technology to next-generation wireless technology (4G).[24–33] OWCs offer several attractive features, compared to radio, such as the following[24]: (1) They have a huge available unregulated bandwidth, (2) these systems do not introduce electromagnetic interference to existing radio systems, (3) optical wireless signals do not penetrate through walls so that the room represents an optical wireless cell, (4) the same equipment can be reused in other cells without any intercell interference, (5) a simple design of high-capacity LANs because coordination between neighboring cells is not needed, and (6) inherent applicability in environments in which radio systems are not desired (airplanes, hospitals, etc.).

The majority of OWCs are designed to operate in the near infrared (IR) region at wavelengths of either 850 or 1550 nm. Research in this area is increasing, and there are a number of possible point-to-point applications, including short-range links described in the IrDA standard.

In this section, we describe the basic concepts of IR OWCs, different techniques to improve the link performance such as diversity and adaptive signal processing, and potential use of visible light communications (VLCs). We believe that OFDM will become an important multiplexing and modulation technology for future IR OWCs and VLCs because of many advantages, including immunity to multipath interference, capability of supporting different services and different modulation formats, and adaptivity to time-varying channel conditions.

11.5.1 Infrared Optical Wireless Communications

The intensity modulation and direct detection are considered by many authors and they are the only viable technologies for both IR OWCs and VLCs. A typical OFDM-based OWC link is shown in Figure 11.16. Because of unguided propagation, the OWC link faces significant free-space loss. Similarly as for wireless indoor links, the most important factor in performance degradation is the multipath effect. Namely, in addition to the LOS component, several diffuse reflection-based components arrive at the receiver side. Because direct detection is used, we do not worry about co-phasing of different multipath components.

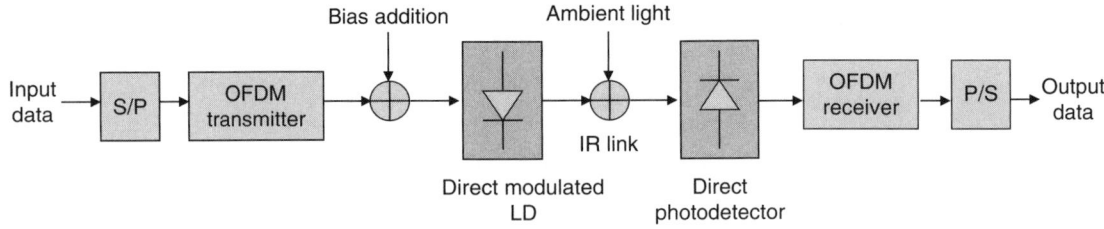

Figure 11.16: A typical OFDM-based optical wireless communication link with direct modulation and direct detection. LD, laser diode.

However, since high-speed signals are transmitted over an IR link, different multipath components arrive in time slots of neighboring symbols causing ISI. To deal with the ISI effect, different methods can be classified into three broad categories: (1) equalization, (2) multicarrier modulation (e.g., OFDM), and (3) spread spectrum. Multicarrier modulation can also be combined with diversity and multiple-input multiple-output techniques. The dominant noise source in IR links is the ambient light, which is responsible for the shot noise component. Artificial light sources also introduce harmonics up to 1 MHz in the transmitted signal. When the ambient light is weak, the receiver preamplifier is the predominant noise source. Fortunately, most of the ambient noise can be removed by proper optical filtering. A typical receiver for OWC consists of an optical front end, with an optical concentrator being used to collect the incoming radiation, an optical filter (to reduce the ambient light), a photodetector, and an electrical front end. The OFDM-based OWC receiver also contains an OFDM receiver, whose operational principle with direct detection was explained previously (e.g., see Chapter 6).

Different link types used in OWC systems, shown in Figure 11.17, can be classified as follows[25]: point-to-point links (there exists an LOS component), diffuse links (the transmitter emits optical power over a wide solid angle, and emitted light is reflected from the ceiling so that the receiver is illuminated from diffused light), and quasi-diffuse links (the transmitter illuminates the ceiling with a series of slowly diverging beam sources). The point-to-point links may contain both an LOS component and diffused reflected multipath components (see Figure 11.17d). The channel transfer function can be written as follows:

$$H(f) = H_{\text{LOS}} + H_{\text{diff}}(f) \tag{11.11}$$

where H_{LOS} corresponds to the LOS path amplitude gain, and $H_{\text{diff}}(f)$ represents the contribution of diffuse reflected multipath components, which can be approximated as a low-pass filtering function.[26] The capacity of the OFDM-based OWC with N independent subcarriers of bandwidth B_N and subcarrier gain $\{g_i, i = 0, \ldots, N-1\}$ can be evaluated by

$$C = \max_{P_i:\ \sum P_i = P} \sum_{i=0}^{N-1} B_N \log_2\left(1 + \frac{g_i(P_i + P_{\text{bias}})}{N_0 B_N}\right) \tag{11.12}$$

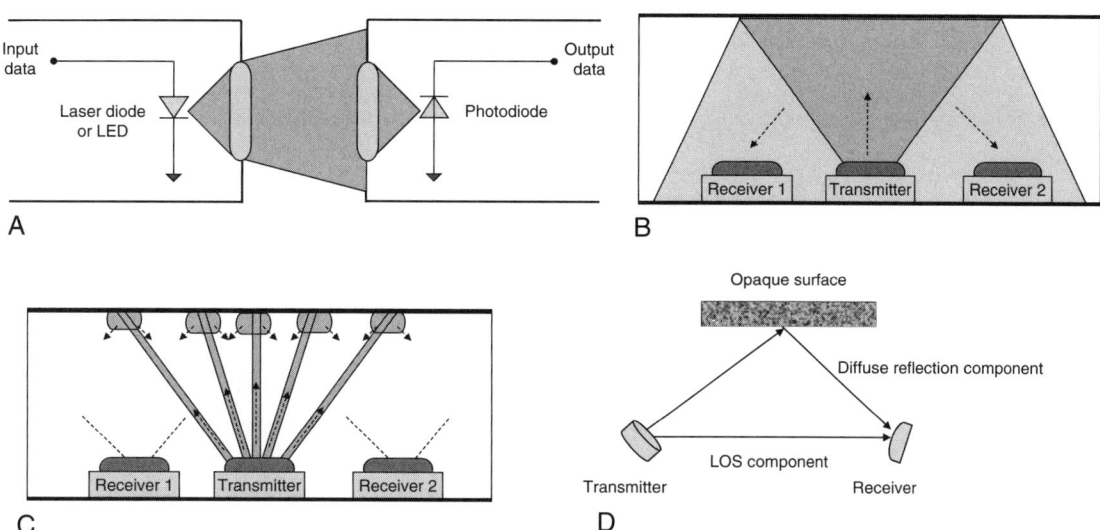

Figure 11.17: **Different link types in OWCs: (a) point-to-point link, (b) diffuse link, and (c) quasi-diffuse link. (d) In point-to-point links, the received signal may contain both an LOS component and diffusion reflected components.**

where P_i is the optical power allocated to the ith subcarrier, P_{bias} is the optical power wasted on DC bias (see Figure 11.16), and N_0 is the power spectral density of the transimpedance amplifier. The subcarrier gain g_i can be calculated by $g_i = R^2 |H_i|^2$, with R being the photodiode responsivity and H_i being the channel transfer function amplitude of the ith subcarrier (see Eq. 11.11). The signal bias is used to convert negative values of the OFDM signal to positive. Because of noncoherent detection, Eq. (11.12) represents essentially a lower bound on channel capacity. Nevertheless, it can be used as an initial figure of merit, with the lack of exact channel capacity expression. Equation (11.12) assumes that ambient noise is properly filtered out by an optical filter. By carefully designing the OFDM subsystem and providing that the guard interval is smaller than the delay spread due to diffuse reflective components, we can eliminate ISI due to the multipath effect. In some practical situations, the guard interval length could be too long compared to effective OFDM symbol length, leading to large overhead. To reduce the overhead, we have to redesign the OFDM subsystem such that ISI due to the multipath effect is only partially eliminated and use either adaptive loading[26] or diversity to improve the tolerance to the multipath effect.[24]

It can be shown using the Lagrangian method that the optimum power allocation policy is the water-filling over frequency[27]:

$$\frac{P_i + P_{bias}}{P} = \begin{cases} 1/\gamma_c - 1/\gamma_i, \ \gamma_i \geq \gamma_c \\ 0, \ \text{otherwise} \end{cases}, \ \gamma_i = g_i(P + P_{bias})/N_0 B_N \qquad (11.13)$$

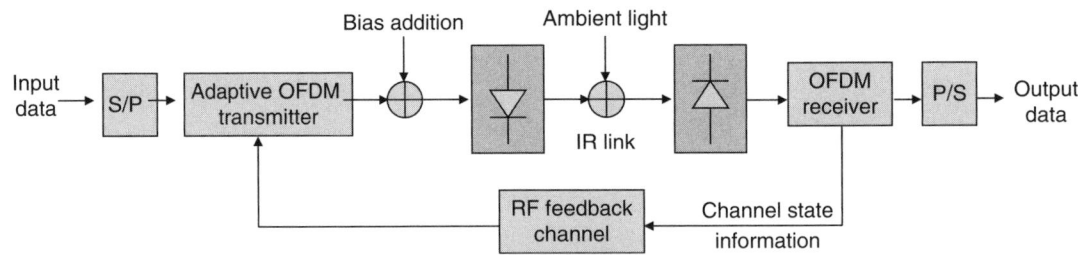

Figure 11.18: Adaptive OFDM-based OWC systems with RF feedback channel.

where γ_i is the SNR ratio of the ith subcarrier, and γ_c is the threshold SNR. By substituting Eq. (11.13) into Eq. (11.12), the following channel capacity expression is obtained:

$$C = \sum_{i:\, \gamma_i > \gamma_c} B_N \log_2 \left(\frac{\gamma_i}{\gamma_c} \right) \tag{11.14}$$

The ith subcarrier is used when corresponding SNR is above threshold. The number of bits per ith subcarrier is determined by $m_i = \lfloor B_N \log_2(\gamma_i/\gamma_c) \rfloor$, where $\lfloor \ \rfloor$ denotes the largest integer smaller than the enclosed number. This approach assumes that the RF feedback channel is used to transmit the channel state information back to the transmitter, which is illustrated in Figure 11.18. Therefore, signal constellation size and power per subcarrier are determined based on IR channel conditions. When channel conditions are favorable, larger constellation sizes are used and power per subcarrier is chosen as given by Eq. (11.13); smaller constellation sizes are used when channel conditions are poor, and nothing is transmitted when SNR per subcarrier falls below a certain threshold.

Some numerical results on potential improvements by adaptive loading are provided by Grubor et al.[26] The authors define the K-factor in a similar manner as the Rician K-parameter in wireless communications, $K = H_{\text{LOS}}/H_{\text{diff}}$, assuming that only one diffuse reflected component is dominant. If the diffuse component is dominant or comparable to the LOS component, the adaptive loading scheme performs comparably to the nonadaptive one. However, for K-factor above 10 dB, significant improvement from the adaptive loading scheme is obtained.

11.5.2 Visible Light Communications

VLCs represent a relatively new research area,[24,28–30] with operating wavelengths of 380–700 nm employing white-light LEDs with available bandwidth of approximately 20 MHz. With visible-light LEDs, it would be possible to broadcast broadband signals in various environments such as offices in which the lighting is already used. Power line communication systems or plastic optical fiber systems can be used as feeders. VLCs can also be used in car-to-car communications by employing LED-based head- and taillights, and probably in airplanes to provide high-speed Internet. It has been shown[30] that transmission with data rates above 100 Mb/s is possible by employing OFDM and VLC. The VLC systems model is similar to that shown in Figure 11.16, except that white-light LED is used instead of laser diode. Different approaches described in Section 11.5.1 are also applicable in VLCs. The VLC channel model can be represented by

$$H(f) = \sum_i H_{\text{LOS},i} \exp(-j2\pi f \Delta\tau_{\text{LOS},i}) + H_{\text{diff}} \frac{\exp(-j2\pi f \Delta\tau_{\text{diff}})}{1 + jf/f_0} \tag{11.15}$$

where $H_{\text{LOS},i}$ and H_{diff} represent the channel gain of the ith LOS component (coming from the ith LED) and diffuse signal gain, respectively. $\Delta\tau_{\text{LOS},i}$ and $\Delta\tau_{\text{diff}}$ represent the corresponding delays of the ith LOS component and diffuse signal components. f_0 is the cutoff frequency of diffuse channel frequency response, and for a medium-sized room it is approximately 10 MHz.[30] The LOS gain from the ith LED chip is related to the luminous flux $\Phi = 683 \int_{380 \text{ nm}}^{720 \text{ nm}} p(\lambda)V(\lambda)d\lambda$, where $p(\lambda)$ is the source power distribution, and $V(\lambda)$ is the eye sensitivity function, by[30]

$$H_{\text{LOS},i} = A_R(m+1)\cos^m \Phi_i \cos\psi_i / (2\pi r_i^2) \tag{11.16}$$

where A_R is the effective receiver surface; ψ_i and r_i are the angle of irradiance and the distance to illuminated surface, respectively; and m is the Lambert index related to the source radiation semiangle by $m = -1/\log_2(\cos\theta_{\text{max}})$. The diffuse signal can commonly be modeled as being constant anywhere in the room, and it is related to the effective receiver surface A_R, the room size A_{room}, and average reflectivity ρ by[30]

$$H_{\text{LOS}} = A_R\rho / A_{\text{room}}(1-\rho) \tag{11.17}$$

11.6 Summary

In this chapter, we described the use of OFDM in optical access networks. In Section 11.1, we described the use of OFDM in RoF systems, both the concepts and possible applications. Possible applications include use in WiMAX to reduce the deployment and maintenance cost and extend the coverage, in cell telephony systems to establish connection between MTSO and BSs as well as to avoid so-called dead zones, and in hybrid fiber coaxial systems to reduce the cost and extend the distance. System designers of OFDM RoF systems have to consider both multipath wireless fading and multimode dispersion in MMFs.

In Section 11.2, we described an OFDMA-PON scheme suitable to deliver different services to end users. This offers many advantages over conventional PONs, such as improved bandwidth efficiency, flexibility in dealing with bandwidth resource sharing, protocol independence and service transparency, scalability, cost-effectiveness, and low-overhead MAC.

In Section 11.3, we described the transmission of UWB signal over fiber optical links as a cost-effective solution to extend the coverage of UWB radios to hundreds of meters. This approach has several advantages, such as transparent electro-optical signal conversion regardless of the UWB's modulation methods, avoidance of the use of expensive high-speed electronic components required during signal transmission, and the opto-electronic technologies integration can potentially reduce the size and power consumption of the optical UWB transceivers. We also described the MB-OFDM UWB proposal and how to transmit this kind of signal over different fiber links.

In Section 11.4, we discussed the possibility of 10 Gb/s transmission over GI-POF links using biased-, unclipped-, and clipped-OFDM SSB transmission and LDPC codes. It has been shown that the 16-QAM OFDM system (of an aggregate rate of 10 Gb/s) supplemented with LDPC codes is able to operate with BER below 10^{-9}. The data rates can further be increased to 40 Gb/s or even 100 Gb/s through the use of adaptive modulation and coding.

Finally, in Section 11.5 we described the basic concepts of IR OWCs, different techniques to improve the link performance such as diversity and adaptive signal processing, and the potential use of VLCs. OWCs offer several attractive features, compared to radio, such as the following: They have a huge available unregulated bandwidth, optical wireless signals are contained within the room (OWC cell), the same equipment can be reused in other OWC cells without any intercell interference, the coordination between neighboring OWC cells is not needed, and OWCs are applicable in environments in which radio systems are not desired (airplanes, hospitals, etc.).

References

1. Ghavami M, Michael LB, Kohno R. *Ultra wideband signals and systems in communication engineering.* Chichester, UK: Wiley; 2007.
2. Kim YH, Kim Y. 60 GHz wireless communication systems with radio-over-fiber links for indoor wireless LANs. *IEEE Trans Consum Electron* 2004;**50**(5):517–20.
3. Wu Y, Caron B. Digital television terrestrial broadcasting. *IEEE Commun Mag* 1994;**32**(5):46–52.
4. Singh G, Alphones A. OFDM modulation study for a radio-over-fiber system for wireless LAN (IEEE 802.11a). In: *Proc. ICICS-PCM*. Singapore; 2003. p. 1460–4.
5. Dang BL, Larrode MG, Prasad RV, Niemegeers I, Kooned AMJ. Radio-over-fiber based architecture for seamless wireless indoor communication in the 60 GHz band. *Computer Commun* 2007;**30**:3598–613.
6. Song JB, Islam AHMR. Distortion of OFDM signals on radio-over-fiber links integrated with an RF amplifier and active/passive electroabsorption modulators. *J Lightwave Technol* 2008;**26**(5):467–77.
7. Xu L, Qian D, Hu J, Wei W, Wang T. OFDMA-based passive optical networks (PON), In: *2008 Digest IEEE/ LEOS Summer Topical Meetings*; 2008. p. 159–60.
8. Qian D, Hu J, Ji P, Wang T, Cvijetic M. 10-Gb/s OFDMA-PON for delivery of heterogeneous services, In: *Opt. Fiber Commun. Conf. Exposition* and *Natl. Fiber Opt. Engineers Conf.*, OSA Technical Digest (CD), paper no. OWH4. Optical Society of America; 2008.

9. Wei W, et al. Resource provisioning for orthogonal frequency division multiple access (OFDMA)-based virtual passive optical networks (VPON), In: *Opt. Fiber Commun. Conf. Exposition* and *Natl. Fiber Opt. Engineers Conf.*, OSA Technical Digest (CD), paper no. OTUI1; Optical Society of America; 2008.

10. Ramaswami R, Sivarajan KN. *Optical networks: A practical perspective*. Morgan Kaufmann; 2002.

11. Verizon Fiber Optic Service (FiOS). Available at: http://www22.verizon.com/Residential/FiOSInternet.

12. Lowery AJ, Armstrong J. 10 Gbit/s multimode fiber link using power-efficient orthogonal-frequency-division multiplexing. *Opt Express* 2005;**13**(25):10003–9.

13. Djordjevic IB. LDPC-coded OFDM transmission over graded-index plastic optical fiber links. *IEEE Photon Technol Lett* 2007;**19**(12):871–3.

14. Jazayerifar M, Cabon B, Salehi JA. Transmission of multi-band OFDM and impulse radio ultra-wideband signals over single mode fiber. *J Lightwave Technol* 2008;**26**(15):2594–602.

15. Kurt T, Yongaçoğlu A, Choinard J-Y. OFDM and externally modulated multi-mode fibers in radio over fiber systems. *IEEE Trans Wireless Commun* 2006;**5**(10):2669–74.

16. Multi-band OFDM physical layer proposal for IEEE 802.15 task group 3a. *IEEE P802.15 Working Group for Wireless Personal Area Networks*. 2004.

17. Boon PK, Peng X, Chin F. Performance studies of a multi-band OFDM system using a simplified LDPC code, In: *Proc. 2004 Int. Workshop Ultra Wideband Systems and 2004 Conf. Ultrawideband Systems Technol.* 2004. p. 376–80.

18. Kim S-M, Tang J, Parhi KK. Quasi-cyclic low-density parity-check coded multiband-OFDM UWB systems. In: *Proc. 2005 IEEE Int. Symp. Circuits Systems*. 2005. p. 65–8.

19. Shin B-G, Park J-H, Kim J-J. Low-loss, high-bandwidth graded-index plastic optical fiber fabricated by the centrifugal deposition method. *App Phys Lett* 2003;**82**(26):4645–7.

20. Sibley M. Analysis of multiple pulse position modulation when operating over graded-index plastic optical fibre. *IEEE Proc-Optoelectron* 2005;**151**:469–75.

21. Randel S, et al. 1 Gbit/s transmission with 6.3 bits/s/Hz spectral efficiency in a 100-m standard 1-mm step-index plastic optical fibre link using adaptive multiple sub-carrier modulation. In: *Proc. Post-Deadline Papers ECOC*, paper no. Th4.4.1; 2006.

22. Yang H, et al. 40-Gb/s transmission over 100 m graded-index plastic optical fiber based on discrete multitone modulation. In: *Proc. OFC/NFOEC 2009 Postdeadline Papers*, paper no. PDPD8. San Diego; 2009.

23. Lin C-C, Lin K-L, Chang H-Ch, Lee C-Y. A 3.33Gb/s (1200, 720) low-density parity check code decoder. In: *Proc. ESSCIRC*. Grenoble, France; 2005. p. 211–4.

24. Langer K-D, Grubor J. Recent developments in optical wireless communications using infrared and visible light. In: *Proc. ICTON*. 2007. p. 146–52.

25. Hranilovic S. *Wireless optical communication systems*. New York: Springer; 2004.

26. Grubor J, Jungnickel V, Langer K-D. Capacity analysis in indoor wireless infrared communication using adaptive multiple subcarrier transmission. In: *Proc ICTON*. 2005. p. 171–4.

27. Goldsmith A. *Wireless communications*. Cambridge, UK: Cambridge University Press; 2005.

28. Tanaka Y, et al. Indoor visible communication utilizing plural white LEDs as lighting. In: *Proc. Personal Indoor Mobile Radio Commun (PIMRC)*. 2001. p. F81–5.

29. Komine T, Nakagawa M. Fundamental analysis for visible-light communication system using LED lightings. *IEEE Trans Consumer Electron* 2004;**15**:100–7.

30. Grubor J, Gaete Jamett OC, Walewski JW, Randel S, Langer K-D. High-speed wireless indoor communication via visible light. In: *ITG Fachbericht 198*. Berlin: VDE Verlag; 2007. p. 203–8.

Uncited References

31. Gonzalez O, Perez-Jimenez R, Rodrıguez S, Rabadan J, Ayala A. OFDM over indoor wireless optical channel. *IEE Proc Optoelectron* 2005;**152**(4):199–204.

32. Gonzalez O, Perez-Jimenez R, Rodrıguez S, Rabadan J, Ayala A. Adaptive OFDM system for communications over the indoor wireless optical channel. *IEE Proc Optoelectron* 2006;**153**(4):139–44.

33. Grubor J, Jungnickel V, Langer K. Rate-adaptive multiple-subcarrier-based transmission for broadband infrared wireless communication, In: *Opt. Fiber Commun. Conf. Exposition* and *Natl. Fiber Opt. Engineers Conf.*, Technical Digest (CD), paper no. NThG2; Optical Society of America; 2006.

34. Prasad R. *OFDM for wireless communications systems*. Boston: Artech House; 2004.

35. Kim A, et al. 60 GHz wireless communication systems with radio-over-fiber links for indoor wireless LANs. *IEEE Trans Consum Electron* 2004;**50**(2):517–20.

36. Djordjevic IB, Vasic B, Neifeld MA. LDPC coded orthogonal frequency division multiplexing over the atmospheric turbulence channel. In: *Proc. CLEO/QELS 2006*. paper no. CMDD5; 2006.

37. Djordjevic IB, Vasic B. Orthogonal frequency-division multiplexing for high-speed optical transmission. *Opt Express* 2006;**14**:3767–75.

38. Djordjevic IB, Vasic B. 100 Gb/s transmission using orthogonal frequency-division multiplexing. *IEEE Photon Technol Lett* 2006;**18**(15):1576–8.

39. Djordjevic IB, et al. Low-density parity-check codes for 40 Gb/s optical transmission systems. *IEEE J Select Topics Quantum Electron* 2006;**12**(4):555–62.

40. Djordjevic IB, Vasic B. MacNeish–Mann theorem based iteratively decodable codes for optical communication systems. *IEEE Commun Lett* 2004;**8**:538–40.

41. Anderson I. *Combinatorial designs and tournaments*. Oxford: Oxford University Press; 1997.

42. Fan JL. Array codes as low-density parity-check codes. In: *Proc. 2nd Int. Symp. Turbo Codes Related Topics*. Brest, France; 2000. p. 543–6.

43. Milenkovic O, Djordjevic IB, Vasic B. Block-circulant low-density parity-check codes for optical communication systems. *IEEE/LEOS J Select Topics Quantum Electron* 2004;**10**(2):294–9.

44. Djordjevic IB, Sankaranarayanan S, Vasic B. Projective plane iteratively decodable block codes for WDM high-speed long-haul transmission systems. *J Lightwave Technol* 2004;**22**:695–702.

45. Lin S, Costello DJ. *Error control coding: Fundamentals and applications*. Upper Saddle River, NJ: Prentice-Hall; 2004.

46. Xiao-Yu H, et al. Efficient implementations of the sum-product algorithm for decoding of LDPC codes. *Proc IEEE Globecom* 2001;**2**:1036–1036E.

CHAPTER 12

Future Research Directions

12.1 Introduction

Optical orthogonal frequency-division multiplexing (OFDM) has received significant attention from the optical communication community, and it has certainly demonstrated its potential to permeate broad ranges of applications across every level of the optical networks, from long-haul to metro, access, and home networks. There may be some sideline debate about the superiority of the optical OFDM versus the single-carrier system, but the rapidly increasing interest in optical OFDM is ushering in a new era of software-defined optical communication (SDOT) in which various functionalities, such as optical dispersion mitigation, channel estimation, phase estimation, performance monitoring, bandwidth provisioning, data rate adaptation, and even modulation format, can be performed via software without human intervention. Within the next decade, the digital signal processing (DSP) enabled transmission system is expected to fundamentally change the way we see and operate optical networks. The ongoing study of optical OFDM also heralds a trend that the fundamental communication theory has begun to be seriously pursued in the field of optical communications, which has its birth from pure physics and traditionally places its focus on device-oriented photonics. In particular, optical communications have been incorporating rather straightforward modulation and reception methods for a much static fiber optical channel, whereas wireless communications in need of adapting to usually fast deep fading environments have employed sophisticated modulation, reception, and coding concepts that form the foundation of modern communication theory. However, the recent drive toward dynamically reconfigurable optical networks with the transmission speed beyond 100 Gb/s has brought the two forms of communications closer together. DSP for optical communications and that for wireless communications are fundamentally similar because both can be modeled as bandpass systems. DSP has played a vital role in wireless communications and may enable the ultimately flexible software-defined radio, once only theoretically possible.[1,2] On the other hand, the application of DSP to optical transmissions was hampered due to the enormous computation power required for optical communications. Advances in microelectronics, such as analog-to-digital converter (ADC), digital-to-analog converter (DAC), and digital signal processor, have resulted in the development of many novel optical subsystems and systems via electronic DSP up to 40 Gb/s.[3] It is anticipated that increasing DSP functionalities will be applied to optical communications, fueled by the ever-advancing silicon technology predicted by Moore's law.

In this chapter, we discuss some interesting research problems in the emerging field of optical OFDM as well as touch upon the intriguing issue of standardization of optical OFDM technology. As the industry is embracing the imminent commercial rollout of 100 Gb/s Ethernet (100 GbE), the feasibility of 1 Tb/s Ethernet is the next logical step. In Section 12.2, coherent optical OFDM (CO-OFDM) is investigated as the promising pathway toward 1 Tb/s Ethernet (1 TbE) transport, through a three-layer multiplexing architecture. In Section 12.3, multimode fiber in conjunction with multiple-input multiple-output OFDM (MIMO-OFDM) is proposed as a technology to achieve 100 Tb/s per fiber that takes advantage of mode multiplexing in the optical fiber. Optical OFDM, CO-OFDM in particular, hinges heavily on the success of photonic integrated circuits (PICs). In Section 12.4, various integration scenarios for high-speed optical OFDM are discussed. Due to its superior scalability to higher capacity, optical OFDM holds promise to be the future-proof technology. The traditional optical networks only support fixed link data rate. OFDM provides many enabling functionalities for future dynamically reconfigurable networks, such as a channel rate adaptable to the channel condition. In Section 12.5, we discuss various adaptable techniques, such as rate-compatible low-density parity-check (LDPC) codes and adaptive loading, to optimize the link capacity. Orthogonal frequency-division multiple access (OFDMA) is an interesting multiuser access technique in which the subsets of subcarriers are assigned to individual users. It enables flexible time and frequency domain resource partitioning. In Section 12.6, we discuss OFDM as a technique that can seamlessly bridge wireless and optical access networks, such as in radio-over-fiber (RoF) systems. OFDMA will also be shown to be promising in offering resource management for passive optical networks (PONs). The migration to the next-generation design in OFDM systems allows for the reuse of the same or similar software or hardware design, which presents an excellent opportunity for the standardization of optical OFDM technology. In Section 12.7, we discuss the prospect of optical OFDM standardization in optical communications and its impact on the optical communication industry.

12.2 Optical OFDM for 1 Tb/s Ethernet Transport

As the 100 Gb/s Ethernet has become increasingly a commercial reality, the next logical pressing issue is the migration path toward 1 Tb/s Ethernet transport to cope with ever-increasing Internet traffic. In fact, some industry experts believe that standardization of 1 TbE should be available by 2012 or 2013.[4] However, the electronic switching speed of the silicon IC as well as the optoelectronic devices will not be able to advance fast enough to support a serial 1 Tb/s interface by that time. Therefore, some form of dividing the entire 1 Tb/s data into multiple parallel pipes is the only viable option for the foreseeable future.

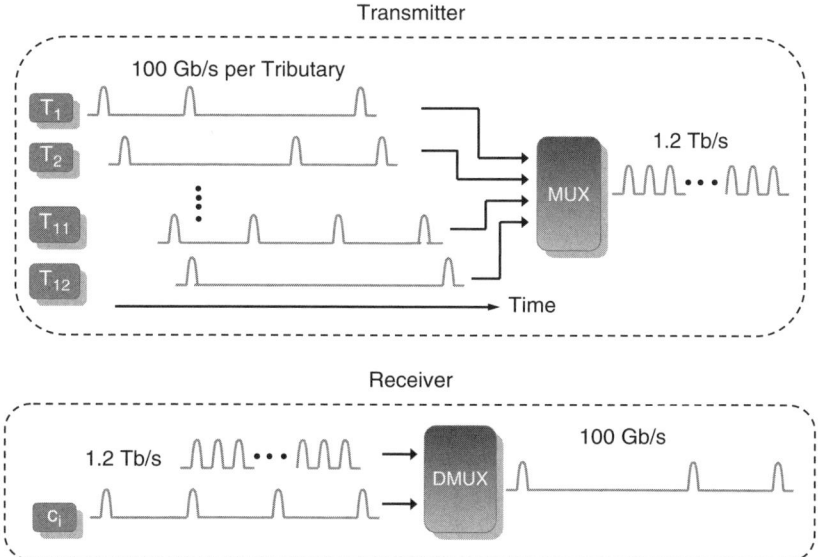

Figure 12.1: Conceptual diagram of multiplexing and demultiplexing architecture for 1 Tb/s OTDM systems. A 1.2 Tb/s OTDM signal comprising 12 tributaries (T_1–T_{12}) is shown as an example. C_i is the optical clock signal for the *i*th tributary.

Optical time-division multiplexing (OTDM) is one of the popular approaches to divide the high-speed channel into lower bit rate in the time domain. Figure 12.1 shows the conceptual diagram of multiplexing and demultiplexing architecture for the OTDM systems, where a 1.2 Tb/s high-speed signal is demultiplexed into 12 tributaries of 100 Gb/s. The OTDM approach had been adopted in several pioneering Tb/s transmission experiments.[5,6] However, there are three key problems that have not been addressed well by the OTDM approach. First, OTDM relies on precise timing alignment of many tributaries for multiplexing and demultiplexing. Whether the TDM is performed in the optical domain now, or perhaps in the electronic domain in the future, the timing accuracy on the order of fetoseconds is challenging if not impossible. Second, the demonstrated transmission reach of the Tb/s OTDM system is limited to 240 km[6] due to its extreme sensitivity to chromatic and polarization dispersion that calls for high-order optical compensation for either dispersion, which is costly if not infeasible. Third, because of the short pulse employed for each tributary, it is questionable whether migration to an OTDM-based Tb/s system will lead to any optical spectral efficiency improvement. All these challenges have cast doubt on the viability of OTDM-based Tb/s transport.

On the other hand, CO-OFDM may offer a promising alternative pathway toward Tb/s transport that possesses high spectral efficiency, resilience to tributary timing alignment, and, most important, channel dispersion. Figure 12.2 shows the multiplexing and demultiplexing architecture of CO-OFDM, where 1.2 Tb/s is demultiplexed into 12 frequency domain tributaries.

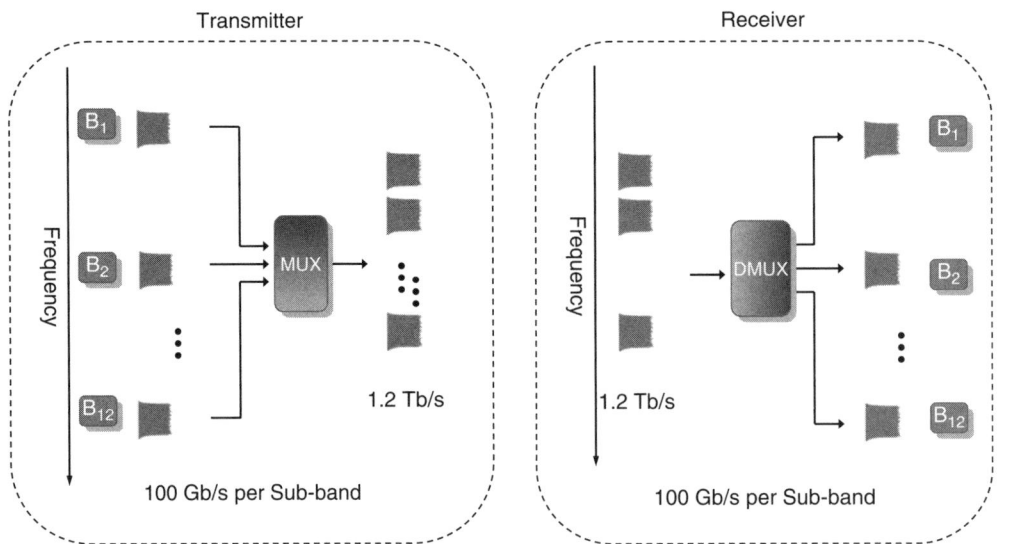

Figure 12.2: Conceptual diagram of multiplexing and demultiplexing architecture for 1 Tb/s CO-OFDM systems. A 1.2 Tb/s CO-OFDM signal comprising 12 bands (B$_1$–B$_{12}$) is shown as an example.

Using this principle, the orthogonal band multiplexing (OBM) scheme was proposed to sub-band the OFDM without sacrificing spectral efficiency and computational complexity.[7] Nevertheless, migrating from 100 Gb/s to 1 Tb/s poses a major challenge because it requires 10 times more OFDM bands. To solely rely on the optical OBM-OFDM entails many coherent optical transmitter and receiver pairs, which is cost-prohibitive. It is thus imperative to introduce an immediate integration process in the RF layer. Figure 12.3 illustrates the concept of such a three-layer IC architecture for the Tb/s CO-OFDM system. For exhibitory clarity, the CO-OFDM is assumed at the rate of 1.2 Tb/s partitioned into 12 bands. The entire mixed-circuit IC design involves three layers of integration as follows: (1) In the baseband layer, baseband OFDM is

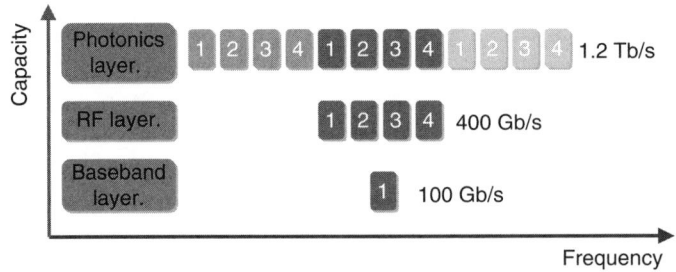

Figure 12.3: Conceptual diagram of three-layer optoelectronic IC hierarchy for 1 Tb/s CO-OFDM transceiver. Each box represents an OFDM band at 100 Gb/s. Multiplexing is performed from the bottom layer to the top layer, whereas demultiplexing is in the reverse order.

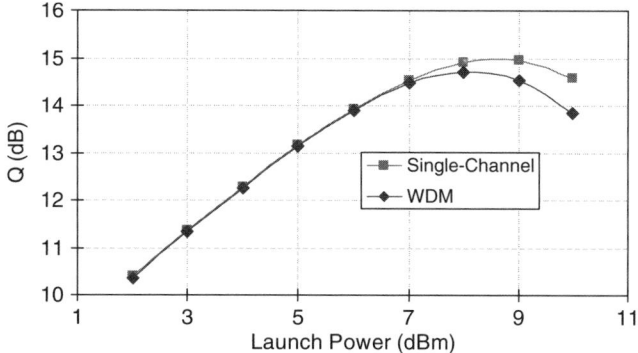

Figure 12.4: System performance of 1 Tb/s CO-OFDM as a function of launch power for a reach of 1000 km standard single-mode fiber.

generated at the rate of 100 Gb/s; (2) in the RF layer, each of the four basebands is multiplexed up to an RF carrier and combined electrically to 400 Gb/s RF signal; and (3) in the photonics layer, the output signal from the RF layer will modulate three phase-locked wavelengths and combine them optically into a 1.2 Tb/s optical signal. The baseband layer and RF layer are ideally implemented in mixed-circuit CMOS ASICs, leveraging the recent progress in millimeter-wave CMOS technology.[8,9] The PICs within the photonics layer or the integration of photonics and CMOS is of major importance for the 1 Tb/s transport. Compared with OTDM, the CO-OFDM-based Tb/s system employs the multiplexing and demultiplexing of each OFDM band with RF and wavelength combiners, eliminating the need for extremely tight timing alignment. Furthermore, the CO-OFDM has shown extreme dispersion resilience,[10–12] without using complicated higher order optical dispersion compensation.

Because of its high spectral efficiency, one of the interesting questions regarding the 1 Tb/s CO-OFDM system concerns its transmission performance. Figure 12.4 shows the system performance for 1 single Tb/s channel and three Tb/s wavelength-division multiplexing (WDM) channels after a transmission of 1000 km (10×100 km) standard single-mode fiber (SMF). It can be seen from Figure 12.4 that the Q-factor can reach 13 dB. With more powerful error-correction coding, such as LDPC[13] or Raman amplification,[14,15] there could be sufficient margin for the 1 Tb/s system for long-haul transmission. Because of the wide spectrum of 1 Tb/s CO-OFDM signal, we do not expect significant performance degradation if the CO-OFDM increases to multiple Tb/s.

12.3 Multimode Fiber for High Spectral Efficiency Long-Haul Transmission

Multimode fiber (MMF) has long been perceived as the medium for short-reach systems, although it can achieve very high capacity.[16,17] Recent experiments with 20 Gb/s CO-OFDM transmission over 200-km MMF fiber may change that stereotype and spur research interest

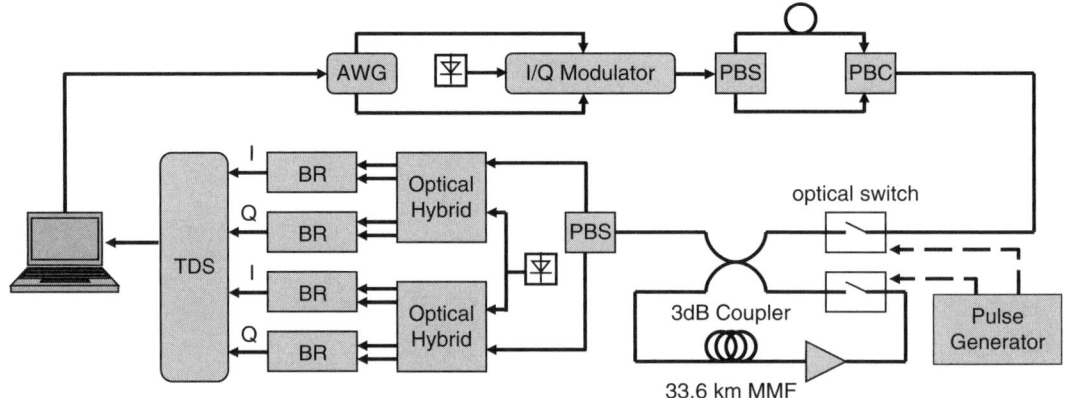

AWG: Arbitrary Waveform Generator, PBS(C): Polarization Beam Splitter (Combiner),
TDS: Time Domain-sampling Scope, BR: Balanced Detector

Figure 12.5: Experimental setup for 20 Gb/s CO-OFDM over MMF link.

in MMF-based long-haul transmission.[18] Figure 12.5 shows the experimental setup for polarization multiplexed 20-Gb/s CO-OFDM signal transmission over optically amplified MMF systems. The MMF link is emulated with a recirculation loop comprising 33.6-km 50/125 μm MMF and an erbium-doped fiber amplifier to compensate for the loss of 16 dB. The optical signal is launched from 9-μm core-diameter SMF into the 33.6 km of 50-μm core-diameter graded-index MMF. SMF-to-MMF launch introduces spatial mode filtering effect, resulting in only a small number of lower order modes excited in the MMF link.[16] The OFDM signal is coupled out from the recirculation loop and fed into a polarization diversity coherent receiver. The signal is sampled with a time domain sampling scope and processed off-line. Figure 12.6 shows the bit error ratio (BER) performance for the back-to-back transmission and 200-km MMF transmission at a launch power of 2.5 dBm. A 1-dB penalty is observed for MMF transmission at the BER of 10^{-3}. The 200-km MMF fiber transmission

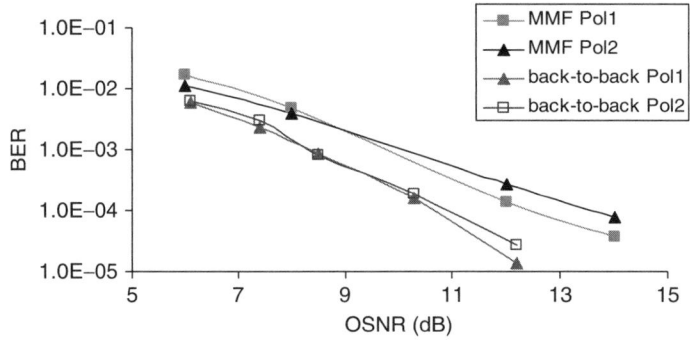

Figure 12.6: Transmission performance of 20 Gb/s CO-OFDM over MMF link at back-to-back and 200-km MMF.

of 20 Gb/s CO-OFDM represents more than one order of magnitude reach extension over that previously reported.[19]

Although the previous experimental demonstration is instrumental in setting a new record of data rate and reach product of 4 Tb·km/s for the MMF link, it does not indicate that the installed MMF is an ideal long-haul transmission medium. In fact, the appropriate MMF needs to be redesigned and installed. The problems with current MMF are its high loss and excessive number of modes that may overwhelm the computation capacity of silicon chips. Therefore, the ideal MMF for long-haul transmission may be the few-mode MMF, such as dual-mode MMF. Interestingly, such a dual-mode transmission fiber was investigated three decades ago for quite a different purpose of reducing coupling loss due to the larger core size of dual-mode fiber.[20] The MMF-based long-haul systems not only entail massive MIMO-OFDM signal processing but also require many critical devices that are not employed in the conventional optical communications. Figure 12.7 shows the conceptual diagram of an MMF-based long-haul system, including some critical devices detailed as follows:

Transmitter/receiver laser array with low laser linewidth: Because of the relatively large dispersion arising from the MMF, the OFDM symbol length for MMF link is to be one or two orders of magnitude longer than that for the SMF link. This implies a need for laser linewidth on the order of 1–10 kHz. It may be a major challenge to achieve such a low linewidth for distributed feedback or distributed Bragg reflector lasers, but it may be feasible for external cavity lasers or fiber ring lasers. Because of the enormous capacity that each wavelength can carry, on the order of terabytes per second, the cost of these low-linewidth lasers can be absorbed very well into the overall transponder cost.

Multimode compatible passive optical components: Current passive optical components, including optical multiplexer/demultiplexer and RODAM, are single-mode based and have relatively tight wavelength accuracy in passband center wavelength. Subsequently, they will not work properly for the multimode signal. This is because the center frequency of these optical filters is expected to have mode dependence, and the mismatch

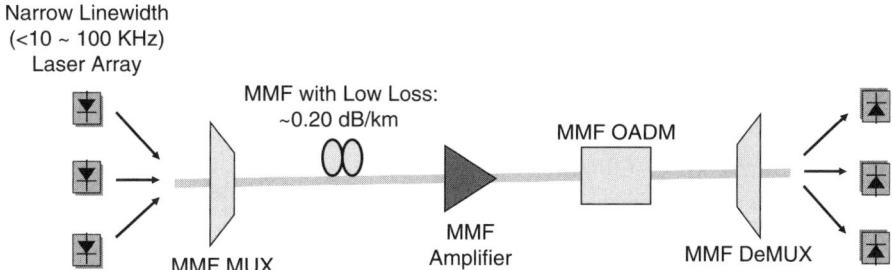

Figure 12.7: Conceptual diagram of MMF-supported communication systems. DeMUX, demultiplexer; MUX, multiplexer.

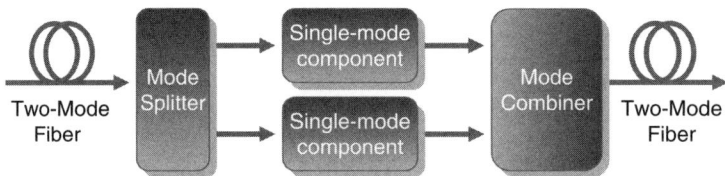

Figure 12.8: Conceptual diagram of an multimode compatible optical component.

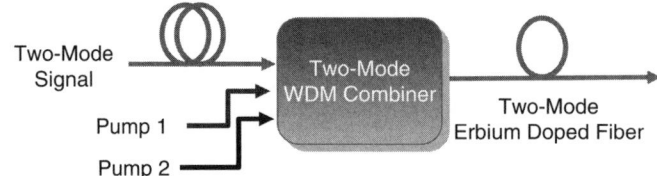

Figure 12.9: Architecture of a dual-mode optical amplifier using two separate pumps, each controlling its corresponding mode gain.

of the center frequencies for the two modes will cause penalty for multimode signals. Figure 12.8 shows a simple solution to this mode-dependence problem using a two-mode signal as an example. The signal is first separated into two modes, with each mode going through SMF components, and then the modes are combined back into the two-mode signal.

Multimode optical amplifier: The optical amplifier is one of the most critical components for optical networking. Erbium-doped multimode fiber can be used as the amplification medium, but the challenge is that mode-dependent gain (MDG) will inevitably arise due to the uncertainty of the doping concentration and the waveguide profile. Figure 12.9 shows the two-mode optical amplifier in which the dual-mode signal and pumps are combined by using the multimode wavelength multiplexer discussed previously. What is unique here is that there are two optical pumps, with each pump controlling effectively the gain of its corresponding mode. In doing so, there is no need for a separate erbium-doped fiber for each mode. MDG can be minimized by dynamically adjusting the power ratio between these two optical pumps.

12.4 Optoelectronic Integrated Circuits for Optical OFDM

The notion that optoelectronic integrated circuits (OEICs) will enable the placement of a large number of optical and electronic devices onto a single chip can be traced back approximately four decades.[21] It is a dream that has not been fulfilled for the photonics community, whereas during the same period, electronic silicon digital signal progressing has advanced in leaps and bounds and has profoundly impacted many fields, from telecommunications to photography. However, in the past decade there has been a dramatic resurgence of interest in OEICs from both industry and academia.[22–26] In 2004, the Defense Advanced Research Projects Agency's (DARPA's) microelectronic technology office initiated a 4-year project on electronic and photonic integrated circuits in silicon, aiming for monolithic integration of silicon very large-scale integration electronics with silicon nanophotonics on a single silicon chip in a commercial state-of-the-art CMOS silicon-on-insulator production plant (foundry). Another notable event is that Infinera has successfully developed and commercially deployed monolithic InP photonic integrated circuits, each having more than 50 integrated components on a single chip.[26]

The fundamental reasons for the of OEICs are twofold[26,27]:

Supporting software-defined photonics: Through incorporation of the electronic signal processing, the photonic chip can be reconfigured for multiple functionalities in a number of aspects of transmission, reception, and filtering.

Performance enhancement: The speed and noise performance of optoelectronic devices can be significantly improved by integration due to the reduction of parasitic reactance, an almost perfect matching condition for balanced devices, and mechanical stability.

Because of the extensive signal processing involved in optical OFDM, it is natural to expect the silicon technology to be the platform to monolithically integrate the electronic DSPs and photonic components onto a single chip. Figure 12.10 shows a CO-OFDM transceiver architecture that includes four functional blocks: RF OFDM transmitter, RF-to-optical up-converter (RTO), optical-to-RF down-converter (OTR), and RF OFDM receiver. Experience in the wireless domain has shown that the OFDM transceiver can be placed on one silicon chip[8,9] and probably can be implemented at the 40-Gb/s baseband rate. By using the band-multiplexing architecture shown in Figure 12.3, and a mixed analog and digital circuit,[8,9] the 100-Gb/s CO-OFDM transceiver integrated silicon chip is not a remote possibility. The RTO up-converter includes an optical IQ modulator composed of dual-arm Mach–Zehnder modulators (MZMs) with 90° shift. The optical silicon MZM based on the free carrier plasma dispersion effect has been demonstrated.[28] The OTR down-converter includes polarization splitter, 90° hybrid, and balanced photodetectors. The polarization splitter based on a two-dimensional grating etched in a silicon-on-insulator waveguide is reported in De la Rue et al.[29] Although the crystalline Si is highly transparent within the 1.2- to 1.6-μm infrared wavelength window, the use of silicon–germanium heterostructures permits the realization of Si-based optoelectronic detectors within this window.[30–32] The relatively large number of balanced photodiodes required in CO-OFDM receivers could ideally be realized in OEICs for cost reduction and improved matching within the balanced photodetectors. It seems that there is great potential for optical OFDM to be implemented in OEICs with incorporation of powerful electronic DSP, ADC/DAC, RF amplifier, RF filter, silicon optical modulator, silicon-based 90 hybrid, and silicon balanced photodetector. We believe that advances in silicon OEICs open up new avenues for multicarrier technology to

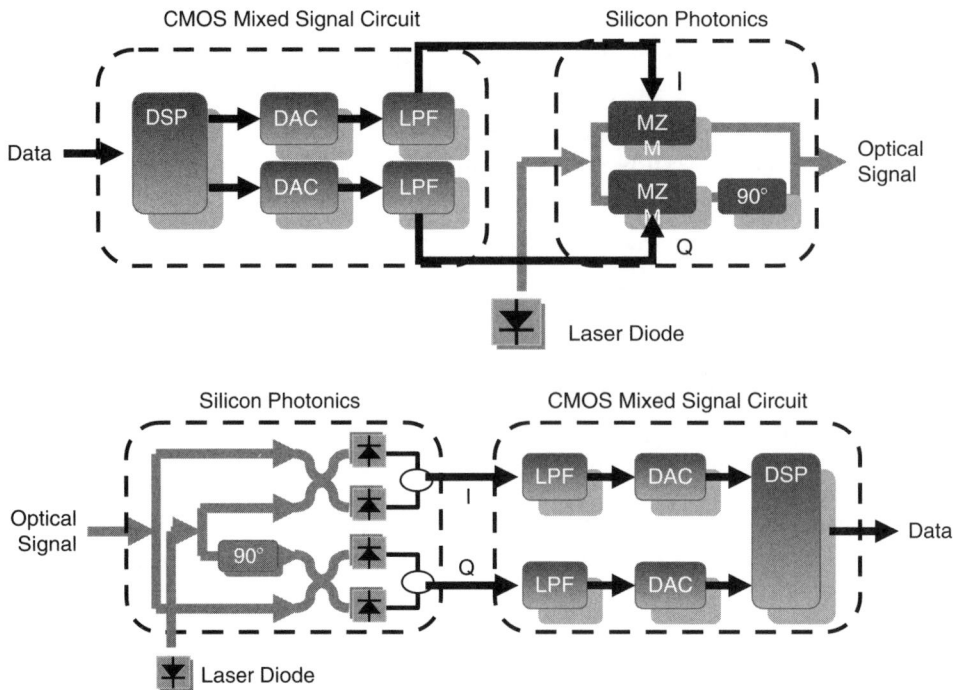

Figure 12.10: Functional blocks of a CO-OFDM transceiver and its corresponding mapping to an integrated silicon chip.

make inroads into the broad range of optical communication applications from access to core optical networks. We anticipate that the optical multicarrier subsystems/systems based on OEICs will present rich research opportunities for the optical communications community.

12.5 Adaptive Coding in Optical OFDM

Codes on graphs, such as LDPC codes and turbo product codes, have generated much research interest. It has been shown that LDPC codes are able to closely approach Shannon's capacity limits. OFDM is an excellent modulation technique to be used for multiuser access, known as OFDMA. In OFDMA, subsets of subcarriers are assigned to individual users. OFDMA enables time and frequency domain resource partitioning. In the time domain, it can accommodate for the burst traffic (packet data) and enables multiuser diversity. In the frequency domain, it provides further granularity and channel-dependent scheduling. In OFDMA, different numbers of subcarriers can be assigned to different users to support differentiated quality of service (QoS). Each subset of subcarriers can have different kinds of modulation formats and can carry different types of data. The differentiated QoS can be achieved by employing the LDPC codes of different error correction capabilities. Therefore, OFDMA represents an excellent interface between wireless/wireline and optical

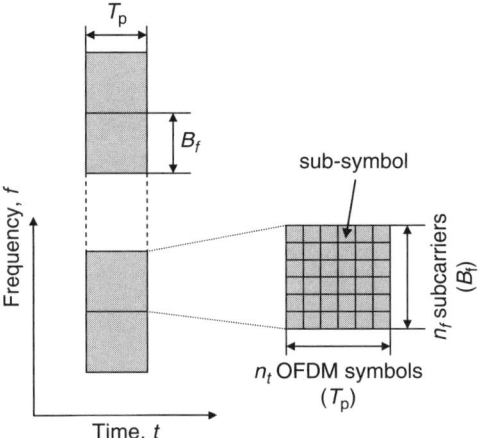

Figure 12.11: OFDM packets for rate-adaptive coded modulation.

technologies. Because of their low complexity of decoders and highly regular structure of parity-check matrices, LDPC codes are considered as excellent candidates for rate-adaptive coded OFDM applications. An example of rate-adaptive coding is shown in Figure 12.11. The transmitted data are organized in OFDM packets of duration $T_{\rm p}$, each containing n_t OFDM symbols, and every OFDM symbol has n_f subcarriers and occupies the frequency bandwidth of B_f Hz. The cyclic extension guard interval of OFDM is sufficiently long to compensate for chromatic dispersion and maximum differential group delay. The number of OFDM packets and constellation size are determined by channel conditions. When channel conditions are favorable, larger constellations are used and higher data rates are transmitted.

The quasi-cyclic LDPC codes described in Chapter 6 are excellent candidates for use in rate-adaptive codes. The parity-check matrix of these codes is given as

$$
H = \begin{bmatrix}
I & I & I & \ldots & I \\
I & P^{S[1]} & P^{S[2]} & \ldots & P^{S[c-1]} \\
I & P^{2S[1]} & P^{2S[2]} & \ldots & P^{2S[c-1]} \\
\ldots & \ldots & \ldots & \ldots & \ldots \\
I & P^{(r-1)S[1]} & P^{(r-1)S[2]} & \ldots & P^{(r-1)S[c-1]}
\end{bmatrix}
\tag{12.1}
$$

where P is a $p \times p$ permutation matrix, r is the column weight (the number of rows in Eq. (12.1) being used), c is the row weight, and p is a prime. The codeword length is determined by pc, whereas the code rate is lower bounded by

$$
R(r) \geq \frac{pc - rp}{pc} = 1 - r/c
\tag{12.2}
$$

For code rate adaptation, we have several options: (1) keep the codeword length fixed (or equivalently keep the row weight c fixed) but change the number of rows r being used, (2) keep r constant but change c, and (3) let both parameters be variables. Note that the parity-check matrix stays quasi-cyclic as the code rate changes. The rate-compatible convolutional codes are already in use in wireless communications.[33,34] However, the code rates of these codes are usually small, puncturing is commonly random, and excessive puncturing yields performance degradation. On the other hand, rate-compatible LDPC codes can be used in a bit-interleaved coded modulation manner,[35] and code rate adaptation can be performed by maximizing the channel capacity. The channel capacity of this scheme can be calculated by employing Ungerboeck's approach[36]:

$$
C_{\mathrm{u}} = I(X;Y) = \log_2 M - E_{X,Y}\left[\log_2 \frac{\sum\limits_{z\in X} p(y|z)}{p(y|x)}\right] \tag{12.3}
$$

which represents the channel capacity for uniform input distribution (or achievable information rate), where $E_{X,Y}$ denotes the expectation operator with respect to input and output symbols, and M is the signal constellation size. Transition probabilities $p(y|z)$ are evaluated by employing the training sequences, whereas for the amplified spontaneous emission noise-dominated scenario the transition probability is given by

$$
p(y|z) = \frac{1}{2\pi\sigma^2} \exp\left[-\frac{|y-z|^2}{2\sigma^2}\right] \tag{12.4}
$$

where σ^2 is the noise variance. One possible scenario is shown in Figure 12.12 (see Chapter 6 for full description of this scheme). The number of bits per symbol $m = \log_2 M$ is determined to maximize the channel capacity expression (Eq. 12.3). The same mapping is employed for the duration of OFDM packets shown in Figure 12.11. Based on optical signal-to-noise ratio estimate, we can determine which code rate is to be used.

Another approach described here is based on adaptive loading, which is already in use in wireless communications.[35] The key idea is to vary the data rate and power assigned to each OFDM subcarrier relative to the subcarrier gain. In adaptive loading, power and rate on each subcarrier are adapted to maximize the total rate of the system using adaptive modulation, such as variable-rate variable-power M-ary QAM or M-ary phase-shift keying. The capacity of the OFDM system with N independent subcarriers of bandwidth B_N and subcarrier gain $\{g_i, i = 0, \ldots, N-1\}$ can be evaluated by

$$
C = \max_{P_i:\sum P_i = P} \sum_{i=0}^{N-1} B_N \ \log_2\left(1 + \frac{g_i^2 P_i}{N_0 B_N}\right) \tag{12.5}
$$

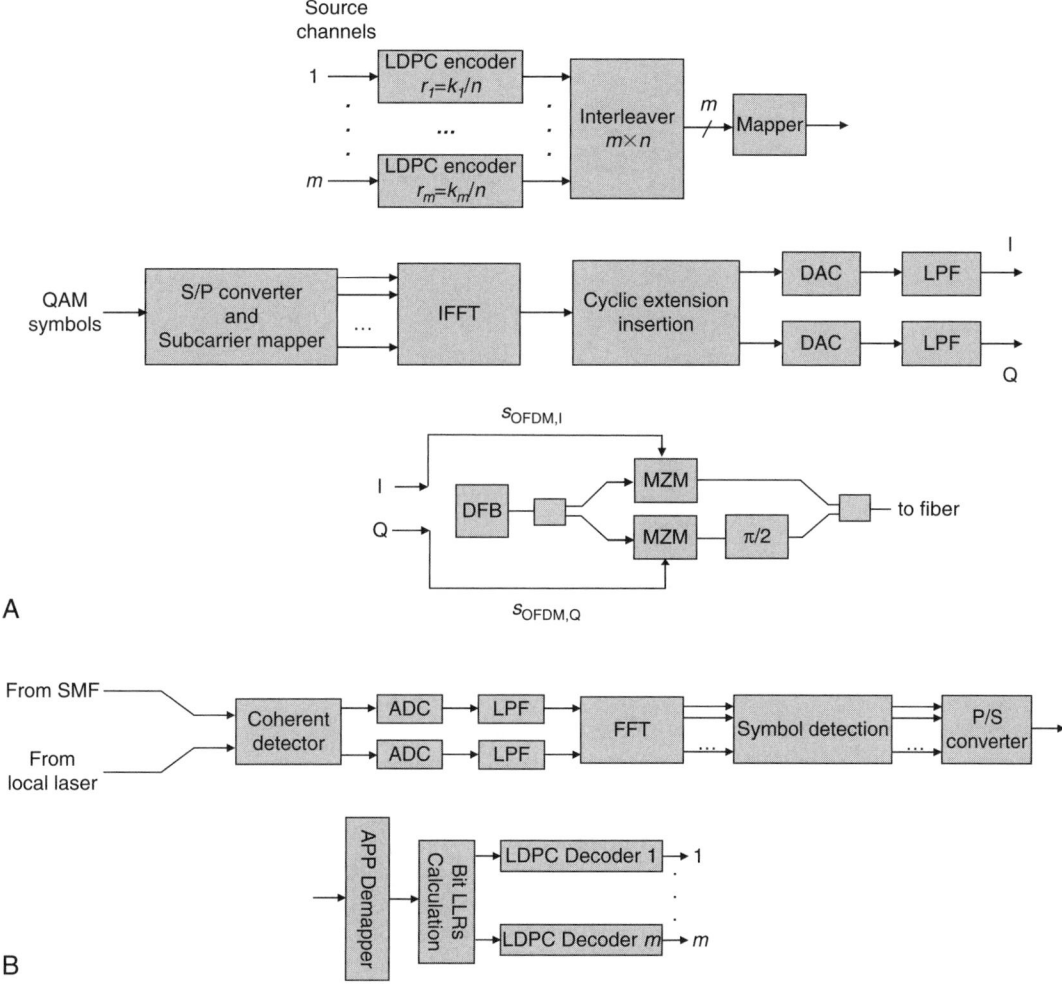

Figure 12.12: Code rate adaptive optical OFDM scenario based on multilevel coding and bit-interleaved coded modulation.

where P_i is the power allocated to ith subcarrier. It can be shown by using the Lagrangian method that the optimum power allocation policy is the water-filling over frequency[37]:

$$\frac{P_i}{P} = \begin{cases} 1/\gamma_c - 1/\gamma_i, & \gamma_i \geq \gamma_c \\ 0, & \text{otherwise} \end{cases}, \quad \gamma_i = g_i P / N_0 B_N \tag{12.6}$$

where γ_i is the signal-to-noise ratio (SNR) of the ith subcarrier, and γ_c is the threshold SNR. By substituting Eq. (12.6) into Eq. (12.5), the following channel capacity expression is obtained:

$$C = \sum_{i:\gamma_i > \gamma_c} B_N \log_2\left(\frac{\gamma_i}{\gamma_c}\right) \tag{12.7}$$

The ith subcarrier is used when corresponding SNR is above threshold. The number of bits per ith subcarrier is determined by $m_i = \lfloor B_N \log_2(\gamma_i/\gamma_c) \rfloor$, where $\lfloor \rfloor$ denotes the largest integer smaller than the enclosed number.

12.6 Optical OFDM-Based Access Networks

OFDM is an important technology that enables high-speed, high-capacity, and robust transmission over both wireless and optical channels. It has been used in many communication standards, such as wireless local area networks (also known as Wi-Fi), digital video and audio broadcasting standards, digital subscriber loop (DSL), and Worldwide Interoperability for Microwave Access (WiMAX, or IEEE 802.16). For reducing the deployment and maintenance costs of wireless networks while providing low power consumption and large bandwidth, the RoF system seems to be a promising candidate.[38–41] In RoF systems, the fiber is used to distribute the RF signal from a central station to remote antenna units. Different types of fibers, such as SMFs, MMFs, and plastic optical fibers, can be used.[38–44] Possible applications include (1) in cellular systems to establish the connection between a mobile telephone switching office and base stations (BSs), (2) in WiMAX to extend coverage and reliability by connecting WiMAX BSs and remote antenna units, and (3) in ultra wideband communications to extend the wireless coverage range.[45] A typical example of RoF systems is shown in Figure 12.13a. The data for a given end user are generated in a central station, imposed on a set of OFDM subcarriers assigned to that particular user, transmitted over optical fiber upon modulation in an MZM, and converted into the electrical domain by an optical receiver in the BS. From the BS, the signal is transmitted over wireless channel to the end user. An example of downlink transmission is shown in Figure 12.13b.

In these systems, system designers have to deal not only with multipath fading present in wireless links but also with dispersion effects present in fiber links. In previous chapters, we showed that chromatic dispersion and PMD can be compensated by employing optical OFDM, provided that the guard interval is longer than the total delay spread due to chromatic dispersion and maximum differential group delay. In wireless networks, the number of subcarriers in OFDM is chosen in such a way that the bandwidth of the signal per subcarrier is smaller than the coherence bandwidth of the wireless channel so that each subcarrier experiences the flat fading. Therefore, for simultaneous compensation of multipath fading in wireless channels and dispersion effects in optical fiber links, we have to provide that the cyclic prefix is longer than the total delay spread due to multipath spreading and dispersion effects.

OFDM is also an excellent candidate to be used for indoor optical wireless applications[46–49] and PON applications.[42] Currently, PON is being deployed to replace conventional cable-based access networks. With optical fiber used as the transmission medium, PONs offer much higher bandwidth while supporting various communication services. For future PON

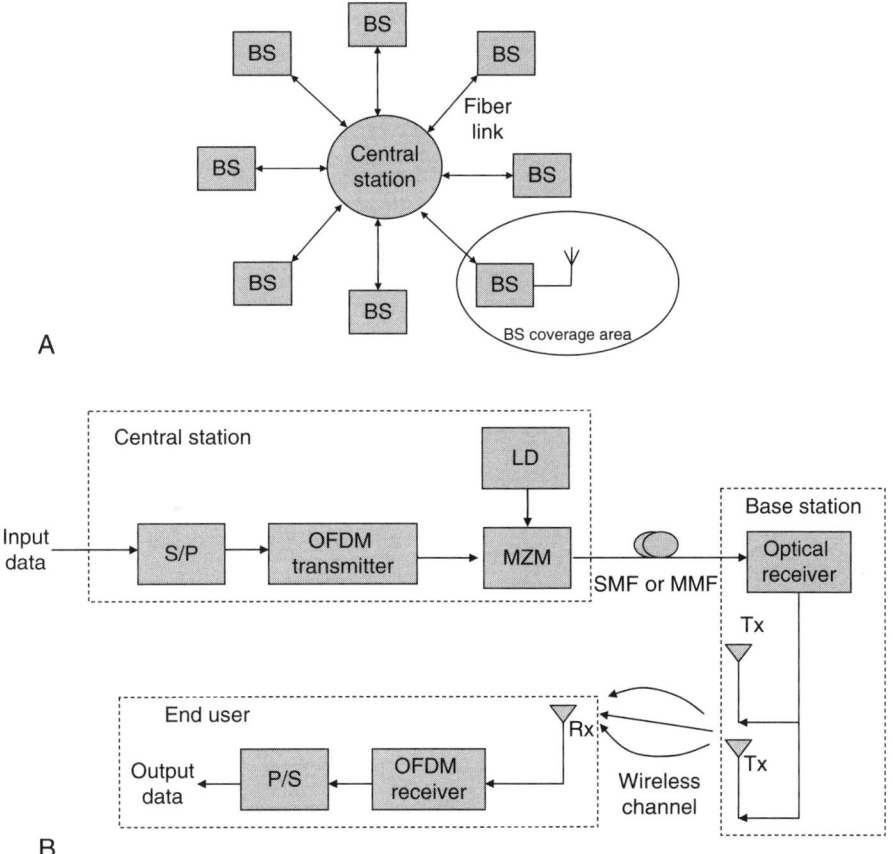

Figure 12.13: (a) An RoF system and (b) downlink transmission. BS, base station; LD, laser diode; MZM, Mach–Zehnder modulator; Rx, wireless receiver; Tx, wireless transmitter.

applications, different PON technologies have been studied, including WDM-PON, SCM-PON, and OCDM-PON. OFDMA-PON has been advocated by Xu et al.[42] OFDMA is a multiple-access technology that allows assigning different subcarriers to multiple users in a dynamic manner. It simultaneously enables time and frequency domain resource partitioning.[42] Compared with TDM-PONs, OFDMA-PONs can be combined with TDM to offer one additional dimension for resource management. For example, in the time domain PON can accommodate burst traffic, and in the frequency domain PON can offer fine granularity and channel-dependent scheduling.[42] The OFDMA-PON principle is illustrated in Figure 12.14. Essentially, PON is commonly implemented as a point-to-multipoint topology, and to reduce overall system cost, all the components between the optical line terminal (OLT) and optical network units (ONUs) are passive. OLT assigns to every particular user (ONU) a subset of subcarriers. Legacy users (denoted as white (E1/T1) and black (wireless

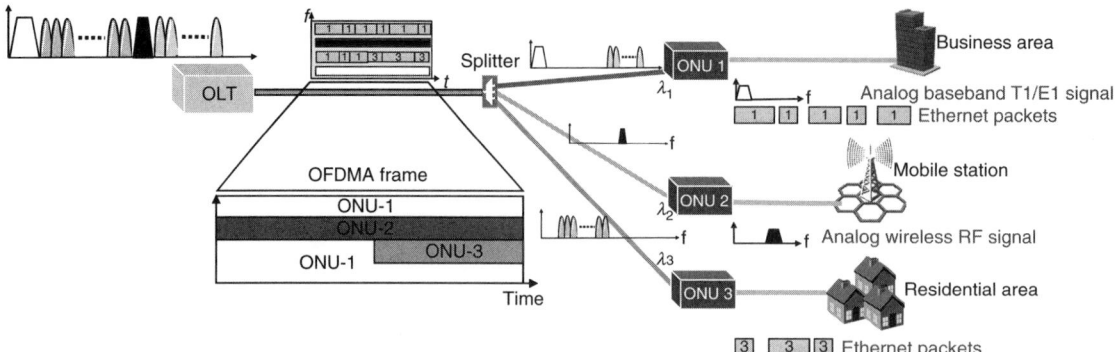

Figure 12.14: An example of OFDMA-PON. OLT, optical line terminal; ONU, optical network unit.

cell) subcarrier frequency slots in Figure 12.14) can coexist with Internet users. In the upstream direction, each ONU modulates the data over the assigned subcarrier set, whereas all the other subcarriers belonging to other ONUs are set to zero.

12.7 Standardization Aspects of Optical OFDM

The standardization in optical networks often ushers in a deployment of advanced communication technology. The introduction of SONET/SDH in 1980 led to the worldwide rollout the high-speed synchronous digital systems that still serve as the backbone of today's modern communication networks.[50] The completion of the GPON (ITU-T G.984) and EPON (IEEE 802.3ah) standard saw two decades of fiber-to-the-home/fiber-to-the-curb research come to fruition, followed by the explosive growth of the fiber-based broadband access rollout throughout the world. Today, as discussed throughout this book, optical OFDM technology—a mere manifestation of the rapidly advancing electronic DSP—has the potential to profoundly impact optical networks. Standardization of the optical OFDM transport at 100 GbE and beyond is natural to consider.

Despite the standardization effort in other aspects of the physical layer of optical networks, such as the SONET/SDH and optical transport network, telecommunication equipment vendors for core networks have resisted standardization for long-haul transponders. Each supplier relies on its own in-house proprietary technology to develop long-haul transponders from a host of techniques, such as non-return-to-zero (NRZ), return-to-zero (RZ), chirped RZ, duobinary, and differential phase-shift keying. There are major problems with nonstandard transponder approaches. First, each supplier develops its own transponders, resulting in volume reduction for the required optoelectronic components and ASICs, leading to cost increases. Second, the equipment from different vendors cannot be interoperable. Carriers cannot use the transponders and WDM systems from different vendors for the same networks.

Although the popular ROADM is promised to enable transparent optical networks that are agonistic to the underlying lightwave signals, the nonstandardized WDM signals from the different vendors make them very challenging to monitor and maintain. Thus, this calls for standardization of the long-haul dense wavelength division multiplexing transponders at 100 GbE and beyond, which will take advantage of the price reduction on high volume and ease of operation.

In the broader context, SDOT can adopt the different modulation standards, whether RZ or NRZ, or single-carrier or multicarrier. The selection of modulation format can be achieved through the DSP and DAC at the transmitter/receiver and through proper pulse shaping at the transmitter. However, we focus on the software-defined optical OFDM standard and discuss what such an optical OFDM standard will entail:

The OFDM parameters: The standard needs to specify the sampling rate, the number of FFT points, filled/unfilled subcarriers, and pilot tones for the channel and phase estimation, inner/outer error-correction codes.

The modulation format and data rate: The optical OFDM provides a unique opportunity for dynamic data rate adaptation. The channel SNR can be measured through pilot symbols, and the data rate can be adjusted through different levels of QAM coding or the number of subcarriers filled.

OFDM sub-band structure: As alluded to in Section 12.2, sub-banding will be an important technique to scale to higher bit rates. This also offers the opportunity for a low-cost receiver to access a small proportion of the overall bandwidth, or so-called subwavelength bandwidth access. The standard needs to define the number of sub-bands and the grouping of the bands if necessary.

Performance monitoring: The standardization will also benefit the signal monitoring at the optical amplifier site and the ROADM site where the full-fledged optical-to-electronic conversion is cost-prohibitive. The OFDM, a modulation format in the frequency domain, can assign a group of the subcarriers for this purpose. This enables the quality of the signal to be monitored without resorting to the high-speed electronics matched to the entire data rate.

12.8 Conclusions

Optical OFDM is a fast-progressing and vibrant research field in optical communications. It is exciting that the most advanced communication concept and theory in modulation, coding, reception, and channel capacity is being applied in the optical domain, as has taken place in the wireless counterpart, but with the major distinction that the signal is processed at a much higher speed that approaches 1 Tb/s. This presents tremendous challenges and opportunities in the field of high-speed electronics and photonics. This chapter discussed several examples of the research-and-development ideas that, in our view, will have significant ramifications in the field of optical OFDM:

1. Tb/s transmission has been traditionally investigated in the context of OTDM. We showed that CO-OFDM can potentially provide a promising alternative toward 1 Tb/s Ethernet transport through implementation of a three-layer optoelectronic multiplexing architecture.

2. As the channel rate approaches 1 Tb/s, the achievable capacity per fiber may become a bottleneck. How to overcome the bottleneck is expected to be an important and interesting research topic. We discussed MMF, more precisely the few-mode fiber, in conjunction with MIMO-OFDM as a technology to achieve 100 Tb/s per fiber that takes advantages of the mode multiplexing in the optical fiber. We described the architecture of the multimode compatible transmission systems as well as some ideas for MMF-compatible components.

3. The traditional optical networks are "rigid," which only supports a fixed link data rate throughout the operation life. The OFDM technique, as one realization of SDOT, provides many enabling functionalities for the future dynamically reconfigurable networks, such as the channel rate being adaptable to the channel condition. We showed that rate-compatible LDPC codes and adaptive loading can be employed to optimize the link capacity.

4. OFDMA has become an attractive multiuser access technique in which the subsets of subcarriers are assigned to individual users. OFDMA enables flexible time and frequency domain resource partitioning. In addition, OFDMA can seamlessly bridge the wireless and optical access networks via RoF systems. OFDMA has also been shown to be a promising approach in offering resource management for PONs.

5. During the past decade, there has been a dramatic resurgence of research interest in OEICs. Considering the extensive digital signal processing involved in optical OFDM, we contemplated the possibility of integration of many digital, RF, and optical components into one silicon IC that can perform the four main functionalities of an optical OFDM transceiver. Without doubt, the success of OEICs will greatly influence the evolution of optical OFDM.

6. Finally, we addressed the problems of product fragmentation in the long-haul transponder market as a result of lack of standards in this space. We discussed the prospect of optical OFDM standardization in optical communications and its impact on the optical communication industry.

References

1. Mitola J. The software radio architecture. *IEEE Commun Mag* 1995;**33**(5):26–38.
2. Abidi AA. The path to the software-defined radio receiver. *IEEE J Solid-State Circuits* 2007;**42**(5):954–66.
3. Sun H, Wu K, Roberts K. Real-time measurements of a 40 Gb/s coherent system. *Opt Express* 2008;**16**:873–9.
4. McDonough J. Moving standards to 100 GbE and beyond. *IEEE Appl Pract* 2007;7–9.
5. Nakazawa M, Yamamoto T, Tamura KR. 1.28 Tbit/s-70 km OTDM transmission using third- and fourth-order simultaneous dispersion compensation with a phase modulator. *Electron Lett* 2000;**36**:2027–9.
6. Weber HG, Ferber S, Kroh M, et al. Single channel 1.28 Tbit/s and 2.56 Tbit/s DQPSK transmission. *Electron Lett* 2006;**42**:178–9.
7. Shieh W, Yang Q, Ma Y. 107 Gb/s coherent optical OFDM transmission over 1000-km SSMF fiber using orthogonal band multiplexing. *Opt Express* 2008;**16**:6378–86.
8. Doan C, Emami S, Niknejad A, Brodersen R. A 60-GHz CMOS receiver front-end. *IEEE J Solid-State Circuits* 2005;**40**:144–55.
9. Razavi B. A 60-GHz CMOS receiver front-end. *IEEE J Solid-State Circuits* 2006;**41**:17–22.
10. Yang Q, Ma Y, Shieh W. 107 Gb/s coherent optical OFDM reception using orthogonal band multiplexing. In: *Opt. Fiber Commun. Conf.*, paper no. PDP 7. San Diego; 2008.
11. Kobayash T, Sano A, Yamada E. Electro-optically subcarrier multiplexed 110 Gb/s OFDM signal transmission over 80 km SMF without dispersion compensation. *Electron Lett* 2008;**44**:225–6.
12. Jansen SL, Morita I, Tanaka H. 10 × 121.9-Gb/s PDM-OFDM transmission with 2-b/s/Hz spectral efficiency over 1000 km of SSMF, In: *Opt. Fiber Commun. Conf.*, paper no. PDP2. San Diego; 2008.
13. Djordjevic IB, Vasic B. LDPC-coded OFDM in fiber-optics communication systems. *J Opt Networking* 2008;**7**:217–26.
14. Bromage J. Raman amplification for fiber communications systems. *J Lightwave Technol* 2004;**22**:79–93.
15. Perlin VE, Winful HG. Distributed Raman amplification for ultrabroad-band long-haul WDM systems. *J Lightwave Technol* 2002;**20**:409–16.
16. Gasulla I, Capmany J. 1 Tb/s km multimode fiber link combining WDM transmission and low-linewidth lasers. *Opt Express* 2008;**16**:8033–8.
17. Tyler EJ, Kourtessis P, Webster M, et al. Toward terabit-per-second capacities over multimode fiber links using SCM/WDM techniques. *J Lightwave Technol* 2003;**21**:3237–43.
18. Tong Z, Ma Y, Shieh W. 21.4 Gbit/s transmission over 200 km multimode fibre using coherent optical OFDM. *Electron Lett* 2008;**44**:1373–4.
19. Panicker RA, Wilde JP, Kahn JM, Welch DF, Lyubomirsky I. 10 × 10 Gb/s DWDM transmission through 2.2-km multimode fiber using adaptive optics. *IEEE Photon Technol Lett* 2007;**19**:1154–6.
20. Sakai J, Kitayama K, Ikeda M, Kato Y, Kimura T. Design considerations of broad-band dual-mode optical fibers. *IEEE Trans Microwave Theory Techniques* 1978;**26**:658–65.
21. Miller SE. Integrated optics: An introduction. *Bell System Technical J* 1969;**48**:2059–69.
22. Jalali B, Yegnanarayanan S, Yoon T, Coppinger F. Advances in silicon-on-insulator optoelectronics. *IEEE J Select Topics Quantum Electron* 1998;**4**:938–47.
23. Soref R. The past, present, and future of silicon photonics. *IEEE J Select Topics Quantum Electron* 2006;**12**:1678–87.
24. Kimerling LC. Electronic–photonic convergence: A roadmap for silicon microphotonics. In: *2nd IEEE Int. Conf. Group IV Photonics*, paper no. WC1. Antwerp, Belgium; 2005.
25. Paniccia M, Koehl S. The silicon solution. *IEEE Spectrum* 2005;**42**:38–44.

26. Nagarajan R, et al. Large-scale photonic integrated circuits for long-haul transmission and switching. *J Opt Networking* 2007;**6**:102–11.

27. Wada O, Sakurai T, Nakagami T. Recent progress in optoelectronic integrated-circuits (OEICS). *IEEE J Quantum Electron* 1986;**22**:805–21.

28. Liu A, Liao L, Rubin D, et al. High-speed optical modulation based on carrier depletion in a silicon waveguide. *Opt Express* 2007;**15**:660–8.

29. De la Rue R, Chong H, Gnan M, et al. Compact two-dimensional grating coupler used as a polarization splitter. *IEEE Photon Technol Lett* 2003;**15**:1249–51.

30. Jalali B, Naval L, Levi AFJ. Si-based receivers for optical data links. *J Lightwave Technol* 1994;**12**:930–5.

31. Giovane LM, Liao L, Lim DR, et al. Si/0.5/Ge/0.5/relaxed buffer photodetectors and low-loss polycrystalline silicon waveguides for integrated optical interconnects at lambda = 1.3 µm. In: *SPIE Int. Soc. Opt. Eng.*, vol. 3007. San Jose, CA; 1997. p. 74–80.

32. Huang FY, Sakamoto K, Wang KL, Jalali B. Epitaxial SiGeC waveguide photodetector grown on Si substrate with response in the 1.3–1.55 µm wavelength range. *IEEE Photon Technol Lett* 1997;**9**:229–31.

33. Prasad R. *OFDM for wireless commun. systems*. Boston: Artech House; 2004.

34. Svensson T, Falahati S, Sternad M. Coding and resource scheduling in packet oriented adaptive TDMA/OFDMA systems, In: *Proc. IEEE Veh. Technol. Conf. VTC06-Spring*. Australia, Melbourne; 2006. p. 1600–4.

35. Caire G, Tarrico G, Biglieri E. Capacity of bit-interleaved channels. *Electron Lett* 1996;**32**:1060–1.

36. Ungerboeck G. Channel coding with multilevel/phase signals. *IEEE Trans Information Theory* 1982;**28**:55–67.

37. Goldsmith A. *Wireless communications*. Cambridge, UK: Cambridge University Press; 2005.

38. Kim YH, Kim Y. 60 GHz wireless communication systems with radio-over-fiber links for indoor wireless LANs. *IEEE Trans Consum Electron* 2004;**50**:517–20.

39. Singh G, Alphones A. OFDM modulation study for a radio-over-fiber system for wireless LAN (IEEE 802.11a). *Proc. ICICS-PCM*. Singapore; 2003. p. 1460–4.

40. Kurt T, Yongaçoğlu A. OFDM and externally modulated multi-mode fibers in radio over fiber systems. *IEEE Trans Wireless Commun* 2006;**5**:2669–74.

41. Song JB, Islam AHMR. Distortion of OFDM signals on radio-over-fiber links integrated with an RF amplifier and active/passive electroabsorption modulators. *J Lightwave Technol* 2008;**26**:467–77.

42. Xu L, Qian D, Hu J, Wei W, Wang T. OFDMA-based passive optical networks (PON). In: *Proc. IEEE/LEOS Summer Topical Meetings*. Acapulco, Mexico; 2008. p. 159–60.

43. Lowery AJ, Armstrong J. 10 Gbit/s multimode fiber link using power-efficient orthogonal-frequency-division multiplexing. *Opt Express* 2005;**13**:10003–9.

44. Djordjevic IB. LDPC-coded OFDM transmission over graded-index plastic optical fiber links. *IEEE Photon Technol Lett* 2007;**19**:871–3.

45. Jazayerifar M, Cabon B, Salehi JA. Transmission of multi-band OFDM and impulse radio ultra-wideband signals over single mode fiber. *J Lightwave Technol* 2008;**26**:2594–602.

46. Gonzalez O, Perez-Jimenez R, Rodriguez S, Rabadan J, Ayala A. OFDM over indoor wireless optical channel. *IEE Proc Optoelectron* 2005;**152**:199–204.

47. Gonzalez O, Perez-Jimenez R, Rodriguez S, Rabadan J, Ayala A. Adaptive OFDM system for communications over the indoor wireless optical channel. *IEE Proc Optoelectron* 2006;**153**:139–44.

48. Grubor J, Jungnickel V, Langer K-D. Capacity analysis in indoor wireless infrared communication using adaptive multiple subcarrier transmission. *Proc ICTON* 2005; Barcelona, Catalonia, Spain; 171–4.

49. Langer K-D, Grubor J. Recent developments in optical wireless communications using infrared and visible light. *Proc ICTON* 2007; Rome, Italy; 146–52.

50. Cavendish D. Evolution of optical transport technologies: From SONET/SDH to WDM. *IEEE Commun Magazine* 2002;**38**:164–72.

Index